Insect Hormones and Bioanalogues

K. Sláma
M. Romaňuk
F. Šorm

Springer-Verlag
New York Wien

Dr. K. Sláma
Institute of Entomology
Ing. Dr. M. Romaňuk
Prof. Ing. Dr. F. Šorm
Institute of Organic Chemistry and Biochemistry
Czechoslovak Academy of Sciences, Prague

Library of Congress Catalog Card Number 72-94418
Printed in Austria by A. Holzhausens Nfg., A-1070 Wien

With 33 Figures

ISBN 0-387-81112-5 Springer-Verlag New York-Wien
ISBN 3-211-81112-5 Springer-Verlag Wien-New York

Introduction

The application of organic insecticides in the agrotechnical praxis resulted in a great and unexpected progress in the control of insect pests, and was of a great economical value all over the world. The widespread application of these agents, however, is also accompanied by negative effects. The principal drawback of classical insecticides consists in the lack of their specificity, the useful insects being killed together with insect pests. Furthermore, the broad-scale application for many years led to the formation of more resistant insect strains requiring higher and higher doses of insecticides. The residues of the mostly used chlorinated compounds accumulate in human and animal foods producing directly or indirectly harmful effects in human subjects. The critical situation led in many developed countries to the restriction in the usage of some types of classical insecticides.

Under these circumstances it is quite natural that novel routes for the control of insect pests are investigated. In this connection, attention has been paid especially to the insect endocrinology and insect hormones which regulate the admirable and in many regards specific development from the egg to the adult insect. The recent successful discoveries in this field are thus in close relation to the practical requirements. Isolation and identification of moulting hormones and juvenile-hormone-like naturally occurring substances not only made possible an exact investigation of their physiological effects but also stimulated the chemical research. In a short period of time numerous biologically active synthetic compounds have been prepared, many of them possessing an enhanced or selective activity. As shown repeatedly by laboratory assays, many of these compounds and especially the juvenile hormone bioanalogues may be used to interfere with the insect development and consequently they are of value as agents acting against some insect pests. The juvenile-hormone-like substances promise to provide the so called "third generation" pesticides which would operate on the physiological basis.

The literature on the insect physiology as well as on the chemistry and biochemistry of insect hormones has undergone very considerable expansion in recent years. We have felt therefore desirable to write this book which should summarize in brief form the present state of knowledge in this field with special respect to the potential use of juvenile-hormone-like compounds as insecticides. The scope of the book is somewhat under influence of our own papers which are completely reviewed here. As far as the patent literature, a selection has been made.

We present this book in the hope that it will stimulate attention in the interesting and gratifying field of insect hormones and their analogues. It is a pleasure to acknowledge the encouragement and help given to us by Springer-Verlag Wien. Furthermore we extend our thanks to all coworkers for their valuable assistance in helping us to prepare this book.

Prague, December 1973 The Authors

Contents

I. Brief Survey of the Neuroendocrine System in Insects 1

A. The Neurosecretory System and the Activation Hormone (AH) (K. SLÁMA) 3

1. Structure of the Neurosecretory Cells and Neurohaemal Organs 4
2. Formation and Release of Neurosecretory Material 7
3. Changes in the Neurosecretory System during Development 9
4. Adenotropic Effects of the Activation Hormone 10
5. Survey of Physiological and Biochemical Effects 13
6. Other Effects of Neurohormones 20
7. Chemical Nature of the AH and other Neurohormones 20
8. Other Proteinaceous Neurohormones with Specific Effects 22

B. Corpora Allata (CA) and the Corpus Allatum Hormone (CAH) (K. SLÁMA) 23

1. External and Internal Anatomy and Morphology of the CA 24
2. Growth and Secretory Activity of the CA during Development 27
3. Action of CAH during Larval Growth and Metamorphosis 30
4. Action of CAH on Reproduction and Adult Diapause 36
5. Physiological and Biochemical Effects of CAH 39
6. Survey of other Effects of CAH 45
7. Chemical Nature of CAH 48

C. Prothoracic Glands (PG) and the Prothoracic Gland Hormone (PGH) (K. SLÁMA) 49

1. External and Internal Morphology of Prothoracic Glands 50
2. Changes in PG during Postembryonic Development 54
3. Effects of PGH on Growth, Differentiation and Moults 58
4. Some Physiological and Biochemical Effects of PGH 62
5. Chemical Nature of the PGH 65

References 67

II. The Chemistry and Physiology of Juvenoids 90

A. Introduction (K. SLÁMA) 90

B. Methods of Testing and Evaluating the Biological Activity of Juvenoids (K. SLÁMA) 93

1 ·Developmental Stages Suitable for the Assays 95

2. Methods of Juvenoid Application 98
3. Evaluation of the Assay Results 104
4. Instructions on Some Biological Assays with Juvenoids 124
5. The Use of Bioassays for Studies on Transport and Stability of Juvenoids 129

C. Chemistry of the Juvenoids 137

1. Introduction (M. ROMAŇUK) 137
2. Acyclic Juvenoids (M. ROMAŇUK) 138
3. Cyclic Juvenoids (F. ŠORM) 170

D. Structure-Activity Relationships (M. ROMAŇUK and K. SLÁMA) 194

1. Physical Properties of Juvenoids 195
2. Molecular Size 196
3. Shape of the Carbon Skeleton in Acyclic Juvenoids 197
4. Double Bonds and Geometrical Isomerism 199
5. Role of Hetero Atoms 203
6. Juvenile Activity of Optical Isomers 212
7. Chemical Structure and Species Specificity 213

E. Physiological and Biochemical Effects of Juvenoids (K. SLÁMA) 217

1. Developmental Periods Sensitive to Juvenoids 218
2. Physiological Conditions for Induction of Supernumerary Instars and Intermediate Forms 221
3. Effect of Juvenoids on Growth and Duration of the Intermoult Cycles 225
4. The Nature of Ecdysial Failures 228
5. The Nature of Juvenoid Effects on Differentiation 232
6. Effects on Reproduction and Adult Diapause 236
7. Action of Juvenoids on Embryonic Development 243
8. Physiological Conditions for Female Sterility 247
9. Relationships in the Effects on Metamorphosis, Embryogenesis and Reproduction 249
10. Synergism and Potentiation of Juvenoid Effects 251
11. Effects of Mitotic and Metabolic Inhibitors 252
12. Effects of Juvenoids on Respiratory Metabolism 255
13. The Effect of Juvenoids on Enzymatic Activity 257
14. The Effect on Carbohydrate, Lipid, and Protein Metabolism 258
15. The Effect on Metabolism of Nucleic Acids and on Puffing Patterns 259
16. The Mode of Action of Juvenoids 261
17. Possibility of Anti-Juvenile or Anti-Juvenoid Effects 263
18. Inactivation of Juvenoids in the Body 264
19. The Effect of Juvenoids on Prothoracic Glands and on Corpora Allata 265
20. Interactions between Plants and Insects by Juvenoids 268
21. Other Effects of Juvenoids, Action on Different Groups of Insects 270

F. Juvenoids as Potential Pesticides (K. SLÁMA) 275

References 281

III. Chemistry and Physiology of Ecdysoids 303

A. The Chemistry of Moulting Hormones (Ecdysones — Ecdysoids) (F. Šorm) 303

B. Relations between the Structure and Activity of Insect Moulting Hormones (F. Šorm) 318

C. Biosynthesis and Catabolism of Ecdysoids (F. Šorm) 320

D. Steroid Compounds Interfering with Ecdysis (F. Šorm) 325

E. Biochemical Effects of Ecdysoids (F. Šorm) 329

F. Bioassays for Ecdysoids (K. Sláma) 332

G. Physiological Effects of Ecdysoids (K. Sláma) 339

1. Developmental Stages Sensitive to Ecdysoid Action 341
2. Physiological Conditions for Ecdysoid Action on Growth and Development 343
3. The Role of Ecdysoids in Puparium Formation 347
4. Excessive versus Small Doses and the Problem of Differentiation 349
5. Effect of Ecdysoids on Reproduction 354
6. The Content, Distribution, and Inhibition of Ecdysoids in the Body 354
7. Developmental Relationships between Ecdysoid and Juvenoid Actions 356
8. Effects of Ecdysoids on Metabolism 358
9. Ecdysoids and Growth of Tissue Cultures 362
10. Ecdysoids and Puffing Patterns on Polytene Chromosomes 364
11. Effect of Ecdysoids in other Animals and Plants 367

References 369

Appendix: Tables (M. Romanuk) **388**

Preface 388
List of Abbreviations 390
Tables 394
Alphabetical List of Juvenoids 458

Addendum 468

Subject Index 470

I. Brief Survey of the Neuroendocrine System in Insects

This introductory chapter represents a condensed review of the components of the neuroendocrine system of insects and the effects of their extirpation or transplantation on development. We have separated these effects from those of plant and animal extracts, other isolation products, or synthetic compounds, which will be described in the following chapters. The recent literature pertaining to insect hormones has been reviewed in several monographs. The results obtained by classical endocrinological techniques are described in detail in the book by PFLUGFELDER [237] which contains detailed descriptions of insect endocrines from the period when morphological studies predominated. An extensive list of the literature pertaining to the endocrine glands of insects with discussions on developmental theories in insect endocrinology is contained in a book by NOVÁK [218]. Comparative data on insects and other invertebrates, especially crustaceans, may be found in a monograph by GERSCH [90] and a brief but very instructive book of comparative endocrinology of invertebrates containing numerous illustrations has been prepared by HIGHNAM and HILL [126]. Comparative data on the structure and function of neurosecretory system may be found in an extensive work by GABE [88]. The basic physiological data for understanding the mode of action of insect hormones are contained in a book by WIGGLESWORTH [370], and a monograph by WIGGLESWORTH [371] on the physiology of metamorphosis of *Rhodnius*. WIGGLESWORTHS most recent review on insect hormones [379] is also highly recommended. Finally, problems concerned with the hormonal control of reproduction in insects have recently been summarized by ENGELMANN [70].

Insects are characterized by an enormous diversity of morphological forms and living habits. Their indirect form of development is manifested by the occurrence of immature larval stages, which in many insect groups live in a completely different environment and possess an entirely different form than those of the adults. The indirect development requires precise regulation of developmental processes and their synchronization with the changing environmental conditions. The same precise regulation occurs during metamorphosis when the transformation of the larva into the mature adult takes place.

The regulation and synchronization of insect development is controled by the neuroendocrine system. It records stimuli from the environment such as photoperiod, temperature, humidity, availability of

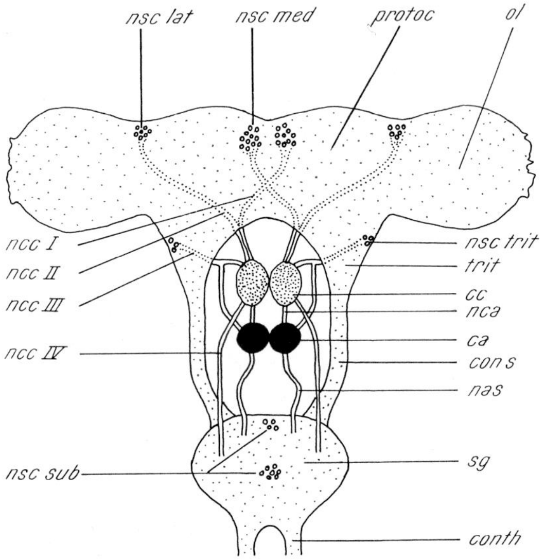

Fig. 1. Schematic illustration of the main neuroendocrine system of an insect. *ca* = corpora allata, *cc* = corpora cardiaca, *con s* = suboesophageal connectives, *con th* = thoracic connectives, *nas* = nervus allatosuboesophagealis, *nca* = nervus corporis allati, *ncc* I–IV = nervi corporis cardiaci, *nsc lat* = neurosecretory cells (nsc) of the lateral part of protocerebrum, *nsc med* = nsc of the pars intercerebralis protocerebri, *nsc sub* = nsc of the suboesophageal ganglion, *nsc trit* = nsc of the tritocerebrum, *ol* = optic lobe, *protoc* = protocerebrum, *sg* = *suboesophageal* ganglion, *trit* = tritocerebrum

food, etc., and transforms these impulses into chemical messengers or hormones which in turn cause functional or developmental changes among various effector organs according to a genetically determined programme. The neuroendocrine system represents a functional unit consisting of the nervous system and the endocrine glands. Its basic structure in insects has many features which are common to higher animals including mammals [289]. The most important part of the system are the neurosecretory cells located in the central nervous system. They are responsible for transmission of neural messages to the endocrine glands or other tissues. Direct effects of their secretions on "target" tissues have the character of first order neuroendocrine functions since they stimulate secretion of hormones by other endocrine glands such as the corpora allata or prothoracic glands. The hormones secreted by these latter endocrine glands represent neuroendocrine functions of the second order [294]. The neurohormones affecting endocrine glands have thus superior position in the whole hierarchy of hormonal regulations. A schematic illustration of the basic neuroendocrine centers in the insect body is given in Fig. 1.

Growth and developmental processes connected with physiological changes in insect tissues are regulated by three hormones collectively referred to as developmental or metamorphosis hormones. One of them is a neurohormone released by the neurosecretory system while the second is released by the corpora allata and the third by the prothoracic glands. It appears difficult to comprehend that the great diversity of developmental patterns in insects is produced by modulating the activities of these three hormones. It is becoming increasingly evident that the responsibility for such diversity depends on inherited genetic factors which determine for a particular species whether at a certain developmental stage individual cells will be dependent or independent of the hormones. Our intention is to summarize the basic facts concerning the insect endocrine system. These facts are important in understanding the mode of action of natural and synthetic compounds possessing hormone activity.

A. The Neurosecretory System and the Activation Hormone (AH)

The neurosecretory system of insects is comprised of a complex of neurosecretory cells located in the cerebral nervous system and in the ganglia of the ventral nerve cord, their axons and special neurohaemal organs engaged in the storage and release of the neurosecretory material into the haemolymph. Differences in histological and anatomical structure, variability in size, location and staining properties as well as specific functional differences suggest that the neurosecretory system may produce several compounds differing in physiological functions and chemical structures. For the regulation of insect development the most important neurosecretory product, or products, of the system appears to be the hormone which stimulates the secretory activity of other endocrine glands. It is assumed that this hormone is secreted by neurosecretory cells of the protocerebrum and released into the haemolymph by the corpora cardiaca. The hormone is often referred to as brain hormone, prothoracotropic hormone or allatotropic hormone but we shall use the term activation hormone which is considered more appropriate [379].

1. Structure of the Neurosecretory Cells and Neurohaemal Organs

The basic unit of the neurosecretory system is the neurosecretory cell. It represents a modified neurone with dual characteristics possessing many features characteristic of a true neurone and also having typical features of an endocrine cell capable of producing secretions which are then released into the haemolymph [88, 285]. In contrast to typical neurones, neurosecretory cells do not establish normal synaptic contacts with other neurones. However, they form vesicular structures (synaptoids) at axon terminals which resemble presynaptic structures and are probably engaged in the release of neurosecretory material [285, 286, 293].

The neurosecretory cells may occur individually or in small groups, e.g. in the ganglia of the ventral nerve cord or may form aggregates as is observed in the central nervous system of some insects. The cells are always situated at determined loci in the nervous system and are sometimes visible under the dissecting microscope or with dark field illumination where they appear as a bluish white reflection.

Each neurosecretory cell (n. s. c.) has an apparent axon and the axons of groups of these cells form bundles leading to neurohaemal organs. Other axons lead to peripheral organs where they may contact parenchymal cells (Fig. 2). There are many types of n. s. c. which are classified mainly according to their histochemical properties, shape, size, and location in the nervous system [88]. The most common n. s. c. which include some of the largest neurosecretory neurones are the so called A type n. s. c. whose granules stain dark purple with paraldehyde fuchsin and deep blue with chrome-haematoxyline-phloxin. The B type n. s. c. contain granules which are phloxinophilic and stain green or bluish-green with paraldehyde-fuchsin and red with chrome-haematoxylin. Finally, it is possible to distinguish cell types which can be classified as C and D on the basis of their staining properties [106, 108, 140, 245, 254, 301]. Some authors have distinguished 8 or more types of n. s. c. by such tinctorial affinities as well as by special functional differences. However, any sort of general description of all neurosecretory cells contained within an insect body is impossible and we must consider individual situation within a particular species [293]. This conclusion also applies to the more or less generalized distinction between the A and B type n. s. c.

The number of n. s. c. in the protocerebrum may be relatively small, i.e. 8 to 50, but very often there are over 100 such cells; in some cases there may be close to 1000 [88, 90, 126, 140, 218, 379]. Distinctive n. s. c. occur also in the tritocerebral part of the brain [250], in the suboesophageal ganglion and in both the thoracic and abdominal ganglia [55, 80, 140, 251, 254]. The most intensively studied n. s. c. are the groups of cells in the pars intercerebralis in each of the hemispheres of the protocerebrum [88]. The axons of these cells form the *nervi corporis cardiaci interni* (NCC I) most of which cross and lead to the corpus cardiacum on the contralateral part of the body. The n. s. c. of the lateral part of the protocerebrum send their axons *(nervi corporis*

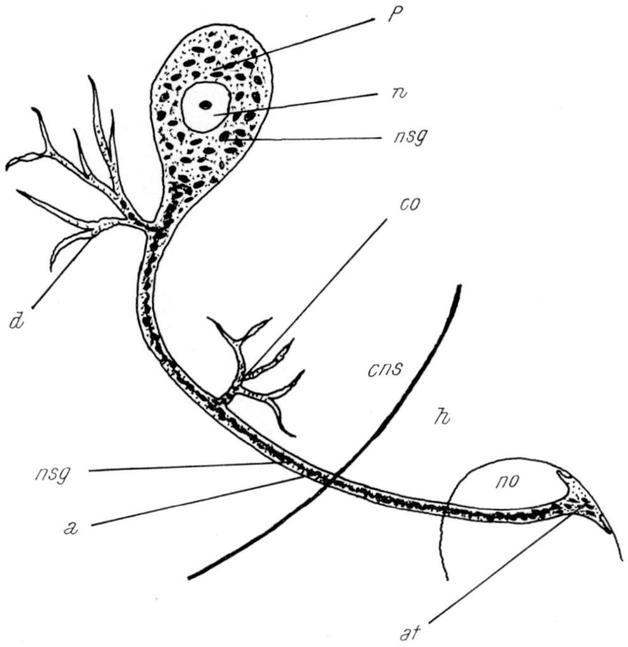

Fig. 2. Schematic illustration of a neurosecretory neuron in the pars intercerebralis of an insect. *a* = axon, *at* = axon terminal, *cns* = central nervous system, *co* = collaterals, *d* = dendrite, *h* = haemolymph, *n* = nucleus, *no* = neurohaemal organ, *nsg* = *neurosecretory granules*, *p* = pericaryon

cardiaci externi) (NCC II) to the ipsilateral corpus cardiacum. The axons of the tritocerebral n. s. c. form the *nervi corporis cardiaci III* leading also to the corpus cardiacum, but in several insect species, these nerves have not been found [301].

The protocerebral n. s. c. occur in all insects including the most primitive Apterygota and the most specialized groups of Endopterygotes [140]. The importance of n. s. c. in the phylogeny of the animals is evident from a fact that similar cells occur in all phyla of animals beginning with primitive worms, the Turbellaria [40, 88, 90, 140, 285, 293].

The main neurohaemal organ in insects is the corpus cardiacum. It is of ectodermal origin and has the character of both nervous tissue and endocrine gland [285, 286]. Macroscopically the corpora cardiaca appear as paired rounded or elongated bluish-white organs located posterior to the brain and usually attached to the wall of the aorta. The morphology of the corpora cardiaca has been modified in different groups of insects. For example in higher Diptera the corpora cardiaca are not distinct paired organs but are fused into the ring gland. The corpora cardiaca form a close connexion with the brain by the cardiacal nerves I to III and also connect via nerves to the hypocerebral ganglion and with the suboesophageal ganglion (*nervus cardiaco-suboesophagealis* or *nervus corporis cardiaci* IV). From the distal end of the corpora cardiaca, nerves lead to the corpora allata *(nervus corporis allati)*. In certain insects it is possible to distinguish more or less separated nervous and glandular parts in the corpora cardiaca [42, 218, 249, 301] (see Figs. 1 to 4).

The corpora cardiaca are surrounded by a connective tissue sheath consisting of acellular stroma. While the surface of the gland may be smoth in some insects it more often shows deep invaginations. The internal structure of the gland is complex and individual structural elements may differ in different species [32, 88, 286]. A large part of the corpora cardiaca consists of nerve bundles and enlarged axon terminals of neurosecretory cells which contain accumulated neurosecretory granules. Many of these axon terminals occur near the outer surface but some are also observed on the stromal invaginations of the organ (Fig. 3). Near the surface it is possible to observe numerous chromophobic interstitial or glial cells having a small nucleus, dense chromatin pattern, little perinuclear cytoplasm and numerous cytoplasmic processes [140, 285, 286, 301, 342]. The glial cells are the main component of the connective tissue inside the gland [42, 280]. Another type of cells present in the corpora cardiaca are the parenchymal cells or intrinsic secretory cells [140, 218, 279, 280, 286, 301, 342] which have a large nucleolus, abundant cytoplasm with short thick process and numerous cytoplasmic granules of Golgi origin. These cells are probably identical with the "chromophilic cells" of the light microscope and are thought to be modified neurons with particularly expressed secretory function [40, 41, 285]. Their axon-like processes contain neurosecretory granules and terminate presumably inside the gland [280, 301, 342]. In addition to these internal structural elements, muscle fibres are also part of the corpora cardiaca in some insects [301].

The extensive work of Raabe and her co-workers has shown that the neurosecretory material of the neurosecretory cells located in the ganglia of the ventral nerve cord is released from the segmentally arranged neurohaemal organs. These organs occur in diverse groups of insects. The neurohaemal organs appear as swellings on the median or transversal visceral nerves and are called perivisceral or perisympathetic neurohaemal organs [251, 253, 254]. Histologically the neurohaemal organs appear similar to the corpus cardiacum. A thin cellular stroma covers

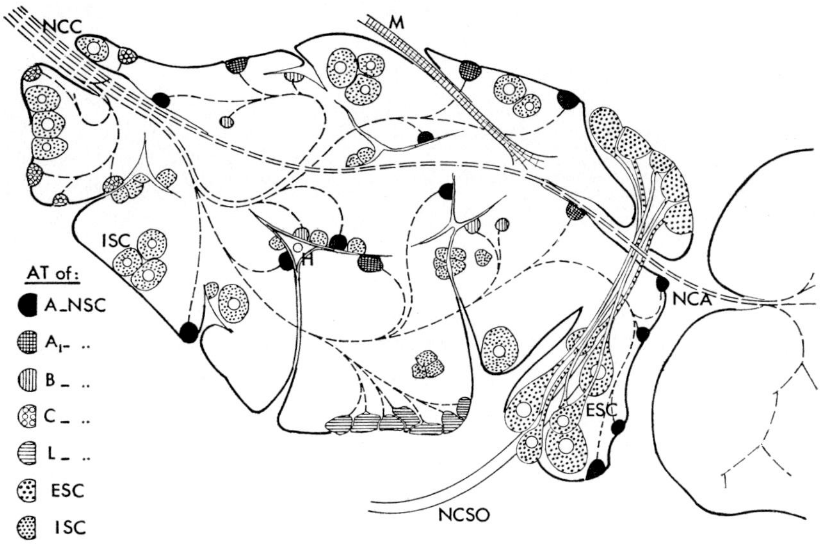

Fig. 3. Diagram of the corpus cardiacum of adult *Leptinotarsa* showing the course of the axons of different neurosecretory cells. AT = axon terminals, ESC = extrinsic secretory cells, H = haemocyte, ISC = intrinsic secretory cells, M = muscle. NCA = nervus corpori allati, NCC = nervi corporis cardiaci, NCSO = nervus cardiaco-suboesophagealis, NSC = neurosecretory cells. From SCHOONEVELD [301]

many axons some of which contain neurosecretory granules. In addition, there are glial cells as in the corpora cardiaca and a well developed lacunar system, but modified neurons comparable to the intrinsic secretory cells of corpus cardiacum with neurosecretory axons have not been found [110, 248, 253, 254, 257].

2. Formation and Release of Neurosecretory Material

The presence of neurosecretory material can be demonstrated in the light microscope by histochemical techniques [245] and selective staining procedures which have already been mentioned. The neurosecretory material of the peptidergic A-type neurosecretory cells can be demonstrated histochemically by the presence of disulphide or sulphydryl groups in polypeptides rich in cysteine or cystine. By means of sensitive fluorescence methods it is possible to demonstrate in other types of neurosecretory cells the presence of biologically active amines such as noradrenaline, serotonine [293]. According to RAABE and MONJO [256] the granules contained in C-type neurosecretory cells situated in the ventral nerve cord and in the lateral part of the pars intercerebralis do not contain catecholamines or polypeptides rich in disulphide or sulphhydryl groups. They contain basic proteins with indol or pyrrol residues.

Under the electron microscope cytoplasmic granules of high electron density surrounded by the membrane can be observed in the n. s. c. Their size ranges usually from 1000 to 3000 Å [279]. The granules are formed in the perikaryon of the neurosecretory cell by the granular endoplasmic reticulum (ergastoplasm) and are then transported to Golgi bodies where they are packaged and budded off to the axons [23, 108, 126, 246, 285, 286]. Under certain circumstances the granules may accumulate in the perikaryon, in the axons or axon terminals.

The main feature of the neurosecretory cells, in addition to mitochondria, neurotubules, cisternal profiles and multivesicular bodies, is the presence of varying amount of vesicles with electron dense material [286]. It is generally agreed that the neurons containing large vesicles (1000–3000 Å) represent "classical" neurosecretory neurones containing proteinaceous secretory material [288] which consists of phospholipoprotein rich in -S-S- or -SH groups [246]. The neurosecretory product of the axons with smaller vesicles may represent material containing monoamines [285, 286].

The enlarged terminus of an axon which is in contact with the matrix of the extracellular stroma is thought to be the site of release of the neurosecretory material. As pointed out by SCHARRER [286], release can take place also at other preterminal parts of the axon. Moreover, the neurohaemal organs are not the sole place of release. Neurosecretory material can be released directly at the parenchymal gland cells such as those of the corpus allatum, or to connective tissue layers, as is also the case in the prothoracic glands, or in the vicinity of other nerve fibres or glial cells, etc. In some insects such as aphids neurosecretory axons terminate in the musculature or in some visceral organs [141]. In *Oncopeltus* and several other Hemiptera neurosecretory products are released through the wall of the aorta and this is also the place where they eventually accumulate [140]. The release sites are characterized by the presence of numerous small vesicles of low electron density which are judged to be residues of fragmental neurosecretory granule membranes [286].

According to the extracellular pathway of the released chemical mediators we can distinguish between neurohumoral and neurohormonal agents. B. SCHARRER [287, 288] has recently classified neurons along these lines and has proposed a terminology for the biologically important neurosecretions. The neurohumours are engaged in the transfer of informations in the synapses where they cause post-synaptic reactions of very short duration. Hence their action is not of the endocrine type. Various names such as 'chemical transmitters', 'neurotransmitters', or "neurohumours" have been proposed for these materials and typical examples are acetylcholine and noradrenaline. The neurohormonal materials are released by neurosecretory cells without synaptic transmission and are transported by the haemolymph to their target cells where the neurohormones can then act. Their action is therefore very similar to that of other hormones secreted from parenchymal cells of regular endocrine glands. However, there are exceptions. For example, in the case where the secretions reach the peripheral tissues directly via axons, we can not classify this type of activity as neurohormonal since there is no transport of neurosecretory material by the haemolymph.

3. Changes in the Neurosecretory System during Development

Various authors have tried to find some correlations between the amount of stainable material in the neurosecretory system and developmental processes governed by endocrine glands. The scope of this chapter does not permit a detailed analysis of this complicated problem. However, the various aspects of this question are discussed in a number of review articles [8, 66, 67, 88, 90, 92, 121, 123, 124, 126, 140, 190, 218, 303, 379, 384]. Cytological studies have revealed changes in the content of neurosecretory material in the pericarya of neurosecretory cells as well as in the axons, neurohaemal organs, or aorta wall. These changes mainly refer to the relative content of the material in the cells as compared to that in the neurohaemal organ. In many insects it appears that the amount of neurosecretion in the neurohaemal organs is less at the beginning of each developmental cycle. This decrease may occur at the beginning of a larval instar, the onset of metamorphosis, the beginning of a reproduction cycle in adult female, etc. Towards the end of a cycle the neurohaemal organ again accumulates neurosecretory material. There are also instances with no apparent alteration in the amount of neurosecretory material during the developmental period. In addition there are reports suggesting that for some species the relationship described above may actually be reversed [122, 123, 124, 126].

The problem of relationship between the amount of stainable material in the neurosecretory system and the degree of physiological activity of this system has been intensively studied by HIGHNAM [122, 124]. He concluded that the amount of material in the cells was not directly related to the rate of synthesis or release and suggested that both activities were regulated by different mechanisms. Thus, the amount of stainable material present in the n. s. c. is a result of a balance between the synthesis and release of the material. Without information concerning these processes the secretory activity of the system at the time of fixation cannot be determined merely from the amount of neurosecretory material present [122]. Using radioactive tracer methods by measuring the rate of incorporation of labelled cystine, HIGHNAM [123] found that the neurosecretory activity is indeed much higher in females with developing ovaries than in those without. The velocity of axonal transport of neurosecretory material has been estimated to be approximately 0.8 to 1.6 mm per hour. SCHOONEVELD [301] suggested that the axonal transport may be regulated by nervous activity but the release of the material from axon terminals may be regulated by other factors. Selective release of individual factors from the corpora cardiaca has been achieved by electrical stimulation of separate cardiacic nerves [95].

In addition to developmental changes there exist many factors which have more immediate influence on the amount of neurosecretion, especially in the neurohaemal organs. Among these factors are electrical irritations, intensive nervous stress as during mating or oviposition, ingestion of food or water after starvation or thirsting, etc. [126]. Such a sudden release of the neurosecretory material is followed by diuretic or myotropic effects without actual stimulation of development. Thus, the amount of stainable material in the corpora cardiaca may not be directly related to the neurohormone which stimulates development [204].

4. Adenotropic Effects of the Activation Hormone (AH)

Adenotropic (gland-directed) effects represent the most important physiological function of the neurohormones. The common integration of the gland functions by the AH achieves the synchronization of growth with nutritional conditions and a correlation of morphogenetic processes at the time when the feeding is not the necessary stimulus. The tropic influence of the neurosecretory system includes a wide spectrum of glandular tissues. However, it is unknown whether the allatotropic or prothoracotropic hormone is the same as the material which stimulates the intestinal epithelial cells, fat body cells and other tissues. Histochemical and physiological differences among separate groups of neurosecretory cells of the protocerebrum suggest that there is more than one active factor [104, 106, 108].

Recent physiological studies reveal that the adenotropic effects of the AH are not restricted to the endocrine glands. Other tissues with more or less pronounced glandular properties such as intestinal epithelial cells, fat body cells, Malpighian tubule cells, pericardial cells and oenocytes show structural and functional changes which can be correlated with substances released from the corpus cardiacum. For example, intestinal epithelial cells in *Calliphora* undergo structural changes in the cytoplasm in response to the AH which leads to increased rate of RNA synthesis and stimulation of synthesis of proteolytic enzymes [344]. The enzymes are released into the gut lumen but other proteins are synthetized and released into the haemolymph. The function of these cells depends on the AH and also on the presence of nondigested food [68, 235]. Thus, the hormone stimulates biosynthesis of particular enzymes but the actual rate of synthesis and amount of the enzyme produced depends on the feed-back mechanism dependent on the presence of food in the gut [132]. Other types of cells, such as Malpighian tubule cells show increased metabolic activity in response to AH. This increased activity is connected with increased transport and excretion of waste products and water [244].

The fat body can be considered as a large gland which is engaged in synthesis of various materials, especially proteins, which are released into the haemolymph and transported to growing tissues. The AH action on the fat body cells is connected, in addition to specific biochemical changes, with structural changes. In the presence of AH the plasma membrane of the cells becomes extremely thin, the mitochondria increase in number and size and this can be correlated with increased metabolic activity. In the cytoplasm of the fat body cells there appear new proteins, and the cytoplasm becomes filled with products assimilated from food sources [318]. In certain insects the function of fat body cells [379] or intestinal cells [68] represent an adaptive system which is dependent on the presence or absence of food, without direct dependence on hormonal stimulation. In addition to these cells the pericardial cells show striking variations in their content of proteins, especially esterase enzymes [214]. Following stimulation by neurohormones the pericardial cells release cardioaccelerating factors into the haemolymph [53]. In all these glandular or epithelial tissues the AH action results in the stimulation of synthesis of specific cellular products most of which appear to be proteinaceous. The morphogenetic changes, e.g. rearrangement of the intestinal epithelium and formation of new or breakdown of the old body cells, are regulated by other hormones which will be described in chapters B and C.

The most important adenotropic effect of the AH is the "activation" of secretory function in the corpora allata and the prothoracic glands. The early studies of Fukuda [83] and Williams [393] demonstrated the effect of the AH on secretory activity of the prothoracic gland and this effect has been confirmed in all major groups of insects [90, 126, 218, 237, 249, 371, 379].

The AH-dependent stimulation of prothoracic glands is very often synchronized with feeding usually occurring at the beginning of each moult cycle. In a majority of insect species, starved larvae do not develop and do not moult. However, feeding is not the only stimulus for AH release and prothoracic gland activation. During postdiapause development or in metamorphosis of Endopterygotes, the AH release is brought about by determined environmental or internal physiological stimuli. This is long known from the studies on insect diapause.

Conditions which are necessary for the adenotropic effect of the AH on the corpora allata have been reviewed by Johansson [140], Highnam [121], Engelmann [66, 70] and others [126, 218, 301, 340, 379]. Such conditions are generally characterized by a common pathway: food ingestion—AH release—gland activation—allatum hormone release. The dependence of corpus allatum activation on feeding is more pronounced than that of the prothoracic glands. In fact, there is usually never a period during insect development in which a non-feeding stage has active corpora allata. Transplantation experiments indicate that corpus allatum activation is mediated by transport of the AH into the haemolymph [218, 237]. More complex processes of activation have also been suggested [66]. Neurosecretory material can also be transported to the corpora allata through the *nervi allati* [121, 140], but it is not yet clear how important this might be for the activation process. Active glands transplanted into allatectomized insects retain their secretory activity for prolonged periods in the absence of nervous connection to the brain. In some cases, activation of the glands occurs after severing of the nervous connection to the brain [126, 140, 379] suggesting a possibility of nervous inhibition of corpora allata function. However, this phenomenon is not general in insects. According to our experience it does not occur in *Pyrrhocoris* and in few of other Hemipterans.

Activation of the corpora allata is initiated at the beginning of the moult or reproductive cycle. In some species there exist special homeostatic mechanisms which coordinate the AH release, corpus allatum activation, and specific physiological functions connected with diapause [301, 383], initiation of feeding [140, 340] or special conditions of reproduction [66].

Little is known concerning the mechanism by which activity of the corpus allatum is inhibited. Of special interest is the selective inhibition of the gland in the last larval instars. The phenomenon has considerable importance because it determines the number of larval instars and the initiation of the whole metamorphosis process. A simple endocrinological explanation would be an absence of the adenotropic hormone. However, this assumption can only be true provided that the allatotropic and prothoracotropic neurohormones are different. Destruction or removal of the n. s. c.—c. cardiaca complex leads usually to inactivity of both the corpora allata and prothoracic glands. This is not the case in last larval instars where all the functions known to be controlled by neurohormones are in performance except selective inhibition of the corpus allatum. Other possible explanations are based on the concept of hormonal stimulation and nervous inhibition of corpus allatum activity [140, 161, 277, 283], which may also control reproductive activity [66, 121]. Further explanations are based on evidence that individual groups of the neurosecretory cells secrete different materials with selective effects on corpora allata [103, 106, 333].

In conclusion, we may state that the adenotropic effects of the AH depend in inducing functional rather than morphogenetic changes in various glandular target organs. The hormone, or complex of hormones, activates the trophic (nutritive) conditions for tissue growth [208, 312] by an integrated stimulation of all the visceral glandular tissues angaged in assimilation of nutrients and their metabolism. Through its effects on endocrine glands, the AH synchronizes morphogenesis and reproduction with the nutritional and homeostatic state of the insect according to the particular genetic programme of the organism.

5. Survey of Physiological and Biochemical Effects

Involvement of the central nervous system in regulation of insect metamorphosis and development was demonstrated in the early studies by Kopeč [171, 172] which were followed by further observations [28, 177, 365]. These findings led to the explanation of the role of the neurosecretory cells of the brain [367]. From the studies of Fukuda, Williams, and Wigglesworth it became clear that most developmental processes were influenced indirectly by the neurosecretory system through its action on the endocrine glands. Hence, many developmental effects ascribed to direct control by the neurosecretory system appeared later to be regulated by hormones from the corpora allata and prothoracic glands. On the other hand, recent physiological and biochemical advances have revealed that the neurosecretory system causes many physiological and biochemical changes in the body which are not dependent on the presence of endocrine glands. Our intention is to give a brief description of these effects of the neurosecretory system. The complexity of this problem has been well stated by Scharrer and Weitzman [293]: "there is hardly a physiological process in existence for which control by neurosecretory mediators cannot be claimed in at least one or another invertebrate animal". Physiological effects of the neurosecretory system have been reviewed by JOLY and CAZAL [147], see also [126].

a) The Effect on Growth and Diapause

Animals with cauterized or extirpated neurosecretory cells—corpora cardiaca complex consume less than normal amount of food and growth is arrested [126, 312, 340]. After implantation of corpora cardiaca feeding and growth resume [312]. This suggests that the neurohormones from corpus cardiacum influence feeding and consequently growth of the body. This effect on feeding and growth is most noticeable in larvae and adult females, whereas only small variations in size occur in adult males [208, 313]. The release of hormones from the neurosecretory system is stimulated by initiation of feeding, however, subsequent intake of food or increase in body size are mutually interdependent on further release of the neurohormone [67, 140]. Furthermore, the quality of food may also be important in initiating release of the neurosecretory material from corpus cardiacum. For example, in Diptera the hormone which activates corpus allatum is released only in response to proteinaceous food [140, 331, 340]. In some insects, like the seed-feeding Hemipterans, the release of neurohormones is dependent on ingestion of determined amounts of water before feeding.

Diapause is a programmed arrest of growth, development, or reproduction, resulting in reduction of active physiological functions. The onset of diapause is directly related to a hormonal defficience [182, 183]. Involved in the regulation of diapause are mainly the neurohormones with adenotropic effects [126, 383, 384, 393, 397]. These relationships between the hormones and diapause have been several times reviewed [90, 126, 218, 379, 383]. The environmental stimuli leading to hormonal inhibition at the onset of diapause are occasionally similar, but more often different from the stimuli resulting in restoration of hormonal activity and termination of diapause.

The larval and pupal diapause are characterized by inactivity of the AH-prothoracic gland hormone system. Induction of this diapause is determined by environmental stimuli such as photoperiod and temperature while termination of diapause is affected by prolonged exposures of the diapausing stages to low temperature [126, 182, 183, 388, 393, 394]. Adult diapause, on the other hand, results as a consequence of secretory inactivity of the system AH-corpus allatum hormone [384]. Finally, the egg or embryonic diapause as in *Bombyx* is initiated by neurohormone released from the suboesophageal ganglion [87, 115, 116]. All these examples of diapause are controled by the neurosecretory system and the absence of AH is a primary cause of all types of postembryonic diapause. However, one must be aware that most of the physiological and biochemical changes associated with diapause result from inactivity of the target endocrine glands, i.e., corpora allata and prothoracic glands.

b) The Effect on Respiratory Metabolism

All the data concerning the effects of the AH on respiratory metabolism confirm that it has a stimulatory effect on oxygen consumption of whole body or isolated body parts [72, 157, 200, 208, 316, 339]. Removal of the source of AH, i.e., neurosecretory cells or corpora cardiaca usually results in a decreased metabolic rate. In particular, the most affected metabolic functions are directly or indirectly associated with food digestion and utilization [208, 312]. SLÁMA [312] has distinguished three types of respiratory metabolism which are responsive to changes in hormonal activity: i) basal cellular metabolism which occurs after removal of all the known neuroendocrine centers or in diapause; ii) digestive metabolism which is dependent on AH; iii) reproductive metabolism which is dependent on the corpus allatum hormone and growth of the ovaries in adult females. The last two examples of hormone stimulated respiratory activity can occur independently [318]. Removal of the corpora cardiaca from larvae caused a decrease in respiratory metabolism to a low level, while removal of the corpora allata had a relatively minor effect on respiration [315]. Removal of the prothoracic glands from larvae of *Pyrrhocoris* does not inhibit digestive metabolism but after prolonged periods during which ecdysis does not occur there is a continuous decrease in respiratory rate due to hypertrophy of the body and inability for larval instars to continue growing beyond certain limits. Extracts prepared from corpora cardiaca stimulate respiratory metabolism of the whole body [158] and also *in vitro* preparations of the fat body [200, 403].

In contrast to the rather high intensity of respiratory metabolism associated with assimilation of food, morphogenetic process may occur during the period of relatively low intensity of respiration. This is especially apparent during metamorphosis and during postdiapause development. In this case it is rather difficult to determine what proportion of the total metabolism depends on the AH and what is due to the prothoracic gland hormone action.

c) The Effect on Enzymatic Activity

As already mentioned, the main physiological function of the AH seems to depend upon stimulation of secretion in various organs which are completely or partly glandular in nature. Many of the specific secretions synthesized by activated gland cells appear to be enzymes.

The AH has a pronounced effect on the stimulation of synthesis of hydrolytic enzymes in tissues engaged with food utilization, e.g. the effect on the proteolytic enzyme synthesis in the epithelial intestinal cells which has already been mentioned [69, 132, 137, 235, 344]. Esterase enzymes present in the epithelial intestinal cells and the Malpighian tubules also exhibit changes which can be correlated with periods of AH release [214, 341]. Furthermore, many proteins with esterase activity accumulate in the haemolymph in the absence of corpora allata, provided that the AH releasing system remains intact [214]. Some enzymes abundant in the intestine or Malpighian tubules such as phosphatases remain very active even in the complete absence of the AH. As shown by NOHEL and SLÁMA [216] the enzymes of catabolic metabolism, such as transaminases are more active in insects in which AH sources have been removed.

In many glandular tissues which are presumptive targets of the AH, i.e., intestine, fat body, Malpighian tubules, pericardial cells, etc., it is possible to observe many changes in the mitochondrial cytochrome enzyme system which are associated with AH release [318]. Extracts of corpora cardiaca increase phosphorylase activity in the fat body [328, 364] and it has been suggested that the hormone from corpora cardiaca may affect the enzymes of carbohydrate metabolism by inhibition of glycolysis at the level of phosphofructokinase and by activation of the trehalose-6-phosphate synthetase [363]. These data suggest that increased enzyme synthesis may be a general pattern associated with the AH action on its glandular target cells. The hormone does not seem to stimulate the "*de novo*" synthesis of new enzymes but more probably it stimulates the amount of gene sets already available and this in turn increases the synthesis of enzymes without re-programming of the genome.

d) The AH and Metabolism of Carbohydrates, Fat, and Protein

Towards the end of the larval stage and during the course of the reproductive cycles, considerable amounts of carbohydrate, fat, and protein reserves accumulate in the insect body. The amount that is accumulated is dependent to a large extent on the amount of food intake and the intensity of digestive processes. Insects in which the neurosecretory cells—corpus cardiacum system is inhibited do not accumulate reserves at the same rate as normal animals. Furthermore, insects in which other endocrine glands are removed but with intact sources of AH are characterized by pronounced hypertrophy of the organs which contain reserve materials and by accumulation of glycogen, fat, and protein [67, 126, 140, 208, 218, 340].

It appears that the extracts of corpora cardiaca influence the carbohydrate metabolism *in vivo* as well as *in vitro*. The action of the extracts depends in lowering the amount of glycogen and increasing the haemolymph trehalose concentration [327]. This has been confirmed by series of observations in different species [31, 44, 147, 164, 211, 212, 262, 328, 364]. In mosquitoes, the neurosecretory system causes a decrease in glycogen synthesis while it stimulates the synthesis of the triglycerides [112]. Despite the fact that these "hyperglycaemic" factors from the neurosecretory system have been intensively studied, it is not yet known whether AH or some other neurohormones are responsible for these effects on carbohydrate metabolism. Some of the effects may be nonspecific because, as found by Wyatt and his coworkers [400, 402], considerable increase in haemolymph trehalose may occur after epidermal injury. In addition, there are many nonspecific factors which activate trehalose activity in insect tissues.

The effect of the AH on protein synthesis in its target cells is manifested by an increase of cytoplasmic volume and development of various cytoplasmic structures such as mitochondria and ribosomes. Furthermore, AH stimulation of protein synthesis is based on its effects on enzyme synthesis. For example, there are reports that the extracts of corpora cardiaca can influence the rate of protein synthesis in isolated fat body cells [403]. Moreover, large variations in the protein pattern of insect haemolymph occur after electrocautery or extirpation of various parts of the neurosecretory system. This has been most intensively studied in connexion with ovarian growth in adult females. Since changes in haemolymph protein patterns are also dependent on other hormones and represent a rather complex problem with respect to species diversity, more detailed information is contained in special papers on this subject [18, 127, 208].

In response to AH, the stimulated target cells show an increased rate of nucleic acid synthesis. DNA replication does not seem to be a necessary condition for this hormone action, but increased rate of RNA synthesis seems to be a general phenomenon [48, 174, 379] related to increase in enzyme synthesis. In addition, there is a considerable increase in the rate of RNA synthesis in the corpora allata and in the prothoracic glands following stimulation with AH [98, 163, 222].

e) The Effect on Diuresis

Vertebrate endocrinologists assume that the effects of neurohormones on diuresis represent one of the oldest means of hormone mechanisms. This seems true for insects as well. These preferentially terrestrial Arthropods have no internal skeleton and diuretic functions are therefore very important as pressoric regulators. For example, the overall shape of the body of caterpillars is largely dependent on the internal pressure of the coelomic fluid.

We have already pointed out that changes connected with water uptake or thirsting are often related to the amount of neurosecretory material [213] present in the neurosecretory system. Also, it has been shown by many investigators that extracts prepared from the central nervous system or from the neurohaemal organs can affect water balance in the body, water intake or excretion. This problem has been reviewed by Raabe [249], Gersch [90], Highnam and Hill [126], Wigglesworth [379] and others. A formidable problem in understanding the regulation of diuresis is associated with diversity of insect material [92]. Different parts of the neurosecretory system may contain both diuretic and anti-diuretic substances, or, the same parts of the system may yield extracts which may have different functions in different species [249]. In some insects it has not been possible to prove that the neurosecretory system plays a role in regulating diuresis [41]. Moreover, some authors have found that extracts from the corpora allata also affect diuresis [5]. This effect may be due to their content of certain amounts of the neuro-secretory material [218].

It has been reported that the extracts from the pars intercerebralis or from the corpora cardiaca have pronounced diuretic effects [105, 138, 244, 249]. In other species extracts from the same tissues cause anti-diuretic effects as measured by water excretion by Malpighian tubules and resorption by the rectum [22, 249]. Extracts of the perisympathetic neurohaemal organs have also antidiuretic properties [22]. As found by Maddrell [202, 203] in *Rhodnius*, the axon terminals of the neuro-secretory cells located in the thoracic ganglionic mass release a very efficient diuretic hormone. On the other hand, Gersch [91] has found an antidiuretic factor in the thoracic ganglia and a diuretic factor in the abdominal ganglia of *Chaoborus*. It has also been reported that extracts from corpora cardiaca of insects have an antidiuretic effect when tested in mice [334] or measured by the frog urinary bladder assay [96].

It is thus evident that the insect neurosecretory system contains factors which affect diuresis in both a positive and a negative way. Provided that the assays for diuresis were specific for the active material, this suggests the presence of antagonistic diuretic and antidiuretic systems.

f) Myotropic Effects of the Neurosecretory System

The neurohormones of insects, like the vertebrate hormones oxytocin and vasopressin, are known to induce rhythmical contractions or affect tonicity in determined groups of muscles. The muscles which are mainly affected are those of the visceral organs [153]. The myotropic effects of the hormones are most apparent on pulsations of the dorsal aorta, peristaltic movements of the intestine, movements of the Malpighian tubules, contractions of the common oviduct during oviposition, and rhythmical contractions of the intersegmental muscles which occur during ecdysis. Various aspects of these hormone effects have been reviewed by GERSCH [90, 92], HIGHNAM and HILL [126], JOLY and CAZAL [147], RAABE [249], and WIGGLESWORTH [379].

The most intensively studied myotropic effect has been the changes in rate of beating and amplitude of contraction of the aorta produced by extracts from the corpora cardiaca. CAMERON [39] has suggested that the active factor is an orthodiphenol, although it is not identical with adrenaline. DAVEY [51, 52, 53] concluded that the active factor released from the corpora cardiaca is a peptide which acts on pericardial cells. These cells respond by releasing an indolalkylamine which is responsible for the heart accelerating activity. Several factors acting on the insect heart beat were isolated by GERSCH *et al.* [92, 95, 350] from the central nervous system. One of them was identified as acetylcholine and the others were peptidic fractions with antagonistic activity. Similar factors with heart accelerating and decelerating activity were obtained not only from the central nervous system and corpora cardiaca but also from other nervous ganglia [212, 260]. RAABE *et al.* [255] have also demonstrated the presence of active factors in the perisympathetic neurohaemal organs.

Neurohormones extracted from various parts of the insect neuroendocrine system can influence intestinal peristaltics [90], movement of the Malpighian tubules [39, 170], contractions of the oviducts [63, 147], and ecdysis [348]. The myotropic effects, especially contractions of the aorta, have been commonly used for the bioassays in isolations of neurohormones. In the *in vivo* system the heart beat is influenced by developmental stage and age of the animal, factors such as the rate of feeding, temperature, ionic changes among the tissues and haemolymph, and concentration of oxygen [275]. It should be noted, however, that in some cases removal or implantation of corpora cardiaca has no effect on the heart *in vivo*, although the cardiacum extracts have strong effects in the isolated aorta [275]. Further, these myotropic effects are influenced by various drugs and salts [153], amino acids such as glutamic or aspartic acids [131], and some toxic compounds [275]. The nervous system of insects contains the compounds like acetylcholine [170, 350], catecholamines [226], or 5-hydroxytryptamine (serotonine) [93, 212], with potent myotropic effects in addition to the mentioned peptidic fractions. It is clear that the problem of specificity in bioassays is a major obstacle in isolation of the neurohormones. Hence, it appears extremely difficult to determine which of the above factors have a real physiological significance in insect body and which are merely pharmacological factors. It seems reasonable to assume, however, that the myotropic factors in the insect neuroendocrine system, as well as factors affecting diuresis, are not identical chemically with the adenotropic neurohormone (AH).

6. Other Effects of Neurohormones

Further effects of the neurohormones include colour change, control of circadian rhythms in locomotory activity, and various aspects of insect behaviour [90, 218, 249]. With regard to the colour change we can distinguish a relatively rapid physiological change which depends on the displacement and migration of the epidermal pigments, and a morphological colour change which remains more or less constant during a developmental instar. The first response appears to be influenced by neurohormones, as has been shown by RAABE [249] in *Dixippus* and GERSCH [90] in *Chaoborus*. A significant influence on colour change has been ascribed to a neurohormone originating in the lateral neurosecretory cells of the protocerebrum [107] or in the neurosecretory cells of the suboesophageal ganglion [252].

The circadian rhythms in locomotory activity are also regulated by the neurohormones from the neurosecretory cells of the suboesophageal ganglion in *Periplaneta* [113] and eclosion of some adult moths is synchronized by day-night regimes. The mechanism responsible for this effect is located in the neurosecretory system of the brain [347]. Finally, some behavioural changes may also be associated with neurosecretory activity, such as preecdysial or ovipositing behaviour [349]. Furthermore, neurohormones can also influence behaviour indirectly through stimulation of female pheromone production [264].

7. Chemical Nature of the AH and Other Neurohormones

The neurosecretory system of insects appears to contain several factors which differ in chemical nature and physiological activity. In general, there are two basic types of neurohormones present in the neurosecretory system. One type is represented by compounds of relatively low molecular weight such as catecholamines (dopamine, adrenaline, noradrenaline), or biogenic amines like 5-hydroxytryptamine, or acetylcholine [92]. The other type is characterized by compounds of relatively large molecular weight such as peptides or proteins. The low molecular weight compounds have especially pronounced myotropic, diuretic, hyperglycaemic and neuro-excitatory effects while the peptides produce in addition to this effect adenotropic responses of AH. Since these peptidic compounds are most important we can concentrate our discussion on procedures to isolate them. For a discussion of the low molecular substances, see the reviews by GERSCH [90, 92] and WIGGLESWORTH [415].

The problems concerned with isolation of neurohormones with adenotropic effects in insects have been reviewed several times [90, 100, 162, 218, 299, 303, 396]. Assays for the neurohormones are based on the fact that decerebrated pupae of Lepidoptera do not develop until supplied with neurohormone which activates the prothoracic glands [165]. Positive result is then judged according to developmental changes induced by the prothoracic gland hormone. Thus, the assay is based on a second order neuroendocrine influence and this is less specific. The nonspecificity of the assay became apparent soon after KOBAYASHI *et al.* [166] identified an active principle from brain extracts to be cholesterol, and when positive responses were also obtained with several phytosterols [166, 167, 299], as well as some catecholamines such as noradrenaline [167].

Recently it has been observed [215] that water extracts from brains, pituitary, pancreas and other organs of vertebrates contain highly active compounds which are temperature resistant, nondialysable and precipitate by acetone or ammonium sulphate.

WILLIAMS [396] increased sensitivity of the assay by using partially cooled diapausing pupae in which some endogenous AH was already present in the haemolymph. In addition to increased sensitivity, there was a simultaneous decrease of specificity. In fact, positive responses were obtained by using salts of bivalent metal ions.

ICHIKAWA and ISHIZAKI [133, 134] obtained active extracts from insect brains which contained water soluble, temperature-resistant and nondialysable materials. The materials were resistant against enzymatic hydrolysis by pepsin and trypsin but the activity was lost after treatment with potent bacterial peptidases nagase and pronase. The authors have concluded that the active material is a protein with molecular weight of 9,000 to 31,000 [136].

At the same time GERSCH and his coworkers suggested on the basis of their earlier work with extracts from *Periplaneta* that the AH was a protein or peptide [89]. More recently, they have identified several active fractions of which the so called "activation factor I and II" had a molecular weight between 20,000 and 40,000 and possessed considerable AH activity [97]. In further studies KOBAYASHI *et al.* [169] also confirmed the proteinic nature of AH and succeeded in achieving a 2,000-fold purification obtaining a protein (probably glycoprotein) of 20,000 molecular weight and an isoelectric point of approximately 8.5 [406].

Finally, it may be noted that according to an earlier emphasis AH may be a mucopolysaccharide [396].

Analysis of water extracts from corpora cardiaca revealed the presence of at least six neurohormones which differed in physiological activities but were peptides in nature [36, 212]. So far, nothing definite can be said about the chemical nature of the AH except that it seems to be a protein with molecular weight between 10,000–50,000.

8. Other Proteinaceous Neurohormones with Specific Effects

In addition to the neurohormones which regulate some general physiological processes there are also those which are involved in regulation of specific developmental or physiological events in certain groups of species of insects. Among these hormones are the diapause hormone, bursicon, and the accelerator of pupariation in flies.

The diapause hormone is secreted by the neurosecretory cells of the suboesophageal ganglion of *Bombyx mori* pupae and its release into the haemolymph is regulated from the brain via connectives to the suboesophageal ganglion [87]. The presence of the diapause hormone in the female insect during ovarian maturation results in deposition of diapausing eggs [115, 116]. Like some other neurohormones, the diapause hormone influences carbohydrate metabolism causing an increase in glycogen deposition in the ovaries and accumulation of trehalose in the haemolymph due to increased metabolism of fat body glycogen. Other physiological effects of the hormone depend on its influence on trehalose activity in the ovaries and on the concentration of 3-hydroxykynurenine associated with changes in serosal pigments [117, 404, 405].

Postecdysial hardening and tanning of the cuticle is in some insects stimulated by a sudden release of neurohormone [50, 77]. FRAENKEL and HSIAO [78] subsequently concluded that this hormone, bursicon, was not identical with any of the known hormones. In the newly emerged adult fly, for example, the hormonal release is initiated when the insect starts to dig its way out the soil. The stimulus for bursicon release depends on the neuromuscular activity associated with intensive locomotory activity. The hormone appears to be a protein with molecular weight of 40,000 [79]. This suggestion was confirmed also in studies on *Periplaneta* [206]. Bursicon originates in the neurosecretory cells of the brain and is released into the haemolymph by the corpora cardiaca but its presence has also been demonstrated in the ganglia of the ventral nerve cord [353]. Biochemical studies on *Periplaneta* revealed that there was no correlation between the presence of the hormone and polyphenoloxidase activity, which are important enzymes in cuticular hardening [205]. In flies, bursicon affects tyrosine hydroxylation and its effect results in accumulation of dihydroxyphenylalanine (DOPA) following inhibition of DOPA decarboxylase [307]. Bursicon also stimulates melanization of the outer cuticular layers and formation of postecdysial endocuticle and this requires RNA and protein synthesis [75]. According to Mills and Whitehead [207] certain properties of bursicon bear resemblance to factors with diuretic effects. Both have similar molecular weights and both increase permeability of the haemocytes for tyrosine which enhances the formation of N-acetyldopamine. Gersch [92] has observed that chromatographic fractions with bursicon activity occur close to the materials with AH activity.

Recently a neurohormone has been described which accelerates formation and tanning of the puparium in cyclorrhaphous flies. The active factor, or factors, appear few hours before pupariation in larval haemolymph and hasten muscular contractions, cuticular shrinkage and tanning coordinating thus the complicated morphogenetic process of conversion of a soft larva into a hard sclerotized puparium. The active principle occurs in water extracts from the central nervous system or corpora cardiaca of various insect species. It is not identical with bursicon, though it is also a protein with similar properties. The function of the hormone is different from that of the AH. It acts after ecdysone had already potentiated the onset of metamorphosis [410].

B. Corpora Allata (CA) and the Corpus Allatum Hormone (CAH)

The main aspect of this book is consideration of the natural and synthetic materials which mimic the action of juvenile hormone. Although these substances are commonly referred to as juvenile hormone analogues, we will use the term juvenoids. In order to facilitate comparisons with different compounds, we have briefly summarized in this chapter the effects on development and metamorphosis of the endogenous hormone obtained by transplantations and extirpations of the CA. We have used the term corpus allatum hormone instead of juvenile hormone to avoid possible confusions with the endogenous hormone and the mimetic substances. Special attention has been paid to the morphological changes associated with the CA activity and to the physiological processes stimulated or inhibited by secretion of the hormone from the glands.

The presence of organs known as corpora allata was observed by insect anatomists as early as in 1762 [237]. Histological evidence suggesting their endocrine activity began to accumulate at the beginning of 20th century [218]. Finally, their physiological action which depended on the inhibition of metamorphosis and the stimulation of ovarian development was discovered by WIGGLESWORTH in 1935 [365, 366]. Since then, morphological and physiological studies on the CA and its hormone made rapid progress. A more detailed review of these studies is outlined in the monographs [90, 126, 218, 237, 370, 379] and is also briefly summarized below.

1. External and Internal Anatomy and Morphology of the CA

Most insects have paired CA which are usually spherical or slightly elongated and located in the hind part of the head or in the cervical region close to the aorta or oesophagus. The CA are connected to the distal part of the corpora cardiaca (CC) to which they are linked by the nerve, *nervus corporis allati* (see Figs. 1 and 4). Further nervous connections lead usually to the suboesophageal ganglion. Differences in the anatomical structure of the CA in different species are most apparent in the degree of fusion of the originally paired organs [218]. In some insects the CA and CC are separated, while in other insects the CA and CC are paired. In some species fusion of one or the other gland occurs. Thus in these insects the CA can be separated with fused CC or fused into one compact gland. Finally, as in the case of the higher Diptera, the CC, CA and also the prothoracic gland may be fused into one complex gland, the ring gland (Fig. 4).

The CA originate as ectodermal evaginations of the mandibular and maxillar pleurite. Their presence has been confirmed in all groups of Pterygota and homologous glands have also been observed in several groups of Apterygotes [218, 237, 379].

Histologically distinct CA have been observed. The most primitive is the vesicular type of CA with parenchyme cells distributed in the peripheral zone. These parenchymal cells may form concentric lamellar structures with connective tissue or liquid inside. The second morphological type is the compact or massive type represented by parenchymal cells distributed throughout the whole volume of the organ. Slight structural differences may also occur between peripheral and centrally located parenchymal cells. In some species the CA cells contain only relatively small amount of cytoplasm resembling lymphocyte cells of vertebrates. Another, more common morphological type, is characterized by large number of relatively small parenchymal cells. Finally, the cells corresponding to CA portion of the ring gland in Diptera are relatively few but considerably larger [40].

Light microscope studies have revealed the presence of rather thick extracellular connective sheath with numerous invaginations extending into the gland. The cell membranes of the parenchymal cells are irregular and poorly resolved, giving an impression of a syncy-

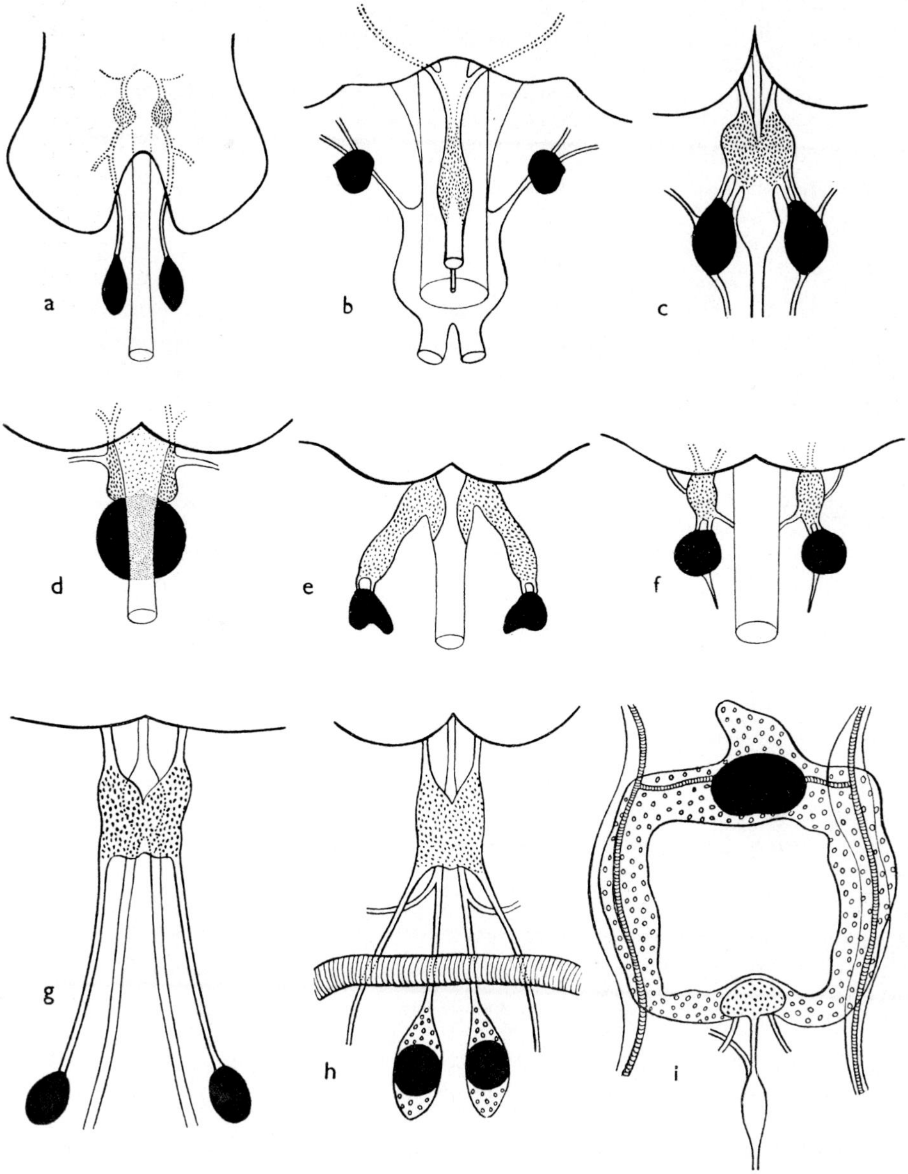

Fig. 4. Diagram with structures of the retrocerebral endocrine glands in various groups of insects. Dotted = corpora cardiaca, black = corpora allata, small rings = peritracheal or pericardial glands. *a* = *Japyx* sp., *b* = *Ephemera vulgata*, *c* = *Blatta orientalis*, *d* = *Pyrrhocoris apterus*, *e* = *Hydrous piceus*, *f* = *Sphinx ligustri*, *g* = *Sialis lutaria*, *h* = *Culex* sp., *i* = *Calliphora erythrocephala*. From Novák [218]

tium. The nuclei have irregular shape or are ovoid, slightly elongated with distinct nucleoli. Cytoplasm contains many vacuoles and inclusions [85, 291, 292, 301].

Electron microscopic studies have shown that cells of active CA are characterized by increased cytoplasmic and nucleolar volume as compared to cells of inactive glands. SCHARRER [278, 282, 290] has studied the ultrastructure of CA in *Leucophaea*, WAKU and GILBERT [358] in *Hyalophora*, KING *et al.* [159, 160] in *Drosophila*, PANOV and BASURMANOVA [231] in *Eurygaster*, JOLY et al [150, 152] in *Locusta*, ODHIAMBO [223] in *Schistocerca*, FUKUDA *et al.* [85] in *Bombyx* and THOMSEN and THOMSEN [345, 346] in *Calliphora*. Most of the investigators agreed that the spherical, ovoid, or irregular nuclei with smooth membranes were either larger or more numerous in the active glands. These nuclei contained two or more nucleoli which were characterized by a round cortical layer forming a thick ring around the nucleolar margin [231]. The cytoplasm of these cells contained numerous mitochondria, which were elongated, with well developed cristae [148, 159, 160, 358). The presence of well developed cristae suggested high respiratory metabolism. Furthermore, the cytoplasm also contained more free and membrane bound ribosomes than cytoplasm of cells in inactive glands [150, 358]. The Golgi complex was usually inconspicuous and did not show evidence for intensive secretory activity [85, 231, 346, 358]. The plasma membranes were folded and the cell membranes interdigitated. Between the cells intercellular spaces were observed and the cells appeared to communicate by thin processes [278]. In some other insects these intercellular spaces were absent [231]. The acellular stromal sheath on the surface of the gland was similar to that observed on other glandular organs [278].

In most insect species the cells of the CA have well developed agranular or smooth surfaced endoplasmic reticulum [85, 150, 152, 159, 160, 225, 278]. But in *Eurygaster* [231] and *Locusta* [150, 152] the granular endoplasmic reticulum was highly developed. The most conspicuous cytoplasmic structures were the systems of small vesicles arranged in the form of concentric lamellae or spirally oriented whorls. In certain species these structures were found only in the active glands [278, 358], while in other insects they occurred in both active and inactive glands [85] or in some cases the whorls occurred only in the inactive glands [150, 159, 160, 346]. The cytoplasm also contained osmiophilic bodies which represented lipid droplets and have been observed only in active glands [85, 358]. It has been suggested that these lipid droplets may originate from the vacuoles released by the microtubules of the smooth surfaced endoplasmic reticulum [346]. Inside the CA there are also axons of neurosecretory cells containing granules of high electron density [278, 358] and the presence of muscle fiber inside the CA has also been reported [278].

B. Scharrer [290] has described the general cellular structures of CA of *Leucophaea* as viewed with the electron microscope. She has discovered special electron dense bodies (c-bodies) originating in the Golgi zone. The distribution and fate of these bodies during various periods of the animals life cycle suggested relationships with secretory functions of the CA cells.

It has been suggested that CAH is synthesized in the tubules of the endoplasmic reticulum [223], transformed into vesicles, taken up and dissolved by the lipid droplets and finally released in lipid droplets into the haemolymph [346]. This would be consistent with the assumption of the lipoid nature of the hormone [159]. In *Leucophaea* it has not been possible to determine the character of the secretion according to the ultrastructural patterns, however, the characteristic patterns of protein secreting structures have not been found [278]. In contrast to *Leucophaea* a well developed ergastoplasm in *Eurygaster* [231] suggested the possibility of protein secretion. Earlier cytological studies on CA of *Chrysocornis* revealed certain similarities with the cells of anterior pituitary and the hormone of the CA was also assumed to be a protein [210].

2. Growth and Secretory Activity of the CA during Development

Changes in the volume of CA during larval growth have been studied and reviewed by Pflugfelder [237] and Novák [218]. In each larval instar the volume is approximately doubled. The largest increase usually occurs between the second and third larval instars. During this time, however, there occurs also greatest increment of the whole body. The pattern of growth in the CA differs according to particular insect studied. In cells of CA of some insects both nuclei and cytoplasm enlarge after each ecdysis. Following the cellular increase glandular volume becomes larger and mitoses begin. In species like bees, the nuclear and cellular size increases during the whole larval period [237]. In *Bombyx* the number of cells in the CA remains constant but gland volume increases because of cell enlargement [185]. We may observe in general that the CA cells of Endopterygote insects grow by cell enlargement as do many other glandular tissues. Among Exopterygote insects the gland volume enlarges due to an increased number of cells.

Some authors have recorded maximal increase in gland volume during the middle of the instar, followed by a decrease in volume near the time of ecdysis [218]. In some insects, however, a maximal increase in volume occurs at the time of ecdysis [185] while in other species there is a gradual enlargement of the gland volume throughout the larval stage [146]. Hence, with respect to gland volume there are many variations in size of the gland during a moult cycle. It is assumed that the CA are very active in all larval instars except the final larval one. NOVÁK [217] noticed that enlargement of the CA was not proportional to the increase in volume of the central nervous system or whole body. The disproportionate increase in volume or surface of CA was used to explain the physiological inactivity of the CA during the last larval instar. Recently the problem of CA inactivation in last instar larvae has been reviewed by WIGGLESWORTH [379]. Available evidence suggests that the CA are inactive at the beginning of the larval instar. However, when dissected from these larvae and transplanted into other last instar larvae the glands exhibit perfect hormonal activity [146, 218, 230, 379]. In certain species such as *Leucophaea* the CA of the last instar larvae may become active after cutting their nervous connections with brain [199]. LÜSCHER and ENGELMANN [199] attempted to use a morphological criterion for determining CA activity by assuming that a real secretory activity could be measured by an "activity volume" as a difference between the gland volume and a certain minimum CA volume with equal number of nuclei. The "activity volume" data indicated that in *Leucophaea* the larval CA were active with exception of the last instar when the "activity volume" dropped to zero.

During metamorphosis, the CA usually exhibit all morphological signs of secretory inactivity, such as decreased total volume, reduced amount of cytoplasm in the parenchymal cells, and reduced chondriome. The signs of activity begin to appear again at the end of the pharate adult period [84, 218] or, most commonly at the beginning of adult stage. Exceptionally, the glands of diapausing metamorphosis stages, such as prepupae and pupae of Lepidoptera seem to be present in active form even during diapause. These signs of activity of the CA disappear with diapause termination [81, 119, 357]. It has been assumed that in these cases an active neurosecretory system and CA are essential for diapause while an active neurosecretory system and prothoracic glands but an inactive CA are essential for postdiapause development [81]. In contrast to this, CA implanted into decerebrated pupae of the silkworm stimulate adult development. Apparently, the glands contained neurosecretory material with prothoracotropic activity [168].

Changes in the volume of CA have been most intensively studied in adult insects in connexion with reproduction and adult diapause (cf. ADAMS [1], DOANE [57], ENGELMANN [66, 67, 70], HIGHNAM [121], HIGHNAM and HILL (126], JOHANSSON [141], and DE WILDE [384]). In general, the glands of freshly emerged adults of both sexes are relatively small and according to histological observations the glands are inactive. When the adult insects start to feed, the gland volume rapidly increases and the amount of cytoplasm and cytoplasmic inclusions also increase. Now the gland acquires the ability to secrete the hormone. Several examples illustrate the relationships between CA activation and changes in neurosecretory system [66, 340] which are dependent upon feeding [140, 331].

Following the initial increase in CA volume the glands of certain adult insects continue to enlarge and eventually, as is the case in bees, old adults contain extremely large CA. This enlargement is usually associated with certain degenerative changes such as amitotic divisions, or pyknosis [237], suggesting that these extremely large glands need not be the most active ones. Such enlargement of CA may occur in adults of both sexes but seems to be more common in males.

In a large number of insect species, the female CA undergo cyclic changes of their volumes which can be correlated with the cyclic patterns occuring during ovarian development, feeding periods, and oviposition [1, 331]. In cockroaches, for example, the cyclic changes in CA volume are causally connected with general cyclicity of reproduction [70]. However, in other female insects reproduction maintains a cyclic nature even after transplantation of many nonsynchronized CA. In this case the less pronounced cyclical changes of CA volume would not be a cause but rather a consequence of physiological changes associated with reproduction. Finally, there are females of many species where reproduction and oviposition proceed more or less continuously and the CA volume does not undergo cyclical changes in volume. As already mentioned, stimulation of growth in the CA is dependent on AH from neurosecretory system. In reproduction, there occur special interactions between developmental state of the ovaries, presence of spermatophore, mating, feeding, etc., and the CA activating mechanisms. The proprioceptive stimuli taking part in these interactions act by nervous or hormonal pathways and it is not understood whether these stimuli act through the neurosecretory system or directly on the CA [66].

The main cause of adult diapause is a lack of AH which results in lack of CA activation. This is followed by suppression of behavioural and physiological patterns prerequisite for reproduction. In diapausing adults CA are always subnormal in size. Hormonal control of adult diapause in *Leptinotarsa* as studied by DE WILDE and his associates [384] is a good example of the role of CA in adult insects.

Distinct differences in size of the CA are associated with caste dimorphism in social insects like bees [237] and termites [197]. In addition, in some insects there occurs sexual dimorphism with respect to CA size and structure. For example, in adult Cecropia the CA of males is 6–8times greater in size than the CA of females [102]. Although similar relations also exist in a few other insects, e.g. *Ephestia*, in the majority of insects including also other Lepidoptera the female CA are larger [237]. In

Bombyx silkmoths the male CA in contrast to the female CA are not only smaller but also inactive [85]. In general, CA are larger in females where growth of the ovaries is stimulated by action of CAH.

Following castration the CA may enlarge tremendously [292] which may indicate some feedback relationships between the functioning gland and the physiological state of the ovaries. Furthermore, there are clear indications that such a feedback exists between the haemolymph titre of CAH and growth of the CA. For example, implantation of active CA into freshly emerged adult females prevent growth of the *in situ* CA, which is known as a "compensation phenomenon" [56, 218, 229, 237] (see also p. 265 and Fig. 31). The implanted glands may become extremely large [59]. When implantation is performed at the time when the host's CA are maximum active, the compensation effect is exerted on the implanted CA which cease to grow. ADAMS [1] has assumed that the ovaries release an oöstatic hormone which has a feedback effect on CA activity. In many insect species, nevertheless, the ovarian function continues uninterrupted even in the presence of an excess of CAH for prolonged time. In these cases the presence of such ovarian hormone is not implied [379].

In addition to the effects of CAH already mentioned, active CA may also influence size of the neurosecretory cell nuclei and it has been observed that the volume of CA may not be always proportional to the actual rate of secretory activity [343].

3. Action of CAH during Larval Growth and Metamorphosis

The spectacular changes in postembryonic development which occur as a consequence of CAH activity may be easily recognized macroscopically as alterations in the size and pattern of the epidermal structures. As a result of these morphological changes the role of CAH in metamorphosis was recognized early in developmental studies on insect hormones by WIGGLESWORTH in 1935 and since then the basic developmental effects of CAH have been reviewed several times [90, 126, 218, 237, 370, 379]. Because a major part of biological research on synthetic juvenoids has been based on their effects on metamorphosis, we decided to summarize the conditions necessary for CAH action in: i) young larval instars, ii) last larval instars of Exopterygotes, iii) Endopterygotes, and iv) in the pupal stage.

a) Action of CAH in Young Larval Instars

Young larval instars are the developmental periods from hatching of the larvae until ecdysis into the last larval instar. We have mentioned already that during this time the CA increase in size and secrete the hormone. WIGGLESWORTH [371] elegantly demonstrated in a series of experiments involving parabiosis and transplantation of larval CA into allatectomized larvae [365] or adults [366] of *Rhodnius* that the CA of young instars are really secreting CAH.

The action of CAH in young larval instars results in the maintenance of the larval stage in development until the insect has accumulated enough reserve materials to construct an adult of determined size and shape with the ability to reproduce. Hence, the most apparent consequence of CAH action is the prevention of metamorphosis. This was clearly demonstrated in the early experiments of BOUNHIOL [29] who induced precocious metamorphosis in young larval instars of Lepidoptera by ablations of CA.

The young larval period can be divided into successive developmental cycles represented by individual instars and separated by ecdyses. The larval character of each developmental cycle is determined by presence of CAH and all such larval cycles have many physiological features in common. One of them is isometric growth. Another pattern is the relatively short duration of the instar, which can be one third or one quarter shorter than the last larval instar in Exopterygotes or even three quarters shorter in some Endopterygotes. Further developmental patterns are represented by specific distribution of mitoses among the epidermal cells [367], origin and distribution of small epidermal organs [371] or epidermal pigments [218]. Other distinguishing characteristics of these larval cycles are the growth patterns on the wing lobes in Exopterygotes [151, 310] or wing imaginal discs in Endopterygotes [302]. In addition, special physiological patterns in the rate of feeding and metabolic activity occur during the young larval period.

Dependence of all these complex processes on CAH is evident after removal of CA from young larvae. As a consequence of allatectomy the larval isometric growth cycle is replaced by allometric growth, which normally occurs only in the last larval instar. Following ecdysis miniature adults [310] or miniature pupae [29] are observed. Implantation of extra CA into young larvae has no apparent effect on development of young larval instars, however, the period of isometric growth is extended and more larval developmental cycles are produced with formations of supernumerary larval instars. WIGGLESWORTH [366] suggested earlier a possibility that the wing lobe cells of *Rhodnius* might develop differently in young larval instars, depending on the titre of CAH. But later [374] he abandoned this suggestion. At present there is no reliable evidence that an excess titer of CAH can alter the normal course of development in young larval instars.

Secretory activity of the CA is inhibited at the last larval ecdysis. The physiological mechanism by which the CA is turned off has not been completely elaborated but appears to be located in the nervous or neurosecretory system. This mechanism represents a hormonal "switch" which allows morphogenesis to proceed again after its suppression by CAH at the end of embryonic development. Indeed, both the periods of intensive morphogenesis, i.e. embryonic development and metamorphosis have many common features [218]. In some insects the initial processes of morphogenesis prerequisite for metamorphosis occur already in the penultimate larval instar. For example, the last larval instars of some Coleoptera (*Dermestes*) or sawflies possess new structural patterns never present in the young larval instars, such as new small epidermal organs, pigmentation, different structure of the mouth organs and many other morphological patterns. These structures have differentiated during penultimate larval instar and indicate that the process of metamorphosis began here before the last larval instar. Finally it has to be mentioned that in a primitive group of Apterygota the morphological difference between young larval instars and adults is very small. The adults are recognized according to reproduction activity and the process of larval-adult transformation in these insects is also conditioned by CAH [263].

b) *Action of CAH in Last Larval Instars of Exopterygotes*

Like adults, the larvae of Exopterygotes occur at the oligopod developmental stage. Metamorphosis occurs predominantly in the last larval instar and the amount of morphological and physiological change that animal undergoes during the transformation from larval to adult stage is much the less than the changes which occur in Endopterygotes. The most apparent metamorphic transformations in the Exopterygotes include development of the wings and external genitalia. Internally, changes occur in the development of indirect flight musculature, gonads, sexual ducts, and accessory sexual glands. Furthermore, significant alterations in the structure of the epidermal cells occur, e.g. formation of new pigment patterns or small epidermal organs such as hairs, bristles, and dermal glands.

Thus it is evident that the problem of CAH action is associated primarily with the morphological changes characteristic for metamorphosis. Our intention is to describe briefly the basic phenomena of CAH action in the metamorphosis of Exopterygote insects while a more complete description can be found in the monographs [126, 218, 237, 379] and is also presented in connexion with the action of juvenoids (Part II). We feel that such a brief recapitulation of the effects of CAH may aid in understanding of the mode of action of the juvenoids.

The effects of CAH on morphology of Exopterygote insects are easily recognized. When the CAH is present at the beginning of the last larval instar, the morphological changes associated with metamorphosis are prevented and at the next ecdysis a giant larva occurs which is approximately the same size and weight as the adult. The giant larva is commonly called a supernumerary larva or an extra-larva. When larvae are provided with a source of CAH a few days after the last larval ecdysis, larval-adult intermediates or adultoids are produced at the next moult. Finally, when a source of CAH is provided in the second-half-period of the last larval instar the CAH is ineffective and normal adults are produced. Thus, the difference between the presence and absence of CAH in the last larval instar of Exopterygotes is exemplified by the above morphological differences between larval and adult structures with all the physiological consequences resulting from structural changes among the organs or tissues. The original observations on the action of CAH were made by WIGGLESWORTH [365] on *Rhodnius* and later they have been confirmed with CA implantations in all major groups of Exopterygote insects [218, 237].

The supernumerary larvae produced by CA implantations were able to feed, grow, and eventually to moult into other extra-larval instars or adultoids. Most of extralarvae and especially adultoids showed certain physiological disturbances during ecdysis and were unable to escape from the old exuviae. The relatively small differences in size between extra-larvae and adults of the same instar suggested that CAH had no direct effect on the rate of nutrition and digestion. The length of the instar in which CAH was supplied was proportional to that of previous larval-larval instars, but generally shorter than the larval-adult instar. The epidermis of extra-larvae produced by CA implantations may be completely larval in most species. However, some epidermal structures such as wing lobes had sometimes partially metamorphosed into adult wings. In order to be affected these structures require the presence of CAH very early after or even before the last larval ecdysis. Growth of extra-larval structures follows the normal growth curve. Distribution of small epidermal larval organs on the body of extra-larvae follows the same distribution relationships which occur in young larval instars. Thus, the development of an extra-larva is a physiological continuation of previous larval developmental cycles. The preservation of larval patterns is considered as one out of several developmental abnormalities called metathetely [218] in distinction to precocious appearance of adult patterns on larvae which is called prothetely.

Implantations of CA into last instar larvae reveal that the sensitive period for CAH action occurs at the beginning of the instar. The critical period of sensitivity is dependent on the activity of AH-prothoracic gland system. Starvation or exposure to low temperatures which inhibit hormonal release from the AH-prothoracic glands system can delay the sensitive period for many days or weeks. The time at which cells or tissue become insensitive to CAH and are committed to metamorphosis varies with the cells and the tissue. In most insects the cells of the wing lobes or external genitalia pass their critical period sooner than epidermal cells of the abdomen. For example, implantation of CA within the sensitive period led to appearance of adultoids with adult patterns on the wings but larval patterns on the abdominal epidermis. These rather common results of CA implantations suggest that CAH has not acted since the very beginning of the sensitive period. In contrast to CA implantations allatectomy has no apparent effect on morphological and physiological changes which occur during the last larval instar [315]. The effects of allatectomy may appear later in adult stage in form of arrested reproduction, behavioural changes, etc. Because similar responses with respect to the sensitive periods occur following the application of juvenoids, we are discussing these problems in more details in Part II.

Several authors have attempted to determine the cytological differences between the development of larval and adult instars in Exopterygote insects. WIGGLESWORTH [371] discovered certain differences in distribution of mitoses among the abdominal epidermal cells in *Rhodnius*. JOLY *et al.* [151] have also observed differences in the distribution of mitoses and orientation of the cells in the wing lobes of *Locusta*. By combinations of allatectomy and CA implantations SLÁMA [310] succeeded in inducing both larval and adult type of development in several successive developmental instars of *Pyrrhocoris*. Cytological differences between larval and adult development were observed in the time of epidermal cell apolysis, timing and orientation of mitoses, extensive breakdown of cells and nuclei and formation of chromatic droplets during metamorphosis. The formation of chromatic droplets indicating a breakdown of cells were earlier observed by WIGGLESWORTH [368]. Further differences were found in the amount of cellular differentiation, formation of small epidermal organs and finally differences in structure of the newly formed cuticles [310]. Cytological observations are in agreement with the assumptions that the sensitive cells to CAH should be in the interphase state. The hormone seems to be ineffective on cells which were already undergoing differentiative divisions [310]. One of the major differences between larval and adult developmental programmes depends in the amount of cell differentiation. The cellular differentiation process is very intensive in adult development while it is restricted to minimum in larval development. There occurs also some differentiation in presence of CAH during young larval instars, as demonstrated on dermal glands of *Rhodnius* [371, 378]. However, the amount of such differentiation among the cells is far less when compared with metamorphosis.

c) Action of CAH in Last Instar Larvae of Endopterygotes

The hormonal conditions which determine the initial processes of metamorphosis in Endopterygotes are more complex and diversified than those in Exopterygota. In certain Endopterygote insects it is possible to produce larval-pupal intermediates or extra-larval instars by implanting active CA into last larval instars. The experimental results are similar to those described for Exopterygote insects. Thus, *Galleria* larvae can undergo one or more extra-larval moults after transplantation of brain-cardiaca-allata complexes [239, 306]. The transplantation of the complex was still effective when performed in the middle of the last larval instar [302]. This suggested that the sensitive period for CAH action in *Galleria* was not restricted to the time immediately after ecdysis. Moreover as in Exopterygotes the extra-larvae of *Galleria* showed all the characteristic morphological and physiological patterns found in young larval instars [306]. Cytological differences in epidermal cells between larval-larval and larval-pupal developmental cycle have many common features with larval-adult cycle [305]. Extra-larvae were also obtained after CA implantations in *Bombyx* [84] and larval-pupal intermediates in *Tenebrio* after transplantations of brain-cardiaca-allata complexes into larvae [258].

In certain Endopterygote insects it is extremely difficult or impossible to produce extra-larval instars or larval-pupal intermediates by implantations of CA. This is true even for some species where implantations of CA into pupae produce a second pupal moult. The usual effect of CA implantations into last instar larvae is merely an inhibition of metamorphosis as indicated by a delay or absence of pupation. This phenomenon has been described in more detail for higher Diptera by Possompès [247]. A delay in metamorphosis, e. g. in larval-pupal transformation, after implantation of CA also occurs in Saturniids, in Coleoptera such as *Dermestes*, and in Hymenoptera. Most species responding to CA implantations into last larval instars in this manner have larvae which develop rapidly. The ability of CAH to cause a delay in the onset of metamorphosis and its inability to cause extra-larval instars or pupal intermediates suggest that a possible selective advantage of this adaptation might depend in prevention of spontaneous, lethal, metathetelic forms. This might possibly occur in species with rapid larval development after incomplete degradation or excretion of CAH residues from young larval instars. These adaptions are pointing out to the presence of physiological anti-juvenile mechanisms discussed later in connexion with juvenoids.

d) Action of CAH in the Pupal Instar

The action of CAH on pupal-adult development is quite similar to that on larval-adult development in that it inhibits formation of adult structures and leads to repetition of pupal structures. For example, PIEPHO [241] demonstrated that implantation of the CA into pupae of *Galleria* resulted in formation of second pupae or pupal-adult intermediates. In *Galleria*, as well as in many other Endopterygote insects, the sensitive period for CAH action occurred at the very beginning of the pupal instar. In certain pupae, the sensitive period has passed within a few hours after larval-pupal ecdysis. Therefore, in order to obtain positive results the CA must be implanted into freshly ecdysed pupae or into prepupae. The sensitive period in insects having a pupal diapause is restricted to the first few days of postdiapause development [392]. However, implantations of CA near the end of diapause may also result in formation of nearly perfect extra-pupae [395]. Adultoids or extra-pupal instars produced by CA implantations are unable to shedd off the old exuviae. Although the new cuticle is formed, absorption of the old cuticle more or less complete, and the ecdysial line ruptured, specific synchronous movements of the body characteristic for adult eclosion are usually lacking. Thus the failure to ecdyse may result from disturbances in somatic muscle development.

Some pupae contain specific structures which are absent in other stages, e.g. thoracic tubercles and other small epidermal organs. These structures not only persist on the body of extra-pupae, but also new structures differentiate with each extra-pupal moult (Fig. 24). This suggests that CAH has no general differentiation inhibiting effect in pupal stage. The hormone is interfering only with the differentiation associated with a new programme of pupal-adult morphogenesis.

4. Action of CAH on Reproduction and Adult Diapause

The effects of CAH on insect reproduction have been extensively reviewed. In addition to the monograph by ENGELMANN [70] and his earlier reviews [66, 67] there are also reviews by DOANE [57], HIGHNAM [121], JOHANSSON [140], WIGGLESWORTH [371], DE WILDE [382], and others [90, 218, 237, 370]. The effects of CAH on adult diapause have been reviewed by DE WILDE [383, 384].

Stimulation of ovarian development in *Rhodnius* by CAH was discovered by WIGGLESWORTH [366]. Extensive studies on *Oncopeltus* [140], *Dytiscus* [143], *Melanoplus* [236], *Leucophaea* [277], *Calliphora* [340] and on many other unrelated species or groups of insects [70] confirmed rather common occurrence of the gonadotropic effects of CAH. In the young larval instars the gonads are not completely differentiated and the effect of CAH on this immature tissue is similar to that on other larval tissues. The gonads grow in size but do not complete differentiation until metamorphosis. Thus, the CAH acts differently on immature gonads than on fully differentiated adult gonads. In many insects, such as Lepidoptera, Coleoptera and Hymenoptera, the reserve materials required for development of the eggs are accumulated by larval stages. Ovarian development and formation of mature eggs in these insects take place concurrently with metamorphosis of other tissues. Furthermore, these processes are stimulated by AH-prothoracic gland hormone. Extirpations of CA from last instar larvae or pupae of these insects had no effect on egg development suggesting that CAH was not required for the process [314]. Moreover, in certain adults of Diptera the CAH does not stimulate ovarian development [180] which is under the control of the neurosecretory system [387].

Control of reproduction by CAH is quite common in other groups of insects. The CAH stimulates ovarian development and formation of mature eggs in insects such as Apterygota [268], various orders of Exopterygota, e.g. Orthoptera, Dermaptera, Dictyoptera, Hemiptera (with exception of Phasmida); among Endopterygotes it acts in some Diptera, Coleoptera, and occasionally in some Lepidoptera [70].

Feeding is a common stimulus for release of hormones with the gonadotropic effects. Allatectomy of adult females inhibits ovarian development but does not prevent feeding and digestion. This results in enormous hypertrophy of the reserve organs or finally of the whole body. Observations on alltectomized females of Orthoptera [236], Dictyoptera [66, 67], Hemiptera [139, 140, 312, 371], Diptera [340] and others [218] suggested that feeding can take place independently from reproduction and is probably controled by hormone different than CAH. Implantations of active CA into hypertrophic allatectomized females induced ovarian development and egg formation but did not stimulate feeding [312] indicating the independence of both processes.

The main target of CAH in adult female are the ovarian follicle cells. Following stimulation by the hormone these follicle cells divide mitotically, endomitotically, or amitotically according to species, and their nuclei and cytoplasm increase tremendously in size. After their division the normal morphological and physiological events associated with vitellogenesis occur. The activated follicular epithelium of each oöcyte undergoes a developmental cycle similar to that of epidermal cells during a moult cycle. At the end of the cycles the follicular cells deposit a chorion. In ovaries with well separated germarium and vitellarium parts it is the vitellarium which responds to CAH while the germarium develops spontaneously or is coordinated with the development of the vitellarium by the neurosecretory system. A simple criterion of CAH action on ovaries is the size of the oöcytes and their content of yolk.

In contrast to its action on immature insects, the action of CAH on ovaries of adult females does not require the presence of other hormones. Stimulation of growth and divisions of the ovarian follicle cells by CAH is in many respects similar to the action of the prothoracic gland hormone on the epidermal cells. With regard to the number of eggs produced there is no direct relationship to the titre of CAH in the body. Since feeding is not under general control of CAH and the number of eggs to be produced depends on the state of nutrition in the body, the CAH apparently only stimulated vitellogenesis but the number of eggs depended on the available nutrients. During vitellogenesis considerable amounts of reserve materials are consumed and transformed into egg yolk. This has profound effect on the nutritional state of the body and both the processes of vitellogenesis and nutrition are synchronized by specific feed-back mechanisms.

Physiological systems leading to activation and inactivation of CA during reproduction represent a complex problem discussed in more detail by Engelmann [70], Highnam [121], Johansson [140], and Schooneveld [301]. Environmental and internal stimuli are received by the nervous system and adenotropic neurohormone is released which in most insects activates the CA. Among the environmental stimuli which can lead to CA activation are initiation of feeding, long-day photoperiod, optimum temperature changes, gregarious or solitary conditions, mating, perception of sexual pheromones, and various other stimuli which may be species specific.

Implantation experiments revealed that CA of young larvae were not always able to substitute the action of adult CA when implanted into adult females. Theories suggesting that larval and adult CA might release different hormones have been proposed. However, the volume of the larval CA may be more than 30 times smaller than that of adult glands [132], and hence the level of CAH after their implantation may not have been sufficient. This may account for the difference mentioned above. At present, there is a general agreement that both larval and adult CA release identical hormone [cf. 70, 126, 218, 237, 379].

Another tissue in adults of both sexes which responds to CAH by pronounced growth and morphological changes is the epithelium of the accessory sexual glands. Activated cells synthesize and transport specific materials into the lumen of the gland. In females the gland develops concurrently with the ovaries. The chemical nature of the product by the accessory glands is different in different species. In females it is essential for specific properties of the egg chorion. The effects of CAH on accessory secual gland functions in males are generally less pronounced and in certain species the male gland can function without CAH.

Moreover, the influence of CAH on male reproduction is generally less apparent. The hormone is not the main stimulus for spermiogenesis which in many insects occurs already during metamorphosis in absence of CAH. The newly emerged males of these insects possess mature spermatozoids in the testicular vesicles or spermaducts and the effect of CAH on reproduction is here limited to stimulation of accessory sexual gland activity or to changes in behaviour.

Reproduction may be suspended for prolonged periods during unfavourable environmental conditions. This is commonly represented by adult diapause which is caused by physiological inactivity of the neuroendocrine system. Conditions regulating the onset and termination of adult diapause have most intensively been studied in Coleoptera [143, 383, 384], Diptera [179, 340], Hemiptera [140, 312, 313], and others [384]. Significance of neuroendocrine system in regulation of adult diapause is evident from the fact that transplantations of the active brain-cardiaca-allata complexes lead to termination of diapause and restoration of reproduction. Implantations of CA alone may stimulate ovarian growth [37] but do not remove all the diapause symptoms [381, 385]. One of the most important diapause features in adult females is the suppressed ovarian growth. This is caused by a lack of CAH as a part of general hormonal defficiency.

5. Physiological and Biochemical Effects of CAH

Any hormone with morphogenetic effects should influence physiological functions and biochemical changes in its target tissues. However, the main obstacle in studying the physiological changes induced by CAH during metamorphosis is that the hormone acts only in cooperation with other hormones of the neuroendocrine system. Thus it is difficult to differentiate specific effects of CAH from those caused by each of the other hormones involved. Insect larvae possess some structures with specifically larval functions, such as abdominal pseudopods, larval hairs and bristles, specific sensory organs, dermal glands, various small epidermal organs, and specifically adapted internal organs for larval life. These structures degenerate in the course of last larval instar when CAH is absent. Consequently, the physiological and biochemical functions prerequisite to these "larval" structures disappear. This illustrates how the morphogenetic changes regulated by CAH can be reflected to physiology and biochemistry of the animal. Another example which favours this suggestion is inhibition by CAH of the morphogenesis of specific adult structures in insect metamorphosis. It is apparent that adultoids with incompletely differentiated adult organs would be incapable to perform regular physiological functions of the adult stage. On the other hand, the last instar larvae which evidently lack CAH are able to perform all the general physiological and biochemical functions of the larval stage. This suggests that CAH would cause rather specific but not general metabolic effects. The only system where CAH induces more profound physiological and biochemical changes are the ovaries and generally the body of adult females where reproduction activity controled by CAH appears to be the main physiological function of the organism.

a) Growth and Chemical Composition of the Body

Allatectomy or CA implantation seem to have little effect on growth of the body during last larval instars of Exopterygote insects [315]. In Endopterygotes, such as *Galleria,* there is increased accumulation of reserve materials in the last larval instar which can be prevented by implantation of CA [306]. The typical S-shaped growth curve present in all insect larvae is to some extent modified by induction of extra.larval instars by CAH, but the amount of growth is limited by factors other than CAH. For example the growth rate of successive extra-larval instars induced by CAH in *Galleria* gradually decreases and finally stops completely [304]. In other words, the animal is able to produce 5th or 6th supernumerary larval instars with no further increase in size over the preceeding 4th supernumerary instar.

L'HELIAS [187, 190] observed reduced growth of the body in allatectomized larvae of *Dixippus.* The content of glycogen, tissue amino acids, and inorganic phosphate was increased. In *Galleria* [304] glycogen content increased after CA implantation whereas the amount of lipid and protein per dry matter decreased. Chemical analysis of the main body constituents (carbohydrate, lipid, protein) of extra-larvae revealed large similarities with young larval instars. It was assumed that CAH had no direct effect on metabolism of the main nutrients, although the metabolism was significantly influenced by morphogenetical changes induced by CAH.

The females which became hypertrophic after allatectomy contained enormous accumulation of glycogen, lipid, and protein, which were nonutilized precursors of egg yolk [139]. Castrated females had similar body composition as the allatectomized ones except a lower amount of glycogen. The carbohydrate was present as glycoproteins. In addition to common nutrients the content of other materials important as energy sources such as ATP may be influenced by CAH in adult females [359].

MINKS [207] performed a detailed analysis of body composition in alatectomized Locusts. The ovaries stimulated by the CAH were distinguished by rapid growth and increased amount of proteins, free amino acids, lipids, glycogen and RNA, in contrast to the steady low concentration of these materials in inactive ovaries. MINKS concluded that this was due to induced morphogenesis and growth of the organ. Unlike ovaries, the metabolism of the fat body of *Locusta* was not significantly altered by CAH [359]. Some conclusions were made [225] that the action of CAH on adult female depended on its primary action on the fat body metabolism while all other effects were consequences of altered lipid metabolism in the fat body. However, the conclusion cannot be justified with respect to other known functions of the hormone.

b) Action on Respiratory Metabolism

Each of the developmental cycles in insects (larval or pupal moult cycles, reproduction cycles in females) are characterized by distinctive changes in respiratory metabolism [317]. Cyclic changes in oxygen consumption during larval instars and the U-shaped course of oxygen consumption in pupae have long been known [309].

In the young larval instars of most Exopterygotes there occur precisely determined cycles in respiratory metabolism. Such respiratory cycles are altered in the last larval instar. During the period of intensive differentiation there occurs an U-shaped decrease of metabolic rate which is common to metamorphosis in the Endopterygotes. By means of allatectomy it is possible to induce this type of respiratory course in young larval instars while implantations of CA into the last instar produce the characteristic larval respiratory cycle [315], (see Fig. 30). In the initial period of intensive growth at the beginning of an instar the presence or absence of CAH has little effect on the respiratory rate [315].

Similar changes in respiratory cycles were observed in Endopterygotes. In *Galleria* there are 3 types of respiratory metabolism cycles. The first occurs in young larval instars, the second in the last larval instar and the third in the pupal instar. After implantations of endocrine complexes containing CA there appeared several extra-larval instars with several more larval respiratory cycles. It appeared that the course of these metabolic cycles was not dependent on age of the animal or its size. However, the cycles depended on the character of morphogenetical changes [306], which is also the case in Exopterygotes [220].

An increase in CAH titer in the haemolymph of adult insects, especially females is followed by an increase in respiratory metabolism. Conversely, for example allatectomized females have a lower rate of oxygen consumption. This has been demonstrated in such different groups as Diptera [339], Coleoptera [60, 386], Hemiptera [312, 318], Dictyoptera

[276], and Orthoptera [273]. Some authors have suggested that the CAH had a direct metabolic effect [60, 273, 276, 339, 386]. However, it has become apparent that CAH has no general metabolic effect but influences metabolism indirectly by affecting growth and physiological functions of certain organs or tissues [139, 200, 220, 221, 306, 312, 313, 315, 318]. Further, this conclusion is also supported by studies on respiration of isolated body fragments [316].

In adult females of *Pyrrhocoris* the respiratory metabolism of tissues such as nervous system, somatic musculature, and salivary glands may not be influenced by hormones at all. Other tissues such as the fat body, Malpighian tubules, pericardial cells, intestine and corpus allatum show increased respiratory metabolism in response to AH. However, the ovaries and accessory sexual glands are the only tissues in which an increase in the respiratory metabolism is dependent on CAH [318]. According to SLÁMA [312] only a certain portion of the body respiration of adult female *Pyrrhocoris* was dependent on CAH. This CAH dependent respiration was called reproductive metabolism in contradistinction to basal metabolism which did not change under hormonal influence and digestive metabolism which was dependent on AH.

The CAH is assumed responsible for stimulating the reconstruction of indirect flight musculature in postdiapause adults of *Leptinotarsa* [173]. However, in adults of *Gryllus* [45] and *Dysdercus* [52] the CAH was reported to cause reduction of indirect musculature. In contrast to these results in *Pyrrhocoris* CAH has no effect on the somatic muscle morphology and metabolism [312, 316, 318]. This shows that physiological action of CAH on the same tissue type may produce different effects. These different responses to CAH may be due to differences in the genome of various species.

Homogenates of CA or the gland extracts were found to increase the *in vitro* oxidation in isolated mitochondria or sarcosomes [47, 173, 208, 261, 329, 380, 386]. Since these *in vitro* systems are very sensitive to ionic balance and the presence of various endogenous substrates supplied with the CA homogenate, the involvement of CAH in mitochondrial respiration is not clear. These problems are discussed in more details by MINKS [208].

c) *Effect on Haemolymph and Tissue Proteins*

During ovarian growth which is stimulated by CAH there is intensive transport of proteinaceous material through the haemolymph from the reserve organs into the oocytes. These vitellogenic proteins (vitellogenins) are stored in the fat body before or are synthesized during the reproduction process. In certain Lepidoptera and Hymenoptera these vitellogenic proteins are already present in the larval fat body [335, 336].

Allatectomy or cauterization of the neurosecretory cells usually result in the accumulation of haemolymph proteins [125, 127, 128, 311]. Studies performed mainly on adult females demonstrated the effect of hormones, especially the CAH on protein synthesis and its effect on the kind of proteins synthesized. These effects have been confirmed in *Schistocerca* [125, 127, 128, 129], *Locusta* [17, 18, 208], *Leucophaea* [35, 54, 73, 295, 296, 297], *Nauphoeta* [198], *Periplaneta* [2, 3], *Pyrrhocoris* [311], *Rhodnius* [49], *Leptinotarsa* [195], *Polyphemus* [24] and others [70]. Most of these studies confirmed that the haemolymph protein pattern is largely influenced by both the AH and CAH [17, 208]. After removal of the CA the total amount of haemolymph protein increases. However, the most conspicuous changes which occurred were the qualitative changes in vitellogenic proteins, which may be represented by two major fractions and several less abundant ones. These vitellogenins are complex proteins with various prosthetic groups. Lipids are usually present in form of diglycerides [337, 338], phospholipids, phosphatidylcholine or phosphatidylethanolamine [337]. A considerable amount of carbohydrate prosthetic groups is present in form of aminosugars and hexoses [3, 17, 35, 54, 337, 338] and pyrimidine or purine bases can also probably be included. Thus, when these complex proteins are transported into the oocyte they carry with them all the necessary materials for synthesis of nucleic acids, membrane constituents, and other cellular structures in the developing embryo.

The vitellogenins occur solubilized in the haemolymph but after entering oocytes they form more or less insoluble globules [297]. However, their identity with haemolymph proteins has been demonstrated [338]. Histochemical studies have indicated that the material of the oocyte globules has the basic character as has an acid mucopolysaccharide [4]. Thus, the globules have certain similarities with the ovomucoid from avian eggs.

Synthesis of the vitellogenins usually takes place in the fat body [3], and is coordinated with ovarian growth and CA activity [69]. In allatectomized females in which ovarian growth is suppressed, the haemolymph vitellogenins may increase in concentration, or other fractions [295], presumably precursors of the vitellogenins may accumulate. These and other protein fractions appear to be synthesized by intestinal epithelium under the influence of AH.

Although the CAH can initiate the synthesis of vitellogenins, the synthesis of vitellogenins alone does not ensure the development of the ovaries. Vitellogenins may be present in large quantities in some allatectomized or inactive females [16] and the ovaries do not develop. The available data permit the conclusion that in adult female insects the CAH affects formation of specific proteins [208], or at least influences their transport to the ovaries [198]. However, considering the overall accumulation of proteins in allatectomized females and the biochemical changes which take place during metamorphosis in complete absence of the CAH [304] one must accept a view that the hormone has no direct or general control of protein synthesis in the insect body. Such view also applies to sexual differences. The CAH affects in some species metabolism of vitellogenins only in females, whereas in the males there are no such effects and neither there occur any profound morphogenetic changes.

d) Effects on Nucleic Acid Metabolism

Like any other morphogenetic changes, those induced by CAH are also reflected in metabolism of nucleic acids. Possible changes in the DNA replication can simply be deduced from studies of mitosis and endomitosis. One such deduction suggests that CAH has no general effect on DNA replication in immature stages. So, mitosis and DNA replication proceed in young larval instars as well as during metamorphosis in absence of CAH with equal or even increased intensity. There are larval tissues which can be maintained only in the presence of a minimum titer of CAH but even in these tissues DNA replication appears to be stimulated by ecdysone, not by CAH. In pupae of Cecropia, for instance, CAH does not stimulate DNA synthesis in any tissue when other hormones are absent [174]. In fact, it has not been documented so far that CAH would directly stimulate mitosis in any larval or pupal tissue. In *Samia* all the tissues are assumed to require ecdysone for DNA synthesis in all the postembryonic development [175]. As suggested by Schneiderman [298] and others especially in vertebrate literature, DNA replication is essential for reprogramming of the genome. In insect metamorphosis all the cells probably synthesize DNA during larval-pupal transformation [298]. The CAH may act only on the cells which are before or have initiated DNA synthesis and is less active or inactive on cells which have completed DNA replication. Therefore, the mechanism of CAH action on metamorphosis does not depend in a general control over the total rate of DNA synthesis in the tissues.

Metamorphosis is characterized by the appearance of new pupal or adult structures capable of performing new physiological and biochemical functions. This requires formation of new chemical compounds involving whole populations of new protein and RNA molecules. Because CAH prevents metamorphosis it must therefore also suppress the formation of the necessary macromolecules involved in production of the new structures. It has been assumed that one way in which these biochemical changes are regulated is at the level of gene transcription. And, in fact, possible inhibition of specific t-RNA or m-RNA responsible for synthesis of new proteins is one of the expected sites for the action of CAH. Thus, CAH may act by exerting a negative influence on the transcription of certain "master regulatory genes" [398]. Other hypothesis to explain CAH action is the conclusion that CAH may affect the gene expression at the translational level by controlling the production of particular t-RNA's and their activating enzymes [135]. This will be discussed more extensively in section which concerns the action of juvenoids (p. 261).

In the diapausing Lepidopteran pupae RNA and DNA synthesis are highly suppressed [33, 174, 401]. In these insects, as in other immature stages, CAH cannot restore nucleic acid synthesis which is under control of ecdysone [33, 174, 176, 222].

In adult insects the effects of CAH on nucleic acid metabolism appear to be different than in larval or pupal stage. Target tissue for CAH are here less numerous than in larvae and they respond to the hormone differently. For example, ovarian follicle cells grow and divide in the presence of CAH alone. Furthermore, in follicle cells stimulated by CAH there is an increase in rates of synthesis of both RNA and DNA [336]. This has been demonstrated in several insect species such as *Rhodnius* [351], *Calliphora* [20], *Locusta* [208] and in other species [70]. In some other adult tissues associated with reproduction there occurs a significant increase in RNA synthesis which is either directly or indirectly associated with CAH action. However, in no other adult tissue does as intensive DNA synthesis occur as in the stimulated ovarian follicle cells or perhaps in the accessory sexual gland cells, too. It is known that DNA synthesis can be restored in certain adult tissues by ecdysone [176]. But in insects in which reproduction is controled by CAH ecdysone does not initiate DNA replication in ovarian follicles.

The assumptions that DNA synthesis can occur in postembryonic development of insects only in presence of ecdysone [222] is not universal. Obviously, CAH does stimulate DNA synthesis in ovarian follicle cells of adult females without respect to ecdysone presence. Moreover, theories explaining mechanism of CAH action as a negative control of transcription [398] do not actually apply to CAH action on the ovarian follicle cells. According to experience obtained with some vertebrate hormones [58] the mode of action of CAH may remain the same despite differences in the developmental programme taking place in immature tissues on one hand and the ovarian follicle cells on the other hand.

6. Survey of Other Effects of CAH

The morphogenetic changes induced by CAH may be associated with specialized effects characteristic for certain groups or species of insects. Several of these effects may result from feedback control between CAH and other neuroendocrine centers. We have concentrated our attention to the effects on polymorphism, colour change, regeneration and behaviour, which are relevant to the effects of juvenoids.

Polymorphism. The whole ontogeny of insects with the large diversity of immature forms represents *per se* a good example of polymorphism characterized by progressive change of forms [373]. There are special types of polymorphism occurring only in particular species. One of them is a caste polymorphism in social insects. It has been shown that CAH affects the appearance of soldier [197] cast in termites and differentiation of castes in ants [34]. Another type of polymorphism influenced by CAH is phase polymorphism in Locusts. The appearance of the solitary phases which differ by certain biometric and biological patterns from the gregarious phases is influenced by CAH [145, 323, 324, 326] but also by the neurosecretory system [62, 120] and prothoracic gland hormone [120, 323]. Evidently, an interaction of the whole neuroendocrine system may be involved in regulating phase polymorphism.

The CAH may also play a certain role in seasonal polymorphism such as that found in winged and apterous forms or generations in aphids [218]. The apterous forms possess certain neotenic patterns and their CA show some differences in volume [178, 362] which can be correlated with these patterns. There were theories [320] that alary polymorphism in Hemiptera may be under CAH control. However, the reduced wings of the brachypterous specimens always contain an adult cutile and other adult patterns, while CAH causes always appearance of larval cuticle with "larvalized" patterns. As suggested by WIGGLESWORTH [373, 378], genetic factors may be more important than CAH in this type of polymorphism. The alary polymorphism is sometimes associated with sexual dimorphism and possible involvement of CAH in this phenomenon has also been considered [184].

Colour change. Morphogenetic changes initiated by absence of CAH at the beginning of metamorphosis are often associated with a change of pigments. These include both melanin pigments in the cuticle and other pigments such as carotenoids and pteridines in the epidermal cells. All the typical larval pigments are preserved in extra-larval instars which result from implantation of CA and, adultoids with partly larval partly adult epidermal structures may show complete mosaics of larval and adult pigment patterns.

Determination of morphogenetic processes can in some insects proceed at different time than the determination of pigment formation. The pigmentation can be influenced by CAH at the time when the critical period for action on morphogenesis has been terminated. Treated insects show developmental abnormalities including morphological patterns of one stage and pigmentation typical for another stage. This phenomenon has been observed in *Dixippus* [237], *Locusta* [144], *Gryllus* [174] and in others [90, 218, 237]. The CAH may not act alone on pigment formation and the interaction of other hormones of the neuro-endocrine system has been suggested [62, 90].

Regeneration. Extensive regeneration including formation of whole appendages or regeneration of large parts of epidermis requires ecdysone rather than CAH. Localized regeneration of small injuries take place even in diapause and in other stages irrespective of the presence or absence of hormones [21]. In fact, injury can cause many changes in the epidermis which are similar to those which occur after hormonal stimulation. Furthermore, injury to the epidermis can influence some physiological functions of the whole body. This has been shown in studies on diapausing pupae [21, 114] or on *Rhodnius* [375, 378]. There have been suggestions of a possible involvement of CAH in regeneration [90, 218, 237]. However, a detailed study of regeneration in *Locusta* has shown that CAH does not affect regeneration in this species [265]. Moreover, other hormones do not appear to be involved [266, 267].

Behaviour. At the beginning of metamorphosis larvae of some species cease to feed, change their phototropic or geotropic reactions, leave the host plant, start migration, and spin the cocoons. It has been known for a long time that spinning behaviour of caterpillars was influenced by the CA implantation [30, 240]. More recently, Piepho [242] has observed that implantation of CA into spinning larvae of *Galleria* not only influences the spinning behaviour but actually leads to restoration of feeding instincts. Other behaviour patterns such as migration instincts in *Melolontha* [330] seem to be affected by CAH. Allatectomy of adult *Leptinotarsa* leads to changes characteristic of diapause behaviour [385, 386].

The CAH plays significant role in regulation of sexual behaviour. As a part of hormonal regulation of reproduction, the effects of CAH on sexual instincts have recently been reviewed in a monograph by Engelmann [70]. Other factors, e. g. nervous, hormonal, and pheromones also influence sexual behaviour of insects. Significance of each of these factors may differ from species to species and also with respect to sex. The mating instincts involve attraction of the opposite sex from a distance by pheromone, release of certain nervous, tactile, gustatory, acoustic, and other stimuli, after which mating occurs. In many species the mating instincts are more complex, especially in females where attraction may be controled by mechanisms other than receptivity.

The endocrine control of mating behaviour in cockroaches has been reviewed by Barth [8], Engelmann [70] and their co-workers. The CAH may influence the production of female pheromone in certain species and also female receptivity, though it is not the necessary factor for the latter. Allatectomy has no effect on mating behaviour of males in some species, but it may reduce the formation of male sexual attractant. The whole problem of CAH effects on sexual instincts in males is complicated by intricate relationships in regulation of the reproduction cycles in separate species and, therefore, must be studied from special reports [6, 7, 65, 71, 272].

Extensive studies of LOHER, PENER, and associates have provided data on the reproductive cycles in Acridiidae. In *Schistocerca* the CAH stimulates release of the male pheromone which accelerates the sexual maturation of young adults [192]. In certain species like *Schistocerca* [192, 232, 234], *Nomadacris* [234], and *Oedipoda* [37] the male sexual activity is dependent on CAH. In other species such as *Gomphocerus* [193, 194] or *Locusta* [356] CAH has only a small influence. In *Gomphocerus* [194] CAH affects also female receptivity and female stridulation [193].

It has been observed by ŽDÁREK [401] that diapausing adults of *Pyrrhocoris* do not exhibit sexual behaviour. The mating instincts of males are fully under the control of nervous system determined by long photoperiod. Removal of CA or other hormonal sources has no effect on mating instincts. Implantations of active endocrine glands into diapausing males lead to restoration of mating activity after prolonged periods. This suggests that either hormones or physiological changes in the body induced by the hormones have a feed-back effect on the nervous system. On the other hand, the female attractivity and receptivity is more profoundly affected by hormones. The AH seems to be more important than the CAH in controlling the sexual behaviour in females. Extirpations of endocrine glands reduce the sexual activity of females and, gland reimplantations have more immediate effect on restoration of sexual activity in females [407, 409].

The data on the effects of CAH on instincts or nervous activity suggest that the action of the hormone is probably indirect and is therefore dependent on other physiological and biochemical changes in the body. This suggestion agrees with homeostatic regulations. For example, in some cases CAH stimulates mating, but in other cases mating and insemination give stimulus to the initiation of secretory activity of the CA in females [8].

7. Chemical Nature of CAH

Our present knowledge on chemical nature of CAH is based exclusively on methyl-homojuvenate and methyl-dihomojuvenate isolated from male moths of *Hyalophora cecropia*. Since a large number of synthetic analogues of these compounds have been prepared and tested along with other juvenoids we have discussed the problems concerned with isolation and identification of these hormones in chemical section of Part II. All the present evidence suggests that the above indicated compounds, e.g. the homo- and dihomojuvenate are identical with the CAH of Cecropia. These compounds duplicate all the known effects of CAH, are less specific than many juvenoids, and are produced by CA incubated in *in vitro* cultures according to RÖLLER and DAHM [269].

Before identification of Cecropia compounds there were assumptions that the hormones of the insect neuroendocrine system including CAH might be related to pyrimidine or pteridine compounds or related to folic acid derivatives [188, 189]. It was also proposed that the CAH might possibly be a steroid [99], a peptide or protein. The assumption that CAH was a protein was based on certain ultrastructural features of the CA cells and also on morphological and functional analogies between CA and adenohypophysis of vertebrates [218].

For several reasons it appears necessary to attempt to isolate CAH from more insect species. Examples of large amount of juvenoids with relatively different chemical structures suggest that the bioassays used for CAH activity determination are not fully specific. The insect body contains considerable amount of isoprenoids obtained from plant food which have detectable hormonal activity in the assays though inactive in physiological conditions. A typical example is farnesol which can be extracted from the insect body but is not the hormone which is released by CA [299]. There are some insects where the hormones isolated from Cecropia may not be identical with their own CAH. For example the CA of *Pyrrhocoris* appear to be too small to provide an effective concentration of these Cecropia type materials which are little active in *Pyrrhocoris*. As we shall see in part II, there exist synthetic juvenoids with a million times higher activity on *Pyrrhocoris* than the dihomojuvenate, yet there is no proof that the CAH of *Pyrrhocoris* and these synthetic juvenoids would be structurally related.

C. Prothoracic Glands (PG) and the Prothoracic Gland Hormone (PGH)

The overall shape of the insect body is largely dependent on the integumental exoskeleton. Because of the enormous growth of the insect body during postembryonic development, the exoskeleton has to be replaced periodically by the process of ecdysis. Growth of the body is limited in each larval instar to a critical internal volume determined by size and elasticity of the cuticle. The process of periodically repeated moults including ecdyses is essential not only for growth in size, but also for the morphogenetic changes during metamorphosis of Endopterygote insects.

Investigations on the hormonal regulation of the moult cycles begin with the observations of Kopeč [172] and extend to the work on Diptera by Fraenkel [76] and on *Rhodnius* by Wigglesworth [365]. These original studies assumed that the source of the hormone is the head region, but it was later thought that the hormone affecting the moult cycles is produced in the prothoracic region [111] an idea which was confirmed experimentally by Fukuda [82] using prothoracic gland transplantations.

The endocrine glands located in the thoracic region have been found in all groups of insects. They differ in anatomical structure, location, etc., but all have one common feature, i. e. they degenerate at the end of metamorphosis or at the beginning of adult life. Only in primitive groups such as Apterygota do these glands remain active during reproduction and the adults of this group continue to moult [360]. Depending on their structure and location the glands are called prothoracic glands, ventral glands, pericardial glands or peritracheal glands. Since these glands share many developmental and physiological features which are similar it is useful to have some common designation for all these structures. Herman and Gilbert [118] proposed a name "ecdysial glands" but as argued by Wigglesworth [379] this term has already been in use with respect to certain epidermal glands. We decided it would not be inappropriate to use the term prothoracic glands in a broad sense of the word. Hence we have also used the term prothoracic gland hormone (PGH) for the secretory product of these glands.

1. External and Internal Morphology of Prothoracic Glands

The anatomical structure of the prothoracic glands (PG) is described in the monographs on insect endocrines [218, 237, 377]. We have restricted our discussion to a brief description of the various types of prothoracic glands and their distribution in insects. Special attention has been given to the effects of PGH which are related in some way to the effects of ecdysoids as described in Part III.

Prothoracic glands in the strict sense are paired organs formed by groups of secretory cells located in the ventrolateral part of the prothorax. In some species the glands extend anterior to the head and posterior to the mesothoracic region. The cells of the gland form aggregations which are commonly arranged in branches and the individual cells are connected by intercellular fibrils formed by acellular connective tissue (Fig. 5). The prothoracic glands are usually attached to the main tracheal trunks leading from the prothoracic stigmatic valve. This type of arrangement is most frequently found in Lepidoptera [118, 390], and it occurs with certain variations also in other groups of Endopterygota such as Coleoptera [321], Megaloptera [259], and Hymenoptera [46, 196]. Among Exopterygota this type is common in Hemiptera [361] and Dictyoptera [25, 237, 281] where the glands are more compact and less ramified (Fig. 5). The PG are well innervated by several nerves. For example,

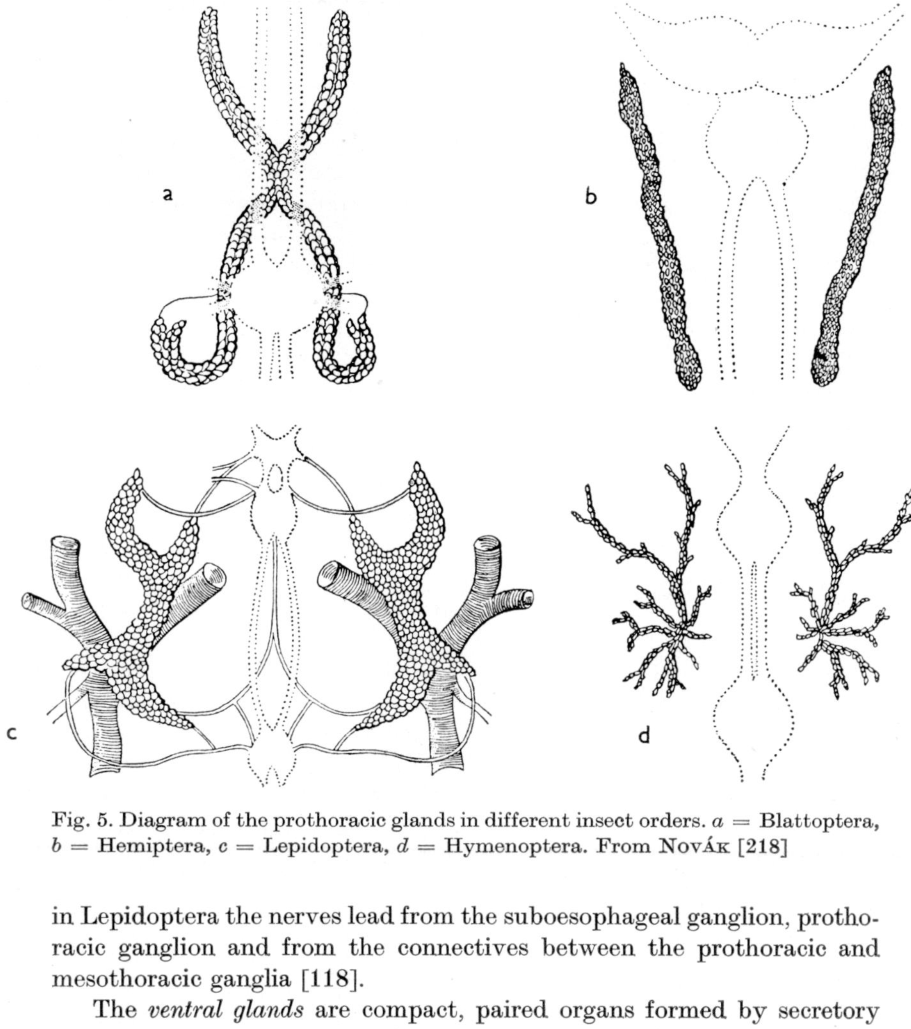

Fig. 5. Diagram of the prothoracic glands in different insect orders. *a* = Blattoptera, *b* = Hemiptera, *c* = Lepidoptera, *d* = Hymenoptera. From Novák [218]

in Lepidoptera the nerves lead from the suboesophageal ganglion, prothoracic ganglion and from the connectives between the prothoracic and mesothoracic ganglia [118].

The *ventral glands* are compact, paired organs formed by secretory cells which are located ventrally in the hind part of the head or between the head and the prothorax. The glands are arranged longitudinally and are slightly branched. They are innervated by nerves from the suboesophageal ganglion. This arrangement is most common in Apterygota and in various groups of Exopterygota, such as Phasmida,

Ephemerida, Plecoptera, Dermaptera, Isoptera, Odonata, and Orthoptera-Saltatoria. The glands are absent in Hemiptera and all Endopterygota [237]. In contrast to the PG of Endopterygotes which grow by endomitosis and have polyploid cells [118], the ventral glands of Exopterygotes enlarge by cytokinesis which takes place during the postembryonic development.

Pericardial and *peritracheal glands.* PFLUGFELDER [237] described groups of secretory cells in Phasmids attached to the dorsal aorta. These glands degenerated at the beginning of adult life and were named pericardial glands. Similar glands have been found in lower Diptera and they may be homologous to the lateral cells of the ring gland of Cyclorrhaphous Diptera. These lateral cells of the ring gland are also homologous to the prothoracic gland and have been named peritracheal glands. The peritracheal part of the ring gland also degenerates at the end of metamorphosis. The ring gland represents fused complex of endocrine glands including corpora cardiaca, corpora allata and prothoracic glands [218, 237].

Histological examination of the PG reveals the presence of parenchymal glandular cells which are similar to those of the parenchymal cells of the CA. The cells are surrounded by a thin membrane formed by an acellular stromal element which actually determines the shape of the whole gland. The cytoplasm communicates with the external stromal membrane by numerous processes with apparent intercellular spaces. The cytoplasm is basophilic and contains a number of granules and vacuoles. The nuclei are ovoid or commonly irregular containing many nucleoli. Near the surface of the gland there are abundant tracheoles, nerve fibers and muscle fibers. In some PG these structures transverse the centre of the glands [118, 281]. Haemocytes are commonly aggregating on the surface or in the vicinity of the glands [369].

Ultrastructural features of the PG of cockroaches have been studied during larval growth and degeneration by SCHARRER [281, 284]. There are long cellular processes separated by extracellular channels and the cell membranes are apposed to an external connective sheath. Micropinocytotic vesicles are present and may possibly be engaged in the release of the secretory product into the haemolymph. The cytoplasm contains membranous structures, the Golgi bodies are inconspicuous, ergastoplasmic elements are little developed. The presence of membrane bound secretory granules typical for cells with proteinaceous secretions, has not been found. The agranular endoplasmic reticulum assumed to be the source of the secretory product. The nerve endings contain numerous vesicles and the neurosecretory granules are also present.

The first sign of the degenerating cells is a breakdown of nucleoli characteristic for pycnosis. The most significant degeneration changes occur in the cytoplasm. They are indicated by appearance of large heterogeneous inclusions which represent autophagic vacuoles engaged in the breakdown of individual cell constituents. Finally, the cell disintegrates and its remnants are absorbed by phagocytic haemocytes. Quite similar ultrastructural features were observed in the PG cells of *Drosophila* [160]. The secretory products were also thought to be manufactured by the smooth endoplasmic reticulum. In contrast JOLY *et al.* [149] found extremely well developed Golgi apparatus, a rich ergastoplasm, abundance of free ribosomes and cytoplasmic vesicles in the PG of *Locusta.* The ergastoplasmic cisternae were organized into long chains of spirals, resembling structures in the CA cells described in the previous section. The authors judged on a proteinaceous character of the secretory product. A detailed analysis of the PG ultrastructure in *Antheraea* and *Bombyx* was made by BEAULATON. In prepupal PG there appeared abundant lysosomes which contained hydrolytic enzymes such as acid phosphatase, thiolactate esterase, or adenosine mono-, di-, and triphosphatases. The lysosomes were also present in the Golgi region [9]. There appeared cytolysomes representing autophagic and autolytic vacuoles originating in the Golgi complex. These vacuoles represented a modified ergastoplasm which engulfed determined parts of the cytoplasm and formed cysts. The above hydrolytic enzymes were also present in these formations [10]. The amount of glycogen in the parenchymal cells of PG increased during the intermoult period, falling down before ecdysis. The presence of a free glycogen in hyaloplasm or in vacuoles pointed out to its connexions with the smooth endoplasmic reticulum. Furthermore, glycogen in the mitochondria accumulated in such an extent that the whole mitochondrion was transformed into a vacuole filled with glycogen [11]. It was noted that similar ultrastructural features were also found in the cells of Amphibian endocrine glands which secrete steroid hormones [238].

BEAULATON further described the origin of annulate lamellae which were absent in the PG cells of young larval instars and appeared in the last larval instar. These structures were transformed into lamellar structures associated with the autolytic process. Apparently the presence of these structures was connected with changes in metabolism of the PG cells. Their appearance may suggest certain transitory form of differentiation in the endoplasmic reticulum [12].

The connective tissue sheath formed processes penetrating inside the cytoplasm of the parenchymal cells. The connective tissue had similar appearance with that surrounding nerves and ovaries, or with that forming a basement membrane. It contained large amount of mucopolysaccharides or mucoproteins. Numerous adipohaemocytes were adhering to the connective sheath [13]. The pericellular spaces delimited by this connective sheath contained interstitial nerve cells and axon terminals of the neurosecretory cells. The axons which ended in the connective sheath contained neurosecretory granules [14]. There occurred large parenchymal cells with numerous diverticula as also found in the cockroaches. The nuclei were relatively large and lobulated resembling the nuclei of the spinning gland cells. Pinocytotic vesicles were seen in the pericellular spaces near to the invaginations of the cytoplasmic membrane. Perinuclear zone contained abundant ribosomes, and ergastoplasmic vesicles. Lipid inclusions were not present in the PG cells of these Lepidopterous insects. The nucleus contained several polymorphic nucleoli which could be digested with ribonuclease. In the female glands there were Feulgen positive intranuclear bodies representing probably a heterochromatin [15].

All these studies on ultrastructure of the PG have revealed many important findings, especially those concerned with degeneration of the gland. Unfortunately, structural patterns which would help to identify probable chemical nature of the secretory product have not been unequivocally determined.

The PG originate embryologically from the second maxillary segment. There exist opinions that they represent serial homologues of CA. It has also been assumed that the PG may be related to the original head nephridia which occur in more primitive animals [218, 237].

2. Changes in PG during Postembryonic Development

In the PG of insects there occur certain structural changes which may indicate secretory activity or inactivity of the glands. These criteria of activity include for example changes in the total volume of the gland, cytological changes in the shape and size of the nucleus, changes in the chromatin patterns, occurrence of cytoplasmic vacuoles, and development of other cytoplasmic organelles. Without respect to the type of growth, which can be endomitotic in most Endopterygota and mitotic in Exopterygota, the gland volume increases successively during the period of larval growth in all insects. The growth increments measured in separate larval instars are roughly proportional to growth of the whole body or growth of some other organs [361]. Maximum size of the gland is usually attained in the last larval instar [118, 160].

Within each larval instar there occur considerable variations in the glandular volume, as well as in the size and number of the parenchymal cells. These changes are precisely synchronized with the course of the intermoult period. Maximum growth and probably also maximum secretory activity occurs usually at the first third of a larval instar. This period is characterized by maximum growth of the whole body and by the beginning of the mitotic divisions in the epidermis.

In Apterygota where the PG function during the adult stage there occurred cyclical changes in the gland volume with maximum growth towards the middle of the intermoult period and minimum at the time of ecdysis. The difference between the maximum and minimum volume was 3 to 1 [360]. In cockroaches which represent Exopterygote insects, there also appeared morphological and physiological changes in the PG during the larval period. The changes had cyclical character and were related to the moult cycles. For example, there were changes in structure and size of the mitochondria, changes in internal structure of the nucleus, etc. However, the size of the nuclei did not change at the corresponding moments of the successive larval instars. Furthermore, mitoses were most common at the middle of the intermoult period [281]. In *Locusta,* mitoses in the PG appeared two days after the ecdysis, e.g. significantly before the appearance of mitoses in the epidermis. The gland volume increased faster than that of the whole body until 4th larval instar. Since then the gland volume still increased but less than that of the body [332].

In *Drosophila,* which represents Endopterygote insects, the nuclei progressively increased in size during the period of larval growth. They became highly polyploid and the secretory activity of the gland was probably increased. It was expected that the gland released more hormone before puparium formation than during the period of larval growth [160]. Comparisons between the volumetric increase of PG and CA during the larval stage have indicated large differences, especially with respect to the last larval instar. Thus, in young larval instars of *Pieris* the ratio of CA to PG volumes was 1 to 2, while in the last larval instar it was as much as 1 to 29. The PG volume was doubled throughout the last larval instar. It was suggested that maximum secretory activity occurred 2 days after each ecdysis in young larval instars [154]. During metamorphosis of *Tenebrio* the PG may attain maximum activity 3 to 4 days before ecdysis, e. g. 2 days before the gland volume begins to decrease [322].

Apparent correlations between structural changes in the PG and the induction of growth in the epidermis were observed in *Rhodnius.* The active PG of this species were always surrounded by numerous haemocytes which may be essential for distribution of the hormone into the haemolymph [372].

These facts indicate that the PG may themselves be a subject of similar growth changes as induced by their hormone in the epidermis or in other larval tissues. This results from a common integration of development by the neurosecretory system, which synchronizes all the growth processes with availability of food and other environmental factors. Further conclusion is that the PG are most active at the beginning or in the middle of an instar but not at the time of ecdysis. This conclusion is in a good agreement with the fact that the PGH stimulates the whole developmental cycle in a given instar which is most apparent on changes in the epidermal cells. The process is initiated very early in an instar by cellular enlargement, apolysis, mitosis, and finally it ends by secretion of the new cuticle. Thus, the final act of ecdysis is not directly stimulated by the PGH. However, some exceptions may exist with respect to the process of sclerotization of the old cuticle which is the case during the puparium formation in higher Diptera. This is discussed in details in connexion with the effects of ecdysoids.

Special attention has been paid to changes in PG activity during metamorphosis. During this period the moments of intensive secretory activity are usually followed by a programmed degeneration of the whole gland. The control of development by PGH during metamorphosis of Cecropia pupae was intensively studied by WILLIAMS and his co-workers [389, 390, 391, 393]. First of all they observed that the PG were inactive during the period of pupal diapause. Reactivation of the glands was dependent on a neurohormone from the brain-corpora cardiaca system. The activated glands released the hormone which stimulated adult development and was responsible for termination of the diapause. During the postdiapause development the PG resumed all signs of secretory activity [93, 118]. These elaborated studies of WILLIAMS represent the best example of the function of PG during insect metamorphosis. Reduction of the PG during diapause and their reactivation during postdiapause development was also found in *Cephus* (Hymenoptera) [46].

Relatively little is known concerning comparisons between development of the PG during an instar with respect to the presence and absence of CAH. There has been some evidence suggesting no basic differences between the PG functions during larval-larval and larval-pupal or pupal-adult moult cycles, although each of these developmental cycles has revealed some minor but characteristic changes in the internal structure of the glands [118]. However, other findings indicated that in the last or in the two terminal cycles of PG activity which took place in absence of CAH, there occurred considerable changes in internal structure of the parenchymal cells [322]. One of such changes is also indicated by appearance of lamellated cytoplasmic structures which are absent during the larval-larval moult cycles [12]. As has been pointed out already, lysosomes and autolytic vacuoles may appear in the cytoplasm long before the glands actually degenerate, but they do not appear in young larval instars which have active CA. Furthermore, the PG of Saturniids synthesized large amounts of DNA during the larval-larval cycles whereas they were unable to synthesize DNA during metamorphosis. In addition, ability of the PG to synthesize DNA was not recovered after their transplantation into larval hosts. In contrast to this, the ability of PG to synthesize RNA and proteins remained preserved during metamorphosis [174]. These data indicate that the PG may themselves undergo structural rearrangements associated with metamorphosis differentiation in absence of CAH. Thus, the PG may be considered as a special tissue which is not programmed for immediate breakdown in the absence of CAH, unlike most of the specifically "larval" tissues or cells. The cells of the PG are programmed for one (in Exopterygota) or more (in Endopterygota) additional developmental cycles in an absence of CAH prior to their final degeneration. Finally, according to analogy with other tissues in immature insect stages, the function of the PG cells may be different in presence and in absence of CAH. However, there is no experimental evidence for such an assumption so far.

In certain nondiapausing pupae of Coleoptera the PG exhibited all signs of secretory activity at the moment of larval-pupal ecdysis. There appeared distinctive degeneration patterns as soon as 16 to 20 hr thereafter. The degeneration of the glands was initiated long before adult emergence [322], suggesting that the PG were not directly involved in the process of ecdysis or adult eclosion.

In general, the PG remain functional throughout the whole postembryonic life only in Apterygota. In most Pterygote insects they degenerate during or soon after metamorphosis. Exceptionally, the PG of certain Pterygotes, such as Locusts remain preserved in reduced anatomical form for different periods of adult life. Most probably these glands do not function [74].

The period of functional activity and survival of the PG is dependent on the CAH. So, the number of the activity cycles in larval or pupal PG is dependent on the period of CAH presence. This is manifested experimentally by a common induction of one or more extra-larval or extra-pupal instars after implantations of active CA. The giant extra-larval instars contain functioning PG for prolonged periods. In certain Exopterygotes such as *Pyrrhocoris* the PG remained active provided that the CA were implanted at the very beginning of the last larval instar. Implantations performed later resulted in formation of adultoids which were unable to undergo additional moult cycles. This suggests that the determination for breakdown of the PG proceeds in this species at the very beginning of metamorphosis just after the ecdysis into the last larval instar. In the pupal instars of Endopterygotes it is also necessary to implant the CA before or very early after the larval-pupal ecdysis, in order to obtain repeated formation of extra-pupal instars. This is discussed in more details in connexion with the effects of CAH and juvenoids (p. 265).

The effect of CAH on degeneration of the PG may be quite different before and after metamorphosis. For example, in some adult insects CAH did not prevent degeneration of the PG as is usual in larvae or pupae, but in contrast to this the CAH stimulated their breakdown. Consequently, allatectomy was followed by persistence of active PG which resulted in formation of extra-adult moults [25]. We have mentioned already that the number of the activity cycles in the PG is variable when CAH is present. On the other hand the number of the activity cycles taking place in metamorphosis, e.g. in the absence of CAH appears to be constant for a given species and is most probably determined genetically. Thus, the PG of Exopterygote insects are determined to undergo only one activity cycle in the absence of CAH; the PG of most Endopterygotes undergo two cycles and, the PG of sawfly larvae may even undergo three cycles of secretory activity all taking place in the absence of CAH [314].

In connexion with the adenotropic effects of the AH (see p. 10) we have described the principal role of the AH in the stimulation of the secretory activity in the PG. A detailed study of the interactions between brain and the PG in Cecropia pupae was made by WILLIAMS [393], and various problems connected with the activation of the PG may be found in literature [46, 83, 90, 99, 218, 219, 237, 299, 303, 371]. In growing larvae, unlike in metamorphosis, the release of the adenotropic factors activating PG is usually dependent on feeding. This arrangement correlates growth of the PG with the growth of the intestine and, consequently with the growth of other organs. Experiments by WIGGLESWORTH [379] pointed out, however, that it was not the feeding alone which stimulated the PG activity. In larvae of a Dermestid beetle *Trogoderma*, the larval moult cycles also occur irrespective of starvation. Atypical cases of PG activation were also reported. For example, the PG of diapausing pupae of Cecropia were activated by implanted CA [222], or, the PGH was itself reported to exert a positive feedback effect on further functions of the PG [393].

3. Effects of PGH on Growth, Differentiation and Moults

The most apparent consequence of the prothoracic gland hormone defficiency in insect development is absence of the moult cycles. This effect was found long ago in the experiments with ligatures [76] or isolations of the abdominal body fragments. It was also a major developmental effect of the prothoracic gland ectomy. For their inability to moult, the larvae with ectomized prothoracic glands were commonly called "permanent larvae" or "dauerlarvae" [227, 237, 247, 266, 332]. Implantations of active prothoracic glands into "permanent larvae" of various insect species resulted in restoration of the moult cycles, and this was eventually associated with induction of the metamorphosis development [83, 227, 237, 247, 369, 371]. When extra prothoracic glands were implanted into larvae at the beginning of a normal larval instar the moult cycle was accelerated so that the next ecdysis occurred sooner [237]. Moreover, in certain cases it was possible to induce also an extra-adult moult by implanting active prothoracic glands into the adults. This specific ability of the PG to induce the adult moult cycles was originally demonstrated by WIGGLESWORTH using parabiosis [371] and PG implantations in the adults of *Rhodnius* [369] but it was later confirmed in several species belonging to various unrelated groups of insects [25, 64, 176, 228].

Large significance of the PGH in the control of metamorphosis by the neuroendocrine system was perhaps best illustrated by WILLIAMS [388—393] on the pupal-adult transformation in Saturniidae. It has been definitely established that the neurohormones originating in the neurosecretory cells of the brain can stimulate functions of the PG but are not *per se* responsible for growth and differentiation of the epidermal tissues in insect metamorphosis. Implantation of active brains was ineffective whereas implantations of the active brains along with the PG induced metamorphosis development even in the isolated pupal abdomens. It is generally agreed that inactivity of the PG combined with absence of the adenotropic neurohormones are major endocrine conditions for larval and pupal diapause. These relationships have been reviewed by DE WILDE [384].

Various authors have observed that the effects of transplanted PG may depend on some internal physiological factors such as the presence of nutritive reserves or developmental changes in the "reactiveness" of the epidermal cells [76, 237, 243, 354]. For example, the implanted PG were ineffective in stimulation of the moult cycles unless the larvae were provided with food and initiated feeding [228]. Furthermore, changes in the sensitivity to PGH in the epidermal cells may be best illustrated on the stimulation of puparium formation in higher Diptera [76, 247]. Here the epidermal cells exhibit different sensitivity to the hormone in the course of the last larval instar [354] which is apparently associated with determination of the whole metamorphosis process [355]. More details concerning changes in the sensitivity of epidermal cells to PGH and ecdysoids can be found in Part III.

We have noticed already that stimulation of growth, especially that in epidermal tissues such as are chitinogeneous epithelium, localized imaginal discs, or variously distributed adult histoblast cells, represents the main function of PGH in insect development. This action is usually associated with multiplication of the epidermal cells or with an increase of the ploidy. In addition, it is followed by synthesis of specific enzymes and substrates which are essential for the normal course of the moult cycles, i.e. formation of the new cuticle and digestion of the old one. However, there is no unequivocal experimental evidence to show that the final act of ecdysis was also stimulated by PGH. Clearly, the hormone stimulates all the foregoing morphological, physiological, and biochemical changes which are absolutely essential for ecdysis to take place, but it is most probable that the timing and release of the ecdysial mechanisms are affected by neurohormones. For illustration, ligated fragments of the body or isolated abdomens can undergo apolysis, formation of the new cuticle, digestion of the old cuticle, and absorption of the moulting fluid in response to the PG implantation. However, there are no coordinated ecdysial movements which are used for shedding off the old cuticle in normal body due to absence in the abdomens of the cerebral neurohormonal sources. Even in the case of puparium formation in flies, which has some features in common with ecdysis, the neurohormones are at least engaged in the contraction to puparium [410].

The above descriptions of the PGH functions apply to growth of some particular tissues within the body. Relatively more complicated relationships occur with respect to the effects of the hormone on growth of the whole body. Novák [218] claims that PGH has no growth promoting effects. He argues that specific "larval" cells are stimulated to grow by juvenile hormone whereas other "imaginal" cells grow due to presence of some hypothetical gradient factor. One must realize that growth of the whole body in insects is limited in each developmental instar by certain capacity of the sclerotized integument. Thus, prevention of moult cycles after PG ectomy and inability of these experimental animals to exchange periodically the old integumental cover for a new enlarged one (as it occurs in normal development), must invariably result in suppression of growth of the whole body. With respect to a single larval instar growth of the whole body may not depend on the activity of PG for as long as the above indicated limitation of the internal volume by capacity of the integument is reached. It is generally known that insect larvae feed and increase the internal body volume at the beginning of each instar prior to the neurohormonal activation of the PG.

The last instar larvae of *Pyrrhocoris* with ectomized PG could feed and increase their body size much over the usual level without exhibiting a moult cycle. The body of these "permanent larvae" was loaded with nutritional reserves in the same way as in the hypertrophic allatectomized females. This suggests that larvae without PG can feed and utilize the food products and are able to increase the size of the body in a single instar irrespective to the development associated with the moult cycle in the epidermal cells. This conclusion that PGH has no direct effect on food utilization and synthesis of reserve products in the fat body is supported by observations on the regulation of the moult cycle in *Rhodnius* [375, 379]. Furthermore it is generally known that adult females of most insects perform all the nutritive physiological functions in complete absence of the active PG.

Whereas digestive and nutritional processes can proceed independently from the PGH stimulated development in the epidermal cells the reciprocal relationships are not always possible. First of all, PGH can induce moult cycles in the metamorphosis stages of Endopterygote insects without respect to feeding as all the necessary substrates for tissue growth have already been accumulated in the larval body. However, in larvae of Exopterygote insects or in young larval instars of all insects (perhaps with one exception of *Trogoderma* larvae) we find considerable limitations of growth of the epidermal tissues due to insufficient supply of nutrients in the body. A regular moult cycle in these larvae is usually associated with two to four times increase of the epidermal cell number or size. Therefore, it seems quite obvious that such cellular growth cannot take place in the absence of essential substrates, which are frequently defficient at the beginning of each larval instar. These relationships constitute an important part of physiological interactions between the neurohormonally integrated trophic functions and the PGH plus CAH integrated morphogenetic changes in insect development, as has been discussed by SLÁMA [312, 315, 317, 318].

Responses of the epidermal cells to PGH stimulation as well as the regulation of PG functions by neurohormones may differ from species to species according to specific genetical instructions. This is reflected into considerable differences involving developmental regulations in various species, especially with respect to metamorphosis of Endopterygote insects. In certain Lepidoptera, for instance, the PG require neurohormonal stimulus for their control of metamorphosis at first in the last larval instar and then again in the pupa [393]. In contrast, many non-diapausing Lepidoptera listed by WILLIAMS [393] and PFLUGFELDER [237] undergo metamorphosis in complete absence of the neurosecretory system of the brain. Also, in certain species of sawflies the whole metamorphosis, including several moult cycles, can proceed normally in decapitated larvae [314]. Not only that, but even the isolated larval abdomens which are apparently lacking PG can undergo several moult cycles and develop progressively into pupal and adult abdominal fragments in *Dermestes vulpinus* (see p. 352). This may suggest that the larval epidermal cells of this species require only one instantaneous stimulation by PGH for the whole process of metamorphosis. It has also been reported that roaches deprived of PG can undergo a moult cycle [43].

Although PGH is a common stimulus for growth and multiplication among the epidermal tissue it is not a general and obligatory factor for any kind of cell division in insects [375]. Supporting evidence to this interpretation is derived from regulation of embryogenesis, reproduction, and regeneration in diapausing pupae [21], in *Rhodnius* [375], and in *Locusta* [266]. Hence, the PGH, like certain hormones of other animals which have also morphogenetic effects [58], would intermediate chemical signals for initiation and termination of development at various parts of the body. It seems to affect only some determined target cells at some determined ontogenetic stages according to predetermined genetical instructions.

With respect to development of the reproductive system the PGH stimulates growth of the imaginal discs of the gonads in young larval instars. During metamorphosis it is also essential for growth and differentiation of the internal reproductive system. Moreover, during metamorphosis of some Lepidoptera, Hymenoptera, and in certain other Endopterygote insects (see p. 238) the PGH is essential for development of the ovaries including vitellogenesis and formation of the ripe eggs. In this case the hormone controls virtually the same developmental process which is usually under control of the CAH in adult females of other insect species. In adult females of Apterygota the moult cycles are alternating with ovipositions in regular intervals [360]. In adult females of *Rhodnius* it was possible to induce a moult cycle without remarcable effect on growth of the ovaries [369]. In *Locusta* [142] implantation of the prothoracic glands affected growth and development of the accessory sexual glands in both sexes, while in *Leucophaea* implantations of active PG into adult females inhibited the activity of CA and so it inhibited ovarian growth. Further problems concerned with the action of PGH on growth have been discussed by Novák [219]. Finally it may be added that the effects on insect development produced by the PG transplantations are neither species nor genus specific [218, 237].

4. Some Physiological and Biochemical Effects of PGH

Experiments which involved extirpations and transplantations of the PG appeared extremely difficult due to complicated anatomical structure of the glands. Therefore, usual experimental techniques in studies of PG functions included ligatures and decapitations which may seriously intervene with many physiological functions, especially with feeding and digestion, nervous activity, and with the action of the retrocerebral neuroendocrine complex in general. Because of these difficulties our present knowledge on the physiological effects of the PGH is mainly based on an indirect evidence. However, as seen in Part III, this lack of our knowledge is partly counterbalanced by extensive amount of data obtained with the ecdysoids.

One such conclusion which is based on an indirect evidence suggests that the PGH, similarly like other insect hormones has no direct and general influence on the crucial physiological and biochemical functions of the organism, such as digestion, excretion, muscle activity, nerve activity, or the total respiration rate. This is substantiated by the fact that all these functions can intensively proceed in the adult insects which have completely degenerated PG. In addition, WIGGLESWORTH [375) has found that the fat body cells of *Rhodnius*, in contrast to the epidermal cells, can perform all their intensive physiological functions during the moult cycle in a complete absence of the PGH. According to our recent observations, larvae of the last larval instar of *Pyrrhocoris* with the removed PG are able to feed, grow, and perform all the main physiological functions with exception that they cannot moult. They become hypertrophic and accumulate reserve materials in a similar way as do also allatectomized adult females. The assimilated food products are not utilized. This suggests that the role of the PGH in the last larval instar of this insect may be in some physiological respects similar to that of CAH in adult females. The main difference depends perhaps in the character and the developmental programme of the target tissue, e.g. ovaries in adult females and larval tissues.

Physiological effects of hormones are generally dependent on nature and ontogenetic state of their target tissues. The main target of the PGH are the epidermal tissue of ectodermal origin, i.e. chitinogenous epithelium, imaginal disc cells or imaginal histoblast cells. More complex is the situation with internal organs such as the nerve system, intestine, fat body cells, undifferentiated gonads, etc. We have already pointed out that as to the physiological functions is concerned, most of these organs are targets for the neurohormones. However, with respect to the control of morphogenetical changes, these tissues appear to be targets of the PGH as well. Evidence supporting this assumption is derived from the effects of PGH on growth and differentiation of the internal organs during insect metamorphosis.

Although the PGH has not the direct effect on the main physiological functions, many changes in specific physiological processes appear as an indirect result of morphological changes induced. For example, during the post diapause development stimulated by PGH in Lepidopteran pupae the developing tissues and organs reconstitute and resynthesize the mitochondrial cytochrome enzyme systems [101, 300, 308, 401]. Further examples suggesting that morphological changes associated with PGH activity may be reflected to certain specific physiological and biochemical functions are also discussed together with the effects of ecdysoids in Part III.

Stimulation by PGH of tissue growth and cell divisions has a profound influence on metabolism of nucleic acids. KRISHNAKUMARAN *et al.* [174, 175] demonstrated that virtually all epidermal cells synthesized DNA during the larval-larval moult cycles. The period of DNA synthesis was closely correlated with the activity cycles in the PG. In an absence of the PGH, growth and DNA synthesis was suppressed. Thus, we may observe that unlike the action of AH which is followed by increase in RNA and protein synthesis, the action of PGH is in addition to this also followed by the DNA replication. The increased rate of RNA and protein synthesis are rather nonspecific biochemical phenomena which usually occur in target organs of most animal hormones [58]. However, the induction of cell divisions or DNA replication by PGH in insects is a more specific process which undoubtedly represents one of the main physiological functions of the hormone.

Some physiological changes associated with PGH action on the epidermal cells of *Rhodnius* were described by WIGGLESWORTH [376]. At the beginning of the moult cycle the epidermal cells occurred in a dormant state. During further development they were transformed into mitotic state after which they secreted a new cuticle. In presence of the PGH the transfer from dormancy into "activated" state proceeded in few hours. The nuclei and nucleoli of the "activated" cells were enlarged, the amount of cytoplasm was also enlarged, there appeared increased amounts of ribosomal RNA and enlarged mitochondria. This period of cell growth and proteosynthesis was 4 to 5 days later followed by mitoses. Under normal physiological conditions the described activities of the cells proceeded simultaneously in the epidermal cells and in the fat body cells. When the larvae obtained food and hormonal sources were removed, the fat body cells did react normally including mitosis, while the epidermal cells remained dormant. In other words, the epidermal cells required nutrition and PGH for their development, whereas the fat body cells required only nutrition. The amount of mitoses among the epidermal cells was not directly dependent on the PGH. This was under the influence of homeostatic mechanisms related to cell density. Thus, the PGH gives only the stimulus for development while the actual amount and extent of developmental changes may be subjected to specific feed-back regulations depending on factors such as the presence of nutrients, expansion of the internal body volume, or possibly other homeostatic reactions. We have already pointed out

(p. 38) that similar relationships exist in the action of CAH on the ovaries.

Some insects, such as higher Diptera, are characterized by rapid development during the larval stage. The larval tissues grow endomitotically and this process is also stimulated by the PGH. The epidermal cells of Exopterygote larvae usually grow by regular mitotic divisions as do also the cells of some internal organs. It is very common that some tissues in the same larval body develop mitotically whereas other tissues develop by cryptomitosis, both being stimulated by PGH. This indicates that the hormone induces DNA replication while the type of the cell development which follows depends on the genetic programme. Hinton [130] has pointed out that in certain cases the epidermal cells which have released the adult cuticle may be the same or are direct descendants of the cells which have previously released the larval cuticle. In many Endopterygotes, however, the cells which secrete adult cuticle are completely different and may not even be descendants of the cells which have released the larval cuticle. It is known that the cells of various internal organs, such as the fat body can remain functional from larval to adult stage without undergoing cell divisions [174]. In contrast to this, mitosis seems to be quite necessary in metamorphosis of the epidermis of Lepidoptera [174—176]. It appears rather difficult at present to relate these developmental changes with the PGH activity because the targets of the hormone may change from tissue to tissue and from species to species with respect to a general problem of insect diversity.

Other specific physiological processes can be directly or indirectly influenced by PGH activity. For example, in *Cerura* it was possible to influence epidermal colour change at the beginning of metamorphosis by ligatures behind the PG [38]. In addition, in the solitary and gregarious phases of *Locusta* and *Schistocerca* partial ectomy of the PG also caused a change in the pigmentation [61]. Large injuries, such as amputations of appendages led to considerable prolongation of the intermoult period with delayed ecdysis [224, 352]. It was assumed that PGH could influence regeneration but was not considered as an essential factor for this process in general [266]. Finally, histolysis of specific larval structures which occurs at the beginning of metamorphosis depends on the PGH, although a necessary condition for histolysis is also an absence of CAH. Further physiological effects of the PGH are mentioned in Part III.

5. Chemical Nature of the PGH

Compounds with PGH activity isolated from insect material are the polyhydroxylated sterols ecdysone and ecdysterone. For similar reasons as with CAH and juvenoids we have placed all the problematics concerning extractions, assays, isolation procedures and identification of their chemical structure in Part III. In this section we have discussed some arguments with regard to the identity of ecdysone with the PGH.

As already indicated, the ultrastructural analysis of the PG cells did not provide satisfactory solution of whether the secretory product of PG would be steroidal or not. Some ultrastructural patterns, such as the rich perinuclear zone with abundant ribosomes, appearance of the Golgi complex and a well developed ergastoplasm in some species indicated secretion of proteinaceous material but other features, such as the well developed smooth endoplasmic reticulum, vesicles released into intercellular spaces, lamellated structures, or abundance of glycogen in the chondriome as found in some other species suggested that the ultrastructural signs of steroid secretion are also present.

LOCKE [191] found certain ultrastructural features of steroid secretion in the oenocytes. He assumed that ecdysone might not be released from the PG cells but from the oenocytes after stimulation with PGH. This assumption was accepted by BONNER-WEIR [27] who found that anterior parts of *Calpodes* containing the source of the PGH did not pupate without presence of at least a part of the abdomen containing oenocytes. Finally, ROMER [271] obtained extracts from oenocytes with higher ecdysone activity per gramm of tissue than had the extracts prepared from the PG.

Further doubts on the real identity of ecdysone with the PGH were conditioned by some atypical physiological effects. For example, extracts with ecdysone activity were obtained from adult Locusts. Furthermore, fragments of the body of certain insects were able to moult and develop without respect to thoracic centers which were the source of the PGH. The first argument can be explained by findings of FAIN-MAUREL and CASSIER [74] who found that the PG of *Locusta* remain present for long periods of adult life including the reproductive period. The second argument can be explained by possible presence of small amounts of the endogenous hormone in the haemolymph, for it is known that relatively small amounts of ecdysone can induce the whole process of adult development following only one application. In addition, the suppressed development of the isolated thoracic parts as observed in *Calpodes* may be caused by a lack of reserve materials and substrates for growth of the thoracic muscles since most of these reserves are present in the abdominal fat body.

The oenocytes have rather a character of a decentralized peripheral endocrine organ which is most probably engaged in regulation of some extensive metabolic processes. This assumption is supported by considerably large mass of the organ in contrast to relatively small volume of the principal endocrine glands. Further, the oenocytes exhibit perfect developmental cycles in adults with the fully degenerated PG. According to a working hypothesis suggested by LOCKE [191], (i.e. neurohormone –PG–PGH–oenocytes–ecdysone–epidermal targets) this system would represent a neuroendocrine function of the third order. The presence of such a system has not been reported to occur in insects so far.

WILLIG *et al.* [399] have recently made an attempt to confirm the occurrence of ecdysone synthesis within the ring gland of Diptera. They were able to detect formation of polar fractions from the labelled cholesterol. However, repeated recrystallizations of these materials with ecdysone and ecdysterone revealed a loss of radioactivity, suggesting that the polar fractions obtained from the labelled cholesterol were not identical with the ecdysone or ecdysterone. The data contained in the Part II on ecdysoids are indicating that ecdysone, ecdysterone, and most other ecdysoids isolated from plants are able to induce all the known physiological and morphological effects of the PGH. Similar compounds do occur in the insect body and their identity with the PGH is very probable.

References

[1] ADAMS, T. S.: Ovarian regulation of the corpus allatum in the housefly, *Musca domestica*. J. Insect Physiol. *16*, 349–360 (1970).

[2] ADIYODI, K. G., and K. K. NAYAR: Haemolymph proteins and reproduction in *Periplaneta americana*. Current Sci. *35*, 587–588 (1966).

[3] ADIYODI, K. G., and K. K. NAYAR: Haemolymph proteins and reproduction in *Periplaneta americana*: The nature of conjugated proteins and the effect of cardiac-allat-ectomy on protein metabolism. Biol. Bull. Woods Hole *133*, 271–286 (1967).

[4] AGGARWAL, S. K.: Histochemistry of vitellogenesis in the adult mealworm, *Tenebrio molitor* L. (Coleoptera, Tenebrionidae). La Cellule *64*, 371–379 (1963).

[5] ALTMANN, G. G.: Hormonale Steuerung des Wasserhaushaltes der Honigbiene. Zool. Anz. Suppl. *19*, 107–112 (1956).

[6] BARTH, R. H., JR.: The endocrine control of mating behaviour in the cockroach, *Byrsotria fumigata* (Guérin). Gen. Comp. Endocrinol. *2*, 53–69 (1962).

[7] BARTH, R. H., JR.: Insect mating behaviour: endocrine control of a chemical communication system. Science *149*, 882–883 (1965).

[8] BARTH, R. H., JR.: The comparative physiology of reproductive processes in cockroaches. Part I. Mating behaviour and its endocrine control. Advances Reprod. Physiol. *3*, 167–207 (1968).

[9] BEAULATON, J.: Localisation d'activités lytiques dans la glande prothoracique du Ver à soie du chene (*Antheraea pernyi* Guér.) au stade prénymphal. I. Structures lysosomiques appareil de Golgi et ergastoplasme. J. Microscopie *6*, 179–200 (1967).

[10] Beaulaton, J.: Localisation d'activités lytiques dans la glande prothoracique du Ver à soie du chene (*Antheraea pernyi* Guér.) au stade prénymphal. II. Les vacuoles autolytiques (cytolysomes). J. Microscopie *6*, 349–370 (1967).

[11] Beaulaton, J.: Modifications ultrastructurales des cellules sécrétrices de la glande prothoracique de Vers à soie au cours des deux derniers ages larvaires. II. Le glycogéne, ses relations avec le chondriome et le réticulum endoplasmique. J. Microscopie *7*, 673–692 (1968).

[12] Beaulaton, J.: Modifications ultrastructurales des cellules sécrétrices de la glande prothoracique de Vers à soie au cours des deux derniers ages larvaires. III. Les lamelles annelées et leur dégradation. J. Microscopie *7*, 895–906 (1968).

[13] Beaulaton, J.: Étude ultrastructurale et cytochimique des glandes prothoraciques de Vers à soie aux quatrième et cinquième ages larvaires. I. La tunica propria et ses relations avec les fibres conjonctives et les hémocytes. J. Ultrastruct. Res. *23*, 474–498 (1968).

[14] Beaulaton, J.: Étude ultrastructurale et cytochimique des glandes prothoraciques de Vers à soie aux quatrième et cinquième ages larvaires. II. Les cellules interstitielles et les fibres nerveuses. J. Ultrastruct. Res. *23*, 499–515 (1968).

[15] Beaulaton, J.: Étude ultrastructurale et cytochimique des glandes prothoraciques de Vers à soie aux quatrième et cinquième ages larvaires. III. Les cellules sécrétrices. J. Ultrastruct. Res. *23*, 516–536 (1968).

[16] Bell, W. J.: Starvation induced oöcyte resorption and yolk protein salvage in *Periplaneta americana.* J. Insect Physiol. *17*, 1099–1111 (1971).

[17] Bentz, F., and A. Girardie: Action de la pars intercerebralis et des corpora allata sur les protéines ovariennes de *Locusta migratoria migratoroides* (Orthoptére) au cours de la vitellogenése. C. R. Acad. Sci. (Paris) *269*, 2014–2017 (1969).

[18] Bentz, F., A. Girardie, and M. Cazal: Étude éléctrophoretique des variations de la proteinémie chez *Locusta migratoria* pendant la maturation sexuelle. J. Insect Physiol. *16*, 2257–2270 (1970).

[19] Bentz, F., L. Joly, and P. Joly: Action des corpora allata sur les mitochondries d'un muscle alaire chez *Locusta migratoria.* Bull. Soc. Zool. France *94*, 195–199 (1969).

[20] Berreur, P.: Effects de l'ablation et de l'implantation de l'anneau de Weismann sur l'évolution des acides ribonucléiques au cours de la métamorphose de *Calliphora erythrocephala* Meig. (Diptére). C. R. Acad. Sci. (Paris) *252*, 4201–4203 (1961).

[21] Berry, S. J., A. Krishnakumaran, H. Oberlander, and H. A. Schneidermann: Effects of hormones and injury on RNA synthesis in saturniid moths. J. Insect Physiol. *13*, 1511–1537 (1967).

[22] Bessé, N. de, and M. Cazal: Action des extraits d'organs périsympathiques et de corpora cardiaca sur les ovaries de quelques insects. C. R. Acad. Sci. (Paris) *266*, 615–618 (1968).

[23] Bloch, B., E. Thomsen, and M. Thomsen: The neurosecretory system of the adult *Calliphora erythrocephala.* III. Electron microscopy of the medial neurosecretory cells of the brain and some adjacent cells. Z. Zellforsch. *70*, 185–208 (1966).

[24] Blumenfeld, M., and H. A. Schneidermann: Effect of juvenile hormone on the synthesis and accumulation of a sex-limited blood protein in the *Polyphemus* silkmoth. Biol. Bull. Woods Hole *135*, 466–475 (1968).

[25] Bodenstein, D.: Studies on the humoral mechanisms in growth and metamorphosis of the cockroach, *Periplaneta americana*. J. Exp. Zool. *123*, 413–431 (1953).

[26] Bodenstein, D., and B. Sacktor: Cytochrome-c-oxidase activity during the metamorphosis of *Drosophila virilis*. Science *116*, 299–300 (1952).

[27] Bonner-Weir, S.: Control of moulting in an insect. Nature (London) *228*, 580–581 (1970).

[28] Bounhiol, J. J.: Dans quelques limits l'écérebration des larves de Lépidoptéres est-elle compatible avec leur nymphose. C. R. Acad. Sci. (Paris) *203*, 1182–1184 (1936).

[29] Bounhiol, J. J.: Métamorphose prématurée par ablation des corpora allata chez la jeune vers à soie. C. R. Acad. Sci. (Paris) *205*, 175–177 (1937).

[30] Bounhiol, J. J.: Recherches expérimentelles sur le déterminisme de la métamorphose chez les Lépidoptéres. Bull. Biol. France Belg. *24*, 1–199 (1938).

[31] Bowers, W. S., and S. Friedmann: Mobilization of fat body glycogen by an extract of corpus cardiacum. Nature (London) *198*, 685 (1963).

[32] Bowers, B., and B. Johnson: An electron microscope study of the corpora cardiaca and secretory neurons in the aphid, *Myzus persicae* (Sulz.) Gen. Comp. Endocrinol. *6*, 213–230 (1966).

[33] Bowers, B., and C. M. Williams: Physiology of insect diapause. XIII. DNA synthesis during the metamorphosis of the cecropia silkworm. Biol. Bull. Woods Hole *126*, 205–219 (1964).

[34] Brian, M. V.: The neurosecretory cells of the brain, the corpora cardiaca and corpora allata during caste differentiation in an ant. Acta Symp. Evol. Ins. Praha *1959*, 167–171.

[35] Brookes, V. J., and R. K. Dejmal: Yolk protein: Structural changes during vitellogenesis in the cockroach *Leucophaea maderae*. Science *160*, 999–1001 (1968).

[36] Brown, B. E.: Pharmacologically active constituents of the cockroach corpus cardiacum: resolution and some characteristics. Gen. Comp. Endocrinol. *5*, 387–401 (1965).

[37] Broza, M., and M. P. Pener: Hormonal control of the reproductive diapause in the grasshopper *Oedipoda miniata*. Experientia *25*, 414–415 (1969).

[38] Bückmann, D.: Die Auslösung der Umfärbung durch das Häutungshormon bei *Cerura vinula* L. (Lepidoptera, Notodontidae). J. Insect Physiol. *3*, 159–189 (1959).

[39] Cameron, M.: Secretion of an orthodiphenol in the corpus cardiacum of the insect. Nature (London) *172*, 349 (1953).

[40] Cazal, P.: Les glandes endocrines rétro-cérébrales des insectes. Bull. Biol. France Belg. suppl. *32*, 1–10 (1948).

[41] Cazal, P.: Role des corpora cardiaca sur le teneur en eau chez quelques Orthoptéres. C. R. Acad. Sci. (Paris) *261*, 3895–3898 (1965).

[42] Cazal, P., L. Joly, and A. Porte: Étude ultrastructurelle des corpora cardiaca et de quelques formations annexes chez *Locusta migratoria* L. Z. Zellforsch. *114*, 61–72 (1971).

[43] Chadwick, L. E.: Removal of prothoracic glands from the nymphal cockroach. J. Exp. Zool. *131*, 291–305 (1956).

[44] Chalaye, D.: La tréhalosémie et sous controle neuroendocrine chez le criquet migrateur, *Locusta migratoria migratoroides*. 2. Role des corpora cardiaca et des organes périsympathiques. C. R. Acad. Sci. (Paris) *268*, 3111–3114 (1969).

[45] Chudakova, I. V., and O. M. Bocharova-Messner: Endocrine regulation of the condition of the wing musculature in the imago of the house cricket (*Acheta domestica* L.) Proc. Acad. Sci. U.S.S.R. *179*, 157–159 (1968).

[46] Church, N. S.: Hormones and the termination of reproduction and diapause in *Cephus cinctus* Nort. Canad. J. Zool. *33*, 339–369 (1955).

[47] Clarke, K. U., and R. W. Baldwin: The effect of insect hormones and of 2,4-dinitrophenol on the mitochondrion of *Locusta migratoria* L. J. Insect Physiol. *5*, 37–46 (1960).

[48] Coles, G. C.: Some effects of decapitation on metabolism in *Rhodnius prolixus* Stål. Nature (London) *203*, 323 (1964).

[49] Coles, G. C.: Studies on the hormonal control of metabolism in *Rhodnius prolixus* Stål. I. The adult female. J. Insect Physiol. *11*, 1325–1330 (1965).

[50] Cottrel, C. B.: The imaginal ecdysis in blowflies. Detection of the blood-borne darkening factor and determination of some of its properties. J. Exp. Biol. *39*, 431–448 (1962).

[51] Davey, K. G.: Substances controlling the rate of beating of the heart of *Periplaneta*. Nature (London) *192*, 284 (1961).

[52] Davey, K. G.: Changes in the pericardial cells of *Periplaneta americana* induced by exposure to homogenates of the corpus cardiacum. Quart. J. Microscop. Sci. *103*, 349–358 (1962).

[53] Davey, K. G.: The release by feeding of a pharmacologically active factor from the corpus cardiacum of *Periplaneta americana*. J. Insect Physiol. *8*, 205–208 (1962).

[54] Dejmal, R. G., and V. J. Brookes: Solubility and electrophoretic properties of ovarial protein of the cockroach *Leucophaea maderae*. J. Insect Physiol. *14*, 371–381 (1968).

[55] Delépine, V.: Recherches sur la neurosécrétion dans l'ensemble du systéme nerveux central d'un Lépidoptére *Galleria mellonella*. Bull. Soc. Zool. France *90*, 525–540 (1965).

[56] Doane, W. W.: Effect of implanted adult corpora allata on that of host in two species of *Drosophila*. Science *134*, 838–839 (1961).

[57] Doane, W. W.: Endocrine control of reproduction. Insect Physiology. 23rd Biol. Congr. Oregon State University Press, pp. 65–88 (1962).

[58] Dubovský, J.: Molekulární endokrinologie, pp. 1–235. Praha: Státní Zdravotnické Nakladatelstvi. 1968.

[59] Edwards, F. J.: Endocrinecontrol of flight muscle histolysis in *Dysdercus intermedius*. J. Insect Physiol. *16*, 2027–2031 (1970).

[60] El-Ibrashy, M. T.: A comparative study of metabolic effects of the corpus allatum in two adult Coleoptera, in relation to diapause. Meded. Landbouwhogeschool Wageningen *65*, 1–65 (1965).

[61] Ellis, P. E., and D. B. Carlisle: The prothoracic gland and colour change in Locusts. Nature (London) *190*, 368–369 (1961).

[62] Ellis, P. E., and V. J. A. Novák: Metamorphosis hormones and phase dimorphism in *Schistocerca gregaria*. Endocrinol. exp. *5*, 13–18 (1971).

[63] Enders, E.: Die hormonale Steuerung rhythmischer Bewegungen von Insekten-Ovidukten. Zool. Anz. suppl. *19*, 113–116 (1956).

[64] Engelmann, F.: Über die Einwirkung implantierter Prothoraxdrüsen im adulten Weibchen von *Leucophaea maderae* (Blattaria). Z. Vergl. Physiol. *41*, 456–470 (1959).

[65] Engelmann, F.: Hormonal control of mating behaviour in an insect. Experientia *16*, 69–70 (1960).

[66] Engelmann, F.: The mode of regulation of the corpus allatum in adult insects. Arch. Anat. Microscop. *54*, 387–404 (1965).

[67] Engelmann, F.: Endocrine control of reproduction in insects. Annu. Rev. Entomol. *13*, 1–26 (1968).

[68] Engelmann, F.: Food-stimulated synthesis of intestinal protoeolytic enzymes in the cockroach *Leucophaea maderae*. J. Insect. Physiol. *15*, 217–235 (1969).

[69] Engelmann, F.: Female specific proteins: Biosynthesis controled by corpus allatum in *Leucophaea maderae*. Science *165*, 407–409 (1969).

[70] Engelmann, F.: The Physiology of Insect Reproduction, 307 pp. Pergamon Press Inc. 1970.

[71] Engelmann, F., and R. H. Barth, Jr.: Endocrine control of female receptivity in *Leucophaea maderae* (Blattaria). Ann. Ent. Soc. Amer. *61*, 503–505 (1968).

[72] Engelmann, F., and H. P. Mueller: Fat body respiration as influenced by previously isolated corpora cardiaca. Naturwissenschaften *53*, 388–389 (1966).

[73] Engelmann, F., and D. Penney: Studies on the endocrine control of metabolism in *Leucophaea maderae* (Blattaria). I. The haemolymph proteins during egg maturation. Gen. Comp. Endocrinol. *7*, 314–325 (1966).

[74] Fain-Maurel, M. A., and P. Cassier: Rôle de l'appareil de Golgi dans l'involution des glandes de mue du criquet migrateur (*Locusta migratoria migratoroides* R. et F.). IVth Europ. Reg. Conf. Electron Microscop. Rome, p. 233 (1968).

[75] Fajal, W., and G. Fraenkel: The role of bursicon in melanization and endocuticule formation in the adult fleshfly, *Sarcophaga bullata*. J. Insect Physiol. *15*, 1235–1247 (1969).

[76] Fraenkel, G.: Puparium of flies initiated by a hormone. Nature (London) *133*, 834 (1934).

[77] Fraenkel, G., and C. Hsiao: Tanning in the adult fly: a new function of neurosecretion in the brain. Science *141*, 1057–1058 (1963).

[78] Fraenkel, G., and C. Hsiao: Bursicon, a hormone which mediates tanning of the cuticle in the adult fly and other insects. J. Insect Physiol. *11*, 513–556 (1965).

[79] Fraenkel, G., C. Hsiao, and M. Seligman: Properties of bursicon: An insect protein hormone that controls cuticular tanning. Science *151*, 553–561 (1969).

[80] Fréon, G.: Contribution à l'étude de la neurosécrétion dans la chaîne nerveuse ventrale du criquet migrateur, *Locusta migratoria* L. Bull. Soc. Zool. France *89*, 819–830 (1964).

[81] Fukaya, M., and J. Mitsuhashi: The hormonal control of larval diapause in the rice stem borer, *Chilo suppresalis*. II. The activity of the corpora allata during the diapause. Jap. J. Appl. Zool. *2*, 223–226 (1958).

[82] Fukuda, S.: Role of the prothoracic glands in differentiation of the imaginal character in the silkworm pupae. Annot. Zool. Jap. *20*, 9–13 (1941).

[83] Fukuda, S.: The hormonal mechanism of larval molting and metamorphosis in the silkworm. J. Fac. Sci. Tokyo Univ. *6*, 477–537 (1944).

[84] Fukuda, S.: Secretion of juvenile hormone by the corpora allata in pupae and moths of the silkworm, *Bombyx mori*. Annot. Zool. Jap. *35*, 199–212 (1962).

[85] Fukuda, S., G. Eguchi, and S. Takeuchi: Histological and electron microscopical studies on sexual differences in structure of the corpora allata of the moth of the silkworm, *Bombyx mori*. Embryologia *9*, 123–158 (1966).

[86] Fukuda, S., and K. Endo: Hormonal control of the development of seasonal forms in the butterfly, *Polygonia c-aureum* L. Proc. Jap. Acad. *42*, 1082–1087 (1966).

[87] Fukuda, S., and S. Takeuchi: Studies on the diapause factor-producing cells in the subeosophageal ganglion of the silkworm, *Bombyx mori*. Embryologia *9*, 333–353 (1967).

[88] Gabe, M.: Neurosecretion. Internat. Ser. Monogr. Biol. *28*, 872 pp. Oxford-London-New York: Pergamon Press Inc. 1966.

[89] Gersch, M.: Insect metamorphosis and the activation hormone. Am. Zoologist *1*, 53–57 (1961).

[90] Gersch, M.: Vergleichende Endokrinologie der wirbellosen Tiere, 535 pp. Leipzig: Akademische Verlagsgesellschaft, Geest & Portig K.-G. 1964.

[91] Gersch, M.: Experimentelle Untersuchungen zur endokrinen Regulation des Wasserhaushaltes der Larve von *Corethra (Chaoborus)*. Zool. Jb. Physiol. *75*, 1–16 (1969).

[92] Gersch, M.: Generelle Probleme der Neuroendokrinologie wirbelloser Tiere. Biol. Rundschau *8*, 77–90 (1970).

[93] Gersch, M., F. Fischer, H. Unger, and W. Kapitza: Vorkommen von Serotonin im Nervensystem von *Periplaneta americana*. L. (Insecta). Z. Naturforsch. *16б*, 351–352 (1961).

[94] Gersch, M., F. Fischer, H. Unger, and H. Koch: Die Isolierung neurohormonaler Faktoren aus dem Nervensystem der Küchenschabe *Periplaneta americana*. Z. Naturforsch. *15 b*, 319–322 (1960).

[95] Gersch, M., K. Richter, G.-A. Böhm, and J. Stürzebecher: Selektive Ausschüttung von Neurohormonen nach elektrischer Reizung der Corpora Cardiaca von *Periplaneta americana in vitro*. J. Insect Physiol. *16*, 1991–2013 (1970).

[96] Gersch, M., K. Richter, J. Stürzebecher, and B. Fabian: Experimentelle Untersuchungen zum Wirkungsmechanismus des Neurohormons D von *Periplaneta americana* (L.) am „biologischen Modell" der Harnblase von *Bufo bufo* (L.). Gen. Comp. Endocrinol. *12*, 40–50 (1969).

[97] Gersch, M., and J. Stürzebecher: Weitere Untersuchungen zur Kennzeichnung des Aktivationshormons der Insektenhäutung. J. Insect Physiol. *14*, 87–96 (1968).

[98] Gersch, M., and J. Stürzebecher: Experimentelle Stimulierung der zellulären Aktivität der Prothorakaldrüsen von *Periplaneta americana* durch den Aktivationsfaktor. J. Insect Physiol. *16*, 1813–1826 (1970).

[99] Gilbert, L. I.: Physiology of growth and development. Endocrine aspects. In: Rockstein, M., ed.: Physiology of Insects, pp. 198—263. New York: Academic Press. 1964.

[100] Gilbert, L. I.: The chemistry of insect hormones. Progress in Endocrinology, Excerpta Med. Int. Congr. Ser. *184*, 340–346 (1968).

[101] Gilbert, L. I., and H. A. Schneidermann: Some biochemical aspects of insect metamorphosis. Am. Zoologist *1*, 11–51 (1961).

[102] Gilbert, L. I., and H. A. Schneidermann: The content of juvenile hormone and lipid in Lepidoptera: sexual differences and developmental changes. Gen. Comp. Endocrinol. *1*, 453–472 (1961).

[103] Girardie, A.: Contribution à l'étude du contrôle de l'activité des corpora allata par le pars intercerebralis chez *Locusta migratoria* L. C. R. Séanc. Acad. Sci. (Paris) (0) *261*, 4876–4878 (1965).

[104] Girardie, A.: Contrôle de l'activité génitale chez *Locusta migratoria*. Mise en évidence d'un facteur gonadotrope et d'un facteur allatotrope dans la pars intercerebralis. Bull. Soc. Zool. France *91*, 423–424 (1966).

[105] Girardie, A.: Mise en évidence, dans la protocérébron de cellules neurosécrétrices contrôlant le métabolisme hydrique. C. R. Acad. Sci. (Paris) *271*, 504–507 (1970).

[106] Girardie, A.: Neurosécrétions cérébrales chez les Acridiens. Bull. Soc. Zool. France *95*, 783–802 (1970).

[107] Girardie, A., and M. Cazal: Rôle de la pars intercerebralis et des corpora cardiaca sur la mélanisation chez *Locusta migratoria* (L). C. R. Acad. Sci. (Paris) *261*, 4525–4527 (1965).

[108] Girardie, A., and J. Girardie: Étude histologique, histochimique et ultrastructurale de la pars intercerebralis chez *Locusta migratoria* L. (Orthoptére). Z. Zellforsch. *78*, 54–75 (1967).

[109] Gnatzy, W.: Struktur und Entwicklung des Integuments und der Oenocyten von *Culex pipiens* L. (Dipt.). Z. Zellforsch. *110*, 401–443 (1970).

[110] Grillot, J.-P.: Recherches sur la localisation et la structure des organes neurohémaux métamériques associés à la chaine nerveuse ventrale chez les Coléoptéres. C. R. Acad. Sci. (Paris) *270*, 847–850 (1970).

[111] Hachlow, W.: Zur Entwicklungsmechanik der Schmetterlinge. Arch. Entw. Mech. Org. *125*, 26–49 (1931).

[112] Handel, E. van, and A. O. Lea: Medial neurosecretory cells as regulators of glycogen and triglyceride synthesis. Science *149*, 298–300 (1965).

[113] Harker, J. E.: Internal factors controlling the suboesophageal ganglion neurosecretory cycle in *Periplaneta americana* L. J. Exp. Biol. *37*, 164–170 (1960).

[114] Harvey, W. R., and C. M. Williams: The injury metabolism of the cecropia silkworm. I. Biological amplification of the effects of localized injury. J. Insect Physiol. *7*, 81–99 (1961).

[115] Hasegawa, K.: The diapause hormone of the silkworm, *Bombyx mori*. Nature (London) *179*, 1300–1301 (1957).

[116] HASEGAWA, K.: Studies on the mode of action of the diapause hormone in the silkworm, *Bombyx mori*. I. The action of diapause hormone injected into pupae of different ages. J. Exp. Biol. *40*, 517–529 (1963).

[117] HASEGAWA, K., and O. YAMASHITA: Mode d'action de l'hormone de diapause dans l'métabolisme glucidique du Ver à soie *Bombyx mori* L. Ann. Endocrinol. (Paris) *31*, 631–636 (1970).

[118] HERMAN, W. S., and L. I. GILBERT: The neuroendocrine system of *Hyalophora cecropia* (L.). (Lepidoptera: Saturniidae). I. The anatomy and histology of the ecdysial glands. Gen. Comp. Endocrinol. *7*, 291 (1966).

[119] HIGHNAM, K. C.: Activity of the corpora allata during pupal diapause in *Mimas tiliae* (Lepidoptera). Quart. J. Micr. Sci *99*, 171–180 (1958).

[120] HIGHNAM, K. C.: Neurosecretory control of ovarian development in *Schistocerca gregaria*, and its relation to phase differences. Coll. Int. Centr. Natn. Rech. Scient. *114*, 107–121 (1962).

[121] HIGHNAM, K. C.: Endocrine relationships in insect reproduction. In: HIGHNAM, K. C., ed.: Insect Reproduction. Roy. Entomol. Soc. London Symp. *2*, 26–42 (1964).

[122] HIGHNAM, K. C.: Insect Hormones. J. Endocrinol. *39*, 123–150 (1967).

[123] HIGHNAM, K. C.: Neurosecretion in Insects. Progress in Endocrinology. Excerpta Medica International Congress Series *184*, 351–355 (1968).

[124] HIGHNAM, K. C.: Estimates of neurosecretory activity during maturation in female locusts. Insect Endocrines. Academia Praha *1971*, 81–90.

[125] HIGHNAM, K. C., and L. HILL: The role of the corpora allata during oöcyte growth in the desert locust, *Schistocerca gregaria* Forsk. J. Insect Physiol. *9*, 587–596 (1963).

[126] HIGHNAM, K. C., and L. HILL: The Comparative Endocrinology of the Invertebrates. Contemporary Biology, 270 pp. London: Edward Arnold Ltd. 1969.

[127] HILL, L.: Neurosecretory control of haemolymph protein concentration during ovarian development in the desert locust. J. Insect Physiol. *8*, 609–619 (1962).

[128] HILL, L.: Endocrine control of protein synthesis in female desert locusts (*Schistocerca gregaria*). J. Endocrinol. *26*, 17–18 (1963).

[129] HILL, L.: The incorporation of C^{14}-glycine into the proteins of the fat body of the desert locust during ovarian development. J. Insect Physiol. *11*, 1605–1615 (1965).

[130] HINTON, H. E.: Metamorphosis of the epidermis and hormone mimetic substances. Sci. Progress *51*, 306–322 (1963).

[131] HOLMAN, G. M., and B. J. COOK: Pharmacological properties of excitatory neuromuscular transmission in the hindgut of the cockroach, *Leucophaea maderae*. J. Insect Physiol. *16*, 1891–1907 (1970).

[132] HRUBEŠOVÁ, H., and K. SLÁMA: The effect of hormones on the intestinal proteinase activity of adult *Pyrrhocoris apterus* L. (Hemiptera). Acta entom. bohemoslov. *64*, 175–183 (1967).

[133] ICHIKAWA, M., and H. ISHIZAKI: Brain hormone of the silkworm, *Bombyx mori*. Nature (London) *191*, 933–934 (1961).

[134] ICHIKAWA, M., and H. ISHIZAKI: Protein nature of the brain hormone of insects. Nature (London) *198*, 308–309 (1963).

[135] Ilan, J., J. Ilan, and N. Patel: Mechanism of gene expression in *Tenebrio molitor*. J. Biol. Chem. *245*, 1275–1281 (1970).

[136] Ishizaki, H., and M. Ichikawa: Purification of the brain hormone of the silkworm *Bombyx mori*. Biol. Bull. Woods Hole *133*, 355–368 (1967).

[137] Janda, V., Jr., and P. Krieg: Proteolytische Aktivität des Mitteldarms von *Galleria mellonella* in Zusammenhang mit Wachstum und Metamorphose. Z. vergl. Physiol. *64*, 288–300 (1969).

[138] Janda, V., Jr., and F. Sehnal: The influence of juvenile hormone on glycogen, fat and nitrogen metabolism in *Galleria mellonella* L. Endocrinol. exp. *5*, 79–83 (1971).

[139] Janda V., Jr., and K. Sláma: Über den Einfluß von Hormonen auf den Glykogen-, Fett- und Stickstoffmetabolismus bei den Imagines von *Pyrrhocoris apterus* L. (Hemiptera). Zool. Jb. Physiol. *71*, 345–358 (1965).

[140] Johansson, A. S.: Relation of nutrition to endocrine-reproductive functions in the milkweed bug *Oncopeltus fasciatus* (Dallas) (Heteroptera: Lygaeidae). Nytt Mag. Zool. *7*, 1–132 (1958).

[141] Johnson, B.: A histological study of neurosecretion in aphids. J. Insect Physiol. *9*, 727–739 (1963).

[142] Joly, L.: Action de l'implantation de glandes ventrales sur l'activité génitale des imagos de *Locusta migratoria* L. C. R. Acad. Sci. (Paris) *272*, 2326–2329 (1971).

[143] Joly, P.: Les corrélations humorales chez les insectes. Année biol. *21*, 1–34 (1945).

[144] Joly, P.: Déterminisme de la pigmentation chez *Acria turrita*. C. R. Acad. Sci. (Paris) *235*, 1054–1056 (1952).

[145] Joly, P.: Role joué per les corpora allata dans la realisation du polymorphisme de phase chez *Locusta migratoria*. Coll. Int. Centr. Natn. Rech. Scient. *114*, 77–88 (1962).

[146] Joly, P.: Comparison du volume et de l'activité physiologique des corpora allata de *Locusta migratoria* L. Ann. Soc. Entom. France *3*, 601–608 (1967).

[147] Joly, P., and M. Cazal: Donnée récentes sur les corpora cardiaca. Bull. Soc. Zool. France *94*, 181–194 (1969).

[148] Joly, P., and L. Joly: Comparison des tests cytologiques d'activité des corpora allata des Acridiens. C. R. Acad. Sci. (Paris) *270*, 2560–2562 (1970).

[149] Joly, P., L. Joly, and A. Porte: Remarques sur l'ultrastructure de la glande ventrale de *Locusta migratoria* L. (Orthoptére) en population dense. C. R. Acad. Sci. (Paris) *269*, 917–918 (1969).

[150] Joly, P., L. Joly, A. Porte, and A. Girardie: Analyse ultrastructurale de l'activité des corpora allata de *Locusta migratoria* L. et ses conséquences sur la structure que l'on doit attributer au mécanisme humoral controlant la métamorphose. Arch. Zool. exptl. gén. *110*, 617–628 (1969).

[151] Joly, P., L. Joly, and C. Riotte: La métamorphose des organes du vol chez *Locusta migratoria* L. C. R. 86. Congr. Soc. Savantes Paris (Sec. Sci. Montpellier), pp. 725–733 (1962).

[152] Joly, P., A. Porte, and A. Girardie: Caractéres ultrastructuraux des corpora allata actifs et inactifs chez *Locusta migratoria*. C. R. Acad. Sci. (Paris) *265*, 1633–1635 (1967).

[153] Jones, J. C.: Effects of salts on *Anopheles* heart rates. J. Exp. Zool. *133*, 125–144 (1956).
[154] Kaiser, P.: Histologische Untersuchungen über die Corpora allata und Prothoraxdrüsen der Lepidopteren in bezug auf ihre Funktion. Arch. Entw. Mech. Org. *144*, 99—131 (1949).
[155] Karlinsky, A.: Effects de l'ablation des corpora allata imaginaux sur le développement ovarien de *Pieris brassicae* L. (Lépidoptére). C. R. Acad. Sci. (Paris) *256*, 4101–4103 (1963).
[156] Karlinsky, A.: Reprise de la vitellogénèse aprés implantation de corpora allata chez *Pieris brassicae* L. (Lépidoptére). C. R. Acad. Sci. (Paris) *264*, 1735–1738 (1967).
[157] Keely, L. L., and S. Friedman: Corpus cardiacum as a metabolic regulator in *Blaberus discoidalis* Serville (Blattidae). I. Long-term effects of cardiacectomy on whole body and tissue respiration and on trophic metabolism. Gen. Comp. Endocrinol. *8*, 129–134 (1967).
[158] Keely, L. L., and H. Van Waddill: Insect hormones: evidence for a neuroendocrine factor affecting respiratory metabolism. Life Sciences *10*, 737–745 (1971).
[159] King, R. C., S. K. Aggarwal, and D. Bodenstein: The comparative submicroscopic cytology of the corpus allatum-corpus cardiacum complex of wild type and *fes* adult female *Drosophila melanogaster*. J. Exp. Zool. *161*, 151–176 (1966).
[160] King, R. C., S. K. Aggarwal, and D. Bodenstein: The comparative submicroscopic morphology of the ring gland of *Drosophila melanogaster* during the second and third larval instars. Z. Zellforsch. *73*, 272–285 (1966).
[161] Kloot, W. G. van der: Neurosecretion in insects. Annu. Rev. Entomol. *5*, 35–52 (1960).
[162] Kobayashi, M.: The chemistry and physiology of the brain hormone. Proc. XVIth Int. Congr. Zool. Washington *4*, 226–233 (1963).
[163] Kobayashi, M., Y. Ishitoya, and M. Yamazaki: Action of proteinic brain hormone to the prothoracic gland in an insect, *Bombyx mori* L. (Lepidoptera: Bombycidae). Appl. Entomol. Zool. *3*, 150–152 (1968).
[164] Kobayashi, M., S. Kimura, and M. Yamazaki: Action of insect hormones on the fate of C^{14}-glucose in the diapausing, brainless pupa of *Samia cynthia pyeri* (Lepidoptera: Saturniidae). Appl. Entomol. Zool. *2*, 79–84 (1967).
[165] Kobayashi, M., and J. Kirimura: The "brain" hormone in the silkworm, *Bombyx mori* L. Nature (London) *181*, 1217 (1958).
[166] Kobayashi, M., J. Kirimura, and M. Saito: The "brain" hormone in an insect, *Bombyx mori* L. Mushi *36*, 85–92 (1962).
[167] Kobayashi, M., M. Saito, Y. Ishitoya, and N. Ikekawa: Brain hormone activity in *Bombyx mori* of sterols and physiologically vital active substances. Proc. Soc. Exp. Biol. Med. *114*, 316–318 (1963).
[168] Kobayashi, M., and Y. Yamashita: A function of corpus allatum in neurosecretory system in the silkworm, *Bombyx mori*. J. Sericult. Sci. Japan *28*, 335–339 (1959).
[169] Kobayashi, M., and M. Yamazaki: The proteinic brain hormone in an insect *Bombyx mori* L. (Lepidoptera: Bombycidae). Appl. Entomol. Zool. *1*, 53–60 (1966).
[170] Koller, G.: Rhythmische Bewegung und hormonale Steuerung bei den malpighischen Gefäßen der Insekten. Biol. Zbl. *67*, 201–211 (1948).

[171] Kopeč, S.: Über die Funktionen des Nervensystems der Schmetterlinge während der successiven Stadien der Metamorphose. Zool. Anz. *40*, 353–360 (1912).

[172] Kopeč, S.: Studies on the necessity of the brain for the inception of insect metamorphosis. Biol. Bull. Woods Hole *42*, 323–342 (1922).

[173] Kort, C. A. D. de: Hormones and the structural and biochemical properties of the flight muscles in the colorado beetle. Meded. Landbouwhogeschool Wageningen *69*, 1–63 (1969).

[174] Krishnakumaran, A., S. J. Berry, H. Oberlander, and H. A. Schneidermann: Nucleic acid syntheses during insect development. II. Control of DNA synthesis in the cecropia silkworm and other saturniid moths. J. Insect Physiol. *13*, 1–57 (1967).

[175] Krishnakumaran, A., H. Oberlander, and H. A. Schneidermann: Rates of DNA and RNA synthesis in various tissues during a larval moult cycle of *Samia cynthia ricini* (Lepidoptera). Biol. Bull Woods Hole *125*, 576–581 (1963).

[176] Krishnakumaran, A., and H. A. Schneidermann: Developmental capacities of the cells of an adult moth. J. Exp. Zool. *157*, 293–306 (1964).

[177] Kühn, A., and H. Piepho: Über hormonale Wirkungen bei der Verpuppung der Schmetterlinge. Nachr. Ges. Wiss. (Göttingen) *2*, 141–154 (1936).

[178] Lamb, K. P., and D. F. White: Endocrine aspects of alary polymorphism in *Brevicoryne brassicae* L. Endocrinol. Exp. *5*, 19–22 (1971).

[179] Larsen, J. R., and D. Bodenstein: The humoral control of egg maturation in the mosquito. J. Exp. Zool. *140*, 343–382 (1959).

[180] Lea, A. O.: Egg maturation in mosquitoes not regulated by the corpora allata. J. Insect Physiol. *15*, 537–541 (1969).

[181] Lea, A. O.: Size independent secretion by the corpus allatum of *Calliphora erythrocephala*. J. Insect. Physiol. *15*, 477–482 (1969).

[182] Lees, A. D.: The physiology of diapause in arthropods, I–II. Cambridge University Press. 1955.

[183] Lees, A. D.: The physiology and biochemistry of diapause. Annu. Rev. Entomol. *1*, 1–16 (1956).

[184] Lefeuvre, J. C.: On the precocious determination of cockroaches wing pads. Endocrinol. exp. *5*, 29–33 (1971).

[185] Legay, J. M.: Note sur l'évolution des corpora allata au cours de la vie larvaire de *Bombyx mori*. C. R. Soc. Biol. (Paris) *144*, 512–513 (1950).

[186] L'Hélias, C.: Étude du métabolisme basal chez le phasme *Dixippus morosus* après ablation des corpora allata. C. R. Acad. Sci. (Paris) *239*, 778–780 (1954).

[187] L'Hélias, C.: Variation des métabolismes glucidique, azote et lipidique après ablation des corpora allata chez le phasme *Dixippus morosus* (Br.) Physiol. Comp. Oecol. *4*, 74–88 (1955).

[188] L'Hélias, C.: Biochimie de la métamorphose et role du complexe rétrocérébral chez les insectes. Acta Soc. Linn. (Bordeaux) *97*, 1–11 (1957).

[189] L'Hélias, C.: Le role des ptérines dans le mécanisme hormonal du complexe rétrocérébral chez les insectes. 2nd Int. Symp. Neurosecret. Lund *1957*, 91–95.

[190] L'Hélias, C.: Aspects chimiques des mécanismes endocriniens chez les insectes. Ann. Biol. *3*, 1–31 (1964).

[191] Locke, M.: The ultrastructure of the oenocytes in the molt/intermolt cycle of an insect. Tissue and Cell *1*, 103–154 (1969).

[192] Loher, W.: The chemical acceleration of the maturation process and its hormonal control in the male of the desert locust. Proc. Roy. Soc. (London) *B 153*, 380–397 (1960).

[193] Loher, W.: Die Kontrolle des Weibchengesanges von *Gomphocerus rufus* L., Acridinae durch die corpora allata. Naturwissenschaften *49*, 406–409 (1962).

[194] Loher, W., and F. Huber: Nervous and endocrine control of sexual behaviour in a grasshopper (*Gomphocerus rufus* L., Acridinae). Symp. Soc. Exp. Biol. *20*, 381–400 (1966).

[195] Loof, A. de, and J. de Wilde: The relations between haemolymph proteins and vitellogenesis in the colorado beetle *Leptinotarsa decemlineata*. J. Insect Physiol. *16*, 157–169 (1970).

[196] Lukoschus, F.: Über die Prothoraxdrüse der Honigbiene (*Apis mellifica* L.). Naturwissenschaften *5*, 116 (1952).

[197] Lüscher, M.: Hormonal control of caste differentiation in termites. Ann. N. Y. Acad. Sci. *89*, 549–563 (1960).

[198] Lüscher, M.: Hormonal control of respiration and protein synthesis in the fat body of the cockroach *Nauphoeta cinerea* during oöcyte growth. J. Insect Physiol. *14*, 499–511 (1968).

[199] Lüscher, M., and F. Engelmann: Histologische und Experimentelle Untersuchungen über die Auslösung der Metamorphose bei *Leucophaea maderae* (Orthoptera). J. Insect Physiol. *5*, 240–258 (1960).

[200] Lüscher, M., and H. Leuthold: Über die Hormonale Beeinflussung des respiratorischen Stoffwechsels bei der Schabe *Leucophaea maderae* (F.). Rev. Suisse Zool. *72*, 618–622 (1965).

[201] Lüscher, M., and M. Wyss-Huber: Die Adenosin-Nukleotide im Fettkörper des adulten Weibchens von *Leucophaea maderae* im Laufe des Sexualzyklus. Rev. Suisse Zool. *71*, 183–194 (1964).

[202] Maddrell, S. H. P.: Excretion in the blood sucking bug, *Rhodnius prolixus* Stål. I. The control of diuresis. J. Exp. Biol. *40*, 247–256 (1963).

[203] Maddrell, S. H. P.: The site of release of the diuretic hormone in *Rhodnius*–a new neurohaemal system in insects. J. Exp. Biol. *45*, 499–508 (1966).

[204] Meola, R., and A. O. Lea: Interdependence of paraldehyde-fuchsine staining of the corpus cardiacum and the presence of the neurosecretory hormone required for egg development in the mosquito. Gen. Comp. Endocrinol. *16*, 105–111 (1971).

[205] Mills, R. R., S. Androuny, and R. R. Fox: Correlation of phenoloxidase activity with ecdysis and tanning hormone release in the american cockroach. J. Insect Physiol. *14*, 603–611 (1968).

[206] Mills, R. R., and D. J. Nielsen: Hormonal control of tanning in the american cockroach–V. Some properties of the purified hormone. J. Insect Physiol. *13*, 273–280 (1967).

[207] Mills, R. R., and D. L. Whitehead: Hormonal control of tanning in the american cockroach: changes in blood cell permeability during ecdysis. J. Insect Physiol. *16*, 331–340 (1970).

[208] Minks, A. K.: Biochemical aspects of juvenile hormone action in the adult *Locusta migratoria*. Archs. Neerl. Zool. *17*, 175–258 (1967).

[209] Mordue, W.: The hormonal control of excretion and water balance in Locusts. Endocrinol. exp. *5*, 79–83 (1971).

[210] Nair, K. M.: The localization of ribonucleic acid and hormone production in the corpus allatum of *Chrysocornis purpureus* (Westw.). J. Histochem. Cytochem. *11*, 495–499 (1963).

[211] Natalizi, G., and N. Frontali: Purification of insect hyperglycaemic and heart accelerating hormones. J. Insect. Physiol. *12*, 1279–1287 (1966).

[212] Natalizi, G., M. Pansa, V. D'Ajello, O. Casaglia, S. Bettini, and N. Frontali: Physiologically active factors from corpora cardiaca of *Periplaneta americana*. J. Insect Physiol. *16*, 1827–1836 (1970).

[213] Nayar, K. K.: Water content and release of neurosecretory products in *Iphita limbata* Stal. Current Sci. *26*, 25 (1957).

[214] Němec, V., K. Sláma, and H. Hrubešová: The effect of hormones on the intestinal proteinase activity of adult *Pyrrhocoris apterus* L. (Hemiptera). Acta entomol. bohemoslov. *64*, 175–183 (1967).

[215] Nishiitsutsuji-Uwo, J.: An insect brain hormone–activity from mammalian tissues. Botyu-Kagaku *36*, 66–77 (1971).

[216] Nohel, P., and K. Sláma: Effect of a juvenile hormone analogue on glutamate-pyruvate transaminase in the bug *Pyrrhocoris apterus*. Insect Biochem. *2*, 58–66 (1972).

[217] Novák, V. J. A.: The growth of the corpora allata during the postembryonal development in insects. Acta Soc. Zool. Čsl. *18*, 98–133 (1954).

[218] Novák, V. J. A.: Insect Hormones, 478 pp. London: Methuen & Co. Ltd. 1966.

[219] Novák, V. J. A.: Hormonal control of the molting process in Arthropods. Gen. Comp. Endocrinol., suppl. *2*, 439–450 (1969).

[220] Novák, V. J. A., and K. Sláma: The influence of juvenile hormone on the oxygen consumption of the last larval instar of *Pyrrhocoris apterus* L. J. Insect Physiol. *8*, 145–153 (1962).

[221] Novák, V. J. A., K. Sláma, and K. Wenig: Influence of implantation of corpus allatum on oxygen consumption of *Pyrrhocoris apterus*. Acta Symp. Evol. Insects (Prague) *1959*, 147–153.

[222] Oberlander, H., S. J. Berry, A. Krishnakumaran, and H. A. Schneidermann: RNA and DNA synthesis during activation and secretion of the prothoracic glands of saturniid moths. J. Exp. Zool. *159*, 15–32 (1965).

[223] Odhiambo, T. R.: The fine structure of the corpus allatum of the sexually mature male of the desert locust. J. Insect Physiol. *12*, 819–828 (1966).

[224] O'Farell, M., and A. Stock: Influence of illumination on molting and regeneration in the cockroach *Blatella germanica* L. Nature (London) *195*, 1122–1123 (1962).

[225] Orr, C. W. M.: The influence of nutritional and hormonal factors on the chemistry of the fat body, blood, and ovaries of the blowfly *Phormia regina* Meig. J. Insect Physiol. *10*, 103–119 (1964).

[226] Östlund, E.: The distribution of catecholamines in lower animals and their effect on the heart. Acta Physiol. Scand. *31*, suppl. *112*, 1–67 (1954).

[227] Ozeki, K.: Secretion of molting hormone from the ventral glands of the earwig, *Anisolabis maritima*. Sci. Papers Coll. Gen. Educ. Univ. Tokyo *9*, 255–262 (1959).

[228] Ozeki, K.: Hormonal control of molting in the earwig *Anisolabis maritima*. Sci. Papers Coll. Gen. Educ. Univ. Tokyo *10*, 87–97 (1960).

[229] Ozeki, K.: Secretion of the juvenile hormone in the imago of the earwig, *Anisolabis maritima*. Sci. Papers Coll. Gen. Educ. Univ. Tokyo *11*, 101–107 (1961).

[230] Ozeki, K.: Studies on the function of the corpus allatum during the last nymphal stage in the earwig, *Anisolabis maritima*. Sci. Papers Coll. Gen. Educ. Univ. Tokyo *15*, 149–156 (1965).

[231] Panov, A. A., and O. K. Basurmanova: Fine structure of the gland cells in inactive and active corpus allatum of the bug, *Eurygaster integriceps*. J. Insect Physiol. *16*, 1265–1281 (1970).

[232] Pener, M. P.: On the influence of corpora allata on maturation and sexual behaviour of *Schistocerca gregaria*. J. Zool. *147*, 119–136 (1965).

[233] Pener, M. P.: Effects of allatectomy and sectioning of the nerves of the corpora allata on oöcyte growth, male sexual behaviour, and colour change in adults of *Schistocerca gregaria*. J. Insect Physiol. *13*, 665–684 (1967).

[234] Pener, M. P.: The effect of corpora allata on sexual behaviour and "adult diapause" in males of the red locust. Entomol. Exp. and Appl. *11*, 94–100 (1968).

[235] Persaud, C. E., and K. G. Davey: The control of protease synthesis in the intestine of adults of *Rhodnius prolixus*. J. Insect Physiol. *17*, 1429–1440 (1971).

[236] Pfeiffer-Weed, I.: Effects of the corpora allata on the metabolism of adult female grasshoppers. J. Exp. Zool. *99*, 183–233 (1945).

[237] Pflugfelder, O.: Entwicklungsphysiologie der Insekten, 490 pp. Leipzig: Akademische Verlagsgesellschaft, Geest & Portig K. G. 1958.

[238] Picheral, B.: Les tissues élaborateurs d'hormones stéroides chez les Amphibiens urodéles. II. Aspects ultrastructuraux de la glande interrénale de *Salamandra salamandra* (L.) Étude particulière du glycogène. J. Microscopie *7*, 907–926 (1968).

[239] Piepho, H.: Über die Hemmung der Verpuppung durch corpora allata. Untersuchungen an der Wachsmotte *Galleria mellonella* L. Biol. Zbl. *60*, 367–393 (1940).

[240] Piepho, H.: Hormonale Grundlagen der Spinntätigkeit bei Schmetterlingsraupen. Z. Tierphysiol. *7*, 424–434 (1950).

[241] Piepho, H.: Über die Lenkung der Insektenmetamorphose durch Hormone. Verh. deutsch. Zool. Ges. *1951*, 62–76.

[242] Piepho, H.: Juvenilhormon und Verhalten bei der Wachsmotte. Naturwissenschaften *54*, 50 (1967).

[243] Piepho, H., and H. Meyer: Reaktionen der Schmetterlingshaut auf Häutungshormone. Biol. Zbl. *70*, 252–260 (1951).

[244] Pilcher, D. E. M.: Hormonal control of the malpighian tubules of the stick insect, *Carausius morosus*. J. Exp. Biol. *52*, 653–665 (1970).

[245] Pipa, R. L.: Studies on the hexapod nervous system. IV. A cytological and cytochemical study of neurons and their inclusions in the brain of a cockroach, *Periplaneta americana* (L.). Biol. Bull. Woods Hole *121*, 521–534 (1961).

[246] PIPA, R. L.: A cytochemical study of neurosecretory and other neoplasmic inclusions in *Periplaneta americana*. Gen. Comp. Endocrinol. *2*, 44–52 (1962).

[247] POSSOMPÈS, B.: Recherches expérimentelles sur le déterminisme de la métamorphose chez *Calliphora erythrocephala* Meig. Arch. Zool. Exp. Gén. *89*, 203–364 (1953).

[248] PROVANSAL, A.: Mise en évidence d'organes neurohémaux métamériques associés à la chaine nerveuse ventrale chez *Vespa crabro* L. et *Vespula germanica* Fabr. (Hyménoptéres Vespidae). C. R. Acad. Sci. (Paris) *267*, 864–867 (1968).

[249] RAABE, M.: Neurohormones chez les insectes. Bull. Soc. Zool. France *90*, 631–654 (1959).

[250] RAABE, M.: Nouvelles recherches sur la neurosécretion chez les insectes. Ann. Endocrinol. (Paris) *25*, suppl., 107–112 (1964).

[251] RAABE, M.: Étude des phénoménes de neurosecretion au niveau de la chaine nerveuse ventrale des Phasmides. Bull. Soc. Zool. France *90*, 631–654 (1965).

[252] RAABE, M.: Recherches sur le neurosécrétion dans la chaîne nerveuse ventrale du Phasme, *Carausius morosus*: Liaison entre l'activité des cellules B, C et pigmentation. C. R. Acad. Sci. (Paris) *263*, 408–411 (1966).

[253] RAABE, M.: Recherches récentes sur la neurosécrétion dans la chaîne nerveuse ventrale des insectes. Bull. Soc. Zool. France *92*, 67–71 (1967).

[254] RAABE, M.: Neurosecretion in the ventral nerve cord of insects. Insect endocrines. Academia Praha *1971*, 105–121.

[255] RAABE, M., M. CAZAL, D. CHALAYE, and N. DE BESSÉ: Action cardioaccélératrice des organes neurohémaux périsympathiques ventraux de quelques insectes. C. R. Acad. Sci. (Paris) *263*, 2002–2005 (1966).

[256] RAABE, M., and D. MONJO: Recherches histologiques et histochimiques sur le neurosécrétion chez le Phasme, *Clitumnus extradentatus*: les neuro-sécrétion du type C. C. R. Acad. Sci. (Paris) *270*, 2021–2024 (1970).

[257] RAABE, M., and F. RAMADE: Observations sur l'ultrastructure des organes périsympathiques des Phasmides. C. R. Acad. Sci. (Paris) *264*, 77–80 (1967).

[258] RADTKE, A.: Hemmung der Verpuppung beim Mehlkäfer, *Tenebrio molitor* L. Naturwissenschaften *30*, 451–452 (1942).

[259] RAHM, U. H.: Die innersekretorische Steuerung der postembryonalen Entwicklung von *Sialis lutaria* L. (Megaloptera). Rev. Suisse Zool. *59*, 173–273 (1952).

[260] RALPH, C. L.: Heart accelerators and decelerators in the nervous system of *Periplaneta americana* (L.). J. Insect Physiol. *8*, 431–439 (1962).

[261] RALPH, C. L., and R. J. MATTA: Evidence for humoral effects on metabolism of cockroaches from studies of tissue homogenates. J. Insect Physiol. *11*, 983–991 (1965).

[262] RALPH, C. L., and C. MAC CARTHY: Effect of brain and corpus cardiacum extracts on haemolymph trehalose of the cockroach *Periplaneta americana*. Nature (London) *203*, 1195 (1964).

[263] RICHTER, A.: Über die Entwicklung der Schuppenorgane und der Genitalanhänge in Abhängigkeit vom Hormonsystem bei *Lepisma saccharina* L. Roux Arch. Entwicklungsmech. *154*, 1–28 (1962).

[264] Riddiford, L. M., and C. M. Williams: Role of the corpora cardiaca in the behavior of saturniid moths. I. Release of sex pheromones. Biol. Bull. Woods Hole *140*, 1–7 (1971).

[265] Rinterknecht, E.: Influence de l'hormone juvénile sur la reconstitution des téguments chez *Locusta migratoria* adulte. Bull. Soc. Zool. France *89*, 723–730 (1964).

[266] Rinterknecht, E.: Influence de l'hormone de la mue sur la reconstitution des téguments chez *Locusta migratoria* larvaire. (stade V). Bull. Soc. Zool. France *91*, 645–654 (1966).

[267] Rinterknecht, E.: Controle hormonale de la cicatrisation chez *Locusta migratoria*. Role de la pars intercerbralis chez les sujets larvaires. Bull. Soc. Zool. France *91*, 789–802 (1966).

[268] Rohdendorf, E.: Der Einfluß der Allatektomie auf adulte Weibchen von *Thermobia domestica* Packard (Lepismatidae, Thysanura). Zool. Jb. Physiol. *8*, 685–693 (1966).

[269] Röller, H., and K. H. Dahm: The identity of juvenile hormone produced by corpora allata *in vitro*. Naturwissenschaften *57*, 454–455 (1970).

[270] Röller, H., H. Piepho, and I. Holz: Zum Problem der Hormonalabhängigkeit des Paarungsverhaltens bei Insekten. Untersuchungen an *Galleria mellonella* L. J. Insect Physiol. *9*, 187–194 (1963).

[271] Romer, F.: Häutungshormone in den Oenocyten des Mehlkäfers. Naturwissenschaften *58*, 324–325 (1971).

[272] Roth, L. M., and R. H. Barth, Jr.: The control of sexual receptivity in female cockroaches. J. Insect Physiol. *10*, 965–975 (1964).

[273] Roussel, J. P.: Consommation d'oxygéne après ablation des corpora allata chez des femelles adultes de *Locusta migratoria*. J. Insect Physiol. *9*, 721–725 (1963).

[274] Roussel, J. P.: Fonctions des corpora allata et contrôle de la pigmentation chez *Gryllus bimaculatus* De Geer. J. Insect Physiol. *13*, 113–130 (1967).

[275] Roussel, J. P.: Rhythme et régulation du coeur chez *Locusta migratoria migratoroides* L. Acrida *1*, 17–39 (1971).

[276] Sägesser, H.: Über die Wirkung der Corpora allata auf den Sauerstoffverbrauch bei der Schabe *Leucophaea maderae* (F.). J. Insect Physiol. *5*, 264–285 (1960).

[277] Scharrer, B.: The relationship between corpora allata and reproductive organs in adult *Leucophaea maderae*. Endocrinology *38*, 46–55 (1946).

[278] Scharrer, B.: Functional analysis of the corpus allatum of the insect *Leucophaea maderae*, with the electron microscope. Biol. Bull. Woods Hole *121*, 370 (1961).

[279] Scharrer, B.: The neurosecretory system of *Leucophaea maderae* and its role in neuroendocrine integration. Gen. Comp. Endocrinol. *2*, 30 (1962).

[280] Scharrer, B.: Neurosecretion. XIII. The ultrastructure of the corpus cardiacum of the insect *Leucophaea maderae*. Z. Zellforsch. *60*, 761–796 (1963).

[281] Scharrer, B.: The fine structure of Blattarian prothoracic glands. Z. Zellforsch. *64*, 301–326 (1964).

[282] Scharrer, B.: Histophysiological studies on the corpus allatum of *Leucophaea maderae*. IV. Ultrastructure during normal activity cycle. Z. Zellforsch. *62*, 125–148 (1964).

[283] Scharrer, B.: Recent progress in the study of neuroendocrine mechanisms in insects. Arch. Anatom. Microscop. *54*, 331–342 (1965).

[284] Scharrer, B.: Ultrastructural study of the regressing prothoracic glands of Blattarian insects. Z. Naturforsch. *69*, 1–21 (1966).

[285] Scharrer, B.: The neurosecretory neuron in neuroendocrine regulatory mechanisms. Amer. Zool. *7*, 161–169 (1967).

[286] Scharrer, B.: Neurosecretion. XIV. Ultrastructural study of sites of release of neurosecretory material in Blattarian insects. Z. Zellforsch. *89*, 1–16 (1968).

[287] Scharrer, B.: Current concepts in the field of neurochemical mediation. MCV Quarterly *5*, 27–31 (1969).

[288] Scharrer, B.: Neurohumors and neurohormones: Definitions and terminology. In: Ariëns Kappers, J., ed.: Neurohormones and Neurohumors (Journal of Neuro-Visceral Relations, Supplementum IX), pp. 1–20. Wien-New York: Springer. 1969.

[289] Scharrer, B.: General principles of neuroendocrine communication. In: Schnitt, F. O., ed.: The Neurosciences, pp. 519–529. New York: Rockefeller Univ. Press. 1970.

[290] Scharrer, B.: Histophysiological studies on the corpus allatum of *Leucophaea maderae*. V. Ultrastructure of sites of origin and release of a distinctive cellular product. Z. Zellforsch. *120*, 1–16 (1971).

[291] Scharrer, B., and M. von Harnack: Histophysiological studies on the corpus allatum of *Leucophaea maderae*. I. Normal life cycle in male and female adults. Biol. Bull. Woods Hole *115*, 508–520 (1958).

[292] Scharrer, B., and M. von Harnack: Histophysiological studies on the corpus allatum of *Leucophaea maderae*. III. The effect of castration. Biol. Bull. Woods Hole *121*, 193–208 (1961).

[293] Scharrer, B., and M. Weitzmann: Current problems in invertebrate neurosecretion. In: Bargmann, W., and B. Scharrer, eds.: Aspects of Neuroendocrinology, pp. 1–23. Berlin-Heidelberg-New York: Springer. 1970.

[294] Scharrer, E.: Photo-neuro-endocrine systems: General concepts. Ann. N. Y. Acad. Sci. *117*, 13–22 (1964).

[295] Scheuer, R.: Endocrine control of protein synthesis during oöcyte maturation in the cockroach *Leucophaea maderae*. J. Insect Physiol. *15*, 1411–1419 (1969).

[296] Scheuer, R., and R. Leuthold: Haemolymph proteins and water uptake in female *Leucophaea maderae* during the sexual cycle. J. Insect Physiol. *15*, 1067–1077 (1969).

[297] Scheuer, R., and M. Lüscher: Nachweis der Synthese eines Dotterproteins unter dem Einfluß der Corpora allata bei *Leucophaea maderae*. Rev. Suisse Zool. *75*, 715–722 (1968).

[298] Schneidermann, H. A.: Control systems in insect development. In: Devons, S., ed.: Biology and the Physical Sciences, pp. 186–208. Columbia Univ. Press. 1969.

[299] Schneidermann, H. A., and L. I. Gilbert: Control of growth and development in insects. Science *143*, 325–333 (1964).

[300] Schneidermann, H. A., and C. M. Williams: The physiology of insect diapause. VII. The respiratory metabolism of the cecropia silkworm during diapause and development. Biol. Bull. Woods Hole *105*, 320–334 (1953).

[301] Schooneveld, H.: Structural aspects of neurosecretory and corpus allatum activity in the adult colorado beetle, *Leptinotarsa decemlineata* Say, as a function of daylength. Netherlands J. Zool. *20*, 151–237 (1970).

[302] Sehnal, F.: Influence of the corpus allatum on the development of intestinal organs in *Galleria mellonella* L. J. Insect Physiol. *14*, 73–85 (1968).

[303] Sehnal, F.: Endocrines of Arthropods. Chem. Zool. *6*, 307–345 (1971).

[304] Sehnal, F.: Juvenile hormone action and insect growth rate. Endocrinol. exp. *5*, 29–33 (1971).

[305] Sehnal, F., and V. J. A. Novák: Morphogenesis of the pupal integument in the waxmoth (*Galleria mellonella*) and its analysis by means of juvenile hormone. Acta entomol. bohemoslov. *66*, 137–145 (1969).

[306] Sehnal, F., and K. Sláma: The effect of corpus allatum hormone on respiratory metabolism during larval development and ecdysis of *Galleria mellonella*. J. Insect Physiol. *12*, 1333–1342 (1966).

[307] Seligman, M., S. Friedman, and G. Fraenkel: Bursicon mediation of tyrosine hydroxylation during tanning of the adult cuticle of the fly, *Sarcophaga bullata*. J. Insect. Physiol. *15*, 553–561 (1969).

[308] Shappirio, D. G., and C. M. Williams: The cytochrome system of the Cecropia silkworm. I. Spectroscopic studies of individual tissues. Proc. Roy. Soc. (London) *B 147*, 218–232 (1957).

[309] Sláma, K.: Oxygen consumption during the postembryonic development of *Pyrrhocoris apterus* (Heterometabola: Heteroptera) and its comparsion with that of Holometabola. Ann. Entomol. Soc. Amer. *53*, 606–610 (1960).

[310] Sláma, K.: Die Einwirkung des Juvenilhormons auf die Epidermiszellen der Flügelanlagen bei künstlich beschleunigter und verzögerter Metamorphose von *Pyrrhocoris apterus* L. Zool. Jb. Physiol. *70*, 427–454 (1964).

[311] Sláma, K.: Hormonal control of haemolymph protein concentration in the adults of *Pyrrhocoris apterus* L. (Hemiptera). J. Insect Physiol. *10*, 773–782 (1964).

[312] Sláma, K.: Hormonal control of respiratory metabolism during growth, reproduction, and diapause in female adults of *Pyrrhocoris apterus* L. (Hemiptera). J. Insect Physiol. *10*, 283–303 (1964).

[313] Sláma, K.: Hormonal control of respiratory metabolism during growth, reproduction, and diapause in male adults of *Pyrrhocoris apterus* L. (Hemiptera). Biol. Bull. Woods Hole *127*, 499–510 (1964).

[314] Sláma, K.: Physiology of sawfly metamorphosis. 2. Hormonal activity during diapause and development. Acta Soc. entomol. bohemoslov. *61*, 210–219 (1964).

[315] Sláma, K.: Effect of hormones on growth and respiratory metabolism in the larvae of *Pyrrhocoris apterus* L. (Hemiptera). J. Insect Physiol. *11*, 113–122 (1965).

[316] Sláma, K.: Effect of hormones on the respiration of body fragments of adult *Pyrrhocoris apterus* L. (Hemiptera). Nature (London) *205*, 416–417 (1965).

[317] SLÁMA, K.: Hormonal control of developmental and metabolic cycles in insects. Gen. Comp. Endocrinol. *9*, 492–493 (1968).

[318] SLÁMA, K.: Hormonal control of metabolism in *Pyrrhocoris*. Endocrinol. exp. *5*, 85–90 (1971).

[319] SLÁMA, K., and H. HRUBEŠOVÁ: Übereinstimmung in der Einwirkung von larvalen und imaginalen Corpora allata auf den Respirationsmetabolismus und die Reproduktion bei *Pyrrhocoris apterus* L.-Weibchen. Zool. Jb. Physiol. *70*, 291–300 (1963).

[320] SOUTHWOOD, T. R. E.: A hormonal theory of wing polymorphism in Heteroptera. Proc. R. Entomol. Soc. *36*, 63–66 (1961).

[321] SRIVASTAVA, P. N.: The prothoracic glands of some Coleopteran larvae. Quart. J. Microscop. Sci. *100*, 51–64 (1959).

[322] SRIVASTAVA, P. N.: Secretory cycle and disappearance of prothoracic glands in *Tenebrio molitor* L. (Coleoptera: Tenebrionidae). Experientia *16*, 445–446 (1960).

[323] STAAL, G. B.: Endocrine effects in phase development in locusts. Acta Symp. Evol. Insect. Praha *1959*, 142–146.

[324] STAAL, G. B.: Studies on the physiology of phase induction in *Locusta migratoria migratoroides* R. et F. Publ. Fonds Land. Exp. Bureau *40*, 1–125 (1961).

[325] STAAL, G. B.: Endocrine aspects of larval development in insects. J. Endocrinol. *37*, 13–14 (1967).

[326] STAAL, G. B., and J. DE WILDE: Endocrine influence on the development of phase characters in *Locusta*. Coll. Int. Centr. Natn. Rech. Scient. *114*, 89–105 (1962).

[327] STEELE, J. E.: Occurence of a hyperglycaemic factor in the corpus cardiacum of an insect. Nature (London) *192*, 680–681 (1961).

[328] STEELE, J. E.: The site of action of insect hyperglycaemic hormone. Gen. Comp. Endocrinol. *3*, 46–52 (1963).

[329] STEGWEE, D.: Metabolic effect of a corpus allatum hormone in diapausing *Leptinotarsa decemlineata* Say. Proc. 11th Int. Congr. Entomol. Symp. *3*, 218–221 (1960).

[330] STENGEL, M., and G. SCHUBERT: Rôle des corpora allata dans le compartement migrateur de la femelle de *Melolontha melolontha* L. (Coleoptera, Scarabaeidae). C. R. Acad. Sci. (Paris) *270*, 181–184 (1970).

[331] STRANGWAYS-DIXON, J.: The relationship between nutrition, hormones, and reproduction in the blowfly *Calliphora erythrocephala*. J. Exp. Biol. *38*, 225–235 (1961).

[332] STRICH, M. C.: Étude de la glande ventrale chez *Locusta migratoria*. Ann. Sci. Nat. Zool. *16*, 399–410 (1954).

[333] STRONG, L.: The relationships between the brain, corpora allata, and oöcyte growth in the central american locust, *Schistocerca* sp. II. The innervation of the corpora allata, the lateral neurosecretory complex, and oöcyte growth. J. Insect Physiol. *11*, 271–280 (1965).

[334] STUTINSKI, F.: Einfluß der Extrakte von Corpora cardiaca und neurosekretorischen Gehirnzellen und die Exkretion des Wassers in der Ratte. Bull. Soc. Zool. France *78*, 202–204 (1952).

[335] TELFER, W. H.: The selective accumulation of blood proteins by the oöcytes of saturniid moths. Biol. Bull. Woods Hole *118*, 338–351 (1960).

[336] TELFER, W. H.: The mechanism and control of yolk formation. Annu. Rev. Entomol. *10*, 161–184 (1965).

[337] THOMAS, K. K., and L. I. GILBERT: Isolation and characterization of the haemolymph lipoproteins of the american silkmoth *Hyalophora cecropia*. Arch. Biochem. Biophys. *127*, 512–521 (1968).

[338] THOMAS, K. K., and L. I. GILBERT: The haemolymph lipoproteins of the silkmoth *Hyalophora gloveri*: studies on lipid composition, origin and function. Physiol. Chem. & Phys. *1*, 293–311 (1969).

[339] THOMSEN, E.: Influence of the corpus allatum on the oxygen consumption of adult *Calliphora erythrocephala* (Meig). J. Exp. Biol. *26*, 137–149 (1949).

[340] THOMSEN, E.: Functional significance of the neurosecretory brain cells and the corpus cardiacum in the female blowfly, *Calliphora erythrocephala* Meig. J. Exp. Biol. *29*, 137–172 (1952).

[341] THOMSEN, E.: Esterase in the cells of the hind midgut of the *Calliphora erythrocephala* female, and its possible dependence on the medial neurosecretory cells of the brain. Z. Naturforsch. *75*, 281–303 (1966).

[342] THOMSEN, M.: The neurosecretory system of the adult *Calliphora erythrocephala*. IV. A histological study of the corpus cardiacum and its connections with the nervous system. Z. Zellforsch. *94*, 205–219 (1969).

[343] THOMSEN, E., and A. O. LEA: Control of the medial neurosecretory cells by the corpus allatum in *Calliphora erythrocephala*. Gen. Comp. Endocrinol. *12*, 51–57 (1969).

[344] THOMSEN, E., and I. MØLLER: Neurosecretion and intestinal proteinase activity in an insect, *Calliphora erythrocephala* Meig. Nature (London) *183*, 1401–1402 (1959).

[345] THOMSEN, E., and M. THOMSEN: Fine structure of the corpus allatum of the female *Calliphora*, with special reference to hormone formation. Gen. Comp. Endocrinol. *13*, 534 (1969).

[346] THOMSEN, E., and M. THOMSEN: Fine structure of the corpus allatum of the female blow-fly *Calliphora erythrocephala*. Z. Zellforsch. *110*, 40–60 (1970).

[347] TRUMAN, J. W.: Hour-glass behavior of the circadian clock controlling eclosion of the silkmoth *Antheraea pernyi*. Proc. Natn. Acad. Sci. Amer. *68*, 595–599 (1971).

[348] TRUMAN, J. W., and L. M. RIDDIFORD: Neuroendocrine control of ecdysis in silkmoths. Science *167*, 1624–1626 (1970).

[349] TRUMAN, J. W., and L. M. RIDDIFORD: Role of the corpora cardiaca in the behavior of saturniid moths. II. Oviposition. Biol. Bull. Woods Hole *140*, 8–14 (1971).

[350] UNGER, H.: Untersuchungen zur neurohormonalen Steuerung der Herztätigkeit bei Schaben. Biol. Zbl. *76*, 204–225 (1967).

[351] VANDENBERG, J. P.: The role of the gonadotropic hormone in the synthesis of protein and RNA in *Rhodnius prolixus* (Hemiptera). Biol. Bull. Woods Hole *125*, 576–581 (1963).

[352] VIETINGHOFF, U., H. PENZLIN, and L. SPANNHOF: Kernvolumetrische Untersuchungen an den Pericardialdrüsen regenerierender Stabheuschrecken *Carausius morosus*. Naturwissenschaften *15*, 370 (1964).

[353] VICZENT, J. F. V.: Effects of bursicon on cuticular properties in *Locusta migratoria migratoroides*. J. Insect Physiol. *17*, 625–636 (1971).

[354] VOGT, M.: Zur Produktion und Bedeutung metamorphosefördernden Hormone während der Larvenentwicklung von *Drosophila*. Biol. Zbl. *63*, 395–446 (1943).

[355] VOGT, M.: Zur hormonalen Förderung imaginaler Differenzierungsprozesse bei *Drosophila*. Naturwissenschaften *32*, 37–39 (1944).

[356] WAJC, E., and M. P. PENER: The effect of the corpora allata on the mating behaviour of the male migratory locust *Locusta migratoria migratoroides* (R & F.). Israel J. Zool. *18*, 179–192 (1969).

[357] WAKU, Y.: Studies on the hibernation and diapause in insects. IV. Histological observations of the endocrine organs in the diapause and nondiapause larvae of the indian meal-moth, *Plodia interpunctella* Hübner. Sci. Rep. Tohoku Univ. Biol. *26*, 327–340 (1960).

[358] WAKU, Y., and L. I. GILBERT: The corpora allata of the silkworm, *Hyalophora cecropia*: An ultrastructural study. J. Morphol. *115*, 69–96 (1964).

[359] WALKER, P. B., and E. BAILEY: Effect of allatectomy on fat body lipid metabolism of the male desert locust during adult development. J. Insect Physiol. *17*, 813–821 (1971).

[360] WATSON, J. A. L.: Moulting and reproduction in the adult firebrat *Thermobia domestica* (Packard) (Thysanura, Lepismatidae). I. The moulting cycle and its control. J. Insect Physiol. *10*, 305–317 (1964).

[361] WELLS, M. J.: The thoracic glands of Hemiptera–Heteroptera. Quart. J. Microscop. Sci. *95*, 231–244 (1954).

[362] WHITE, D. F.: Corpus allatum activity associated with development of wingbuds in cabbage aphid embryos and larvae. J. Insect Physiol. *17*, 761–773 (1971).

[363] WIENS, A. W., and L. I. GILBERT: Regulation of cockroach fat body metabolism by the corpus cardiacum *in vitro*. Science *150*, 614–616 (1965).

[364] WIENS, A. W., and L. I. GILBERT: Regulation of carbohydrate mobilization and utilization in *Leucophaea maderae*. J. Insect Physiol. *13*, 779–794 (1967).

[365] WIGGLESWORTH, V. B.: Function of the corpus allatum in insects. Nature (London) *136*, 338 (1935).

[366] WIGGLESWORTH, V. B.: The function of the corpus allatum in the growth and reproduction of *Rhodnius prolixus*. Quart. J. Microscop. Sci. *79*, 91–119 (1936).

[367] WIGGLESWORTH, V. B.: The determination of characters at metamorphosis in *Rhodnius prolixus*. J. Exp. Biol. *17*, 201–222 (1940).

[368] WIGGLESWORTH, V. B.: The significance of "chromatic droplets" in the growth of insects. Quart. J. Microscop. Sci. *83*, 141–152 (1942).

[369] WIGGLESWORTH, V. B.: The thoracic gland of *Rhodnius* and its role in moulting. J. Exp. Biol. *29*, 620–631 (1952).

[370] WIGGLESWORTH, V. B.: Principles of insect physiology, 3rd ed. London: 1953.

[371] WIGGLESWORTH, V. B.: The physiology of insect metamorphosis, 152 pp. Cambridge: Cambr. Monogr. Exp. Biol. 1. 1954.

[372] WIGGLESWORTH, V. B.: The role of the haemocytes in the growth and moulting of an insect, *Rhodnius prolixus* (Hemiptera). J. Exp. Biol. *32*, 659–663 (1955).

[373] WIGGLESWORTH, V. B.: Insect polymorphism–a tentative synthesis. In: KENNEDY, J. S., ed.: Insect polymorphism, pp. 103–113. Dorking, Surrey: Bartholomew Press. 1961.

[374] WIGGLESWORTH, V. B.: Some observations on the juvenile hormone effect of farnesol in *Rhodnius prolixus* Stål (Hemiptera). J. Insect Physiol. *7*, 73–78 (1961).
[375] WIGGLESWORTH, V. B.: The action of moulting hormone and juvenile hormone at the cellular level in *Rhodnius prolixus*. J. Exp. Biol. *40*, 231–245 (1963).
[376] WIGGLESWORTH, V. B.: Homeostasis in insect growth. Symp. Soc. Exp. Biol. *18*, 265–281 (1964).
[377] WIGGLESWORTH, V. B.: The hormonal regulation of growth and reproduction in insects. In: BEAMENT, J. W. L., *et al.*, eds.: Advances in insect physiology *2*, pp. 248–335. New York: Academic Press. 1964.
[378] WIGGLESWORTH, V. B.: Hormonal regulation of differentiation in insects. In: BEERMANN, W., ed.: Cell differentiation and morphogenesis, pp. 180–209. North Holland Publ. Co. 1966.
[379] WIGGLESWORTH, V. B.: Insect hormones, 159 pp. Edinburgh: Oliver & Boyd, R. & R. Clark Ltd. 1970.
[380] WILDE, J. DE: Diapause in the colorado beetle *Leptinotarsa decemlineata* Say as an endocrine deficiency syndrome of the corpora allata. Acta Symp. Evol. Insects Prague *1959*, 226–230.
[381] WILDE, J. DE: Analysis of the diapause syndrome in the colorado potato beetle (*Leptinotarsa decemlineata* Say); Behaviour and reproduction. Acta Physiol. Pharmacol. Neerl. *11*, 93 (1962).
[382] WILDE, J. DE: Reproduction–endocrine control. In: ROCKSTEIN, M., ed.: The physiology of insecta *1*, pp. 59–90. New York: Academic Press. 1964.
[383] WILDE, J. DE: Photoperiodic control of endocrines in insects. Arch. Anatom. Microscop. *54*, 547–564 (1965).
[384] WILDE, J. DE: Hormones and diapause. Progress in Endocrinology, pp. 351–355. Excerpta Medica International Congress. Series *184* (1968).
[385] WILDE, J. DE, and S. A. BOER: Physiology of diapause in the adult colorado beetle–II. Diapause as a case of pseudo-allatectomy. J. Insect Physiol. *6*, 152–161 (1961).
[386] WILDE, J. DE, and D. STEGWEE: Two major effects of the corpus allatum in the adult colorado beetle *(Leptinotarsa decemlineata)*. Archs. Neerl. Zool. *13*, suppl., 227–289 (1958).
[387] WILKENS, J. L.: The endocrine and nutritional control of egg maturation in the fleshfly *Sarcophaga bullata*. J. Insect Physiol. *14*, 927–943 (1968).
[388] WILLIAMS, C. M.: The function of the brain in terminating pupal diapause in the giant silkworm *Platysamia cecropia*. Anatom. Rec. *99*, 115 (1947).
[389] WILLIAMS, C. M.: Physiology of insect diapause. II. Interaction between the pupal brain and prothoracic glands in the metamorphosis of the giant silkworm *Platysamia cecropia*. Biol. Bull. Woods Hole *93*, 89–98 (1947).
[390] WILLIAMS, C. M.: Physiology of insect diapause. III. The prothoracic glands in the Cecropia silkworm, with special reference to their significance in embryonic and postembryonic development. Biol. Bull. Woods Hole *94*, 60–65 (1948).
[391] WILLIAMS, C. M.: The prothoracic glands of insects in retrospect and in prospect. Biol. Bull. Woods Hole *97*, 111–114 (1949).
[392] WILLIAMS, C. M.: Morphogenesis and metamorphosis of insects. Harvey Lectures *47*, 126–155 (1952).

[393] Williams, C. M.: Physiology of insect diapause. IV. The brain and prothoracic glands as an endocrine system in the cecropia silkworm. Biol. Bull. Woods Hole *103*, 120–138 (1952).
[394] Williams, C. M.: Physiology of insect diapause. X. An endocrine mechanism for the influence of temperature on the diapausing pupa of the cecropia silkworm. Biol. Bull. Woods Hole *110*, 201–218 (1956).
[395] Williams, C. M.: The juvenile hormone. I. Endocrine activity of the corpora allata of the adult cecropia silkworm. Biol. Bull. Woods Hole *116*, 323–338 (1959).
[396] Williams, C. M.: The present status of the brain hormone. In: Beament, J. W. L., and J. E. Treherne, eds.: Insects and physiology, pp. 133–139. Edinburgh-London: Oliver & Boyd. 1967.
[397] Williams, C. M.: Photoperiodism and the endocrine aspects of insect diapause. Symp. Soc. Exp. Biol. *23*, 285–300 (1969).
[398] Williams, C. M., and F. C. Kafatos: Theoretical aspects of the action of juvenile hormone. Bull. Entomol. Soc. Suisse *44*, 151–162 (1971).
[399] Willig, A., H. H. Rees, and T. W. Goodwin: Biosynthesis of insect moulting hormones in isolated ring glands and whole larvae of *Calliphora*. J. Insect Physiol. *17*, 2317–2326 (1971).
[400] Wyatt, G. R.: Effects of experimental injury on carbohydrate metabolism in silkmoth pupae. Fed. Proc. (USA) *20*, 81 (1961).
[401] Wyatt, G. R.: Biochemistry of insect metamorphosis. In: Etkin, W., and L. I. Gilbert, eds.: Metamorphosis: A problem in developmental biology, pp. 143–184. New York: Meredith Corp. 1968.
[402] Wyatt, G. R., A. E. Gussin, and A. R. Gilby: The activation of insect muscle trehalase. Fed. Proc. (USA) *24*, 125 (1965).
[403] Wyss-Huber, M., and M. Lüscher: Über die hormonale Beeinflußbarkeit der Proteinsynthese *in vitro* im Fettkörper von *Leucophaea maderae* (Insecta). Rev. Suisse Zool. *73*, 517–521 (1966).
[404] Yamashita, O., and K. Hasegawa: Further studies on the mode of action of the diapause hormone in the silkworm, *Bombyx mori L.* J. Insect Physiol. *12*, 957–962 (1966).
[405] Yamashita, O., and K. Hasegawa: The effect of the diapause hormone on the trehalase activity in pupal ovaries of the silkworm (A preliminary note). Proc. Japan Acad. Sci. *43*, 547–551 (1967).
[406] Yamazaki, M., and M. Kobayashi: Purification of the proteinic brain hormone of the silkworm, *Bombyx mori*. J. Insect Physiol. *15*, 1981–1990 (1969).
[407] Žďárek, J.: Le comportement d'accouplement à la fin de la diapause imaginale et son controle hormonal dans le cas de la punaise *Pyrrhocoris apterus* L. (Pyrrhocoridae, Heteroptera). Ann. Endocrinol. (Paris) *29*, 703–707 (1968).
[408] Žďárek, J.: Mating behaviour in the bug, *Pyrrhocoris apterus* L. (Heteroptera): ontogeny and its environmental control. Behaviour *37*, 253–268 (1970).
[409] Žďárek, J.: Hormonal control of mating behaviour in *Pyrrhocoris apterus* L. In: Novák, V. J. A., and K. Sláma, eds.: Insect endocrines, pp. 51–61. Praha: Academia. 1971.
[410] Žďárek, J., and G. Fraenkel: Correlated effects of ecdysone and neurosecretion in puparium formation (pupariation) of flies. Proc. Natn. Acad. Sci. Amer. *64*, 565–572 (1969).

II. The Chemistry and Physiology of Juvenoids

This most extensive part of the book deals with compounds generally known as juvenile hormone analogues or juvenile hormone mimics. These materials encompass a wide range of mainly synthetic compounds with a relatively smaller amount of natural products isolated from plants and animals. The compounds have been characterized by certain size and shape of the molecule, low polarity, presence of specific functional groups and substituents, and by some other common structural features. However, the only criterion for classification of a compound as a juvenile hormone analogue is its property to exhibit or mimic some morphological and physiological effects of the hormone secreted by the corpora allata. Since the chemical nature of the endogenous corpus allatum hormone is still incompletely known with respect to diversity of insect species, these compounds cannot be viewed as analogues of the corpus allatum hormone in the chemical sense of the word. It is more appropriate to regard them as its bio-analogues or pharmaco-analogues. To facilitate further discussions we have selected a generic name juvenoids for all the compounds with juvenile hormone activity.

A. Introduction

The history of juvenoids begins with WILLIAMS [368] who was the first to obtain juvenile hormone effects on metamorphosis using lipid extracts prepared from abdomens of adult male Cecropia moths. The presence of juvenile hormone activity in these extracts was soon confirmed by WIGGLESWORTH [355] on *Rhodnius*. Having found a convenient assay, investigations of juvenile activity in other sources than adult Cecropia moths were performed. Surprisingly, WILLIAMS *et al.* [377] found positive responses with lipid extracts prepared from thymus, human placenta, various other vertebrate organs, and even from an ordinary heavy cream. In a series of papers, SCHNEIDERMAN and GILBERT encountered juvenile activity in lipid extracts prepared from adrenal cortex of vertebrates [103], from Crustaceans and almost all phyla of animals from Hydroids to vertebrates [283], and also from microorganisms and plants [287]. They determined the relative changes in the activity of extracts prepared from various developmental stages and various species of Lepidoptera [106]. The problematics were extensively discussed in several reviews by these authors [105, 107, 285, 286]. Juvenile active materials were also found to be produced by insect parasites *Nosema* [96]. For many years the Cecropia extracts were purified and analysed in several laboratories until RÖLLER and his colleagues identified the active compound (see chapter II-C). However, the identity of the chemical substances which were responsible for juvenile hormone activity in the lipid extracts from various other sources is unknown.

Further important steps in juvenoid research were reached when SCHMIALEK [277, 278, 280] identified farnesol as the active substance from *Tenebrio* excrements and yeasts. Farnesol was the first pure compound with definite juvenile hormone activity and this attracted all the attention to the field of terpenoid chemistry. With the first derivatives of farnesol or farnesenic acid begins the history of synthetic juvenoids (see 308 for reference). Curiously, SCHMIALEK [280] was the first to test juvenile hormone activity in the parental alcohol 3,7,11-trimethyl-2,6,10-tridecatrien-1-ol of the second active compound isolated later from the Cecropia extracts. BOWERS *et al.* [31] predicted possible similarity between 10,11-epoxyfarnesoate (juvenoate) and the hormone later be found.

Among some farnesol derivatives, farnesyl alkylethers [30, 280, 289] were widely used in the original investigations on physiological effects of these compounds. Somewhat later, LAW *et al.* [172] described a high activity of hydrochlorination products prepared from the farnesoic acid. These products had higher activity on certain insects than any other farnesoic acid derivative known at that time. Their preparation which is called LAW-WILLIAMS mixture is still commonly used in juvenile hormone research. The most active component of the preparation with especially high activity on certain Hemiptera was identified by ROMAŇUK *et al.* [271] as 7,11-dichlorodihydrofarnesoate. This compound was extensively used in studies on female sterility, ovicidal effects and inhibition of metamorphosis by juvenoids. Identification of dihomojuvenate as an active component of the Cecropia extracts, stimulated new efforts to prepare similar compounds synthetically [16, 213, 242, 265, 274, 351, 358]. Recently we know large series of acyclic juvenoids which are structurally related to farnesoic acid, and all possess juvenile hormone activity (see chapter C).

The industrial interest in juvenoids was connected with the discovery of juvenile hormone activity in American paper products by SLÁMA and WILLIAMS [316, 318, 379]. The fact that the whole insect population was unable to reproduce and perished by merely contact with an ordinary paper attracted attention of many scientists searching for a possibility of insect control by nontoxic agents. The active compounds responsible for these "paper factor" effects were identified as juvabione by BOWERS *et al.* [29] and dehydrojuvabione by ČERNÝ *et al.* [49]. They appeared to be the first alicyclic terpenoid juvenoids. The first aromatic juvenoids were prepared by ŠORM and coworkers [335]. They were structurally related to aromatic juvabione. Further progress in the field of aromatic juvenoid chemistry was stimulated by BOWERS's paper [26] on aromatic terpenoid ethers and insecticide synergists [25]. The findings on high juvenile hormone activity among the aromatic terpenoid ethers opened a way to synthesis of analogous thioethers and terpenoid aniline derivatives [308] which comprise substantial part of the recently known synthetic juvenoids. Finally, the peptidic juvenoids introduced by ZAORAL and SLÁMA [389] were essentially aniline or p-aminobenzoic acid derivatives with side chains containing amino acid residues. They show remote structural resemblances to juvenoids of the terpenoid aniline type.

The biological research on juvenile hormone activity was also initiated with the Cecropia extracts by extensive investigations of WILLIAMS [368, 377], GILBERT and SCHNEIDERMAN [103, 105, 283, 284], and WIGGLESWORTH [355], among others. They studied mainly the effects based on morphological inhibition of metamorphosis. These observations were further extended using farnesol and its derivatives [160, 277, 278, 356]. Most of the studies were performed on few laboratory insects such as *Hyalophora*, *Galleria*, *Tenebrio*, *Rhodnius*, and *Pyrrhocoris*. LARSEN and BODENSTEIN [169] were the first to report ovarian stimulation with Cecropia extracts; this was soon confirmed with farnesol in adult female *Rhodnius* by WIGGLESWORTH [356], and with other farnesol derivatives in cockroaches [31, 52]. Then, SLÁMA and WILLIAMS [316, 317] discovered the effects of juvenoids ("paper factor") on embryonic development which was further extended in the studies by RIDDIFORD [257, 258, 259, 261]. SLÁMA and WILLIAMS also discovered female sterility caused by juvenoids [317] which was connected with the ovicidal effects. These investigations were further extended by studies of MASNER *et al.* [191, 192] who demonstrated that female sterility may be achieved by transfer of juvenoids from treated males to females during sexual contact [192, 193]. BOWERS and BLICKENSTAFF [28] confirmed experimentally the effects of juvenoids on adult diapause. They suggested possibilities for using juvenoids in the control of insects with adult diapause. Following suggestions by WILLIAMS [371] that juvenoids may possibly be used as the "Third-generation pesticides", an increased attention was paid to different methods of application and to the effects of juvenoids on insect populations. It appeared that juvenoids may cause developmental disturbances in insects after their contact with treated plants and, as observed by BABU and SLÁMA [7] certain juvenoids had properties of a systemic agent.

Problems concerned with the chemistry and physiology of juvenoids have been extensively reviewed in recent years. There exist earlier reviews by RÖLLER and DAHM [265] and by BERKOFF [16]. In 1970 there appeared review articles by WILLIAMS [374], EL-IBRASHY [80], and CIZIN and DRABKINA [53]. Among the more recent reviews we may recommend a collection of papers from the Basel Symposium on Juvenile Hormone by BOWERS [27], CRUICKSHANK [66], MORI [213], ŠORM [322], and WAKABAYASHI *et al.* [345]. Furthermore, a review by PFIFFNER is specifically devoted to chemistry and that by SLÁMA [308] is devoted to physiology of juvenoids. Some general problems of juvenoid action are discussed in recent monographs on insect hormones by HIGHNAM and HILL [118], WIGGLESWORTH [361], and WYATT [386].

In Part II we have made a recapitulation of the intensively increasing amount of data concerning bioassays, chemistry, structure-activity relationships and physiology of juvenoids. We have also incorporated considerable amount of unpublished results which have been obtained in our laboratories during several years of intensive research on juvenoids. We have discussed some conflicting problems with new insights on the chemistry, mode of action, and specificity of action of juvenoids, using comparative evidence obtained on representatives of different insect groups. When possible, the action of juvenoids has been confronted with the action of the corpus allatum hormone. We are conscious of the fact that the articulation of the chapters as follows is a compromise between the chemical and biological points of view, which may not be optimal from either one of these views. Finally, we wish to attract the attention of the reader to the extensive list of chemical structures and biological activities of juvenoids contained in the Appendix tables. We hope that this will provide additional information and inspiration for those engaged in juvenoid research.

B. Methods of Testing and Evaluating the Biological Activity of Juvenoids

It is known that the biological activity of hormones in particular is dependent on the mode of application. This is especially true with respect to juvenoids which represent a system with a wide range of chemical compounds to be tested on as many insect species as possible. In the section on structure-activity relationships (p. 194) we have discussed that the nonstandardized bioassay procedures may cause serious difficulties in determination and comparisons of the biological activity of juvenoids. It becomes quite obvious that the recent developments in the field of juvenoid research require such a bioassay system, in which there is only one standardized value of the activity for one juvenoid compound, one species, and one developmental stage. In order to satisfy these demands, we have analysed previous testing methods with an attempt to standardize the conditions so that the values obtained by different authors might be easily compared.

Methods for determination of the juvenile hormone activity underwent rapid development in the last few years. The first demonstration of juvenile hormone activity was achieved by injections of lipid extracts from male Cecropia moths into pupae [368]. Extensive screening for juvenile activity of extracts prepared from various animal and plant sources contributed to the development of a very sensitive local epidermal assay, which anabled to record traces of juvenile activity [103, 283, 284, 377]. These local epidermal assays were based on an increased sensitivity of the regenerating epidermal cells to juvenile hormone. Their typical representative is the so called "wax wound test" on *Galleria* pupae [105, 106, 285, 286]. For its very high sensitivity the assay is still used for determinations of juvenile activity of crude tissue or haemolymph extracts [365, 366]. Similar localized epidermal tests were used for juvenile activity determinations of cecropia extracts and farnesol derivatives by Wigglesworth [355, 356] on *Rhodnius* and by Schmialek [277, 278, 280] on *Tenebrio*. The techniques of topical application were in these studies combined with abrasions of the upper epicuticular layers, or, eventually with epidermal punctures. This facilitated penetration of the active materials through the integument. Karlson and Nachtigal [160] modified the original assay in *Tenebrio* [355] using injections of measured amounts of the compounds diluted with neutral plant oils. This method was further modified by Röller *et al.* [262, 264] who used the presence of localized patches of pupal cuticle around the site of injection as a quantitative measure of juvenile activity. Results of these assays were also influenced by increased sensitivity of the regenerating epidermal cells. This method was successfully used for isolation of the active principle from the Cecropia extracts.

When the more potent farnesol derivatives became available in large quantities, the testing methods for juvenoids underwent further development which may be characterized by a general influence on morphology and physiology of the whole animal [31, 357]. This tendency was further stimulated by the discovery of practically an unlimited supply of juvenile hormone activity in American paper products [316]. Such an availability of juvenoid sources enabled to specify conditions for a simple topical application and also for evaluation of the activity according to the degree of morphological and physiological changes induced [318, 379].

Since 1965, there has been a rapid development of new synthetic juvenoids. This was connected with an involvement of professional chemists in the research and also of the industrial teams which calculated on the possible use of juvenoids as pesticides. This created specific conditions with a special accent on the testing methods. The chemists were interested to know exactly which of the synthetic compounds were more active than the others while economists required exact data for conclusions concerning specificity and possible action of juvenoids in the field conditions. In response to these demands, entomologists have specified the assay conditions and have incorporated some economically important species in addition to a few laboratory insects.

1. Developmental Stages Suitable for the Assays

The biological activity of juvenoids can essentially be assayed on any developmental stage of insects, e.g. on eggs, young larval instars, last larval instars, metamorphosis stages and also on adults. However, the developmental abnormalities caused by juvenoids appear only at some determined developmental periods which depend mainly on absence of the endogenous corpus allatum hormone and on action of other hormones of the neuroendocrine system [308]. Therefore, application of juvenoids on stages with high endogenous hormone titre or on larval or pupal stages with arrested development has little, if any, immediate effect on the stages treated. In this case the effects may appear later, when the endogenous hormone production ceases, or after resumption of the development respectively, provided that the juvenoid compound persists in the body. During metamorphosis or during embryonic development there exist determined periods at which the stages are sensitive to juvenoid action. They usually occur at the beginning of each developmental instar or at the beginning of the egg stage. Thus, after a certain critical period, juvenoids have no effect on the fate of development in the given instar, however, their presence may seriously influence further developmental program at the beginning of the next instar. More details on physiology of these sensitive periods to juvenoids have been described on p. 218. The existence of sensitive periods to juvenoids within each developmental instar indicates that the assay results would be largely dependent on the time of application. And, as a matter of fact, any bioassay of juvenoids is completely lacking a quantitative character without precise timing of the moment of application.

A list of sensitive periods to juvenoids in several insect species is given in Table 1. It shows that in the majority of insect species the presence of the sensitive period is limited to the first-third or first-half of an instar. An exception is found in the last instar larvae of *Galleria* which remain sensitive to juvenoids until the second-half period. Also, in larvae of *Rhodnius* or *Triatoma* the sensitive period occurs after the blood meal but not after ecdysis. Finally, in diapausing pupae of Cecropia the occurrence of the sensitive period depends on the initiation of postdiapause development. However, in the pupal instars of most species the sensitive period is extremely short being restricted to few hours after the larval-pupal ecdysis. And, in certain species of Hymenoptera or in some other insects with rapid development, the sensitive period to juvenoids may be shifted to the pharate pupal period, e.g. before the larval-pupal ecdysis.

When the application of juvenoids is made at the very beginning of the sensitive period, it usually results in a perfect retention of the old larval or pupal epidermal structures. The internal organs do not respond in the same manner, which suggests that the sensitive periods to juvenoids in these tissues do not coincide exactly with those in the epidermal cells. This is documented by physiological disproportions which sometime occur after treatment with juvenoids, such as are morphologically perfect extra-larvae with degenerated prothoracic glands, or pupae with larval structure of some internal organs which has been described in the physiological section. In general, the effects of juvenoids on morphology of internal organs are less conspicuous and more difficult to score than the effects on epidermal structures. For this reason they are not routinely used as an assay measure.

In embryonic development the duration of the sensitive periods to juvenoids is also commonly limited to the first-third of the egg development [317, 347]. Moreover, in insects such as Diptera or aphids, the sensitive period for embryogenesis may sometimes occur before oviposition. In this case the embryonic development cannot be influenced by juvenoid treatment of the freshly deposited eggs. The application of juvenoids to adult females may induce disturbances in embryonic development when applied at any moment of the reproduction cycle, e.g. from initiation of ovarian development to few hours before oviposition [191]. However, this observation cannot be generalized to all adult females and it does not apply to the effects of juvenoids on ovaries which develop during metamorphosis. Induction of ovarian development in diapausing adult females can be achieved by juvenoid treatment at any time of adult diapause although different doses are required to do so at the beginning and towards the end of diapause. According to this brief description of the sensitive periods we may conclude that the most suitable developmental stages for testing juvenoids would be: adult females or freshly deposited eggs for the effects on embryogenesis, freshly

Table 1. *Developmental Periods Sensitive to the Action of Juvenoids during Metamorphosis (Epidermal Structures)*

Insect group and species		Instar or stage	Duration (days)	Sensitive period (days)	Recommended application	Result	References
Hemiptera							
Pyrrhocoridae	*Pyrrhocoris apterus*	last larval	7	0–2.5	0–20 h	extra-larvae	[379]
Pyrrhocoridae	*Dysdercus* (several sp.)	last larval	7	0–2.5	0–20 h	extra-larvae	[315, 335]
Lygaeidae	*Oncopeltus fasciatus*	last larval	6	0–2.5	0–20 h	extra-larvae	[173]
Lygaeidae	*Lygaeus equestris*	last larval	6	0–2.5	0–20 h	extra-larvae	[312, 314]
Pentatomidae	*Graphosoma italicum*	last larval	6	0–2	0–20 h	extra-larvae	[312]
Pentatomidae	*Eurygaster integriceps*	last larval	6	0–2	0–20 h	extra-larvae	[312]
Pentatomidae	*Aelia acuminata*	last larval	6	0–2	0–20 h	extra-larvae	[312]
Reduviidae	*Rhodnius prolixus*	last larval	20[1]	0–9[1]	24 h[1]	extra-larvae	[357, 358]
	Triatoma infestans	last larval	18[1]	0–8[1]	24 h[1]	extra-larvae	
Orthoptera							
Acrididae	*Locusta migratoria*	last larval	9	0–3.5	0–20 h	extra-larvae	[154, 225]
Acrididae	*Schistocerca gregaria*	last larval	9	0–4	0–20–h	extra-larvae	[225]
Gryllidae	*Gryllus bimaculatus*	last larval	9	0–3.5	0–20 h	extra-larvae	[314]
Coleoptera							
Tenebrionidae	*Tenebrio molitor*	pupal	8	0–2.5	0–16 h	second pupae	
Dermestidae	*Dermestes vulpinus*	pupal	8	0–2.5	0–16 h	second pupae	
Lepidoptera							
Pyralidae	*Galleria mellonella*	last larval	8	0–6.5	140 h	extra-larvae	[134]
	Galleria mellonella	pupal	8	0–1.5	0—9 h	2nd pupae	[134]
Saturniidae	*Hyalophora cecropia*	pupal	21[2]	0–4[2]	0–24 h[2]	2nd pupae	[368]
Diptera							
Muscidae	*Stomoxys calcitrans*	puparium	7[3]	0–1[3]	0—3 h[3]	nonemerging adultoids	[385]

[1] The time counted from the moment of feeding. [2] Counted since the initiation of adult development after diapause. [3] Counted from the moment of puparium formation.

ecdysed last instar larvae or pupae for the effects on metamorphosis and, diapausing adult females for the effects on ovarian growth.

2. Methods of Juvenoid Application

The ability of juvenoids to penetrate through the insect integument allows combination of several application methods. The lipoid character and solubility in organic solvents makes it possible to use the application techniques which are similar to the well elaborated procedures used in the work with classical insecticides, or other methods of testing chemicals on insects in general [297, 298].

In principle, there exist two categories of juvenoid application methods. The direct methods, which depend on administration of measured amounts of the tested compounds or extracts directly to the surface or within the body and, the indirect methods, which are based on the treatment of food, water, air, organic or inorganic substrate, or other objects which are then exposed to the insect body. The direct methods are suitable for laboratory studies on structure-activity relationships, whereas the indirect methods which determine effective concentrations of juvenoids in the environment, are more important for practical purposes.

a) Direct Application Methods

Most insects are relatively small and the topical application or injections of juvenoid samples requires equipment for constant delivery of 0.1 to 1 μl amounts with less than 5 per cent standard error. Hamilton microliter syringes or other syringes with micrometric operation [297] are commonly used for this purpose. Water narcosis appears to be the most satisfactory for the injections. The insects to be injected are submerged under water for 10 to 30 min before injecting. Developmental stages with very low gaseous exchange cannot be narcotized this way. They are immobilized by low temperature or gaseous CO_2 or, when necessary, they are narcotized with some common anaesthetics.

With regard to the lipoidic nature of all the juvenoids, injections are performed in oily carriers. Plant oils such as olive oil, sunflower oil or peanut oil are currently used. It is advised to perform control experiments in order to check for possible *per se* juvenile activity of these plant oils, which may be due to their contamination with isoprenoids. Other materials used for dilution of juvenoids are certain nonmetabolizable mineral oils such as paraffine oil. In some cases it is convenient to use inactive but isoprenoid carriers such as squalene [66], methyl geranylate [278] and others. These terpenoid carriers interact with the biological prenol-metabolizing systems and thus inhibit the biological degradation of the active compounds. More polar solvents like ethylene or propylene glycols or free fatty acids are not recommended as juvenoid solvents for injections because of their toxicity to many insects. Water emulsions prepared with the aid of emulsifiers may lead to a complete loss of biological activity [357, 360] in injections.

Neutral plant oils injected into the haemocoele are quickly resorbed in species like *Pyrrhocoris*. However, this is rather exceptional since in many other insects they remain in a form of localized or dispersed droplets for a long time [357]. This is also the case with some juvenoids injected in pure state. For example, the dichloro farnesenic acid esters injected into *Tenebrio* pupae remain as inert lipid droplets in the body cavity for a long time. Thus, we may expect that biological activity of juvenoids injected into the body will be significantly influenced by the ability of the compound to leave the lipid droplets. The idea to use insect's own body lipids as solvents for injections of juvenoids does not solve the problem because most of the lipid extracted from insects is a triglyceride from the fat body with the same properties as the plant oils.

For topical applications the compounds are simply dissolved in some volatile solvent and applied to the selected parts of the body surface. The insect integument is a heterogeneous complex consisting of several layers. Going from inside of the body there is a basement membrane, unicellular epithelium of the epidermal cells, and different layers of the cuticle. It is expected that conditions for transport and permeability of chemical compounds through the basement membrane and epidermal cell layer will be similar to conditions that occur in other tissues and in internal organs. However, the most important parts of the integument with respect to juvenoid transport are the cuticular layers. One of these, a very thin outer epicuticle is of lipoidal nature and most probably quite permeable for all the lipid soluble juvenoids. The remaining exo- and endocuticular layers seem to be most important for transport of juvenoids, as demonstrated by studies on diffusion of non-electrolytes through isolated insect cuticle [339]. Using radioactive tracer technique it was found [177] that oily contaminants may rapidly spread from a localized place over the whole surface of the insect body. After 5–15 min. since the first contact of a fly with the oily deposit, a monomolecular film of oil was established. The film ceased to spread after death of the animal. Assumptions were made that a thin film of oil progresses in the epicuticle as a result of very small elastic movements of the integumental surface in the living insect. Since most of the juvenoids have physical properties of liquid oils, it is possible that a similar film of juvenoid spreads over the insect body after this topical application.

Narcosis is not a necessary condition for topical applications, unless this is required for easier handling of some very mobile forms. For most insects a convenient amount of the solvent to be applied is 1 μl. Devices dispensing the samples should operate rather quickly in order to prevent evaporation of the solvent before it reaches the body. Simple arrangements which are used for delivery of constant 1 μl drops of acetone or similar solvents are microliter pipettes, calibrated glass capillaries, Hamilton microliter syringes and some commercially available automatic microliter dispensors. In routine work with more juvenoid samples care should be taken to prevent contamination. This is achieved by using disposable pieces of polyethylene tubing or exchangeable glass capillaries which are attached to the dispensing devices. The samples are soaked in and dispensed off from the tubings by changes in air pressure between the pistil of the syringe, or a mercury column in the pipette, and the acetone solution. The volume of the air between acetone solution and the calibrated systems has to be reduced to a minimum.

Table 2 shows the effect of various solvents and carriers on the results of some topical assays. The data do not reveal any profound differences in juvenile hormone activity with respect to the nature of the organic solvent used. However, an addition of small amounts of a plant oil to the acetone solutions slightly enhances the activity of the juvenoid used, as was found earlier with other insects [66]. Dioxane is equivalent to or a better solvent than acetone. Pure olive oil is not as good as an organic solvent for topical assays, probably because it causes deposits of the compounds on the body surface. However, less viscous oils, such as methyl oleate, or paraffine oil may be as good carriers as acetone. The Tween emulsions show less activity when compared to organic solvents or oily carriers.

An important criterion of some juvenoid properties is represented by differences in activity between topical applications and injections. Wigglesworth [358] performed such a comparison of the juvenile activity units between topical applications and injections in *Rhodnius*. We observed that a majority of juvenoids were more active in injections, some were equally active in both methods of application and there were also compounds more active in the topical assays. Our data on *Pyrrhocoris* suggest little difference in juvenile activity of various juvenoids between topical applications and injections in this species. This may indicate that the rate of penetration of juvenoids through the integument is not a limiting factor of their activity in this case. Similar conclusion was reached with respect to certain insecticides in other insects [297]. On the other hand, in pupae of *Dermestes* or *Tenebrio*, and probably in pupae of other Endopterygote insects too, there occur significant differences in the assay results between both the methods of application. Here most of the juvenoids are more active in injections [312,

Table 2. *Effect of Various Solvents on Juvenile Activity of the Compounds No. T-251 and T-259 (1:1) in Topical Assays on Dysdercus cingulatus. The ID-50 Morph. Values Were Derived from Individual Dose-Response Curves*

Solvent	Amount applied (µl per spec.)	ID-50	Solvent	Amount applied (µl per spec.)	ID-50
Acetone	0.5	0.01	Dimethylsulphoxyde	0.2	0.04
Acetone + 1% olive oil	0.5	0.008	Citronellylacetate	0.2	0.01
Ethanol	0.5	0.02	Methyl oleate	0.2	0.01
Dioxane	0.5	0.007	Olive oil	0.2	0.05
Heptane	0.5	0.01	Paraffine oil	0.2	0.008
Octane	0.5	0.01	0.1% Tween-80/water	0.5	0.03
Petroleum ether	0.5	0.009	0.1% Tween-80		
Ethyl cellosolve	0.2	0.008	+ 5% olive oil emulsion	0.5	0.08

314]. The observations on *Rhodnius* [358] and other Hemiptera [308, 312, 314] have suggested that a relatively good topical activity, when compared with injections, occurs in the chlorinated juvenoids. This is perhaps related with generally good contact action of chlorinated hydrocarbons on insects as is known from insecticide studies. More details concerning the effects of particular chemical structures on topical activity of juvenoids have been mentioned in the section on structure-activity relationships, and they can also be found in the Appendix Tables.

b) Indirect Application Methods

These methods are designed for measurements of developmental responses in insects exposed to determined concentrations of juvenoids present in the environment. This requires specific procedures for treatment of paper, glass, wood, soil, food plants, water, vapour phase, and various other objects which come in contact with the insect body. Usual procedures involve spraying of the surface by juvenoid solutions in organic solvents, or treatment of known volumes of the substrates with these solutions. The degree of juvenile activity is related to the amounts of juvenoid per unit of surface, volume, or weight, or eventually, it is expressed as the concentration of juvenoid present in the solutions which were used for the treatment. Insects of determined developmental stage are then exposed to the treated objects after complete evaporation of the solvent. When extremely small amounts of juvenoids are used, or when the active compound forms a solid deposit, it is advised to add small amounts of oil in order to facilitate insect contamination. The active surface of the insect body which comes in contact with the contaminated substrate is sometimes rather small and limited to the tips of tarsi and antennae. More uniform results will be obtained with these methods when the insects are exposed slightly before the actual beginning of the sensitive period to juvenoids. However, because of considerable diversity in the breeding methods and feeding habits in insects, selection of the proper methods of indirect application of juvenoids must be made with respect to species specific conditions. Some examples can be found in the literature [43, 241, 242, 342, 347, 379].

Studies on the effects of food application of juvenoids are complicated by their pronounced contact action. In fact, it is almost impossible to differentiate between the effects caused through the digestive system and those caused by contamination through the integument as the food always comes in touch with the epidermis, at least on the mouth organs. We have observed, however, that some juvenoids were considerably more active when administered in food than in topical assays. This suggests that the effect of juvenoids through the digestive tract may exist. Furthermore, some of the peptidic juvenoids were active when administered in drinking water to *Pyrrhocoris* larvae, though they were topically inactive. Some data on the action of juvenoids in food have been reported by CRUICKSHANK [66] and others [174, 235, 278].

Application of juvenoids in water, as used for assays on aquatic insects, such as the larvae of mosquitoes, is usually made by mixing small amounts of acetone-juvenoid solutions with determined amounts of water [325]. Since most juvenoids are almost insoluble in water, special additives or emulsifiers are added [213]. Application of juvenoids in vapour phase is made in closed glass compartments containing determined amounts of the compound adsorbed on materials with a large surface. The samples are separated from direct contacts with the tested insects. The internal atmosphere should contain regulation of constant R. H. and the amount of insect specimens contained within the compartment should be adjusted with respect to their demands for oxygen. For information, the O_2 consumption of insect larvae is roughly in the range of 1000 μl per gram per hour, that of pupae is 200–500, and much over 1000 in flying adults. To minimize the risk of oxygen insufficiency, it is possible to ventilate the closed system after the insects have passed the developmental period sensitive to juvenoids. Measurements of the vapour effects of juvenoids are complicated since fluctuations in temperature and relative humidity may cause condensation of the juvenoid vapours on the walls, food, and insects themselves. Thus, a real significance can be given only to those results which show apparent dose-response dependence. A method for measurements of the juvenoid vapour effects on insect eggs has been described by WALKER and BOWERS [347]. Recently, BABU and SLÁMA [7] have reported a method designed for determinations of systemic activity of juvenoids in plants.

c) *Application of Juvenoids in Field Conditions*

This method represents a combination of direct and indirect applications on determined field areas. The direct contamination of the insect body is in this case less significant than re-contamination by the residues which persist in the treated environment. The application procedures so far used are performed similarly with the application of insecticides. Thus, juvenoids are used in the form of sprays, dusts, aerosols, or ultra-low volume applications of the pure compound. However, unlike most insecticides, the field use of juvenoids requires much higher persistency in the environment, which is connected with some changes in formulation, such as the presence of special additives increasing mechanically or chemically the stability of juvenoids. Also, in contrast to standardized laboratory conditions, the insect population which develops in the field is more or less heterogeneous. Different parts of the population reach the sensitive developmental stage at different time. It is, therefore, quite necessary to coordinate the moment of application with the developmental rate and duration of the sensitive period. An optimal moment for the application occurs when most of the specimens approached the beginning of the sensitive developmental period.

In order to be effective, the compound must persist in active concentrations throughout the sensitive period of most or all members of the population. When this is not possible, repeated applications are more efficient in this respect than an increase of the dose. From this it follows that insect species with faster or seasonally synchronized development are more suitable for a single application of juvenoids, whereas those with slow and asynchronous development require juvenoids with long persistance or repeated applications. Since each of the insect species has specific developmental conditions, it is beyond the scope of this publication to go into further details concerning the field application of juvenoids. These problems are a subject of recent intensive investigations and it is hoped that the results will soon be published.

Although there does not seem to exist any constant proportionality between the effective doses obtained in the laboratory assays and the effective concentrations in the field, we have calculated that a juvenoid compound which is effective in 0.1 μg in laboratory assays should also be effective in the field application in doses of approximately 1 kg per hectare. However, this is derived from field experiments with one type of aromatic terpenoid compound on larvae of Hemipteran insects. It is obvious that in other species with different development and with compounds of different persistence the proportion between laboratory and field effectiveness will be considerably altered. Nevertheless, it is expected that the juvenoid candidates for effective field application should act in less than 0.1 μg per specimen in laboratory assays, irrespective of the animal size.

3. Evaluation of the Assay Results

With few exceptions, the action of juvenoids on insect development is not connected with direct toxicity. Therefore, the common probit system with LD-50 determinations which has been adopted in assays of insecticides has limited application in the juvenoids. These compounds possess hormonal activity with physiological rather than lethal action. Conventionally, their effects are expressed in standardized units of activity which is quite common for hormones. Some of the effects caused by juvenoids are very specific and can be induced only in particular insect species or stage. In such cases their biological activity is expressed in arbitrary "Effective doses -20, -50, -90" (ED-20. -50), which are specifically related to a given species or stage, and to specific developmental conditions. On the other hand, most of the other juvenoid effects are based on the same physiological background in many, sometimes quite unrelated, insect species. This results from a common developmental role of the juvenile hormone in insect evolution. And, in this instance, the juvenile hormone activity of juvenoids can be expressed in standardized units.

One of the most spectacular actions of juvenoids is an inhibition of progressive change of form during the ontogeny of insects. The degree of such inhibition of metamorphosis or embryogenesis transformations is generally used for determination of the biological activity of juvenoids. Under standardized conditions of application there exist clear correlations between the dose of juvenoid applied and the degree of inhibition of the progressive change of structures and functions. In the inhibition of metamorphosis which is most frequently studied, the unit concentrations cause appearance of intermediate forms containing an equal amount of old and new structures. This system of inhibition doses -50 (ID-50) is universal and allows comparisons of the data obtained with one compound on different developmental stages of one species, as well as with many compounds on one species, and eventually, with many compounds on many species and several developmental stages.

Methods have already been developed for evaluation of the biological activity of juvenoids in the whole range of physiological effects caused by juvenile hormone. We have separated these methods with respect to: a) morphological changes induced in metamorphosis, b) effects on embryonic development, and c) effects on ovarian growth. In addition, we have briefly discussed determinations of the biological activity of juvenoids in their assays on whole populations (d), and in the field assays (e).

a) Evaluation of the Effects on Metamorphosis

As pointed out already, the determination of biological activity of juvenoids in insect metamorphosis is made according to a reciprocal change of the epidermal morphological patterns between a larva and adult (in Exopterygotes), and between larva and pupa, or pupa and adult (in Endopterygotes). Among extreme morphological types, e.g. larval, pupal, and adult types, there exist continuous series of intermediary forms (cf. Fig. 8). The type and character of the intermediates are related to two of the main factors, e.g. the time of application and the dose of juvenoid applied. Different authors working with different species described special scoring systems for classification of the types of the intermediates. The activity was originally expressed in terms of concentrations related to a corresponding morphological score. These scoring systems were quite arbitrary and subjective.

Different authors working with the same species used sometimes different scoring scales for exactly the same effects. For instance, in *Tenebrio* some authors used 0 to 4 degree scale [26, 27, 345], while others had 0 to 12 [274, 351]. In Hemiptera 0 to 5 degree scale was commonly used [49, 313, 315, 318, 335, 379], but in *Rhodnius*, for example, 0 to 20 degrees [358]. These old scoring systems, which were satisfactory at the initial phases of juvenile hormone research, do not cope with the tremendously increased number of compounds and species investigated at the present time. Actually, all these old scoring systems indicated nothing more but a 0 to 100 per cent change of morphological form and physiological functions between two different developmental stages. Practically all of the systems can be easily converted into the percentage of morphological change between the two developmental stages concerned. Using this system, a 100 per cent morphological change would correspond to a morphological form of a perfect extra-larval or extra-pupal instar, indicating a complete inhibition of metamorphosis. The formation of normal pupae or normal adults is indicated as zero per cent, whereas 50 per cent morphological change would indicate formation of exactly half-larval or half-pupal adultoids. Furthermore, the scale of 0, 10, 20, 30, . . . per cent morphological change is homologous with all the old 10-degree scoring scales; the 0, 20, 40, 60, . . . per cent scale is homologous with the old 5-degree scale; and the 0, 5, 10, 15, . . . per cent with the 20-degree scale. Thus, all the previous scoring systems can now be transformed into percentages of morphological change which are common to metamorphosis instars irrespective of species or stage.

There are two principal methods for producing a series of intermediate forms between two metamorphosis stages, as required for determination of the degree of morphological change caused by juvenoids. The first method depends on the application of excessive amounts of a juvenoid at various intervals of the sensitive period. The second depends on application of various doses of a juvenoid at constant time at the beginning of the sensitive period. In the next paragraphs we comment on the results which can be obtained using both the methods indicated. The juvenile hormone activity caused by the juvenoids used is expressed in percentages of morphological change induced, as indicated above. The exact relationships between certain type of morphological patterns and a corresponding percentage of the morphological change have been derived from the dose-response curves of several juvenoid compounds, as described below.

When the values of morphological change are plotted diagramatically against the time of application of excessive amounts of a juvenoid, we obtain a curve which shows the time-dependence of the effect. Such curves for the last instar larvae of *Pyrrhocoris* and for the pupae of

Tenebrio are shown in Fig. 6. The course of the curves demonstrates an influence of the time of application on results of these bioassays for juvenoids. The curves have an almost identical shape with many juvenoids at a constant temperature. With the increased temperature the zero per cent effects occur sooner. The moment at which the curves reach zero morphological effect indicates precisely the end of the sensitive period to juvenoids with respect to external morphological patterns. Significant deviations from the usual shape of these curves suggest a change in the properties of the juvenoid tested, such as local action

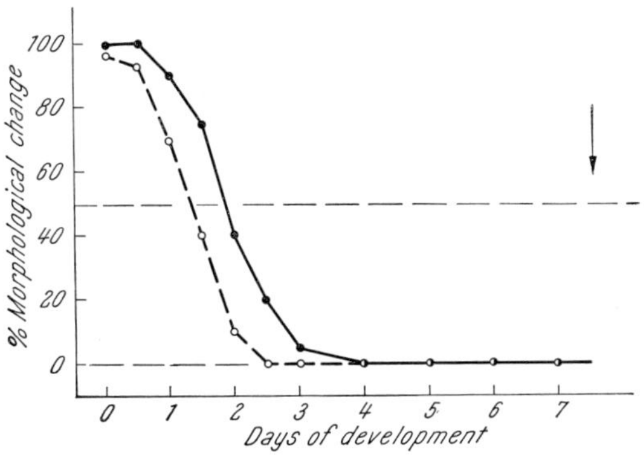

Fig. 6. Effect of excessive amounts of a juvenoid (1000-fold ID-50) applied topically at different time of the intermoult period on the degree of metamorphosis inhibition. Full points = last larval instar of *Pyrrhocoris*; open circles = pupal instar of *Tenebrio*. Arrow indicates ecdysis

on the treated spots of epidermis, limited penetration or limited transport within the body, or a rapid breakdown.

Most important for evaluation of these assays appear to be reversed conditions, e.g. the time of application remains constant while the dose of the juvenoid is varied. The application must be performed before or at the moment when the time-dependent curve (Fig. 6) begins to decline from maximum values of the morphological change. In other words, the application has to be made at the very beginning of the sensitive period. These quite important dose-response relationships can be studied on graphs which are obtained when the percentage of morphological change is plotted against the logarithmic scale of concentrations. Fig. 7 shows two such dose-response curves with two different juvenoids on last larval instars of two different Hemipteran species. The results indicate that the whole morphological change from 0 to 100 per cent effect has been realized within approximately a 10-fold change in concentration. A similar shape of the dose-response curves as shown in

Fig. 7 was found with several other juvenoids which were active between 0.1 ng to 10 μg per spec. However, the curves had, as a rule, a steeper slope with extremely active juvenoids whereas the shape was atypical when less active juvenoids than 10 μg per spec. were used. Also, similar dose-response relationships were observed with topical applications in other Exopterygote insects, such as in other Hemiptera, *Gryllus*, or *Locusta*. We believe that the relationships represented by dose-response curves in Fig. 7 may be general for this large group of insects. A series of larval-adult intermediates which is typical for Hemiptera is illus-

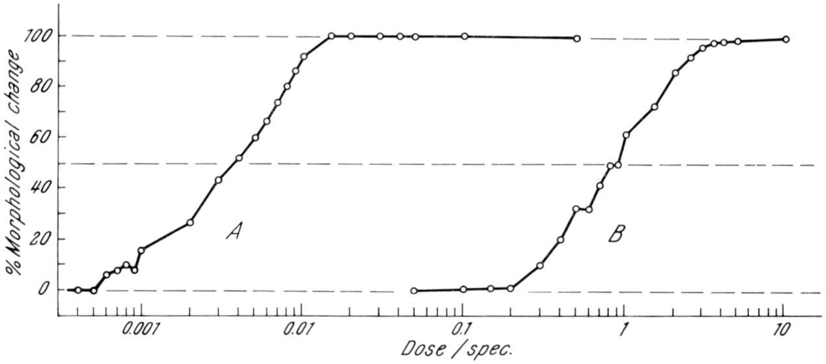

Fig. 7. Dose-response curves obtained with topical application of juvenoids. *A* = *cis*, *trans* mixture of ethyl dichlorodihydrofarnesoate tested on 0 to 20-hr-old last instar larvae of *Pyrrhocoris*. *B* = ± juvabione tested on last instar larvae of *Dysdercus cingulatus*

trated in Fig. 8. With exception of the initial (0 to 20) and terminal (80–100 per cent) values of morphological change, the course of the dose-response curves is more or less linear in the central part, provided that the successive types of intermediate forms were properly selected. In addition, the angles of the straight line portions show rather small deviations throughout a wide range of concentrations. However, when a linear scale of concentrations is used instead of the logarithmical one, the angle of the dose-response curves for compounds with different activity may be different, as observed by WIGGLESWORTH [358].

According to these semi-logarithmic dose-response curves, an optimal position for determination of a standard unit concentration for juvenoid activity should lie within the linear part of the curves, e.g. between 20 and 80 per cent morphological change. For conventional reasons it appeared most useful to define the standard unit of activity by 50 per cent morphological change. This corresponds approximately to the score 3 of the old 5-degree scale, score 6 of the old 12-degree scale, score 10 of the old 20-degree scale, etc. Thus, the unit concentration indicates an amount of a juvenoid (in μg per specimen) which would cause under specified conditions of application, 50 per cent change of developmental morphological patterns. This is manifested by appear-

Fig. 8. The effect of juvenoids on metamorphosis of *Pyrrhocoris* larvae. Normal adults, adultoids of 20, 40, 60, 80 per cent morphological change, and supernumerary larvae. *A* = dorsal view, *B* = ventral view. Note changes in the black cuticular pigment on abdominal sternites

ance of exactly half-larval adultoids in Exopterygota, and half-larval intermediates or half-pupal adultoids in Endopterygota. Such units of juvenile hormone activity which are based on the inhibition of metamorphosis are referred to as "Inhibition doses–50 Morphological" or

(ID-50 Morph.) as distinguished from other ID-50 units which are based on inhibition of embryonic development or adult-diapause. When assayed on insects with an average body weight between 20–80 mg, the juvenoids with ID-50 Morph. values larger than 10 μg per spec. are considered little active or inactive, with 0.1 to 10 as medium active and with less than 0.1 highly active.

In insects such as *Pyrrhocoris*, both injections and topical applications induce similar dose-responses. Moreover, the dose-response curves as shown in Fig. 7 are also quite similar to dose-response re-

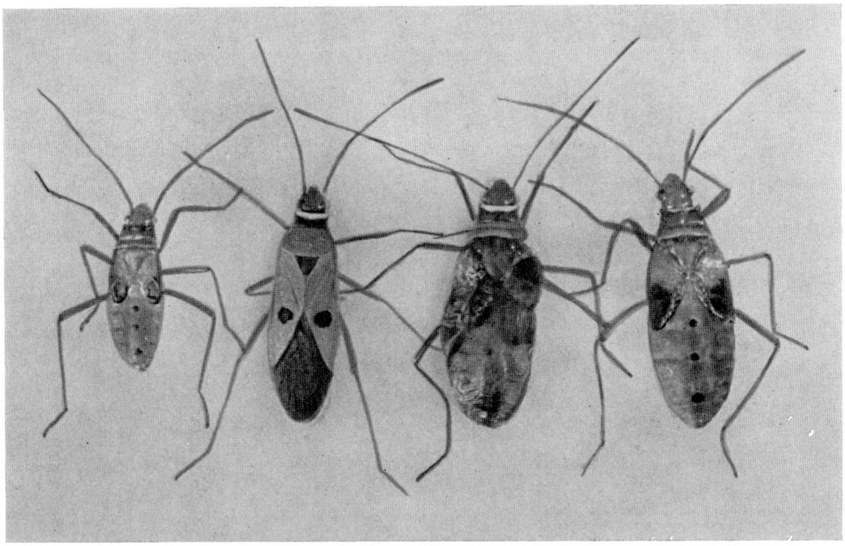

Fig. 9. Effect of juvenoids on metamorphosis of *Dysdercus cingulatus*. Last instar larva, normal adult, adultoid, supernumerary larva

lationships obtained with topical applications of medium or highly active juvenoids on pupae of some Endopterygote insects, such as *Tenebrio* or *Dermestes*. In contrast to this, it was known for a long time that injections in *Tenebrio* pupae induced completely different dose-response effects [274]. The course of the curves was extremely flat when compared with that in Fig. 7, and it was variable with respect to different activity of the juvenoids tested. The concentrations required for induction of minimum and maximum morphological changes differed by as much as three orders of magnitude. In these assays it is rather difficult to determine exactly the standard ID-50 unit concentration. For example, in the injection assay on Tenebrio pupae we have to count with about 30 per cent standard error, whereas the ID-50 values from the topical assays on Exopterygote larvae are reproducible to some $\pm$ 10 per cent.

With respect to physiological changes associated with juvenile hormone action, which have been described in Chapter E, the ID-50 value represents a critical amount of a juvenoid which reaches the effective concentration in the body for only a certain part (usually for the first-half) of the sensitive developmental period. The results are therefore dependent on a balance between the effective concentration and time in which it is present. Obviously, this combination of effective concentration versus time will be influenced by factors such as the rate of penetration, transport, or absorption of juvenoids by the haemolymph. However, our results on biological activity of juvenoids are still empirical

Fig. 10. Effect of juvenoids on metamorphosis of *Graphosoma italicum*. Last instar larva, normal adult, supernumerary larvae with dark and light pigmentation, and giant adultoid obtained from a supernumerary larva

and the significance of the above factors in this respect remains to be resolved.

For better illustration of the effects of juvenoids on insect metamorphosis we have included a series of photographs showing normal developmental stages, intermediates, and extra-larvae or extra-pupae in several Hemipterans, e. g. *Pyrrhocoris* (Fig. 8), *Dysdercus* (Fig. 9), *Graphosoma* (Fig. 10), and *Triatoma* (Fig. 11); in an Orthopteran *Schistocerca* (Fig. 12), in Coleopterans *Tenebrio* (Fig. 13) and *Dermestes* (Fig. 14), and in a Lepidopteran *Pieris* (Fig. 15).

The described methods for evaluation of juvenoids have been related to direct application methods. Similar procedures may be used with indirect application methods. The ID-50 values are then expressed per unit of treated surface, volume or weight. Finally, it has to be added that the described assay conditions apply only to juvenoids which have definite juvenile hormone activity and are used under physiological conditions. Overdosage of insects with excessive amounts of solvents or carriers or using large doses of poorly active compounds may lead to atypical or nonspecific results.

b) Evaluation of Assays Based on Inhibition of Embryonic Development

It is generally assumed that juvenoid action on embryogenesis may have many features in common with that on metamorphosis. This is confirmed by analogous inhibition of embryogenesis by juvenoids which leads to formation of embryos with incompletely differentiated or

Fig. 11. Effect of juvenoids on metamorphosis of *Triatoma infestans*. Last instar larva, normal adult, adultoid, supernumerary larva

completely rudimental larval epidermal structures. Eventually, the action of juvenoids can result in a general suppression of the embryogenesis process from the very beginning [194, 230, 257, 259, 261, 317]. As in the assays on metamorphosis, the dose-response relationships are in embryogenesis also dependent on the moment of juvenoid application.

In contrast to metamorphosis, it would be unrealistic and extremely difficult to evaluate the activity of juvenoids in embryological assays according to the degree of actual morphological change induced. For example, suppression of embryonic development may occur at different stages in different species, or it may bring about species specific patterns of malformations and malfunctions of the embryonic tissues. Furthermore, individual inspections of the affected embryos (which occur sometimes within pigmented or hard sclerotized chorion) may require time consuming dissections and microscopic analysis. However, all the disturbances caused by juvenoids in embryonic development are characterized by one common feature, e.g. there is no hatching of larvae from the eggs. Therefore, for practical reasons, the egg hatchability can be used here as a simple criterion for evaluation of juvenoid effects on the embryos. This criterion would represent a whole complex of symptoms characteristic for action of juvenoids on

Fig. 12. Effect of juvenoids on metamorphosis of *Schistocerca gregaria*. Last instar larva, adultoid, normal adult

embryogenesis. For example, embryos with arrested development at the very early blastoderm stage, as well as the more advanced forms inhibited after blastokinesis, are equally unable to form hatching larvae. In addition, some embryos develop after treatment of the eggs with juvenoids into almost perfect larvae which do not hatch due to physiological disturbances in functions of the internal organs. It thus appears that the effect of juvenoids on the egg hatchability must be considered as an outcome of all possible morphological and physiological deviations from the normal course of embryogenesis development. It is sometimes observed that embryos which resist determined doses of juvenoids and do form hatchable larvae, may suffer from certain physiological disproportions and the larvae may die later in the stage. This particular effect is not included in the usual assays of juvenoids on embryogenesis. It is a subject of special population assays described in further sections.

The standard unit of activity in these ovicidal assays of juvenoids is "Inhibition dose–50 Ovicidal" or (ID-50 Ovic.). It is indicated by an amount of juvenoid in μg per adult female, or per egg, which cause 50 per cent inhibition of larval hatching from the eggs. It is quite common that a certain percentage of the nontreated control eggs will not hatch, either because of lack in fertilization, or due to dessication and various other factors. Therefore, the amount of unhatched eggs used for ID-50

Fig. 13. Effect of juvenoids on metamorphosis of *Tenebrio molitor*. Normal pupa, normal adult, adultoid, second pupa

Ovic. calculations has to be corrected with respect to an actual amount of the hatchable eggs in the control group [308].

The ovicidal dose-response relationships found with topical applications of juvenoids in adult female *Pyrrhocoris* [191] show apparent similarities with the shape of the dose-response curves obtained on the metamorphosis stages. Similarities also exist with respect to specificity. For example, juvenoids which were active in the ovicidal assays had similar activity also in the metamorphosis assays (see p. 249).

Because of the simplified evaluation procedures in the ovicidal assays there exists an increased possibility to obtain all kinds of nonspecific effects. Thus, for example, generally toxic compounds such as normal insecticides or special ovicides can kill the embryo and thus also prevent hatching of larvae from the eggs. However, the true juvenoid effects can be distinguished from the toxic ones according to characteristic morphological malformations, or by a lack of activity and direct toxicity when applied after the sensitive period; sometimes they are identified according to simultaneous activity on metamorphosis.

Evaluation of the ovicidal effects is relatively simple when juvenoids have been applied directly on freshly deposited eggs. It is more complicated when juvenoids have been applied on females. Following treatments of the females, the ovicidal dose-responses are seriously influenced by the time interval between the treatment and oviposition. Some adult females deposit batches of eggs in regular intervals. In this case the ovicidal activity has to be determined separately for each batch of eggs. Eventually it is possible to work with a constant moment of application

Fig. 14. Effect of juvenoids on metamorphosis of *Dermestes vulpinus*. Prepupa, pupa, adult; (lower row) second pupa, predominantly pupal adultoid, and two adultoids with predominating adult patterns

and with one selected batch of eggs, which helps to achieve standardized conditions. Continuous decrease of ovicidal effects recorded in successive batches of eggs may provide additional information to the biological persistence of juvenoids. An optimal juvenoid candidate for practical use causes long lasting ovicidal effects with relatively small doses applied.

In females which oviposit more or less continuously throughout the reproductive period, the ovicidal effects have to be evaluated separately on eggs collected at determined intervals after the treatment. In special cases when the reproduction period is of short duration, or in females with one gonotrophic cycle, the ovicidal effects can be determined for all the eggs produced in one bioassay. With respect to indirect methods of application it has to be realized that there usually occurs simultaneous contamination of both females and eggs. However, it is also possible to assay the ovicidal effects indirectly on freshly deposited eggs which have been exposed to contacts with treated substrate or to juvenoid vapours.

Finally, we should like to mention that ovicidal effects produced by treatment of adult females with juvenoids have been referred to as a special case of female sterility [191].

c) Evaluation of the Assays Based on Stimulation of Ovarian Development

The effects of CAH on ovarian growth in adult female insects have already been described (p. 36). It has been further discussed (p. 239) that this effect of CAH on the ovaries can be duplicated by juvenoids. In the following paragraphs we are going to show that such a gonadotropic activity may be used as a very specific bioassay for juvenoids. In principle, the assay is made on females which have inactive ovaries, and the effect is determined according to deposition of yolk in the oocytes, formation of the ripe eggs in the ovaries, or according to induced oviposition. The females with inactive ovaries, which are used as assay animals, are usually represented by allatectomized adult females [31, 52], or by diapausing females. The former are most suitable with respect to physiological conditions but less convenient for routine assays, since allatectomy performed on a large scale is quite laborious. The latter, diapausing females, are adapted to inactivity of the whole neuroendocrine system, but can be obtained in large numbers by a simple manipulation with the environmental conditions for development, such as photoperiod.

When treated with juvenoids, the allatectomized or diapausing females of some species such as for instance *Pyrrhocoris*, may undergo several reproduction cycles in which ripe eggs are not only formed in the ovaries but also oviposited in regular intervals. But in many other species, the treated females develop ripe eggs which are never oviposited. This phenomenon is probably caused by disproportions in synchronization of the physiological functions which are regulated by neurohormones. In addition, following application of small doses of juvenoids the oocytes begin to accumulate yolk, but after a certain time the yolk is resorbed. Thus, it is necessary in such cases to dissect individual females at different intervals after the moment of application and to perform evaluation of the results on these dissected ovaries [357].

For their economic importance and because of the above mentioned practical reasons, most of these assays are made on diapausing females. Moreover, the main diapause feature in an adult female is actually arrest of ovarian growth, to be reinstalled by juvenoids. Hence, we have assumed that the activity of juvenoids in these diapause assays may be expressed in terms of adult diapause inhibition which is in agreement with the adopted ID-50 unit system. We have suggested that a unit of activity in these assays be an "Inhibition dose–50 Diapause" (ID-50 Diap.). It indicates an amount of juvenoid in μg per female which induces ovarian growth (or oviposition resp.) in 50 per cent diapausing females, e.g. it causes an inhibition of diapause in 50 per cent females. We are conscious, however, that it is more or less paradoxical to make an equation between ovarian stimulation and diapause inhibition. An

Fig. 15. Effect of juvenoids on metamorphosis of *Pieris brassicae*. Normal pupa, adultoid, normal adult at the time of ecdysis

alternative solution, especially when working with the *in vitro* ovarian assays, would be to designate this unit as "Effective dose–50 Reproduction" (ED 50-Reprod.). As has been described later in connexion with physiological effects of juvenoids, there exist apparent relationships between the dose of juvenoid applied and the percentage of females with resumed reproduction or inhibited diapause. However, there is no direct proportionality between the dose of juvenoid applied and the number of eggs produced, although in some cases, such as in *Rhodnius* [357] larger doses stimulate more oocytes to deposit yolk, or prolong duration of the reproductive activity. In general, the number of eggs produced by the treated females may be influenced by juvenoids to some degree, but it is also influenced by other physiological factors, such as the presence of reserve materials in the body, the extent of feeding, and reactivation of the neuroendocrine system. Thus, the number of eggs produced by the treated females can be used as a quantitative criterion of juvenoid activity only when the nutritional and hormonal conditions have been more or less constant and well defined. Since many insects have quite specific endocrine regulations of reproduction, it is important that all the assay conditions used have been specified.

The diapausing males can also be used for juvenoid assays, although evaluation of this activity is based on different criteria than that in the females. The main effect of juvenoids on diapausing males is represented by a reactivation of their sexual instincts. Whether this is a direct or indirect effect of juvenoids is dependent on species. Nevertheless, this action of juvenoids can also be expressed in ID-50 Diap. units, in terms of 50 per cent inhibition of the diapause behavioural syndrome.

d) Evaluation of the Population Assays

With juvenoids, insect endocrinologists are recently facing problems of less academic and more practical significance. For example, the described ID-50 values obtained by direct application methods are satisfactory for laboratory research but are quite insufficient with regard to practical use of juvenoids as pesticides. This statement is further explained by the fact that doses far below the ID-50 values may sometimes completely prevent reproduction thus causing a 100 per cent control with respect to an insect population. Thus, in order to obtain more information on these correlations between the degree of hormonal influence and the ability of insect population to survive and overcome developmental disturbances caused by juvenoids, additional assays of the effects on whole insect populations have been designed. In addition, some insects are too small, too delicate, or too sensitive to allow individual direct treatments; some are living inside grains, cocoons, wood, etc., and here the indirect application methods and studies on the whole populations represent the only reasonable approach to investigations of juvenoid effects.

A common principle for performance of these population assays is derived from the previous methods. The insect population is treated with graded amounts of juvenoids at determined stage. It is then allowed to develop for selected time interval or through the whole generation. The number of healthy specimens in the next generation is counted and related to that obtained in the untreated controls. Except specific assays used for studying the relationships between morphological change and survival of the population, most of the other population assays are based on indirect methods of application. The "Inhibition dose–50 for Population" (ID-50 Popul.) indicates an amount of juvenoid (in concentrations usual for indirect application techniques) which causes a 50 per cent decline of the number of specimens in the next generation in relation to similar untreated control population. It is clear that these results represent an outcome of integrated effects of juvenoids on metamorphosis, reproduction, embryonic development, and on other developmental and physiological processes.

Since the population assays are run for a relatively long time there appears a real problem with respect to persistence of juvenoids in biologically active form throughout the experiment. The results of these population bioassays have, therefore, no absolute value, suggesting that the data can be used only with respect to particular chemical structure, particular species or developmental stage, and with respect to particular conditions of the treatment. It is advised that insects with more or less synchronized development are used and that the colony is exposed to juvenoid treatment at the time when most individuals have approached the sensitive developmental period. Convenient developmental stages are the adults at the beginning of reproduction or larvae of the penultimate instar. The population assays are quite laborious and require statistical evaluation. In addition, they are quite nonspecific, unless a more detailed physiological analysis of the effects has been performed. On the other hand, the population assays enable determination of the significance for insect development of the side effects and other changes caused by juvenoids which are less important in the routine bioassays. For example, as revealed by population assays, factors such as physiological malfunctions of the internal organs, slight malformations of the genital system, ecdysial failures, and other quite inconspicuous physiological and biochemical changes associated with juvenoid action may be followed by drastical decrease of the ability of insect population to reproduce.

With regard to relationships between morphological and population effects of juvenoids, the ID-50 Popul. values may approach the ID-50 Morph. in some species. Thus in *Pyrrhocoris* and in few other Hemipteran species we have found that the ID-50 Popul. is equivalent to some ID-10 or ID-20 Morph. However, in some Endopterygote species, such as *Dermestes*, the ID-50 Popul. values are considerably smaller than that causing any visible morphological effects. We may add finally, that in some stored product insects, aphids, flies, and small size insects in general, our knowledge on juvenoid action is mostly based on these population assays.

e) Evaluation of the Field Assays

Our present knowledge on the effectiveness of juvenoids under field conditions is still insufficient. This lack of information leads to more or less substantiated doubts on the real usefulness of juvenoids for practical insect control; although the laboratory results have been quite stimulating. Insect development is so variable under field conditions that it appears absolutely impossible to elaborate any sort of universal bioassay for juvenoid effectiveness in the field experiments. Each of these assays has to be designed separately, according to specific developmental and external ecological conditions. One of the most difficult problems in the field assays is concerned with finding of appropriate methods for evaluation of the activity. Obviously, we cannot go into special details when discussing these complex problems here. Our intention is to give at least some instructions for determination of the juvenoid activity in these field bioassays, using our own experience with several juvenoids and a few insect species.

In principle, the techniques used for evaluation of the field assays of juvenoids represent an application of one or more of the methods described in the previous sections on a large scale. These methods can be separated into two basic groups. One of them depends on determination and evaluation of certain specific juvenoid effects, such as the effects on metamorphosis, adult diapause, female sterility, or ovicidal effects. The second group of methods is characterized by evaluation procedures which are based on analysis of changes which take place in the whole population after a determined period following the treatment.

The effects of juvenoids on metamorphosis can be best evaluated in field assays on larvae of some Exopterygote insects, such as Hemiptera or Orthoptera. It must be noted, however, that in contrast to insecticides, juvenoids require considerably longer time in order to affect the field population of insects, which is always more or less heterogeneous. During this time the affected stages required for scoring of the morphological effects may become a victim of predators, birds, moulds, or may simply get lost in the soil. For this reason it appears most convenient to collect the material for evaluation of the activity before the final ecdysis, e.g. before the developmental disturbances have occurred. This strategy is based on the fact that stages which have already passed the sensitive developmental period to juvenoids become influenced irrespectively to the contact with juvenoids later in the stage. For example, the last instar larvae of these insects which have initiated feeding and remarkably increased their body size on the treated fields (which is indicative that they have passed the sensitive period to juvenoids) can be collected and put into cages, allowed to develop until ecdysis and then scored for morphological influence in the same manner as in the laboratory assays. This procedure also prevents possible errors which may result from migration, since normal adults of most species fly whereas the affected stages cannot fly. The described sampling of the fully grown larvae after the sensitive period to juvenoids may be repeated in suceessive intervals following the treatment. Relative changes in the activity between the separate proups of samples provide additional information on the persistence of the effect. The collection of samples and prevention of migration is also facilitated by construction of isolation boxes on the treated fields. The field effectiveness of juvenoids is expressed in doses per square of the contamined fields which induce the ID-50 Morph. effect, or in doses related to specific morphological or physiological effects which are critical for further development and reproduction of a given species.

Similar techniques may be used for evaluation of field effects of juvenoids on metamorphosis in some Endopterygote insects. However, scoring of the effects according to the degree of morphological influence is more difficult here, and in many cases it is impossible. Therefore, a simple criterion which is represented by the degree of adult emergence has been adopted. This inhibition of adult emergence is directly correlated with the effects of juvenoids on metamorphosis of Endopterygote insects. In addition, physiological conditions for the effect of juvenoids on adult emergence are quite similar to the inhibition of larval hatching from the eggs, as has been described in earlier sections. It appears very convenient to use such a simple criterion as inhibition of adult emergence for evaluation of juvenoid activity in the field assays, because it can be applied with slight modification to such different insect orders as Diptera, Hymenoptera, Lepidoptera, and Coleoptera. One has to realize, however, that this criterion has different significance in insect control measures with respect to species. For example, some adult insects which survive the juvenoid treatment and do emerge, cannot reproduce as a consequence of failures in internal physiological functions. This effect, which occurs in adult of Endopterygote as well as Exopterygote insects should be included in final calculations of practical effectiveness of juvenoids on insect development and reproduction.

Evaluation of the ovicidal effects of juvenoids in the field experiments is complicated by the difficulty of egg collection. It is relatively easier in species where eggs can be located and collected. In this instance samples of eggs are periodically transferred from the treated fields to the laboratory and are incubated together with the original substrate until hatching. The activity is expressed in ID-50 Ovicidal units.

Insects with asynchronous development are inconvenient for field assays of juvenoid activity. It is expected that in order to achieve efficient control of these insects by juvenoids, the compounds used should be extremely stable and persistent under field conditions. Such compounds are not readily available so far. However, certain modifications of the assay conditions make it possible to test the already available juvenoids on insects with asynchronous development. For instance, selected developmental stages (which were obtained from laboratory breedings or were collected on the fields) can be released on isolated areas of the previously treated fields and the activity is later determined as in the laboratory assays on the homogeneous developmental stages. This method is also useful for rather accurate measurements of persistence of the biological activity of juvenoids in the fields. To this end, the synchronized developmental stages are released in the field at different time intervals following treatment. The decrease of biological activity in successive samples is then proportional to the decrease of juvenile hormone effects.

Field assays of juvenoids in which the activity is determined according to analysis of changes in the whole population have relatively broad application. The techniques for activity evaluation are in this respect basically similar to what has been described for the population assays in general (p. 118). The activity is usually expressed in ID-50 Popul. units which are related particularly to a 50 per cent reduction of the number of specimens of the next generation on the treated area in contrast to control values obtained on several untreated areas. In many cases it appears that the population assay method is the only possible one for evaluation of the biological activity of juvenoids under field conditions. Further advantages of the population analysis method are that it can be used for simultaneous determination of juvenoid activity on several related or unrelated insect species which develop on the same field. This is especially important for distinctions between the effects on insect pests and their parasites and predators. Disadvantages of the method are mostly associated with its low specificity and limited usefulness in species which migrate extensively.

In the foregoing sections we have described some general procedures for performance of the standardized assays used in determination of the biological activity of juvenoids. A list of these assays and activity units is given in Table 3. However, there are many assays pertaining only to certain stage or species which cannot be included in the general schemes, because the methods used for evaluation of these effects are based on quite specific developmental features. These include, for example, assays of juvenoids on accessory sexual glands of adult females, *in vitro* assays on isolated ovaries and other organs, assays based on determination of certain biochemical changes in the body, as well as some special assays on larvae of higher Diptera, on mosquito larvae, aphids, etc. These problems have been discussed in connexion with physiological effects of juvenoids.

Table 3. *Survey of Standardized Assay Methods for Juvenoid Activity*

ID-50 Unit	Stage of application	Method of application	Criteria for the unit determination	No. of spec. for one dilution	Specificity
Morphological	last inst. l. or pupae	direct or indirect	inhibition of metamorph., morphol. change between larva-pupa-adult	5–10	good
Ovicidal	fresh eggs or young females	direct or indirect	inhibition of larval hatching from the eggs	20–100	poor
Diapause (Reproduction)	allatect. or diap. females	direct	stimulation of ovarian development, inhibition of adult diapause	10–20	outstanding
Population	any stage, prefer. young adults or l.	indirect	decline in number of specimens in the next generation	over 50	poor

4. Instructions on Some Biological Assays with Juvenoids

In this section we have briefly described methods which have been used for determination of the juvenile hormone activity of juvenoids. We have selected some special assays which are for some reason significant in juvenoid research, but the main emphasis has been given to those assays which can be performed under more or less standardized conditions.

The Tenebrio Unit Assay. Pupae 1 to 25-hrs-old are narcotized with CO_2 and injected intersegmentally between the 4th and 5th abdominal sternite with 1 μl of olive oil containing the active substance. Adults which emerge after 6 to 8 days at 30 °C and 40 per cent R. H. are inspected; the positive effect is revealed by a patch of pupal cuticle over the area of sample deposition. The *Tenebrio Unit* concentration indicates the amount of juvenoid producing pupal patches in 40 per cent out of the total 50 pupae injected. References: [262, 264]. The assay is very sensitive and can be used for assaying crude extracts or juvenoids with low activity. The T. U. doses are commonly 100 to 1000 times smaller than the ID-50 Morph. units. With correspondingly higher amounts of the active materials the same methods of application can be used for determination of the ID-50 Morph. (injections).

Tenebrio Topical Assay. Pupae 1 to 20-hrs-old are fixed with the ventral side up to plastic adhesive tape on the bottom of Petri-dishes which are supplied with cotton plugged vials with water. Ventral side of the thorax or abdomen is then treated topically with 1 μl of acetone-juvenoid solutions. After 6 to 8 days at 27 °C the percentage of retention of the pupal epidermal patterns is evaluated and used for calculation of the ID-50 Morph. units of activity. (See Fig. 16, references 26, 27, 31, 253, 312, 313, 314, 345.) This assay is less sensitive than the former one but more important for considerations concerning practical use of juvenoids. As seen in the Appendix tables, some juvenoids cause the ID-50 effect in less than nanogram amounts.

Galleria Wax Wound Test. Pupae 1 to 20-hrs-old are narcotized and then a square of epidermis approximately 1 mm^2 is excised from the median part of the mesonotum. The wound is sealed by a mixture of paraffin wax with olive oil 1:1 using slightly warmed needle or electrically heated platinum wire. The olive oil contains graded amounts of the extracts or juvenoids. The activity is evaluated 6 to 8 days later. Positive responses are indicated by patches of cuticle with pupal coloration and architecture in the excised area, surrounded by adult cuticle. One Galleria Unit is indicated by an amount of extract which causes positive responses in 50 per cent pupae tested. References: [105, 285, 365, 366]. This is one of the most sensitive assays which is convenient for determination of juvenile hormone activity in crude extracts. However, it is less specific when compared with assays based on generalized morphological influence.

Galleria Metamorphosis Assay on Larvae and Pupae. The last instar larvae just before spinning the cocoon are immobilized by submerging them under water for 5 to 10 min. The larvae are then injected with 1 μl of olive oil solutions, or they are treated topically with 1 μl of acetone-juvenoid solutions. The ID-50 Morph. values are calculated according to the degree of

Fig. 16. Localized effects of juvenoids on metamorphosis of *Tenebrio molitor*. Adultoids with pupal abdomens and more or less adult head and thorax are shown in the lower row. Above are the acetone treated controls

retention of the old larval epidermal patterns some 3–4 days later. References [134]. The same procedure can be used with 0 to 20-hrs-old pupae, and the ID-50 Morph. values are determined according to retention of the pupal structures. Similarly as in *Tenebrio*, these assays are less sensitive than the above wax wound test, but their results are more important from practical points of view.

Metamorphosis Assay on Hemiptera. The last instar larvae 0 to 20-hrs-old are treated topically with 1 μl of acetone-juvenoid solutions. Alternatively, the larvae are narcotized under water for 15 min. and injected with 0.5 μl of olive oil solutions, or they are exposed to contact with substrate previously treated with juvenoids. After about 5 to 10 days at 27 °C the larvae moult into normal adults, larval-adult intermediates, or extra-larvae, depending on the dose of the juvenoid used. The ID-50 Morph. concentrations are calculated according to the degree of retention of the larval patterns, as has been described in previous sections. The number of specimens per each dilution can be reduced to five, provided that the larvae were not older than 20 hrs after ecdysis at the moment of application. References: [49, 271, 308, 312, 313, 314, 318, 335, 379]. The assay is universal for continuously feeding Exopterygote larvae in which the sensitive period to juvenoids occurs just after the last larval ecdysis. It is convenient for all kinds of indirect application methods. Under constant developmental conditions the standard error is limited to ± 10 per cent and the assay can be used for determinations of slight changes in biological activity of juvenoids, such as are connected with the chemical or biological inactivation, the rate of transport in insect body, and determination of residues. When Pyrrhocorid bugs are used as test animals, the assay is sensitive to all types of juvenoids so far known.

Maximum activity recorded for ID-50 Morph. in topical application was 2 picograms per spec., usual range of activity was between a nanogram and a microgram per spec.

Metamorphosis Assay on Rhodnius. The last instar larvae 2 to 3 weeks after ecdysis are fed and treated with the sample solutions one day after blood meal. For topical application it is recommended to use 1 μl of octane solutions, for injections 1 μl of vegetable oils. The ID-50 Morph. values are determined some 20 days later, after ecdysis, using similar criteria as in the previous assay. References: [358]. The assay is an example of atypical position of the sensitive period to juvenoids. Initiation of the sensitive period is affected by the blood meal, and this allows to obtain experimental animals with synchronized development.

Ovicidal Assay. Adult females of *Pyrrhocoris* 2-day-old are treated topically with 1 μl of acetone-juvenoid solutions. The females are incubated together with active males in glass jars or plastic cylinders with a netting on the bottom. The deposited eggs fall through the netting into prepared Petri-dishes. The eggs are collected daily or in 2 or 3-day intervals. They are fixed to a plastic adhesive tape which facilitates further manipulations and inspections. After 8 to 9 days of incubation at 27 °C the proportion of free egg chorions (indicating hatched larvae) to undeveloped eggs is determined. The ID-50 Ovic. values are calculated from the numbers of hatched larvae related to the egg hatchability in controls. For routine assays only the eggs of the first or first and second reproduction cycle are used for evaluation.

Ovicidal effects produced by treatment of the eggs are evaluated in a similar way as described in the previous paragraph. Treatment is made on freshly deposited eggs, preferably 0 to 15-hrs-old which are fixed to a plastic adhesive tape in groups of 20–100 eggs. Each egg is treated individually with 0.2 μl of acetone-juvenoid solution. For indirect methods of application the eggs are simply laid on a contamined surface (paper, glass, wood, etc.). For testing the vapour effects, the eggs are placed inside closed test-tubes containing graded amounts of a juvenoid adsorbed on some large surface material and isolated from direct contact with the eggs. Finally, for the dipping tests the groups of eggs are glued to paper or glass strips using some inert water soluble and acetone insoluble glue. The treatment is made by dipping the eggs for few seconds into acetone solutions of juvenoids.

With slight modifications in collection of the eggs and incubation conditions the assay has wide applicability on many Exopterygote and other insects.

Diapause Assay on Pyrrhocoris. Diapausing adults are obtained when the colony is kept constantly under 8hr photoperiodic illumination at 25 °C since the young larval period. The diapausing females 10 to 100 days old are topically treated with 1 μl of acetone-juvenoid solutions. A parallel control experiment with untreated diapausing females is made in order to check for spontaneous recovery from diapause. The treated females are incubated individually under the above diapause-maintaining conditions. Positive effects are indicated by oviposition which usually occurs between the 5th and the 10th day after the treatment. The ID-50 Diap. is calculated from percentage of females induced to oviposit. The activated females deposit batches of eggs in regular intervals without respect to insemination. There are no direct relationships between dose of the juvenoid and the number of successive egg batches deposited. After induction of the first reproduction cycle by juvenoids the internal neuroendocrine mechanism is reactivated which stimulates further egg laying. This does not exist in the allatectomized females where the number of egg batches is related to some degree to persistence of juvenoids in the body. However, persistence of juvenoid effects in diapausing females can be calculated from the amount of ovicidal effects on the eggs, provided that the treated females were allowed to mate with active males.

In some other species where diapausing females treated with juvenoids do not oviposit, the positive responses are determined according to growth and yolk deposition in the dissected ovaries. The assay is quite specific for juvenoids. Many toxic compounds, antibiotics, free fatty acids, and other lipids which may mimic the action of juvenoids in metamorphosis or in ovicidal assays appeared inactive in this assay. The doses required for ID-50 Diap. on females are, as a rule, 10 to 50 times larger than ID-50 Morph. on larvae.

Tribolium Population Assay. Male and female adults 1 to 2 weeks old are separated into groups, each containing 50 specimens. Each group is then transferred into breeding containers with 100 g of wheat grains previously sprayed or otherwise treated with determined amounts of juvenoid samples in acetone or in other organic solvents. These colonies are incubated at 27 °C until the adults of the new generation emerge. Their number is counted and the ID-50 Popul. values calculated with respect to the control experiment. With slight modification these assays apply to many insect stored product pests.

Drosophila Population Assay. Five pairs of freshly emerged adults are placed in rearing tubes containing 10 ml of the rearing medium contaminated superficially with juvenoids. The treatment is made in the following way: the samples of juvenoids are dissolved in ethanol and 0.1 ml of the solutions is poured over the surface of the medium. The adults are exposed to the treated medium after evaporation of ethanol. The colonies are incubated under conditions common to *Drosophila* breedings until new adults emerge. The ID-50 Popul. is calculated from percentage of the adults which emerge in the next generation related to that in ethanol treated controls. Continuous inspection of the developmental progress permits determination of the stage in which the effect of juvenoids is most apparent. This may be reduced fertility, failures in egg hatchability, atypical larval growth, inhibition of puparium formation, or failures of adult emergence. It is also possible to inoculate the treated medium with young larval stages to separate ovicidal and metamorphosis effects. The assay is very instructive but little specific, being quite sensitive to various side effects. Characteristic action of juvenoids in the assay is revealed by inhibition of pupariation, formation of atypical puparia, and formation of pupal-adult intermediates.

Special Population Assay. This assay has to determine relationships between the degree of morphological influence caused by juvenoids in metamorphosis and the ability of surviving intermediates to reproduce. The last instar larvae or pupae 0 to 20-hrs-old are individually treated by juvenoids using doses which are equal and lower than ID-50 Morph. Adultoids or adults which emerge after 5 to 10 days from the treated larvae or pupae are incubated under conditions favourable for reproduction. The eggs are collected and allowed to hatch. The amount of hatching larvae is counted and used for determination of the ID-50 Popul. units. The unit values are related to the amount of hatching larvae produced by control groups of larvae or pupae of the same number and sexual ratio. The dose for ID-50 Popul. is given in μg of juvenoid per specimen which was applied on larvae or pupae in the previous generation.

Field Assay on a Hemipteran. Young larvae of *Eurygaster intergriceps* are collected on the wheat fields. The material obtained is subjected to statistical analysis of the developmental stages present. Special attention is paid to relative density of individual larval instars. Then, according to approximative duration of each larval instar under field conditions, it is calculated: a) at what time the first outbreak of last instar larvae would appear, and b) how long it may last from this moment until most or all larvae reach the critical stage at the beginning of the last larval instar. The first calculation is needed to determine the moment of the first treatment, the second reveals requirements for persistence of the activity in the field. Due to a possibility of unexpected seasonal variations, it is necessary to make later corrections of these calculations.

Observations on the wheat fields in Bulgaria revealed that the important time interval (b) in which all larval population traversed the sensitive stage to juvenoids lasted 12 to 18 days, depending on weather conditions and location of the fields. This clearly indicated that requirements for persistence of the juvenoid effectiveness were 18 days for a single treatment or at least 9 days for a doubled treatment.

At the time indicated by appearance of the last larval instars, the localized field areas are sprayed with juvenoids formulated as water emulsions (1 per cent light mineral oil solution of juvenoids with equal amount of commercial emulsifiers used for insecticides, and water), or aerosol sprays using non-phytotoxic oily carriers also used for formulation of insecticides. Both the applications are made in parallel series because they have different effects with different juvenoids.

Evaluation of the results is initiated 5 days after the first treatment by collecting samples of fully grown, large, last instar larvae which have spent the sensitive period to juvenoids in the treated fields. The sampling is periodically repeated in 3 or 5-day intervals on each of the treated areas and on one or more control areas. Each sample consists of 30 to 100 larvae which are incubated in small containers provided with food and water, either under field or in laboratory conditions. The effect of juvenoids becomes visible after the larvae have moulted. It is manifested by the degree of retention of the old larval epidermal patterns which can be scored similarly as in the usual laboratory assays. All specimens showing higher percentage of morphological influence than 30 per cent are unable to reproduce and can be considered as a positive field control, as well as specimens with ecdysial failures which soon die. Differences in activity or eventually disappearance of the activity in samples collected after successive intervals are used retrospectively for calculations of the actual persistence of the juvenoid activity in the field.

It is apparent that the described field assay is based on evaluation of specific effects of juvenoids on metamorphosis. In this case, therefore, the effectiveness of juvenoids can be expressed in concentrations which cause the

ID-50 or ID-30 Morph. effects. It was observed that a juvenoid compound with ID-50 Morph. of 0.1 μg per larva exhibited similar ID-50 Morph. effect on the field population when applied in about 1000 g per Hectare. However, this relationship is profoundly affected by persistence of the juvenoid under field conditions. Description of this assay illustrates a real necessity to know factors such as sensitive periods, developmental rate and dispersion of developmental stages in a given insect, and persistence of the juvenoid to be used before any field assay with juvenoids is initiated.

5. The Use of Bioassays for Studies on Transport and Stability of Juvenoids

Data which are contained in the Appendix tables show that some juvenoids possess enormous biological activity combined eventually with properties to act selectively on certain insect species. This allows one to identify in certain biological assays picogramm amounts of the active compounds with less than 20 per cent standard error. The extremely high sensitivity of these bioassays can compete with radioactive techniques in identification of small amounts of juvenoids. Moreover, the bioassay detects an actual amount of the active compound, being usually insensitive to the breakdown products. Due to these properties, slightly modified biological assays appear very convenient for some specific investigations on the action of juvenoids such as a) determination of their transport through haemolymph, b) studies on their penetration through the integument, and c) distribution of juvenoids in the body, their biological inactivation, and their persistence in active form in the environment.

a) Transport through the Haemolymph

Methods which can be used for determination of juvenoid transport through the haemolymph are derived from some sensitive standard assays. In principle, selected developmental stages are treated with juvenoids; the treated specimens are then incubated separately for a certain time after which the haemolymph samples are taken from the cut leg, antenna, from the tip of the abdomen, etc. The presence of juvenoid in the haemolymph sample is determined biologically as has been described for ID-50 units. For qualitative studies, the haemolymph sample is allowed to dry on the body surface of the test insect. Or, alternatively, it can be injected into the body cavity. For more accurate determinations the haemolymph samples are extracted with acetone and the content of juvenoid present is determined quantitatively as used in methods with topical application of juvenoids. It is evident that these relatively simple techniques may be useful for studying the problems of species specificity. For example, as shown in Table 4, a topical treatment of *Pyrrhocoris* larvae with dichloro-dihydrofarnesoate was followed by almost immediate appearance of the activity in the haemolymph. In about 15 min from the moment of application, one μl of the haemolymph contained enough of the compound to prevent metamorphosis of another larva. In contrast, haemolymph samples taken from *Tenebrio* pupae which had been topically treated or injected with considerable amounts of the same juvenoid compound, revealed no activity when tested on sensitive *Pyrrhocoris* larvae. This suggested that the compound was completely insoluble, and unabsorbed by the haemolymph of *Tenebrio*, on which it was also biologically inactive. It appears necessary to point out that the compound was not decomposed metabolically in *Tenebrio* because lipid extracts made from *Tenebrio* pupae one week after the application contained all the activity.

The described examples show that the solubility of a juvenoid in the haemolymph may represent one of the factors for species specificity. However, we also found that some juvenoids with selective effects on *Pyrrhocoris* occurred in considerable quantities in the haemolymph of other insect species, although these compounds did not have biological activity on any species but *Pyrrhocoris*. Apparently, the haemolymph transport was not the limiting factor here. Nevertheless, the simple haemolymph tests are useful for investigations on transport of juvenoids in different species or stages. Hopefully, they will throw some light upon the problems of species specificity and selectivity which are so little understood so far.

Measurements of radioactivity in the haemolymph samples from larvae treated with tritium labelled dichloro-dihydrofarnesoate were also performed in *Pyrrhocoris* larvae. The obtained results shown in Table 5 are quite consistent with the results obtained by the bioassay methods. The amount of radioactivity in the haemolymph increased during 3 hrs following the treatment. Then the amount of radioactivity in the haemolymph remained more or less constant for certain period of time and it decreased afterwards. The amount of radioactivity in the excrements increased for 24 hrs after treatment, at which time it was about 5-times higher in 1 μl of the rectal content than in 1 μl of the haemolymph. Additional bioassays revealed that the compound was partly present in its biologically active form in the excrements.

b) Penetration of Juvenoids through the Cuticle

Methods for measurements of the rate of penetration through insect cuticle have been described in general by RICHARDS in a monograph on testing chemicals on insects [297]. These methods can be applied to juvenoids in combination with some sensitive bioassay. It was found by TREHERNE [339] that many nonelectrolytes were able to

Table 4. *Juvenile Activity of 1* µl *Samples of Haemolyph Taken at Different Time Intervals from Larvae Treated with Ethyl 3,7,11-Trimethyl-7,11-Dichloro-2-Dodecenoate. The Values Indicate Percentage of Morphological Deviation, Perfect Supernumerary Larvae = 100%. ID-50 Morph. of the Juvenoid Used Was 0.01* µg *in Pyrrhocoris*

Treated insect	Stage	Application	µg per spec.	Haemolymph activity in *Pyrrhocoris* larval assay after: 15 min	30 min	1 h	2 h	4 h	8 h	1 day	2 days	4 days
Pyrrhocoris	last l.	topical	10	35	100	100	100	100	100	90	80	—
Graphosoma	last l.	topical	10	10	60	80	100	100	95	80	30	—
Tenebrio	pupa	topical	500	—	—	0	—	0	0	0	0	0
Tenebrio	pupa	injection	1000	—	—	0	—	0	0	0	30	0

Table 5. *Radioactivity in the Haemolymph and in the Excrements of Pyrrhocoris after Topical Treatment with Tritium Labelled Dihydro-Dichloro Methyl Farnesoate. Stage Treated: Freshly Moulted last Instar Larvae. Specific Activity 10* mC/mMol; *Biological Activity 0.05* µg *(ID-50 Morph.). The Dose per Spec. Corresponded to 18,725 c. p. m.*

	Relative radioactivity counts per *1 µl* of the sample after: 15 min	30 min	1 h	2 h	3 h	4 h	24 h
Haemolymph	22	46	48	66	77	65	50
Rectal content	46	—	152	—	170	—	230

pass through the insect cuticle inside the body. This is most probably true for juvenoids as well, because of their expressed contact action. However, quantitative data on the rate of penetration of different juvenoids through the cuticle are lacking.

We performed some qualitative experiments with dry specimens of cuticle prepared from larvae of *Pyrrhocoris* and pupae of *Tenebrio* and *Galleria*. The results revealed that a wide range of acyclic and aromatic juvenoids were able to pass from the external side of the cuticle to the internal one. It remains to be investigated whether this is also the case under physiological conditions. However, the study is limited by a possible barrier formed by an aqueous milieu in the underlying epidermal cell layer, because most juvenoids are very unpolar compounds almost insoluble in water.

In order to obtain at least some evidence on how much of the topically applied juvenoid remains on the body surface, we have measured distributions of radioactivity after 24 hrs following topical treatment with tritiated dihydro-dichlorofarnesoate. Results of this study which are shown in Table 6, allow the following conclusions: when smaller amounts of the compound were applied, more than half of the total amount of radioactivity was found outside the body. In addition, a larger portion of the remaining radioactivity was found inside the body while less remained on the body surface. Furthermore, when large amounts of the compound were applied, about half of the radioactivity still remained on the body surface, a considerable amount was outside the body, while only about 15 per cent of the total counts was found inside the body. This shows that a topically applied juvenoid can quickly leave the body due to excretion or due to a contact with the substrate or both. Furthermore, the transfer of juvenoids from the surface depots into the body or outside may depend, among other factors, also on the amount and form of the juvenoid applied.

The early experiments with paper extracts [318, 379] revealed that the active compounds were able to get inside the body by contact with any part of the integument. This is in good agreement with some observations that oily materials which were applied on localized spots of insect integument were rapidly spread and formed a thin film over the whole cuticle of a living insect [177]. However, it is obvious that the highly sclerotized parts of the insect integument, such as wing elythrae in beetles, sclerotized pupal cuticle in Lepidoptera, or puparia in flies, may limit the penetration of juvenoids. This is known to occur with respect to insecticides. Consequently, the highly sclerotized parts of the body should never be used for sample application in topical assays with juvenoids.

Table 6. *Distribution of Radioactivity 24 Hours after Topical Application of Tritiated Methyl Dihydro-Dichloro Farnesoate to Freshly Moulted Last Instar Larvae of Dysdercus cingulatus. Values Indicate Percentage Radioactivity from the Total Amount Administered. Specific Activity of the Material Used Was 100* mC/mMol

Radioactivity recovered from	Appl. 3,700 c. p. m. per larva; 30 spec. used	Appl. 23,665 c. p. m. per larva; 75 specimens used
Radioactivity on all objects inside the rearing jars	57	38
Rinsed from the body surface	16	45
Present in body homogenates	25	15.5
Radioactivity lost	2	1.5

Table 7. *The Values of ID-50 Morph. in Pyrrhocoris Obtained after Treatment with Ethyl Dihydro-Dichloro Farnesoate at Different Time of Egg and Larval Stage (Single Treatment). The Treated Groups of Larvae Were either Incubated in One and the Same Jar (A), or Transferred to Fresh Noncontaminated Breeding Jar at the Beginning of Each Larval Instar*

	Developmental stages treated (µg/spec. to cause ID-50)					
	Late eggs	1st instar	2nd instar	3 rd instar	4th instar	5th instar
A	4	4	0.3	0.01	0.005	0.001
B	> 200	> 200	9	0.2	0.04	(0.001)

c) *Distribution, Inactivation, and Persistence of Juvenoids*

The data on biological activity of juvenoids which are obtained from standardized assays are far from satisfactory with respect to contemporary tendencies in juvenoid research. There are requirements for additional information on distribution of juvenoids in the body, stability of these compounds in biological systems and their persistence under environmental conditions. It appears that some of this information may easily be obtained when the bioassay conditions are modified and adjusted to such physiological studies.

The distribution of juvenoids in various insect tissues or organs can be studied with similar methods as has been described for determination of the juvenoid content in haemolymph. A necessary condition is that juvenoids with relatively high biological activity, e.g. ID-50 values around one nanogram, are used. In principle, the methods are based on treatment of some selected developmental stage with about 10 μg of juvenoid per spec. After a determined time of incubation the residues are quickly rinsed out of the surface with acetone, the specimen is dissected in Ringer solution, individual organs are removed, washed in Ringer, and dried. The dry material is extracted with organic solvents and the amount of juvenoid is determined according to some sensitive biological assay. In an insect-juvenoid system where ID-50 Morph. is around 1 nanogram per spec., the treatment with 10 μg gives 10,000 units of activity per specimen. This allows to determine juvenoid content even in such small organs as nervous system, pericardial system, Malpighian tubules, oenocytes, etc. The amount of juvenoid is finally expressed in units of activity or in μg of the compound per mg of dry tissue matter.

Methods designed for measurements of biological stability of juvenoids in insect body are based mostly on the experimental evidence indicating that the effects of juvenoids are realized only in the course of the sensitive period. In other words, effective is only that amount of juvenoid which is present in the body during the sensitive period. Thus, when applied at certain time before the sensitive period, the more stable juvenoids can remain in the body until the sensitive period and will be effective, whereas the biologically unstable juvenoids will not persist until the sensitive stage and will be ineffective. In principle, there are two kinds of methods used for these studies. One of them works with a constant time of application before the sensitive stage, but uses different doses ranging from ID-50 to 100 times of the ID-50 doses. The other method uses a constant dose of ID-50 or larger but the application is made at various moments before the sensitive period. The results are expressed in doses versus time before the sensitive period in which a given juvenoid loses its activity, or, more commonly, in doses which cause the ID-50 effect when applied one, two, or more developmental instars before the sensitive period. Some results which we obtained using a combination of both the above described methods in the larvae of *Pyrrhocoris* are illustrated in Table 7. These results indicate that in order to cause the ID-50 Morph. effect by application of the compound on the young larval instars, it is necessary to use enormously larger doses. These data are also suggesting that extreme care should be taken to prevent re-contamination of the experimental specimens by juvenoid residues left inside the breeding jars or cages.

In *Pyrrhocoris* the application of juvenoids to freshly ecdysed penultimate larval instars (e. g. 5 days before the sensitive period) required about 10-fold ID-50 Morph. doses in order to achieve 50 per cent morphological change in metamorphosis. However, some juvenoids with relatively good biological activity in the standard assay have now appeared virtually inactive. In contrast, there were some juvenoids such as the esters of 3,7,11-trimethyl-2-dodecenic acid which had almost equal activity when applied to both the penultimate and last larval instars.

The above assays for biological stability of juvenoids in the insect body have one weak point. Namely, that duration of development in the young larval instars, and the conditions for growth and ecdyses cannot be fully standardized. It appears however, that under certain circumstances these assays can be performed on the same developmental stage and under more or less constant conditions. These methods are based on findings [379] that initiation of the sensitive period is dependent on feeding in certain stages, such as are the last larval instars of many Exopterygote insects. Moreover, it is observed that the delay of the sensitive period is proportional to the time of starvation. Thus, the compounds can be applied on freshly ecdysed last instar larvae which are then starved for a determined time. In the course of fasting the compounds are inactivated and the effect appears later when the larvae resume feeding and development. Similarly as in the former case, we can use either constant period of fasting with various doses from ID-50 to 10 times ID-50, or, alternatively, constant ID-90 doses with varied time of fasting. The biological stability of juvenoids is expressed in partial or complete loss of activity related to respective time of fasting and dose. This method gives quite consistent and reproducible results, but unfortunately, it is less significant for practical purposes because the growing larvae have relatively more intensive metabolism and excretion, and may thus inactivate juvenoids faster than the starving ones.

With respect to specific developmental conditions in some stages or species it appears possible to study the inactivation of juvenoids in other systems. We have already mentioned, in connexion with the ovicidal assays, that it is possible to judge on biological stability of juvenoids according to disappearance of the ovicidal effects in successive batches of eggs with respect to time of juvenoid application on the females. Another possibility for measurements of juvenoid stability was suggested in *Galleria* [134] where the sensitive period to juvenoids is delayed to the 5th or 6th day of the last larval instar. Here a differential loss of activity between applications performed at the beginning and on the 5th day of the last larval instar is, to a certain degree, indirectly proportional to the biological stability of juvenoids in *Galleria* larvae. Finally, there exist also methods for determination of biological stability of juvenoids which are based on observations that the sensitive period for the effects of juvenoids on morphogenesis may be different from that for the effects on pigmentation. For example, in the last larval instars of *Locusta* and *Schistocerca* the sensitive period for morphological effects proceeds from 0 to 4 days, whereas that for the effects on pigmentation occurs between the 4th and 5th day [154, 225]. It is expected that juvenoids which would cause simultaneously both the effects after being applied on the freshly moulted larvae might be more stable in the body than those which would cause only the morphological effects.

Large series of juvenoid compounds which are made in the chemical laboratories are sometimes stored for some prolonged time. An eventual re-use of the old compounds brings about a risk of their possible spontaneous inactivation. Fortunately, most of the juvenoid compounds retain their biological activity for months or even years when properly stored under refrigeration, in absence of oxygen and humidity. However, there are also compounds which polymerize or decompose under usual conditions of storage and these can be re-used only after fresh chemical or biological analysis. In addition, the problems of chemical stability of juvenoids under field conditions appear very actual in recent time in connexion with the use of juvenoids as pesticides. Solution of all these problems is associated with an availability of some accurate and reproducible biological assay which could distinguish between relatively slight changes in the biological activity of juvenoids. It is generally agreed that such a good bioassay can substitute for laborious chemical analyses, can prevent distribution of inactivated samples or prevent making unsuccessful field experiments with unstable compounds.

For these studies on chemical stability of juvenoids we have adopted a standard method which is based on determination of ID-50 Morph. in Pyrrhocorid bugs. This bioassay has relatively small standard error ($\pm$ 10 per cent) and is quite sensitive to all kinds of juvenoid compounds. Each sample to be analysed consisted of 1 mg of a given juvenoid compound. The samples were dissolved in organic solvents and then applied on determined surface of glas, wood, paper, leaf surface, etc., according to the nature of requested information. Then the samples were exposed to various abiotic environmental conditions, such as intensive light, increased temperature or high relative humidity. The exposure was performed in special boxes under regulated laboratory conditions or in the open under atmospheric conditions. After exposure the samples were quantitatively extracted with organic solvents. The solvent was evaporated and the residue was used for serial dilutions in acetone as required for determination of the ID-50 Morph. units. The loss of activity was calculated from differences between the control ID-50 values and those obtained with the exposed samples. For information we may add that several juvenoid compounds lost 90 per cent of their biological activity from 2 to 10 days under usual outdoor conditions.

C. Chemistry of the Juvenoids

1. Introduction

This chapter describes juvenile-hormone-like substances which have been isolated from naturally occurring sources or prepared by synthesis. The literature on juvenile hormones and compounds with juvenile hormone activity has undergone phenomenal growth in recent years. In the present survey an attempt has been made to select the most important representatives of the many hundreds of juvenile-hormone-active compounds and to point out the structure-activity relationships.

The chapter is divided according to the type of substances described. The physiological properties of the compounds are briefly mentioned in the text while detailed juvenile activity data are listed in tables elsewhere in this book (see the Appendix). Chapter D contains more detailed discussion of the structure-activity relationships of many of the compounds mentioned here.

The compounds are arranged according to their fundamental skeleton along with their derivatives obtained by simple chemical changes. Thus, e.g., some acyclic compounds will be accompanied in the text by heterocyclic derivatives with an oxirane and azirane ring.

The text is illustrated by brief reaction schemes to show the synthetic pathways. These schemes, however, cannot replace the original literature. With few exceptions, the structural formulae are represented in a uniform fashion (cf. the formulae *1* and *2* in Table 8) regardless the geometrical isomerism which, however, is specified in the name of the compound or in the text (cf. the next paragraph). As it is not possible to describe preparations of all compounds mentioned in the text, a selection has been made with special attention to the readily accessible literature which can be consulted in original.

The trivial names have proven useful in many branches of organic chemistry. In the field of juvenile hormones, a system of trivial terms has been proposed by MEYER [205], and the new terms will also be used in our review, especially in the text on Cecropia juvenile hormones. For some examples see Table 8.

2. Acyclic Juvenoids

a) *Farnesol and Derivatives*

The sesquiterpenoid alcohol farnesol (*3*) is undoubtedly one of the first compounds which was shown to posses the insect juvenile hormone activity. This compound occurs in nature in numerous plants, and as a supposed intermediate in the biosynthesis of sesquiterpenoids (in the form of farnesylpyrophosphate), farnesol is potentially present in all tissues where sesquiterpenoids are formed [115, 116].

There are several synthetic procedures for the preparation of farnesol. Commercially available farnesol is usually a mixture of cis-trans isomers, and the highest juvenile activity is shown by the two 6-trans-farnesols [388]. As will be discussed later, the geometrical orientation of farnesol and other compounds is one of the most important factors for insect juvenile hormone activity. For this reason, we mention only some typical stereospecific syntheses leading to the formation of trans,-trans-farnesol.

COREY *et al.* [61] prepared trans,trans-farnesol by a synthesis which included the stereospecific conversion of propargylic alcohol (e.g., *4*) to 3-alkylated allylic alcohol. The stereospecific reduction of the triple bond of propargylic alcohol with lithium aluminium hydride in the presence of sodium methoxide was followed by iodination with excess iodine at a low temperature. For other steps see Scheme 1.

Scheme 1

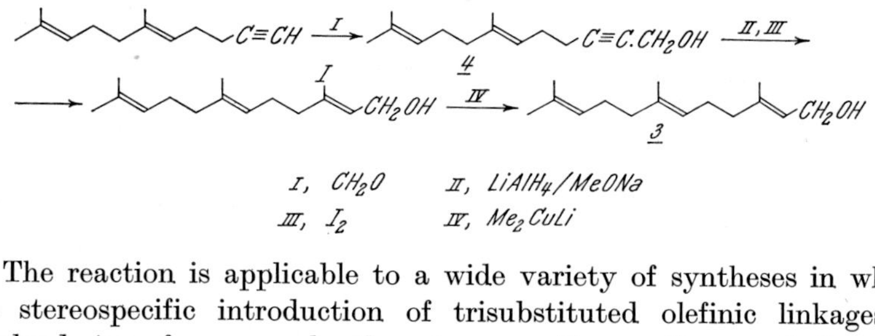

The reaction is applicable to a wide variety of syntheses in which the stereospecific introduction of trisubstituted olefinic linkages is involved (see for example the stereospecific syntheses of Cecropia hormones, p. 150).

A second stereospecific synthesis of trans,trans-farnesol was also

Table 8. *Nomenclature for Substances with Juvenile Hormone Activity*

Compound:	Alternative trivial names:
Substance with juvenile hormone activity	Bio-analog, juvenoid, JH mimic or (hormono)-mimetic compound
Methyl 10,11-epoxy-3,7,11-trimethyl-trans,trans-2,6-dodecadienoate [*1*, $R^1 = R^2 =$ Me]	Methyl juvenate
Methyl cis-10,11-epoxy-3,11-dimethyl-7-ethyl-trans,trans-2,6-tridecadienoate [*1*, $R^1 = R^2 =$ Et]	Cecropia C_{18}-JH, or Methyl 12,14-dihomojuvenate
Methyl cis-10,11-epoxy-3,7,11-trimethyl-trans,trans-2,6-tridecadienoate [*1*, $R^1 =$ Me, $R^2 =$ Et]	Cecropia C_{17}-JH, or Methyl 12-homojuvenate
Methyl 3,7,11-trimethyl-trans,trans-2,6-cis-10-tridecadienoate [*2*, $R^1 =$ Me, $R^2 =$ Et]	Methyl 12-homo-t,t,c-farnesoate
Methyl trans-10,11-epoxy-3,7,11-trimethyl-trans,trans-2,6-tridecadienoate [*1*, $R^1 =$ Me, $R^2 =$ Et]	Methyl 12-homo-t-10-juvenate

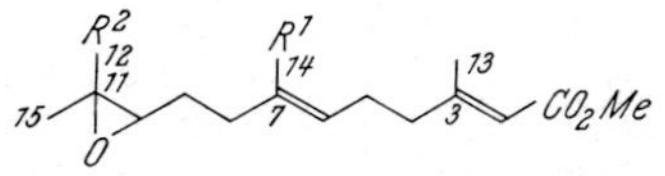

1

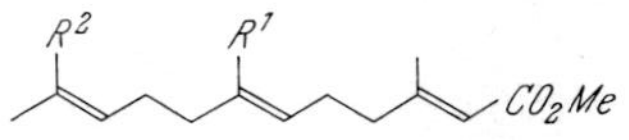

2

described by Corey [64], This synthesis is closely related to the synthesis of Cecropia hormones [57, 63]; (see p. 150) by virtue of the all-trans alcohol of the general formula (5).

In principle, this synthesis of farnesol involves preparation of trisubstituted olefinic systems from β-oxido-phosphonium ylides. The reaction is remarkable in that an intermediate containing three new stereocenters was generated stereospecifically and a directional specificity operates in the final olefin-forming elimination. The hydroxy ether *6* as the key intermediate is converted by Corey into trans,trans-farnesol by means of the disulfate-hydride method [58] (Scheme 2).

Scheme 2

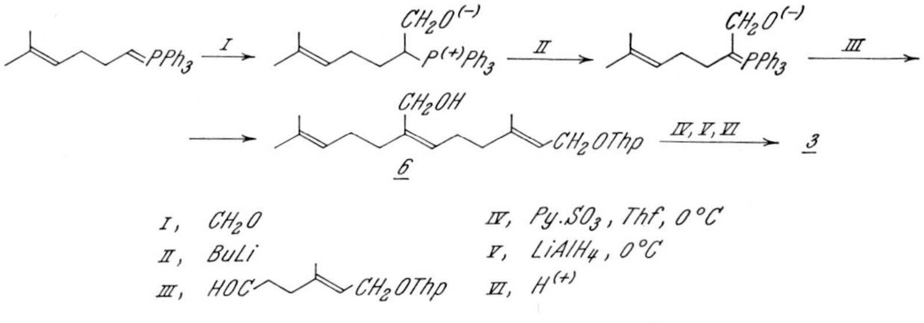

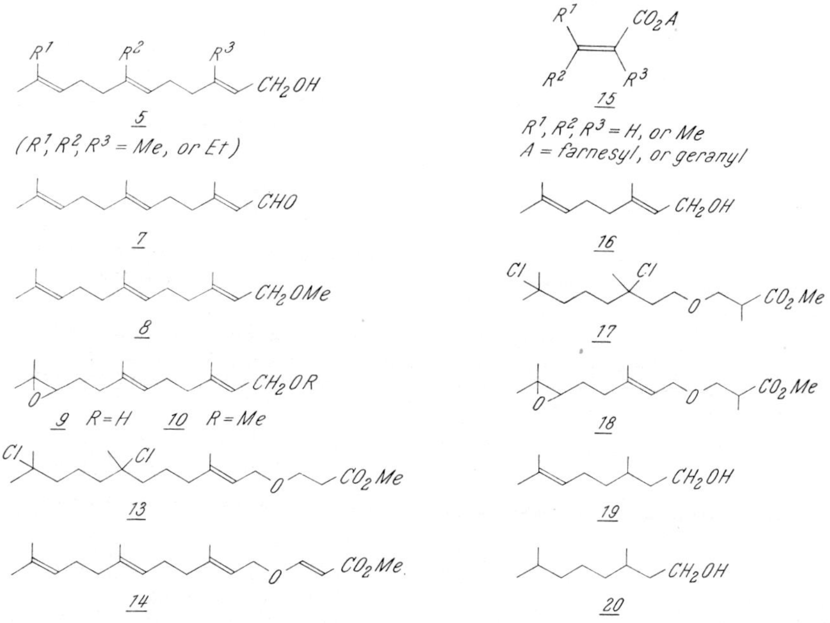

Schmialek discovered farnesol and the aldehyde farnesal (7) in the faeces of the yellow mealworm (*Tenebrio molitor*). The original idea [277,

356] that farnesol constituted a virtual juvenile hormone, was soon abandoned [31, 52, 206, 286, 388]. Nevertheless, some further papers [30, 280, 357] demonstrated a certain juvenile activity of farnesol, farnesal and numerous related compounds in physiological assays, especially when applied to *T. molitor*. Among the active compounds were farnesyl esters, farnesyl ethers and farnesylamines. Both Schmialek [280] and Bowers [30] observed the highest activity with farnesyl methyl ether (*8*) as did Wigglesworth [357] in experiments on the bug *Rhodnius prolixus*. More recently, farnesyl methyl ether has been used as the test substance in several physiological assays and found remarkably active for some insect species. The ether (*8*) is readily accessible by synthesis, e.g., by methylation of farnesol with methyl iodide, as shown by Bowers [31].

Two other compounds, 10,11-epoxyfarnesol (*9*) and 10,11-epoxyfarnesyl methyl ether (*10*) have been prepared by epoxidation of farnesol and farnesyl methyl ether, resp., and are also physiologically active [31].

The preparation of ethers derived from β-hydroxy acids and β-hydroxy nitriles was devised by Ratuský [250]. The synthesis was based on a base catalyzed reaction of alcohols with esters of α,β-unsaturated acids (Scheme 3).

Scheme 3

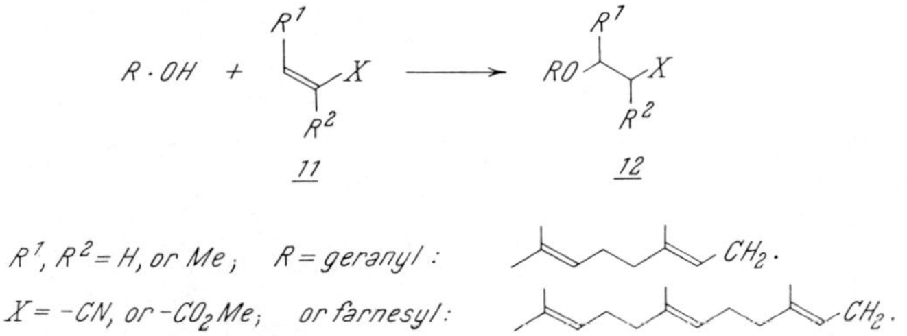

Thus, farnesol (*3*) and methyl acrylate (*11*; $R^1 = R^2 = H$, X = = CO_2Me) have furnished methyl 4-oxa-7,11,15-trimethyl-6,10,14-hexadecatrienoate (*12*; $R^1 = R^2 = H$, X = CO_2Me, R = farnesyl).

Introduction of gaseous hydrogen chloride into solutions of unsaturated compounds at low temperatures results in a ready addition of hydrogen chloride to the isolated double bonds. The 11,15-dichloro derivative (*13*) is thus obtained from the trienoate *12*.

Methyl 4-oxa-7,11,15-trimethyl-2,6,10,14-hecadecatetraenoate (*14*) has been prepared from farnesol and methyl propiolate.

Farnesyl esters of unsaturated acids (*15*) have been prepared by alkaline transesterification of lower esters [249].

The preparation of the above compounds is simple but the insect juvenile hormone activity is rather low.

b) Geraniol and Derivatives

Geraniol (*16*) is a monoterpenic alcohol with the trans-configuration at the double bond. It occurs in numerous plants and is also prepared synthetically together with the cis-isomer, nerol. In contrast to farnesol, geraniol and its lower ethers do not show any juvenile activity [30, 280].

Like farnesol, geraniol has been used in the preparation of ethers of some β-hydroxy acids [250], cf. Scheme 3, p. 141. Thus, methyl 4-oxa-2,7,11-trimethyl-6,10-dodecadienoate (*12*; $R^1 = H$, $R^2 = Me$, $X = CO_2Me$, R = geranyl) has been obtained from geraniol and methyl methacrylate. In tests on *Tenebrio molitor* and the bugs of the family Pyrrhocoridae, compound *12*, the 7,11-dichloro derivative (*17*) and the 10,11-epoxy derivative (*18*) were only poorly active. The activity of geranyl methacrylate (*15*; $R^1 = R^2 = R^3 = Me$, A = geranyl) prepared by transesterification of methyl methacrylate, was also low [249].

The geranyl derivatives of some carbamic acid esters will be mentioned later on (p. 169).

Better juvenile activity was obtained with the 5-oxa esters prepared by Jarolím *et al.* [141]. In principle, the compounds of Jarolím were ethers of geraniol, its dihydro and tetrahydro derivatives, and were synthesized by reaction of the corresponding alcohol with diazoketone in the presence of boron trifluoride etherate [233] followed by the Wittig reaction (Scheme 4):

Scheme 4

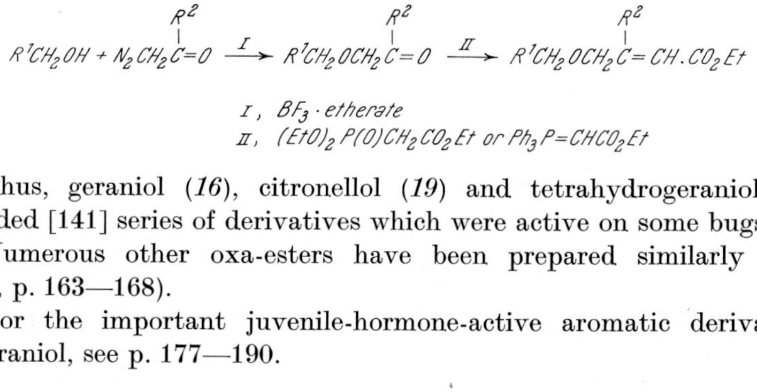

Thus, geraniol (*16*), citronellol (*19*) and tetrahydrogeraniol (*20*) afforded [141] series of derivatives which were active on some bugs.

Numerous other oxa-esters have been prepared similarly (vide infra, p. 163—168).

For the important juvenile-hormone-active aromatic derivatives of geraniol, see p. 177—190.

c) Farnesoic Acid and Derivatives

Farnesoic acid (*22*, R = H) is a sesquiterpenoid acid which does not occur in nature but may be easily prepared e.g. by oxidation of farnesol [41]. In contrast to numerous farnesoic acid derivatives, the juvenile hormone activity of the free acid (*22*) was negligible [313]. However, the methyl and ethyl esters (*22*; R = Me or Et) are considerably more active. The esters are prepared by a mild esterification of farnesoic acid or by a procedure based on the reaction of geranyl-acetone (*21*) with the Wittig reagent [383] or with Wadsworth-Emmons reagent [343].

Scheme 5

21 22

I, $Ph_3P{=}CHCO_2R$, or $(EtO)_2P(O)CH_2CO_2R$

As will be shown later, both these reactions are very often used in the preparation of some juvenoids because of the ready formation of the α, β-unsaturated ester group which is of considerable importance for the juvenile activity especially with acyclic compounds.

This procedure is used inter alia in the preparation of farnesoic acid derivatives differing by the number of double bonds in their molecules.

The low-pressure hydrogenation of methyl farnesoate has afforded the corresponding hexahydro derivative [312, 346] namely, methyl 3,7,11-trimethyldodecanoate (*23*).

Numerous other hydro derivatives have been prepared in a way analogous to the Scheme 5.

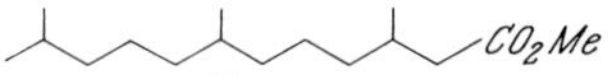

23

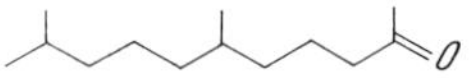

24

25

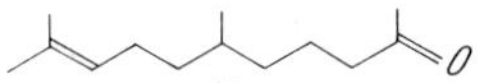

26

27

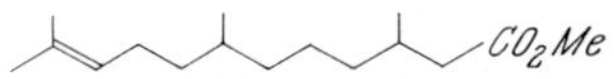

28

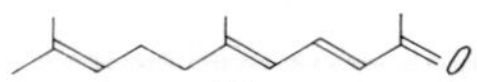

29

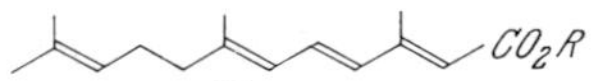

30

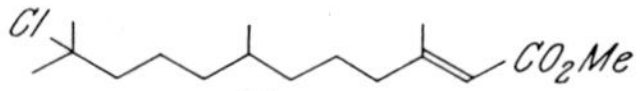

34

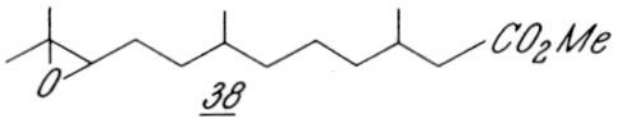

38

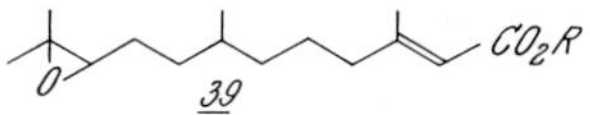

39

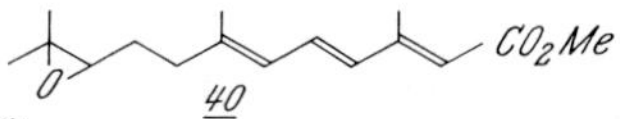

40

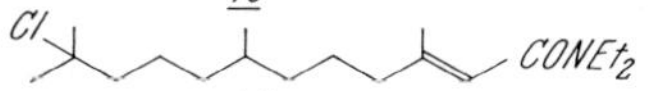

45

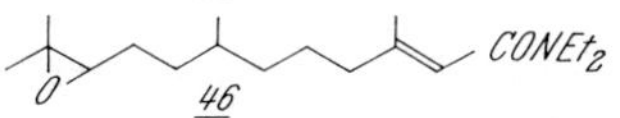

46

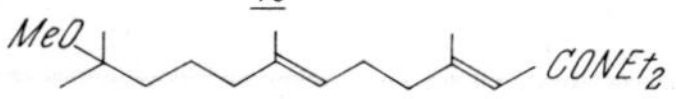

47

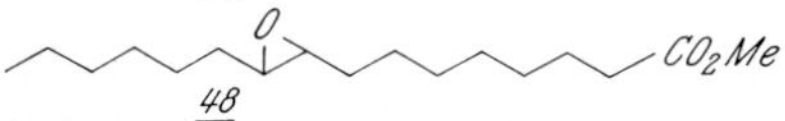

48

Hexahydropseudoionone (*24*) has yielded methyl 3,7,11-trimethyl-2-dodecenoate [114, 312] (*25*), and citronellylacetone (*26*) has furnished methyl 3,7,11-trimethyl-2,10-dodecadienoate [135, 312, 346] (*27*).

Reduction of methyl dihydrofarnesoate (*27*) with sodium in butanol has yielded methyl 3,7,11-trimethyl-10-dodecenoate [346] (*28*).

Pseudoionone (*29*) gives [137] methyl 3,7,11-trimethyl-2,4,6,10-dodecatetraenoate (*30*).

The farnesoic acid derivatives described above, are remarkably active on some beetle species but are much less active on bugs. Juvenile hormone activity is due to the presence of a double bond at position α, β with respect to the carbonyl group [134, 268, 308, 312, 313, 346, 358].

In 1966, LAW *et al.* [172] reported the preparation of a mixture with a high juvenile hormone activity. This mixture was formed by introduction of gaseous hydrogen chloride into a solution of farnesoic acid (*22*; R = H) in lower aliphatic alcohols. The crude product showed morphogenetic activity on butterflies, beetles as well as the bugs of the family Pyrrhocoridae. As shown later by detailed investigations, the introduction of hydrogen chloride into solutions of farnesoic esters was accompanied by addition of hydrogen chloride to the two isolated double bonds [271] (Scheme 6):

Scheme 6

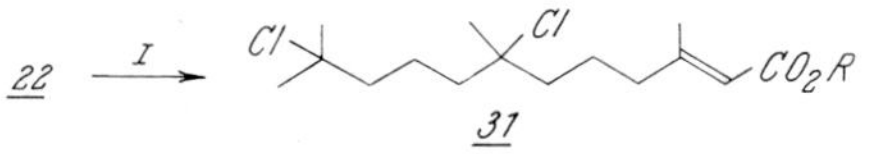

I, HCl

The resulting alkyl 3,7,11-trimethyl-7,11-dichloro-2-dodecenoate (*31*) exhibited an enormously high activity on the above mentioned bugs and obviously constitutes that component which is responsible for the high activity of the "Law-mixture". In the course of further investigations a crystalline acid was isolated [313], namely, 3,7,11-trimethyl-7,11-dichloro-trans-2-dodecenoic acid (*31*; R = H). From this acid, a series of trans-2-dodecenoates [313] was prepared via the dichlorofarnesoyl chloride (*32*): (Scheme 7).

Scheme 7

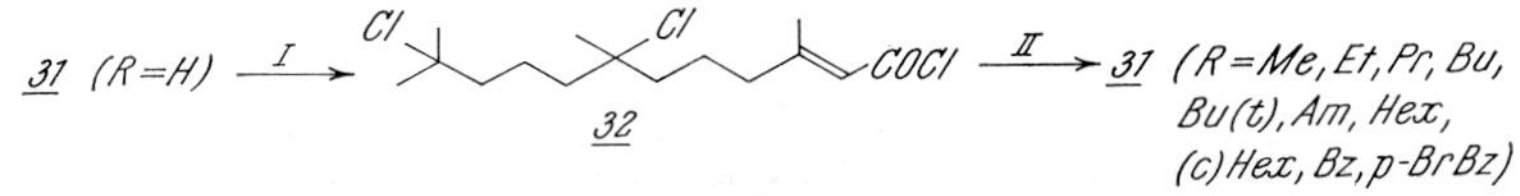

I, $SOCl_2$
II, ROH

Reaction of methyl farnesoate with tritium chloride yields the labelled dichloro derivative [270] (*33*): (Scheme 8).

Scheme 8

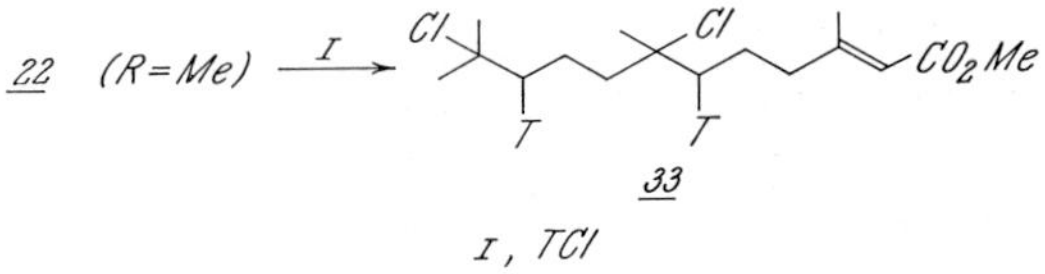

I, TCl

In conclusion, the introduction of a chloro atom into the molecule of farnesoic acid derivatives by means of the addition of hydrogen chloride to one or two double bonds is often accompanied by a considerable increase of juvenile hormone activity. Methyl 3,7,11-trimethyl-11-chloro-2-dodecenoate (*34*) may serve as another example of this effect [312]; compound *34* is prepared by addition of hydrogen chloride to the double bond of the dodecadienoate *27*.

The so called solvomercuration-demercuration [37] has also been used in the preparation of some juvenoids, e.g., by Wakabayashi [344] in the synthesis of methyl 11-methoxyfarnesoate and methyl 11-ethoxyfarnesoate (*35*; R^1 = Me or Et) from methyl farnesoate (*22*; R = Me).

The resulting alkoxy derivatives (*35*) were, however, only slightly more active on *Tenebrio molitor* than the parent esters [344].

Scheme 9

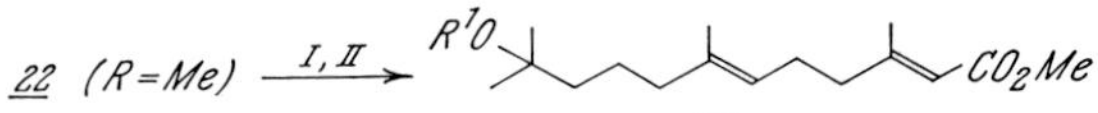

35

I, Hg(OAc)₂, R¹OH
II, NaBH₄, KOH

The presence of the oxirane ring in the molecule of juvenoids plays a very important role. The epoxy derivatives are formed on treatment of the corresponding unsaturated compound with some organic peracids.

As mentioned above (p. 141), the epoxides of farnesol and of its derivatives have been reported by Bowers *et al.* [31]. Epoxidation of methyl trans,trans-farnesoate (*22*; R = Me) has afforded a mixture of 6,7-epoxy and 10,11-epoxy derivatives [31] (*36* and *37*): (Scheme 10).

Scheme 10

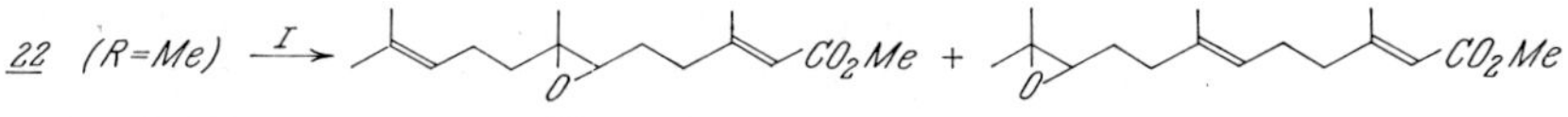

36 37

I, m-Chloroperbenzoic acid

The latter epoxide (*37*) (for the new trivial name methyl juvenate see Table 8, p. 139) was at the time of early investigations a very potent juvenoid. In addition to the treatment with hydrogen chloride, the epoxidation affords an efficient route to increase the morphogenetic activity of parent unsaturated compounds. The 10,11-epoxides *38* and *39* derived from the hydrofarnesoates *28* and *27* and methyl 10,11-epoxy-3,7,11-trimethyl-2,4,6-dodecatrienoate (*40*) prepared from the tetraenoate *30* may serve as further examples [312, 346].

The importance of the experiments of Bowers [31] does not consist merely in the discovery of the morphogenetic activity of epoxy derivatives of farnesol and farnesoic acid esters but also in his suggestion of the close structural relationships of methyl juvenate to the so called Cecropia juvenile hormone (see p. 147) which proved to be entirely correct [205, 340].

The total synthesis of methyl juvenate (*37*) was devised by Cavill and Williams [48]. The Wittig reaction of the starting 5,6-dihydroxy-6-methyl-2-heptanone acetonide (*41*) with (4,4-ethylenedioxy)pentyl-triphenyl-phosphonium iodide (*42*) and the subsequent hydrolysis of the ethyleneketal group yielded a mixture of isomers. The required trans-9,10-dioxy-6,10-dimethyl-5-undecen-2-one acetonide (*43*) was isolated by preparative gas chromatography. The Wittig reaction [383] of compound *43* and hydrolysis led to a diol ester the monomesyl derivative of which (*44*) was converted to the final product *37*, (see the reaction scheme 11):

Scheme 11

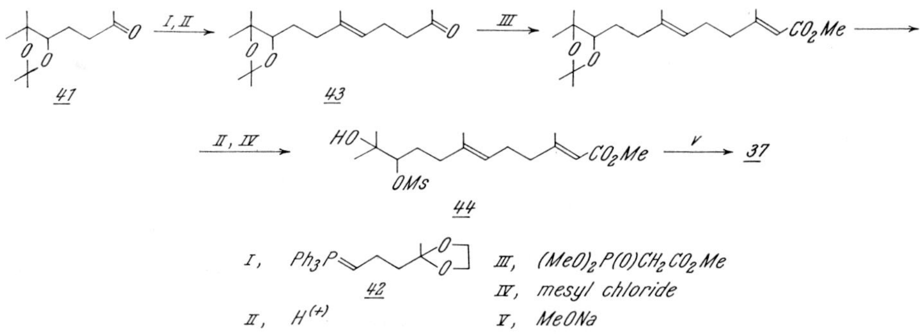

This synthesis is of importance as a model route for the preparation of Cecropia juvenile hormone and its analogues [47] (see p. 154).

Noteworthy also are the amides of farnesoic acid which are usually prepared by the classical route from acid chloride on treatment with corresponding amines. Especially the N-mono- and dialkylamides of some of the farnesoic acid derivatives are highly active or even more active than the corresponding esters, particularly on some Coleoptera, cf. the N,N-dimethylamides (*45*, *46*, and *47*) [66, 67, 268, 308, 322].

d) *Cecropia Juvenile Hormones*

In 1956, Williams [368] discovered that extracts from abdomens of the male moth *Hyalophora cecropia* exhibited strong juvenile hormone activity. The active component of this extract was hypothesized to be methyl 9,10-epoxyhexadecanoate (*48*). Surprisingly enough, both the synthetically prepared pure geometrical isomers of the compound

48 were completely inactive. The activity of the native substance *48* was obviously due to a trace amount of a contaminant the separation of which by chromatography failed.

The discovery of WILLIAMS initiated intensive investigations on the active principle of the Cecropia extract [107, 209, 262]. Finally, the team of ROELLER [262] isolated a very pure specimen of the hormone by a complex isolation procedure consisting of a combination of extraction, low-temperature precipitation, short-path distillation, gas chromatography and column chromatography. To illustrate the difficulty of these operations it should be noted that it was possible to isolate only 0.0008 gram of the pure hormone from 180 g of the crude extract and that the abdomen of one Cecropia moth contains only 0.5–1.0 microgram of the hormone [263, 265].

The structure of the active principle was determined by chemical and physical methods, particularly by a meticulous interpretation of mass spectra and nuclear magnetic resonance spectra of the native compound and products of its partial as well as total hydrogenation. It was observed that spectra and chemical properties of the hormone and methyl juvenate (*37*) were strikingly similar. The total hydrogenation of the Cecropia hormone and methyl juvenate (*37*) gave also very similar products (the total hydrogenation of methyl juvenate furnishes methyl 3,7,11-trimethyldodecanoate *23*). On the basis of the above observations, the Cecropia hormone was ascribed the structure of methyl 10,11-epoxy-3,11-dimethyl-7-ethyl-trans-2,6-tridecadienoate (*49*) [265, 267, 340], i. e. Cecropia C_{18}-JH (see Table 8).

Some data suggested the trans configuration of the $\Delta^{6,7}$ double bond. The configuration of the epoxide ring remained obscure. The stereochemistry of the Cecropia C_{18}-juvenile hormone was finally elucidated on the basis of syntheses [69, 70]. One of these syntheses is illustrated by the Scheme 12:

Scheme 12

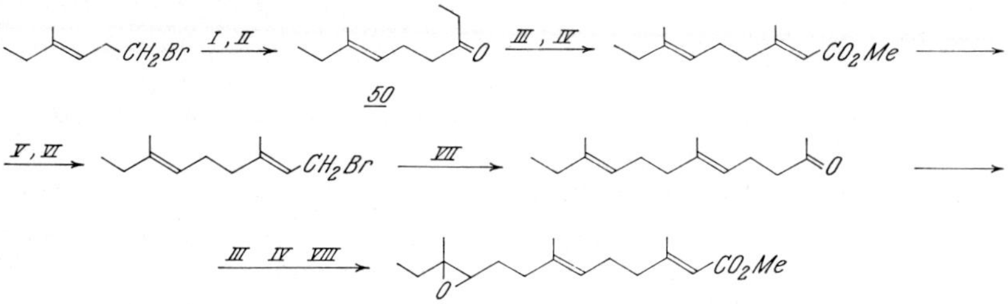

I, Mg, EtCH=O
II, CrO_3
III, $(MeO)_2P(O)CH_2CO_2Me$
IV, Chromatographic separation
V, $LiAlH_4$
VI, PBr_3
VII, $EtCOCH_2CO_2Et$
VIII, m-Chloroperbenzoic acid

The synthesis consisted of the preparation of four geometrical isomers of methyl 3,11-dimethyl-7-ethyl-2,6-trans-10-tridecatrienoate. The pairs of cis-trans isomers formed in the course of the synthesis, were separated by chromatography and processed separately in the subsequent steps. The starting material was trans-1-bromo-4-methyl-3-hexene, obtained according to the method of JULIA and JULIA [156]. A simple reaction sequence (see the Scheme) led to four geometrical isomers of the Cecropia C_{18}-JH with the following configurations at positions 2–3, 6–7, and 10–11: trans,trans,trans; trans,cis,trans; cis, trans, trans, and cis,cis,trans. As shown by gas chromatography, none of these geometrical isomers was identical with the native hormone. Comparison of gas chromatographical data, mass spectra and NMR spectra suggested the trans,trans,cis configuration for the Cecropia C_{18}-juvenile hormone.

This idea was confirmed by another synthesis [70] starting from cis-3-methyl-2-pentenyl bromide (*51*). This bromide was converted by reaction with ethyl 3-oxopentanoate (*52*) to cis-7-methyl-6-nonen-3-one (cf. the corresponding trans-isomer *50*, p. 147), which was processed similarly to the first synthesis [69] to yield after chromatography racemic methyl cis-10,11-epoxy-3,11-dimethyl-7-ethyl-trans-2-trans-6-tridecadienoate (*49*) (= methyl 12,14-dihomojuvenate, cf. Table 8). The physical, chemical and biological properties of the synthetic product were in good agreement [265, 340] with those of the natural Cecropia C_{18}-juvenile hormone.

In the course of investigations on the Cecropia extract [205, 209] MEYER *et al.* reported [208, 210] the isolation of two Cecropia juvenile hormones. The main component of the extract was found to be identical with methyl 12,14-dihomojuvenate (= Cecropia C_{18}-juvenile hormone). The less abundant component, accounting for 13–20% of the juvenile hormone activity in Cecropia moths, was separated by micropreparative gas-liquid chromatography and identified by high-resolution mass spectra and infrared spectra as the lower homologue of Cecropia C_{18}-JH, namely, as methyl cis-10,11-epoxy-3,7,11-trimethyl-trans-2-trans-6-tridecadienoate (*53*) (= methyl 12-homojuvenate or Cecropia C_{17}-JH). When assayed in Lepidoptera, both C_{17} and C_{18} Cecropia juvenile hormones exhibited about the same morphogenetic activity.

The structure of the Cecropia C_{17}-JH was confirmed by a stereospecific synthesis [149] of the racemic form. The synthesis was based on a modified Julia method [156] for introducing trans-substituted double bonds (this method has been also used by the same authors in the synthesis of the racemic Cecropia C_{18}-JH, vide infra).

Methyl-trans-4-bromosenecionate and 1-acetyl-1-methylcyclopropane (*54*; R = H) were used as the starting material. The β-keto ester *54* (R = CO_2Et) was converted to the key intermediate, the keto diester *55* (R = CO_2Et). The subsequent decarboxylation, selective reduction and homoallylic rearrangement [33] gave the bromo ester (*56*). For further reaction steps see the Scheme 13:

Scheme 13

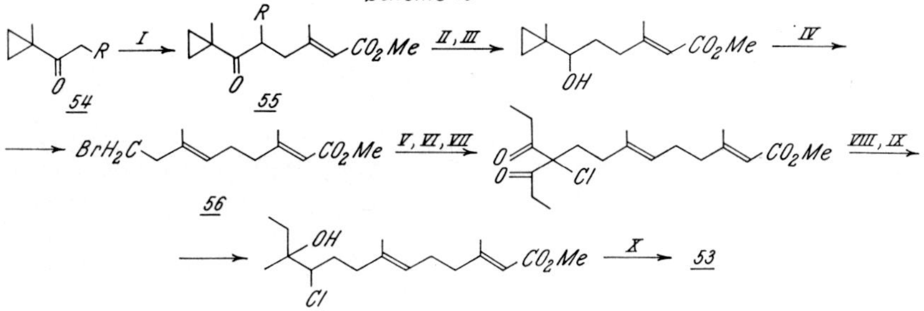

I, Methyl trans-4-bromosenecionate
II, Decarboxylation
III, $NaBH_4$
IV, PBr_3–LiBr, $ZnBr_2$
V, NaI
VI, Li-enolate of heptane-3,5-dione
VII, $CuCl_2$–LiCl
VIII, $Ba(OH)_2$
IX, MeMgCl
X, K_2CO_3

e) Other Syntheses of Cecropia Hormones and Related Compounds

In addition to the above mentioned syntheses, numerous other syntheses have been published in recent years. Many of them are stereospecific or stereoselective or of such a general nature that they may be applied to the preparation of other related compounds. In view of the high biological activity of Cecropia juvenile hormones on Lepidoptera, the aim of some syntheses was to make these hormones commercially available for practical control of insects.

One of the stereoselective syntheses of the racemic Cecropia C_{18}-juvenile hormone was devised by JOHNSON et al. [150]. This synthesis starting from 1-acetyl-1-ethylcyclopropane (*57*) was analogous to that of the racemic C_{17}-homologue starting from 1-acetyl-1-methylcyclopropane (*54*; R = H, cf. Scheme 13). Both these syntheses [149, 150] were very advantageous, the yields of most reaction steps being about 90%.

The ingenious stereospecific synthesis of SIDDALL *et al.* [394] used the ketone *58* as the key intermediate. The synthesis was based on sequential fragmentation of the bicyclic precursor *59*, control of olefine geometry therefore being transposed to a control of relative stereochemistry in cyclic systems. The bicyclic diketone *60* was prepared by Michael addition of 2-ethylcyclopentane-1,3-dione to propyl vinyl ketone and acid catalyzed cyclization of the adduct. Both fragmentations were effected by the action of sodium hydride in tetrahydrofuran on the corresponding p-toluenesulfonates. The stereospecific course of key steps was controlled by NMR spectroscopy. The reaction sequence is shown on the Scheme 14:

Scheme 14

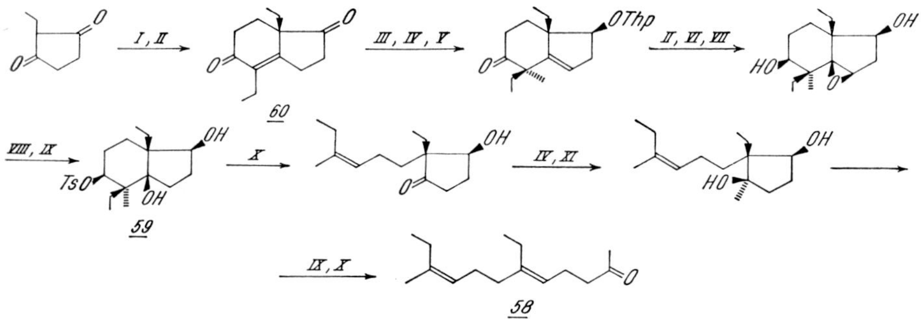

I, Propylvinyl ketone
II, H^+
III, $NaBH_4$
IV, Dihydropyrane
V, MeI
VI, $LiAl(O.tBu)_3H$
VII, m-Chloroperbenzoic acid
VIII, $LiAlH_4$
IX, TsCl
X, NaH
XI, MeLi

The stereospecific total synthesis of the racemic Cecropia C_{18}-juvenile hormone of Corey et al. [60] started from the cis-alcohol *61* which was prepared stereospecifically from p-methoxytoluene. The route of Corey comprised several novel synthetic processes of general applicability, e.g., the conversion of propargylic alcohols to allylic alcohols which has been previously used in the synthesis of farnesol [61] (cf. p. 138). Noteworthy also was the direct transformation of the allylic alcohol *62* to the methyl ester *63* via the acyl cyanide intermediate $=C(CH_3)COCN$. This one-step procedure involved stirring with excess active manganese dioxide in hexane and addition of methanol, sodium cyanide and acetic acid. (Scheme 15, see p. 151).

The syntheses [63, 64] of the two racemic Cecropia hormones (*49* and *53*) and of the biologically active position-isomer *67* of the hormone *53* were very short and simple. The syntheses were based on the same principle as that of trans,trans-farnesol [64]. The racemic Cecropia C_{17}-JH (*53*) was prepared from the hydroxy ether *64* (a homologue of the farnesol precursor *6*) by deoxygenation [58] and oxidation [59], cf. page 140 [64], and 138 [60]. The preparation of the racemic C_{18}-homologue *49* requires an additional reaction of the aldehyde *65* with methylenetriphenylphosphorane to give the vinyl derivative *66* which was

Scheme 15

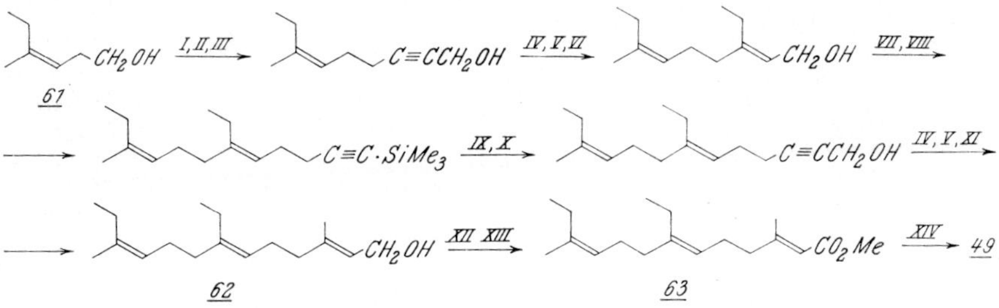

I, TsCl
II, $LiC{\equiv}C\cdot CH_2OThp$
III, $H^{(+)}$
IV, $LiAlH_4-NaOMe$
V, I_2
VI, Et_2CuLi
VII, PBr_3
VIII, $LiCH_2C{\equiv}C.SiMe_3$
IX, $Ag^{(+)}$, then $CN^{(-)}$
X, BuLi, then CH_2O
XI, Me_2CuLi
XII, MnO_2
XIII, MnO_2, $CN^{(-)}$, MeOH
XIV, HOBr, i-$PrO^{(-)}$

then subjected to a selective diimide reduction [62]. The syntheses of the two Cecropia hormones are shown on the Scheme 16:

Scheme 16

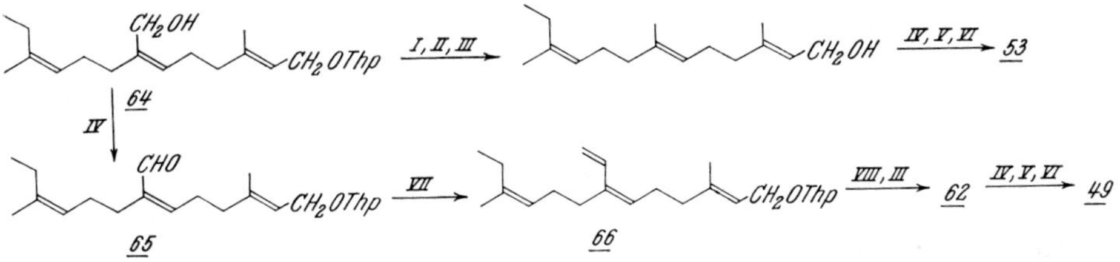

I, $Py.SO_3$, Thf, 0 °C
II, $LiAlH_4$, 0 °C
III, $H^{(+)}$
IV, MnO_2
V, MnO_2, $CN^{(-)}$, MeOH
VI, m-Chloroperbenzoic acid
VII, $Ph_3P{=}CH_2$
VIII, $H_2O_2-N_2H_4$

The position isomer *67* of the hormone *53* was prepared by a similar route from the hydroxy ether *6* [64].

The application of the so called olefinic ketal Claisen reaction for preparing trans-disubstituted as well as trans-trisubstituted olefinic bonds [92, 151] proved also very useful in the synthesis of both racemic Cecropia juvenile hormones. This Claisen reaction involved a thermal rearrangement of the unstable vinyl ether *68* which was produced in situ.

In the synthesis of the racemic C_{18} Cecropia juvenile hormone [148], the starting hydroxy ester *69* was prepared by photooxidation [2] of methyl senecionate. The Claisen rearrangement occured at the stage

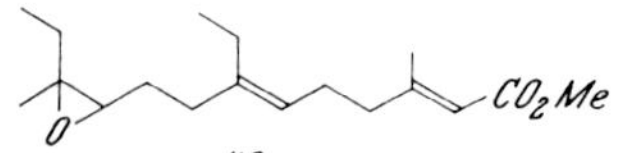

49

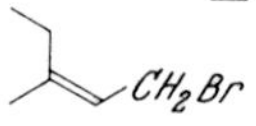

51

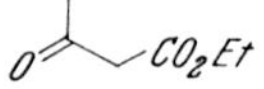

52

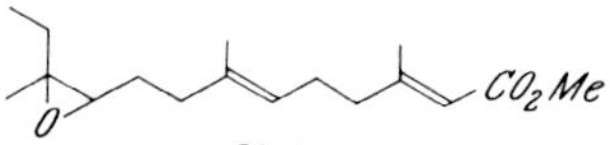

53

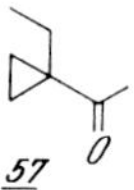

57

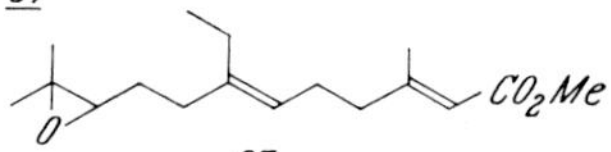

67

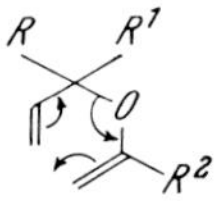

68

R, R¹ = H, alkyl, alkoxyl, alkoxycarbonyl etc.
R² = H, alkyl, alkoxyl, alkenyl etc.

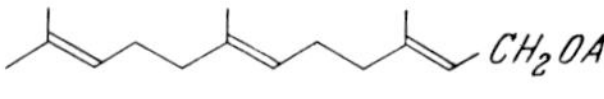

79

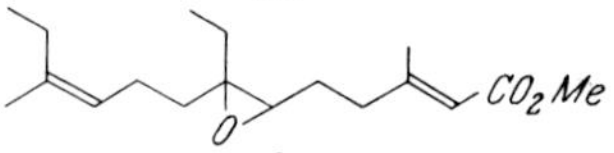

91

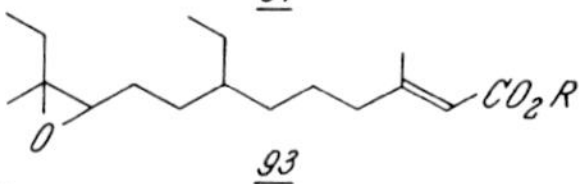

93

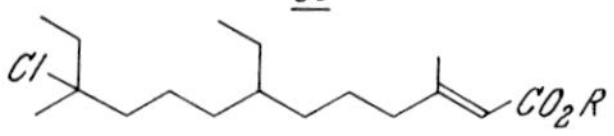

94

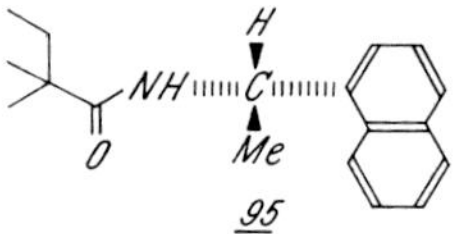

95

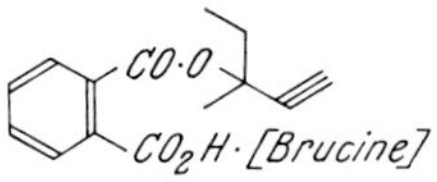

97

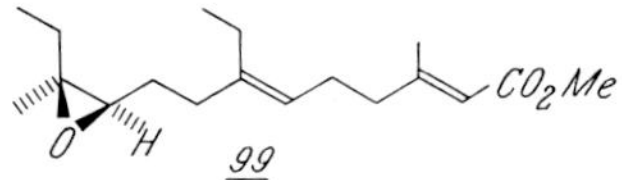

99

when compound *69* (and further *73*) was heated with the ketal *70* at 100 °C under catalysis with organic acids. The S_Ni reaction of the hydroxy ester *71* gave an 85% yield of the chloro ester *72*. The essential steps of the synthesis are shown in the Scheme 17:

Scheme 17

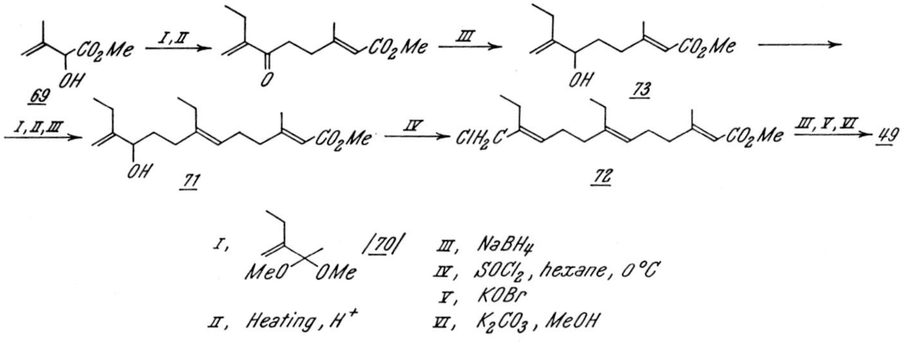

Loew and Siddall [182, 183] shortened considerably the route of Johnson [148] by treating the hydroxy ester *73* directly with the chloro ketal *74*. The resulting chloro keto ester *75* was converted to the final racemic C_{18}-JH (*49*) in two simple steps: (Scheme 18).

Scheme 18

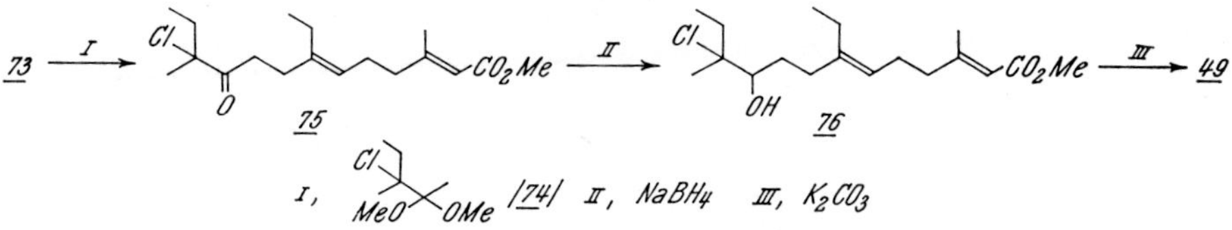

Noteworthy also are the non-stereospecific syntheses leading to mixtures of the racemic Cecropia juvenile hormones with their geometrical isomers.

One of the simple syntheses [291] was based on a series of Wittig reactions using phosphoranes also described by Cavill [48] (see p. 146). (Scheme 19):

Scheme 19

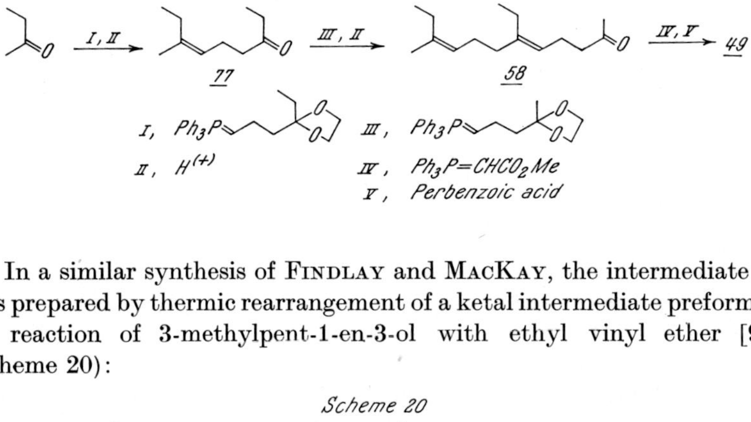

In a similar synthesis of Findlay and MacKay, the intermediate *77* was prepared by thermic rearrangement of a ketal intermediate preformed by reaction of 3-methylpent-1-en-3-ol with ethyl vinyl ether [95] (Scheme 20):

Scheme 20

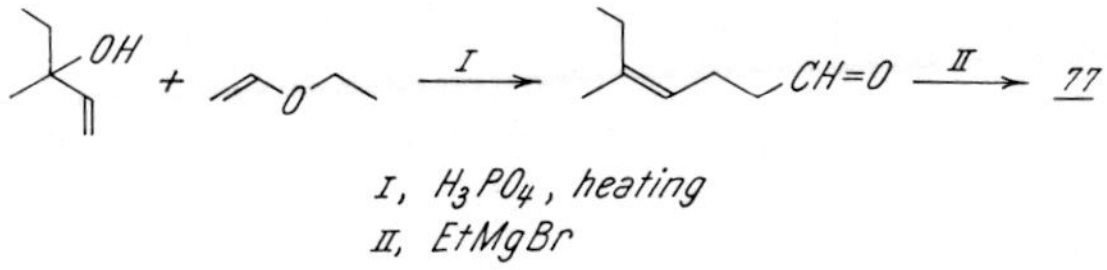

Mori *et al.* [219] made use of the Carroll reaction. This method consisted in a rearrangement of acetoacetates of allylic alcohols [45] (Scheme 21, see p. 154).

The Carroll reaction was also used in another synthesis of racemic isomers of Cecropia juvenile hormones [126]. The acetoacetates were replaced by the corresponding acylmalonates, e.g. ethyl propionyl-

Scheme 21

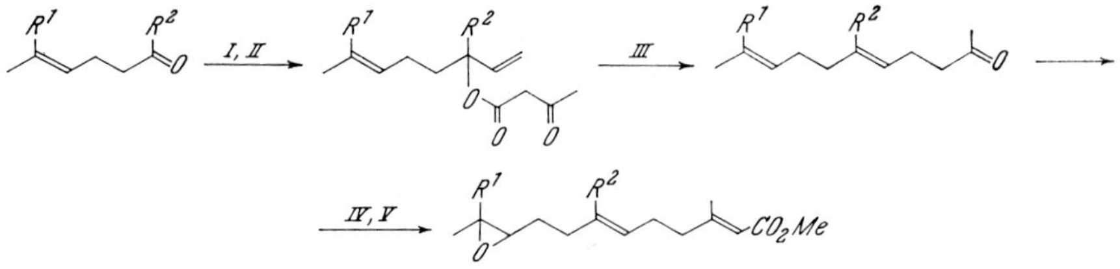

(R^1, R^2 = Me, or Et)

I, Vinylmagnesium bromide
II, Diketene
III, KOAc, heating
IV, $Ph_3P{=}CHCO_2Me$
V, Perbenzoic acid

malonate (*78*), since the reaction of acylmalonates with allylic alcohols is faster (Scheme 22):

Scheme 22

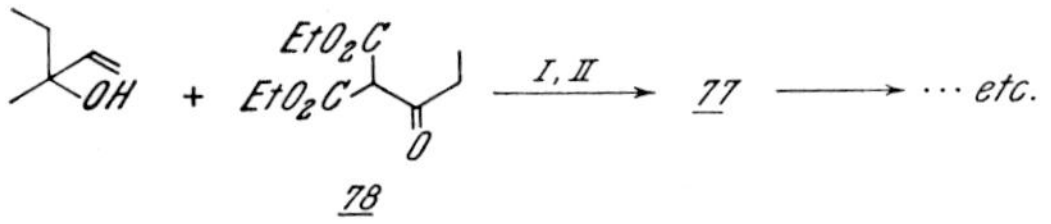

78

I, Heating II, Decarboxylation

The multistep synthesis of the racemic Cecropia C_{18}-JH devised by CAVILL *et al.* [47] was similar to the synthesis of methyl juvenate of the same authors [48] (see p. 146).

Finally, two syntheses were presented which consisted of a bis-homologization of the performed farnesane skeleton.

One of these syntheses [337] started from trans,trans-farnesyl acetate (*79*) which was converted into the 6,7:10,11-diepoxide *80*. By the action of a base, compound *80* was transformed into the trienetriol *81* which was then alkylated and oxidized to afford a mixture of the racemic Cecropia C_{18}-juvenile hormone and its three geometrical isomers in the ratio t,t,c:t,c,t:t,t,t:t,t,c = 1:2:1:1. In the course of the alkylation, the primary hydroxylic function was protected by a trityl group (Scheme 23):

Scheme 23

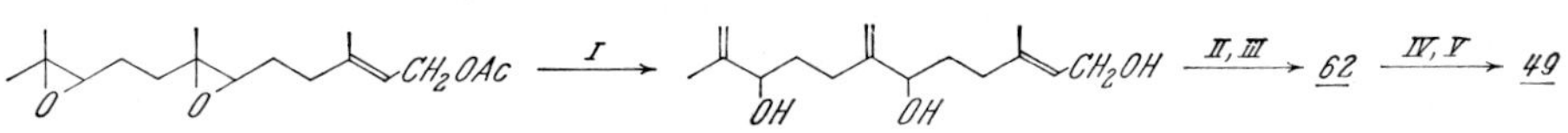

80 81

I, $LiNEt_2$
II, TsCl–LiCl
III $CuLiMe_2$
IV, MnO_2
V, MnO_2, $CN^{(-)}$, MeOH

The other route [2] consisted in alkylation of a similar intermediate *82* but started from ethyl trans,trans-farnesoate (*22*; R = Et). The "one flask conversion" of this ester to the diacetate *82* was comprised of photooxidation, reduction of the corresponding hydroperoxides and acetylation. Methylation of the diacetate *82* with ethereal lithium dimethylcuprate gave a mixture of ethyl dihomofarnesoate (*83*) isomers in the ratio t,t,t:t,c,c:t,c,t = 14:8:76. Methylation of *82* in tetrahydrofuran afforded an appreciable amount of the trans,trans,cis-isomer of *83* [2] (Scheme 24):

Scheme 24

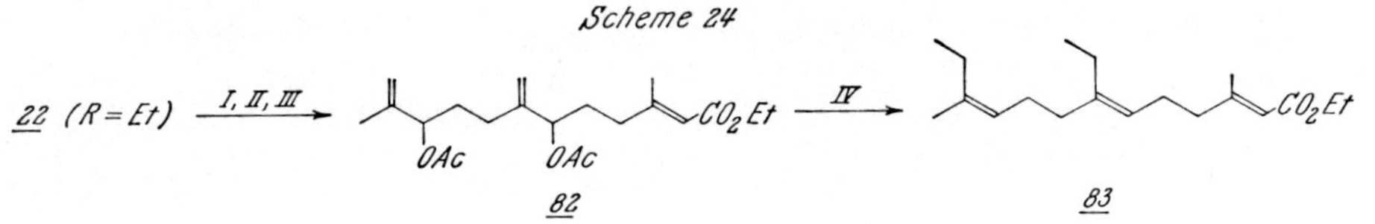

I, $O_2 - h\nu$ III, Ac_2O
II, $(MeO)_3P$ IV, $CuLiMe_2$

Some juvenoids possessing the Cecropia hormone skeleton will be dealt with in the following paragraphs.

Corey [57] reported the preparation of the epoxide *84* via the tetraenoate *85*. The latter compound was obtained from the aldehyde *86* with the use of the reagent of Pattenden and Weedon [238] (Scheme 25): This method is analogous to that which has been used in the synthesis of the Cecropia C_{18}-juvenile hormone [60], cf. p. 150.

Scheme 25

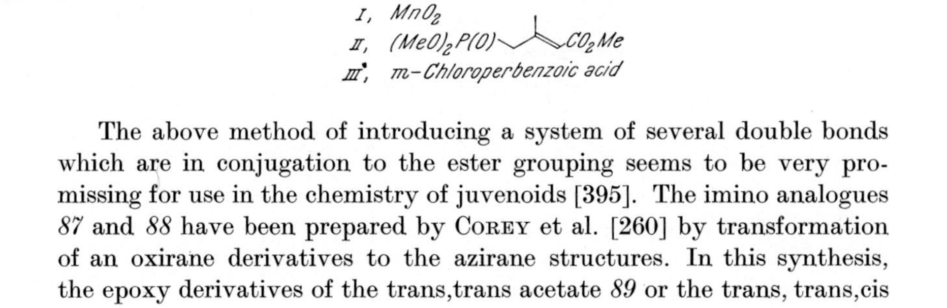

The above method of introducing a system of several double bonds which are in conjugation to the ester grouping seems to be very promissing for use in the chemistry of juvenoids [395]. The imino analogues *87* and *88* have been prepared by Corey et al. [260] by transformation of an oxirane derivatives to the azirane structures. In this synthesis, the epoxy derivatives of the trans,trans acetate *89* or the trans, trans,cis acetate *90* were treated with lithium azide to form a hydroxy azide which was converted to the corresponding tosyloxy azide and this compound was reductively cyclized. The —CH_2OH group of the resulting azirane is transformed to the —CO_2Me group of compounds *87* and *88* by oxidative esterification using the already mentioned [59, 337] manganese dioxide-cyanide-methanol procedure (Scheme 26):

Scheme 26

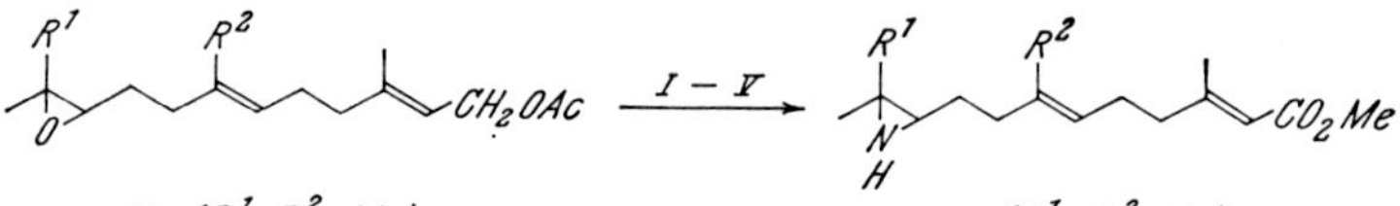

89 ($R^1=R^2=Me$)
90 ($R^1=R^2=Et$)

87 ($R^1=R^2=Me$)
88 ($R^1=R^2=Et$)

I, LiN_3
II, TsCl-Pyridine
III, $LiAlH_4$
IV, MnO_2
V, MnO_2, $CN^{(-)}$, MeOH

The imino analogues *87* and *88* were devoid of juvenile activity but prevented the metabolic destruction of juvenile hormones and thereby potentiated their activity [260].

The position isomers (6,7-epoxy) of the (10,11-epoxy) Cecropia C_{18}-juvenile hormone and their geometrical isomers have been also prepared and tested [242, 358], namely, the geometrical isomers of methyl 6,7-epoxy-3,7,11-trimethyl-2,10-tridecadienoate (*91*).

The esters of 3,11-dimethyl-7-ethyl-2,10-tridecadienoic acid (*92*) were prepared by the classical sequence of reactions [136] shown in the Scheme 27:

Scheme 27

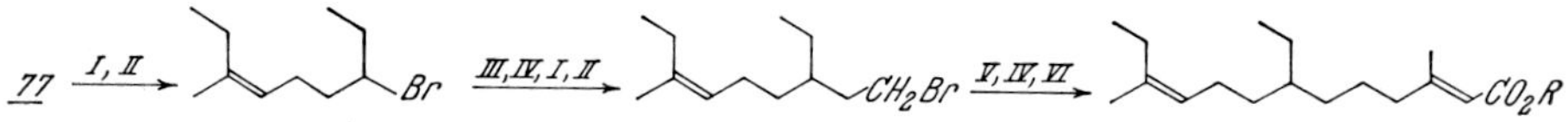

92 (R=Me, or Et)

I, $LiAlH_4$
II, PBr_3-Pyridine
III, Diethyl malonate
IV, Decarboxylation
V, Ethyl acetoacetate
VI, $Ph_3P{=}CHCO_2R$

The 10,11-epoxy (*93*) and 11-chloro (*94*) derivatives of esters *92* have been also prepared [139, 140].

The pure 2-trans-6-cis and 2-cis-6-cis isomers of the methyl and ethyl esters *92* were obtained [136, 138] by chromatographic separations at the appropriate stages of the above reaction sequence.

f) Absolute Configuration of the Cecropia Juvenile Hormones

The asymmetric centers of the two Cecropia juvenile hormones occur at positions C-10 and C-11 of the carbon chain, i.e., at those carbon atoms to which the epoxide group is attached. The spatial orientation of this portion of the molecule remained obscure. The small amount of the available material during early investigations did not allow even the measurement of the optical rotatory power. In 1970, MEYER and HANZMANN [207] observed that a 9:1 mixture of the natural C_{18} and C_{17} Cecropia hormones exhibited a plain positive ORD curve and the optical rotation $[\alpha] \sim +7°$. On the basis of this finding, the synthesis of both enantiomeric forms of the Cecropia hormones gained importance.

LOEW and JOHNSON [181] made use of the reaction sequence previously [182, 183] described for the synthesis of the racemic Cecropia C_{18}-JH (cf. p. 153), except for the separate application of the two enantiomeric forms of the starting chloro ketal *74*. These enantiomers were prepared by a series of especially refined reactions from the corresponding enantiomers of α-chloro-α-methylbutyric acid obtained from the racemic acid by conversion into the acid chloride, treatment of this chloride with (—)-α-(1-naphthyl)ethylamine, separation of the thus-obtained diastereoisomeric amides *95* and hydrolysis.

In the hormone synthesis, each of the enantiomeric chloroketals yielded a pair of diastereoisomeric chlorohydrins which were separated by chromatography. Isomers *76 a* and *76 b* were enantiomeric as were *76 A* and *76 B*. Each of these four chlorohydrins *76* was treated with a base to afford the epoxy compounds *49*. Thus, *76 b* gave the enantiomer *49 b* (dextrorotatory, cis-epoxide) and *76 a* gave the levorotatory enantiomer *49 a*. Similarly, *76 A* gave the levorotatory form of the trans-epoxide *49 A* and *76 B* afforded the dextrorotatory enantiomer *49 B* (Scheme 28):

Scheme 28

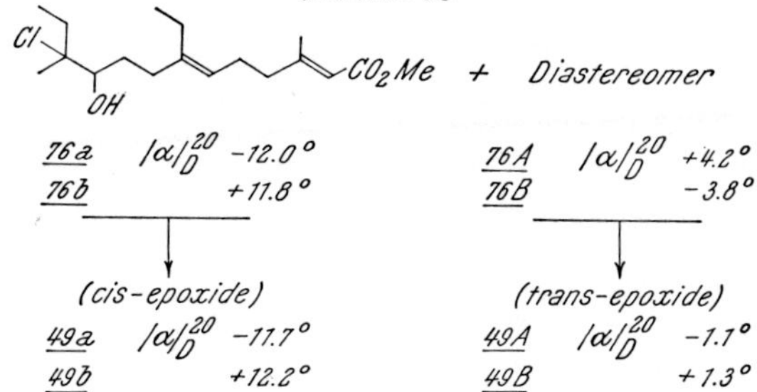

On the basis of the optical rotatory values of the t,t,c-isomers *49 a* and *49 b*, the latter enantiomer *49 b* should represent the Cecropia C_{18}-hormone, as also suggested by biological tests [181] on *Galleria* and

Tenebrio. The low, but unexpected biological activity of the enantiomer *49 a* may have been due to contamination with about 10% of *49 b*.

In spite of the successful synthesis of the optically active forms of the Cecropia C_{18}-JH, it was not possible to determine its absolute configuration because of the unknown absolute configuration of the starting epimeric chloro ketals *74*.

The second synthesis of both enantiomeric forms of the Cecropia C_{18}-juvenile hormone from starting materials of known absolute configuration was reported by FAULKNER and PETERSEN [93]. The route was again analogous to that described by LOEW and SIDDALL [182] for the synthesis of the racemic product. In the present synthesis, the above chloro ketal component *74* of the Claisen reaction was replaced by the hydroxy ketal *96*. The two enantiomers of *96* were obtained via resolution of the brucine salt of 3-methylpent-1-yn-3-yl phthalate half-ester (*97*), hydrolysis and addition of methanol.

As shown by correlation with (—)-2-hydroxy-2-methyl-butyric acid of known absolute configuration, the (—)-hydroxy ketal *96 a* possesses the R configuration and the (+)-hydroxy ketal *96 b* is of the S configuration.

The diastereoisomeric pairs of diols *98 a* + *98 A* (from *96 a*) and *98 b* + *98 B* (from *96 b*) were separated by chromatography. The threo-diols *98 a* and *98 b* gave rise to the required trans,trans,cis-compounds *49 a* and *49 b* while the erythro-diols *98 A* and *98 B* gave the trans,trans-, trans,isomers *49 A* and *49 B*, as it may be seen on the Scheme 29:

Scheme 29

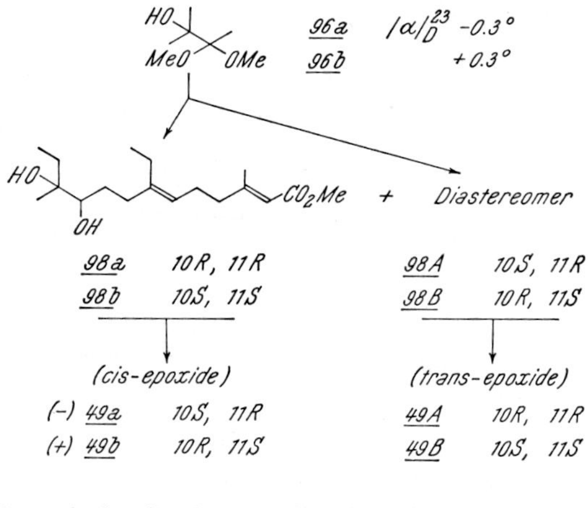

The purity of the final epoxides *49* was about 75–90%. Notwithstanding, the knowledge of the absolute configuration at one center of the reactant *96* made it possible to determine the absolute configuration of both oxirane carbon atoms in the final product. Thus, the natural (+)-Cecropia C_{18}-JH was assigned the 10R,11S configuration (*99*).

The absolute configuration of the Cecropia hormones has been also studied by NAKANISHI *et al.* [223] with the use of the synthetic samples of (+)-C_{18} hormone and its (—)-enantiomer, both prepared by LOEW and JOHNSON [181] (vide supra). The results of NAKANISHI, based on clarification of the mode of epoxide cleavage by NMR spectral analysis and a determination of the absolute configuration of the resultant α-glycol by newly developed circular dichroism method are in full accordance with those of FAULKNER and PETERSEN [93].

g) *Biogenesis of Cecropia Hormones*

ROELLER and DAHM have recently demonstrated [266] that the Cecropia C_{18}-JH is indeed produced *in vitro* by the corpora allata.

The biogenesis of the two Cecropia hormones remains obscure. These hormones differ from farnesol (*3*) and related sesquiterpenoid derivatives by the presence of ethyl groups instead of methyls at positions 7 and 11 (C_{18}-hormone) or at position 11 (C_{17}-hormone) of the carbon chain (cf. Table 8, p. 139).

SCHMIALEK [279] has demonstrated in the case of *Samia cynthia* silk moth that labelled farnesol and farnesal are formed from (2-^{14}C)-mevalonate. Incorporation of both mevalonate and methionine into the carbon chain of Cecropia hormones has not been demonstrated. It seems more likely that the biogenesis of Cecropia hormones in insects involves a simple homologation of farnesol or of a farnesoic acid derivative utilizing adenosyl methionine [340] as outlined on the Scheme 30:

Scheme 30

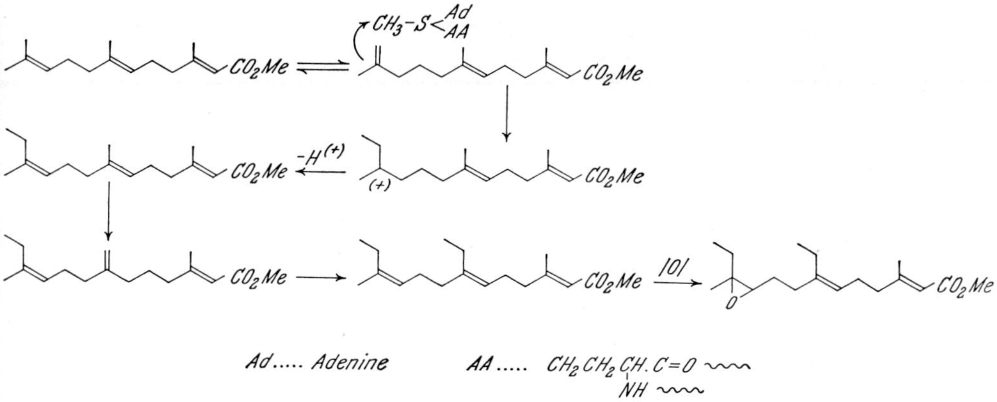

Corroboration of this hypothesis might be seen in the simultaneous occurrence of the C_{17}-hormone and its C_{18}-homologue in silk moths [68, 205, 208].

The recent work of Roeller *et al.* [203, 204] suggests that methionine is involved only in the formation of the ester methyl grouping in the Cecropia C_{18}-hormone but not in the genesis of the carbon chain. Also, the mevalonate does not appear to take part in the genesis of the carbon chain in adult moths. It should be born in mind, however, that in view of the present state of knowledge on the biogenesis of the Cecropia juvenile hormones, the above conclusions are only preliminary.

Whether all insects produce the same hormones as *Hyalophora cecropia* also remains obscure. The isolation of juvenile hormones from another insect species has not yet been achieved. However, one exception is represented by isolation and identification [68] of the C_{17}- and C_{18}-juvenile hormones from the closely related giant silk moth *Hyalophora gloveri*.

h) Other Acyclic Juvenoids

This group comprises all the remaining compounds the carbon skeleton of which more or less differs from the skeleton of the juvenoids described in the previous parts of this chapter. Some of these analogues are closely related to the farnesane type, white others are rather different. The aim of the following syntheses was to prepare juvenile-active compounds which would be more stable and more accessible for practical use in the control of insects than the natural juvenile hormones.

Thus, for example, the compounds carrying a terminal quarternary alkyl group of Romaňuk *et al.* [273] represented more stable analogues of the 11-chloro derivatives of alkyl dihydrofarnesoates (*34*). The compounds were prepared by the Kolbe anodic synthesis from three components. Each of these components contains one of the tertiary carbon atoms or a quarternary carbon atom. This route made it possible to prepare analogues of various chain-length and use them in structure-activity studies. The synthetic route may be outlined as follows (Scheme 31):

Scheme 31

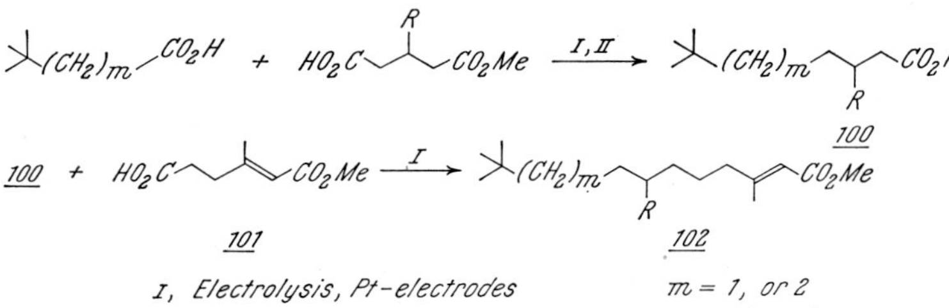

I, Electrolysis, Pt-electrodes
II, KOH

m = 1, or 2
R = Me, or Et

4-Methyl-5-methoxycarbonyl-4-pentenoic acid (*101*) which serves for introducing the important —$C(CH_3){=}CH{\cdot}CO_2Me$ group into the molecule of the product, was formed from levulinic acid according to LINSTEAD *et al.* [180] (Scheme 32):

Scheme 32

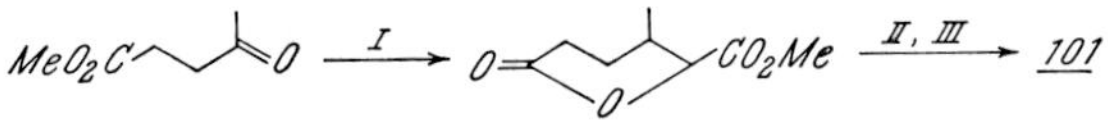

I, $BrCH_2CO_2Me$, Zn, then H_2O, $H^{(+)}$
II, MeONa
III, $H^{(+)}$

In this manner, the following four methyl esters were prepared [273]: methyl 3,7,11,11-tetramethyl-2-dodecenoate (*102*; R = Me, m = 2), methyl 3,11,11-trimethyl-7-ethyl-2-dodecenoate (*102*; R = Et, m = 2), methyl 3,7,10,10-tetramethyl-2-undecenoate (*102*; R = Me, m = 1), and methyl 3,10,10-trimethyl-7-ethyl-2-undecenoate (*102*; R = Et, m = 1). There was no considerable difference in the juvenile hormone activity of these four esters [314]. The esters were considerably active on the cricket *Gryllus bimaculatus*. The activity of the analogously prepared methyl 3,6,9,9-tetramethyl-2-decenoate (*103*) [273] was far lower [314] (Scheme 33):

Scheme 33

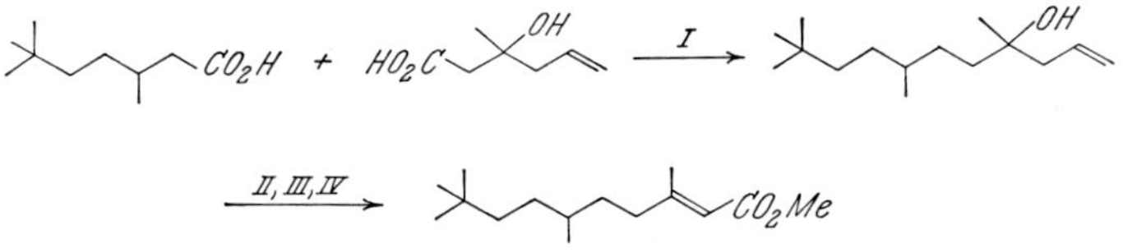

103

I, Electrolysis, Pt-electrodes
II, O_3
III, HCO_2H
IV, CH_2N_2

The preparation [269] of the diesters *104* (R = Me and Et) was based on the Kolbe anodic synthesis with the use of tert-butyl hydrogen β-alkylglutarate and levulinic acid, via the ketoester intermediate *105* (Scheme 34, see p. 162).

tert-Butoxycarbonyl-D,L-β-aminobutyric acid was used as the starting material in the synthesis [269] of the nitrogen analogue *106* of the diester *104* (Scheme 35, see p. 162).

The Wadsworth-Emmons reaction [343] of the ketoester *105* and the corresponding phosphonoacetamides *107 a* and *107 b* afforded the N-ethyl- and N,N-diethylamides *108 a* and *108 b* [269]:

Scheme 34

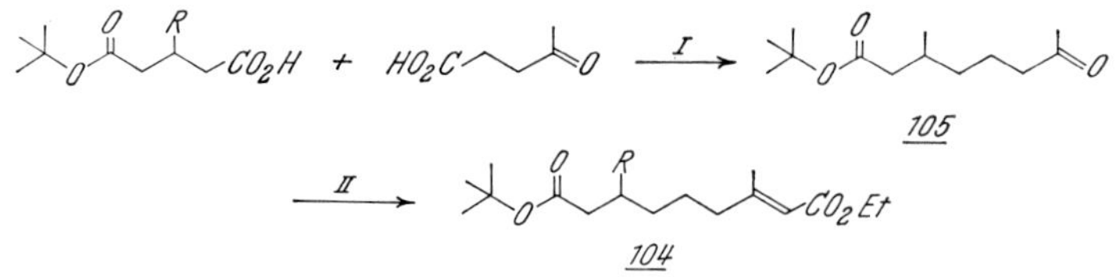

105

104

I, Electrolysis, Pt-electrodes
II, $(EtO)_2P(O)CH_2CO_2Et$

R = Me, or Et

Scheme 35

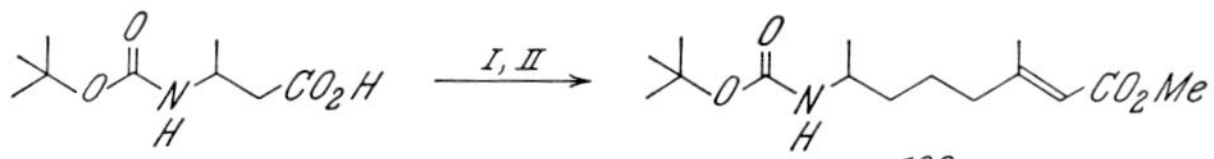

106

I, Laevulinic acid, electrolysis, Pt-electrodes
II, $(EtO)_2P(O)CH_2CO_2Me$

Most of the above mentioned juvenoids carrying the tert-butoxycarbonyl group in their molecule exhibited a high juvenile hormone activity when tested on the bugs of the family Pyrrhocoridae.

Mori *et al.* [213] have prepared a series of aliphatic juvenoids. The non-stereospecific reaction sequence of Mori was very suitable for the preparation of farnesoic acid homologues bearing alkyl or hydrogen at positions 3, 7, and 11 of the carbon chain (cf. Table 9) as well as for the mixture of isomers of the Cecropia C_{18}-JH (*49*). Since most of the compounds prepared by the general route belong to this part of the chapter, the corresponding Scheme is given here. The starting allylic bromide *109* was prepared by a Grignard reaction of the corresponding ketone $R \cdot CO \cdot R^1$ and vinyl bromide followed by treatment of the resulting alkyl vinyl carbinol with hydrobromic acid. The bromide *110* was obtained with the use of the Julia reaction [156] (Scheme 36, see p. 164).

Compounds *111–118* prepared by this route and the corresponding 10,11-epoxy derivatives *119–126* are listed in Table 9). For the activities on *Tenebrio*, and *Tribolium*, see Appendix, tables.

Noteworthy also are the ethers of Jarolím et al. [*322*] These ethers have been prepared from simple and commercially available intermediates such as geraniol (*16*), citronellol (*19*), 6-methyl-5-hepten-2-one (*144*) and the like. The reaction route comprised, inter alia, the reaction of alcohols with diazo ketones [233] and addition of alcohols to methyl vinyl ketone [162]. The most important compounds and the correspond-

Table 9. *Juvenoids Prepared by Mori et al.*[213]

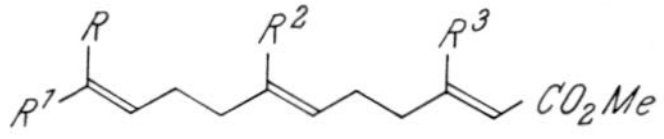	Substituents: R	R^1	R^2	R^3	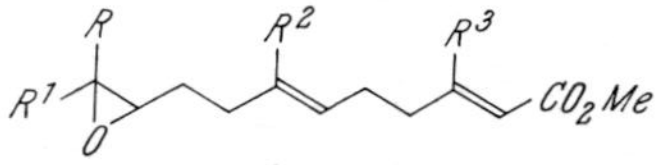
Substance					Substance
111	H	H	Et	Me	*119*
112	H	Me	Et	Me	*120*
113	Et	H	Et	Me	*121*
114	Me	Me	Et	Me	*122*
115	Et	Et	Et	Me	*123*
116	Pr	Me	Et	Me	*124*
117	Me	Me	Et	H	*125*
118	Et	Me	Et	H	*126*

Scheme 36

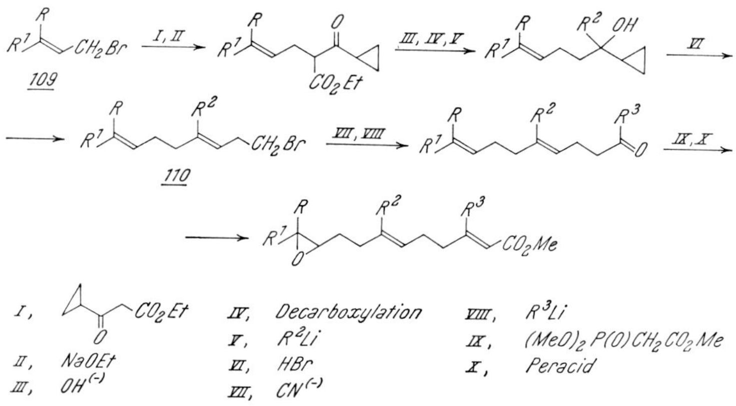

ing reaction Schemes are listed in Table 10 which also contains the geraniol derivatives mentioned on p. 142. 2-Methyl-2-butenol (*145*) has been prepared from methyl ethyl ketone as follows (Scheme 37):

Scheme 37

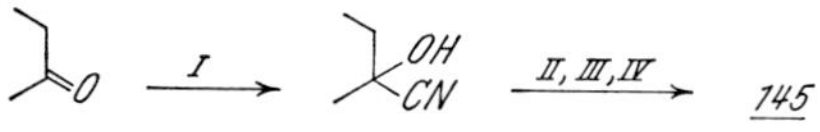

I, HCN III, Dehydration, then CH_2N_2
II, $OH^{(-)}$ IV, $LiAlH_4$

The preparation of 2,6-dimethyl-5-heptenol (*146*) can be illustrated by the following Scheme (Scheme 38):

Scheme 38

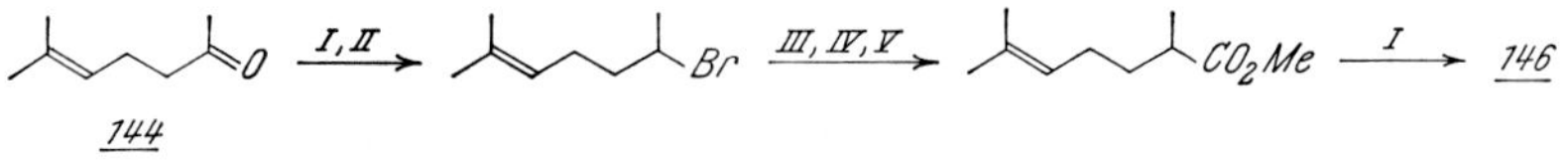

I, $LiAlH_4$ IV, CO_2
II, PBr_3-Pyridine V, CH_2N_2
III, Mg

Numerous juvenoids which are listed in Table 10, especially the 5-oxa, 6-oxa, 5,10-dioxa, 6,9-dioxa, and 6,10-dioxa esters were converted by simple reactions to other juvenile-hormone-active derivatives (see Appendix tables).

The juvenile hormone activity of the above oxa esters was very different and appeared to depend on the chain-length and location of

Table 10. *Oxa-Esters with JH Activity*

Substance:	References	Reaction scheme:	Formula:
2,10-Dodecadienoic acid, 5-oxa-3,7,11-trimethyl-, ethyl ester (**127**)	142 322	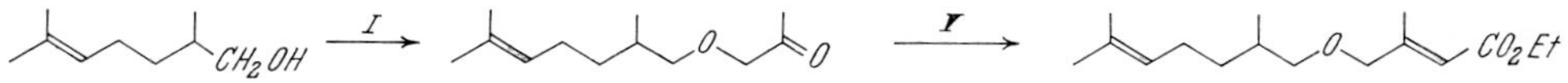 146	127
2-Tridecenoic acid, 5-oxa-3,8,12-trimethyl-, ethyl ester (**128**)	141	 79	128
2,11-Tridecadienoic acid, 5-oxa-3,8,12-trimethyl-, ethyl ester (**129**)	141	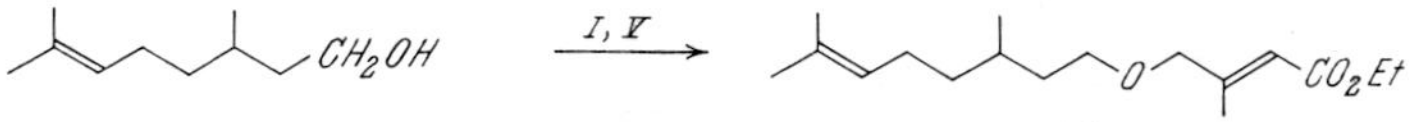	129
2,7,11-Tridecatrienoic acid, 5-oxa-3,8,12-trimethyl-, ethyl ester (**130**)	141 322	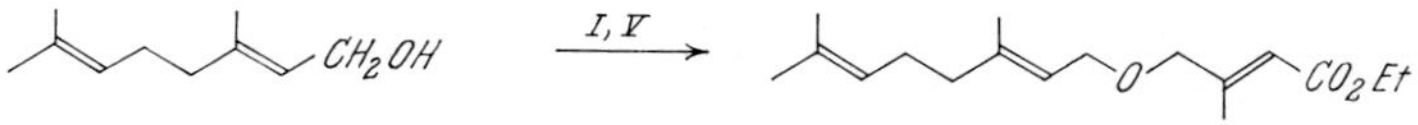 16	130
2,10-Dodecadienoic acid, 6-oxa-3,7,11-trimethyl-, ethyl ester (**131**)	144 145	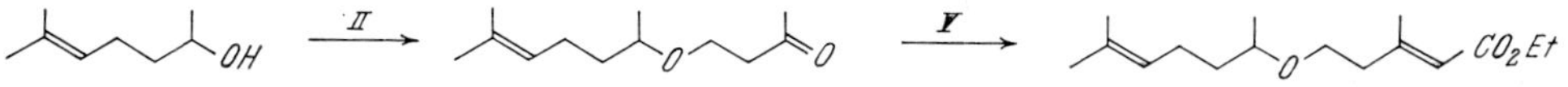	131

(Table 10 — continued)

Substance:	References	Reaction scheme:	Formula:
2-Dodecenoic acid, 9-oxa-3,7,11-trimethyl-, ethyl ester (**132**)	—	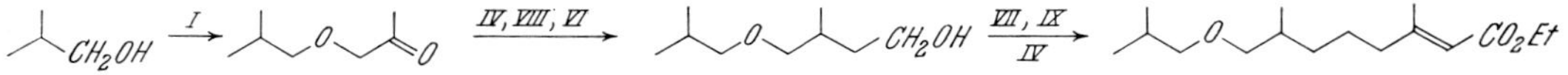	132
2-Dodecenoic acid, 10-oxa-3,7,11,11-tetramethyl-, ethyl ester (**133**)	—	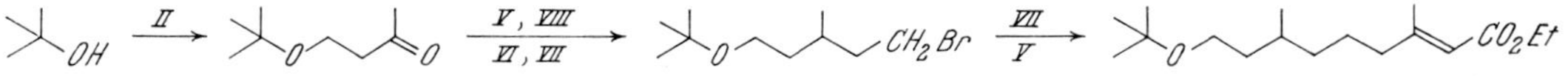	133
2,6-Dodecadienoic acid, 10-oxa-3,7,11,11-tetramethyl-, ethyl ester (**134**)	—	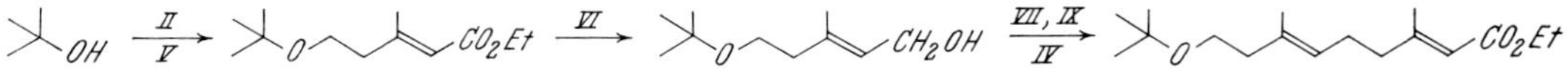	134
2-Tridecenoic acid, 5,10-dioxa-3,8,12-trimethyl-, ethyl ester (**135**)	143	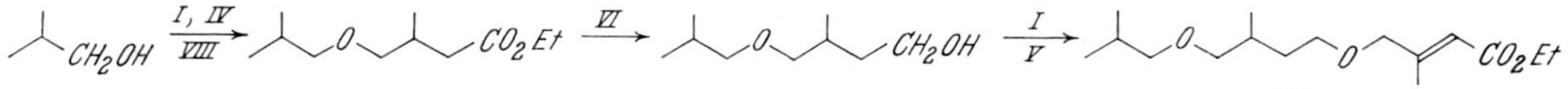	135
2,7-Tridecadienoic acid, 5,10-dioxa-3,8,12,12-tetramethyl-, ethyl ester (**136**)	143	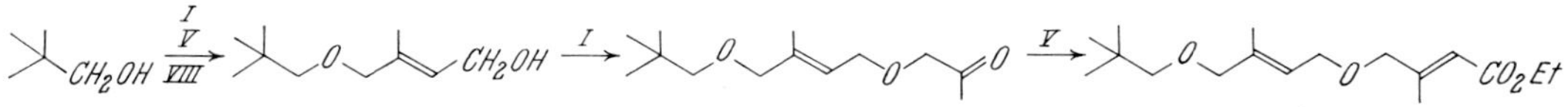	136

(Table 10 — continued)

Substance:	References	Reaction scheme: Formula:
2-Tridecenoic acid, 5,10-dioxa-3,8,12,12-tetramethyl-, ethyl ester (**137**)	143	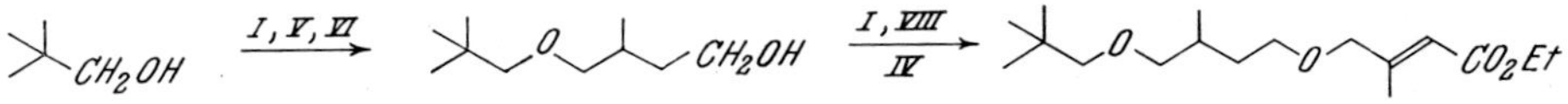 137
2,7,12-Tetradecatrienoic acid, 5,10-dioxa-8-ethyl-3,12-dimethyl-, ethyl ester (**138**)	143	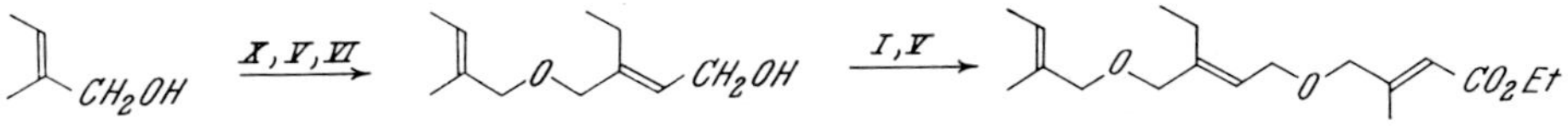 145 138
2,7-Dodecadienoic acid, 5,10-dioxa-3,8,11,11-tetramethyl-, ethyl ester (**139**)	143	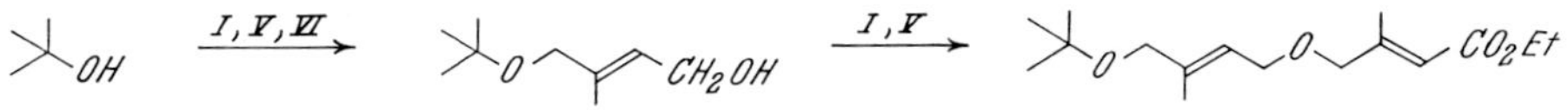 139
2-Dodecenoic acid, 6,9-dioxa-3,7,11-trimethyl-, ethyl ester (**140**)	146	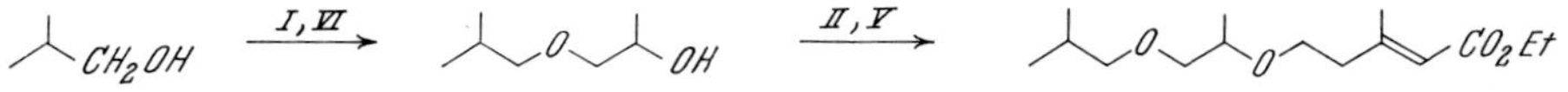 140

(Table 10 — continued)

Substance:	Referen-ces	Reaction scheme: Formula:
2,11-Tridecadienoic acid, 6,9-dioxa-7-ethyl-3,11-dimethyl-, ethyl ester (**141**)	146	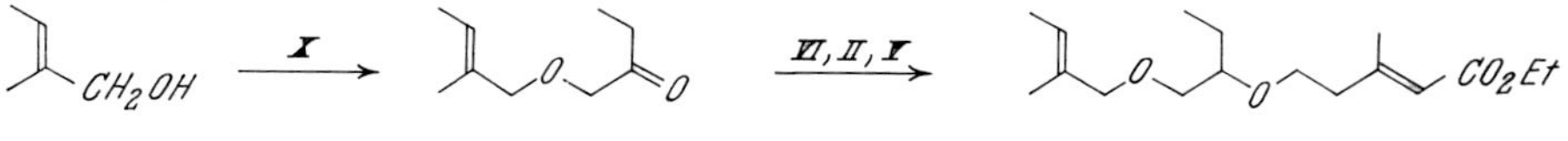 141
2-Dodecenoic acid, 6,10-dioxa-3,7,11-trimethyl-, ethyl ester (**142**)	147	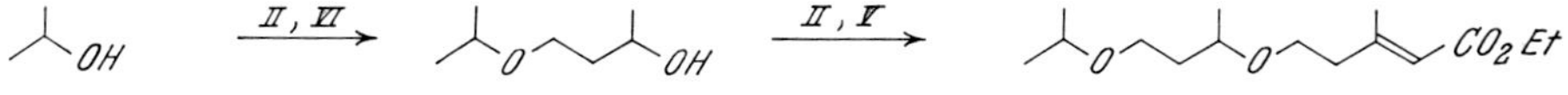 142
2-Dodecenoic acid, 6,10-dioxa-3,7,11,11-tetramethyl-, ethyl ester (**143**)	147	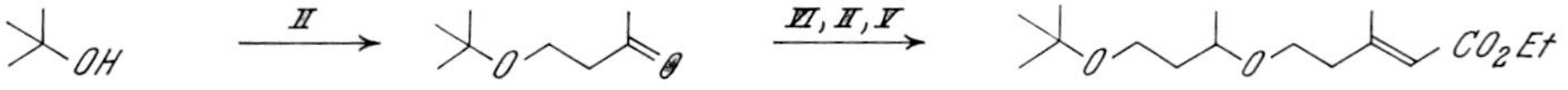 143

Reagents:

i, MeCO · CHN_2, $Et_2O \cdot BF_3$
ii, Methyl vinyl ketone, $Et_2O \cdot BF_3$
iii, $Hg(OAc)_2$, then $NaBH_4$
iv, $(EtO)_2P(O)CH_2CO_2Et$
v, $Ph_3P{=}CHCO_2Et$
vi, $LiAlH_4$
vii, PBr_3—Pyridine
viii, Hydrogenation
ix, Acetoacetic ester, then decarboxylation
x, $EtCO \cdot CHN_2$, $Et_2O \cdot BF_3$

the hetero atom. Some 5-oxa esters showed a remarkably high activity on bugs [308, 322] and some derivatives of 6-oxa esters were highly active on *Tenebrio* and *Dermestes* (cf. Appendix tables).

Replacement of the ester grouping —$C(CH_3){=}CH{\cdot}CO_2R$ by the ketone grouping —$C(CH_3){=}CH{\cdot}CO{\cdot}R$ (R is a lower alkyl) led to substances possessing remarkable juvenile hormone activity. The ketones are usually prepared by reaction of alkyllithium with the corresponding acid, cf. the methyl ketones derived from farnesoic acid and its homologues [292, 385] (*147, 148, 149*): (see T-162, 163, p. 422).

All these compounds exhibit a high activity on *Tenebrio molitor* [292], especially the ketone *149* (T-163).

Esters (*150*) of 2,7,11-trimethyl-3-methylene-4,6,10-dodecatrienoic acid were prepared [112, 113] by reaction of pseudoionone (*29*) with an alkyl α-bromopropionate and the subsequent dehydration of the resulting hydroxy esters to give a mixture of isomers which was separated by chromatography: (Scheme 39)

Scheme 39

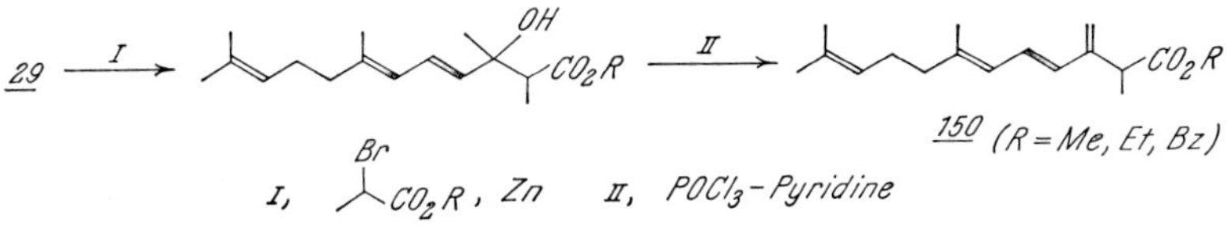

The methyl ester *150* (R = Me) was converted to the 11-chloro derivative *151* on treatment with gaseous hydrogen chloride in methanol [113].

Another type of juvenile hormone analogues [292, 293] is represented by the carbamates *152–160* derived mostly from geranylamine, citronellylamine and their epoxides. These carbamates were prepared from oximes of aldehydes or ketones by reduction with lithium aluminium hydride and the subsequent treatment of the resulting primary amines with alkyl chloroformates, as outlined on the Scheme 40:

Scheme 40

$$R.CH{=}O \xrightarrow{I} R.CH{=}N.OH \xrightarrow{II} R.CH_2NH_2 \xrightarrow{III} R.CH_2NH.CO_2R^7$$

I, NH_2OH II, $LiAlH_4$ III, $ClCO_2R^7$

The same group [293] has also reported without experimental details some related sulfonylamides *161–165*, amides *166–170* and imines *171–173*.

With a few exceptions, the juvenile hormone activities (based on *Tenebrio* assay) of the above nitrogen containing substances, were relatively low [292, 345, 385].

152 $R^1NH.CO_2Me$
153 $R^2NH.CO_2Me$
154 $R^1N(Me).CO_2Et$
155 $R^2NH.CO_2Ph$
156 $R^1NH.CO_2Ph$
157 $R^5NH.CO_2Et$
158 $R^4NH.CO_2Ph$
159 $R^5NH.CO_2Ph$
160 $R^6NH.CO_2CH_2CH_2Cl$

161 $R^2NH.SO_2Me$
162 $R^4NH.SO_2Me$
163 $R^5NH.SO_2Me$
164 $R^2NH.SO_2Ph$
165 $R^6NH.SO_2Me$

166 $R^2NH.COMe$
167 $R^2NH.COEt$
168 $R^2NH.COCMe_3$
169 $R^2NH.COPh$
170 $R^3NH.COMe$

171 $R^1N=CHPh$
172 $R^2N=CHPh$
173 $R^3N=CMe_2$

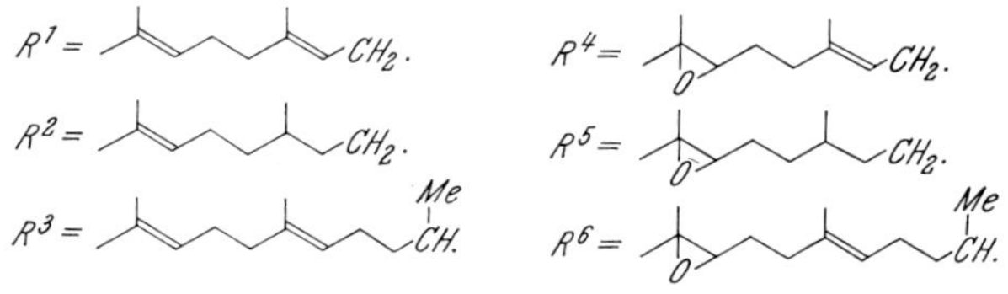

3. Cyclic Juvenoids

Juvabione and Dehydrojuvabione

While working with C. M. Williams in the USA. in 1965, K. Sláma observed that larvae of the hemipteran bug *Pyrrhocoris apterus* failed to metamorphose into adults, while those cultured in Prague developed normally [379]. Later it was shown that this effect was caused by a compound mimicking juvenile hormone and present in the paper with which the larvae came into contact. This compound was most easily detected in paper products manufactured from the Canadian balsam fir (*Abies balsamea*) and was called "paper factor". The active compound, (+)-juvabione (*174*), was independently isolated by two groups of investigators, Bowers *et al.* [29] in Maryland and Černy *et al.* [49] in Prague, and identified as the methylester of todomatuic acid, a previously known sesquiterpenic acid of the bisabolene series [224]. In addition to juvabione, the Czech authors isolated the biologically somewhat more active dehydrojuvabione (*175*) from a Slovak fir.

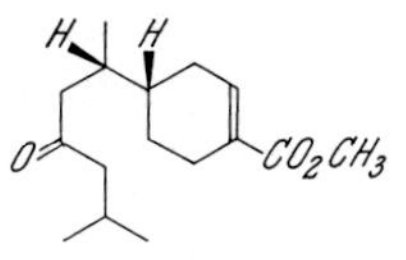

174

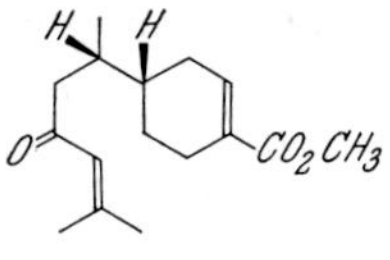

175

As shown by biological assays with pure compounds, both juvabione and dehydrojuvabione exhibit a highly specific juvenile hormone activity since they are active exclusively on hemipterans of the family *Pyrrhocoridae* and practically inactive on other accessible species of experimental insects. Consequently, the initial assumption that "paper factor" could constitute a protective system of plants against insects did not receive support.

The chemical structure of (+)-juvabione and its dehydroderivative was fully established on chemical evidence along with the physical properties of both compounds, particularly by correlation with the naturally occurring todomatuic acid. Later the structure was also confirmed by syntheses. The first synthesis of racemic (+)-juvabione was effected by Mori and Matsui [214, 215] with the use of p-methoxyacetophenone as the starting compound.

Scheme 41

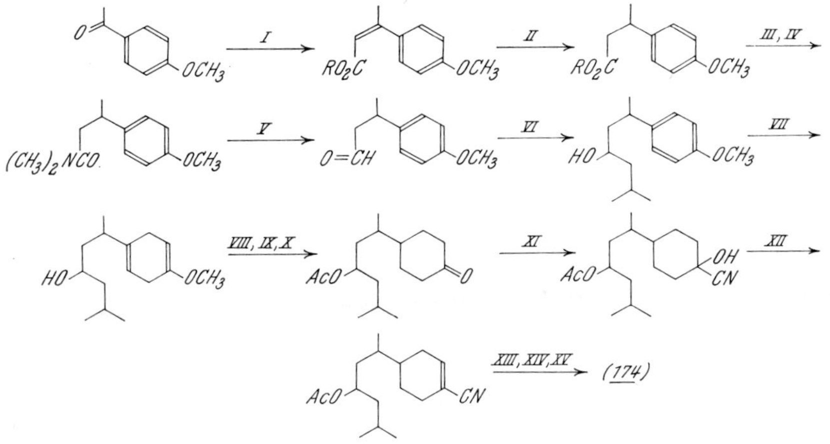

I,	$BrCH_2COOR$	VI,	$(CH_3)_2CHCH_2MgBr$	XI,	HCN
II,	$H_2(RaNi)$	VII,	$Li, (NH_3)$	XII,	$POCl_3$
III,	$SOCl_2$	VIII,	$H^{(+)}$	XIII,	KOH
IV,	$NH(CH_3)_2$	IX,	$H_2(Pd)$	XIV,	Cr_2O_3
V,	$(C_2H_5O)_3LiAlH$	X,	Ac_2O	XV,	CH_2N_2

The series of reactions shown in Scheme 41 led not only to racemic juvabione but also to the racemate of a compound epimeric with juvabione. The synthesis of racemic juvabione and dehydrojuvabione was also reported by AYYAR and RAO [6], and the first stereospecific synthesis was attempted by BIRCH *et al.* [19].

The absolute configuration of naturally occurring (+)-juvabione was originally proposed as R,R at both optically active centres of the molecule, analogous to the configuration of todomatuic acid as described by the Japanese authors [224]. However, PAWSON *et al.* [240] have shown, in connexion with the stereospecific synthesis of all four theoretically possible optical isomers of juvabione, that the original assignment is erroneous and that the naturally occurring (+)-juvabione has an absolute configuration R at position 4 of the menthane skeleton and an absolute configuration S at position 8 of the same system.

Scheme 42

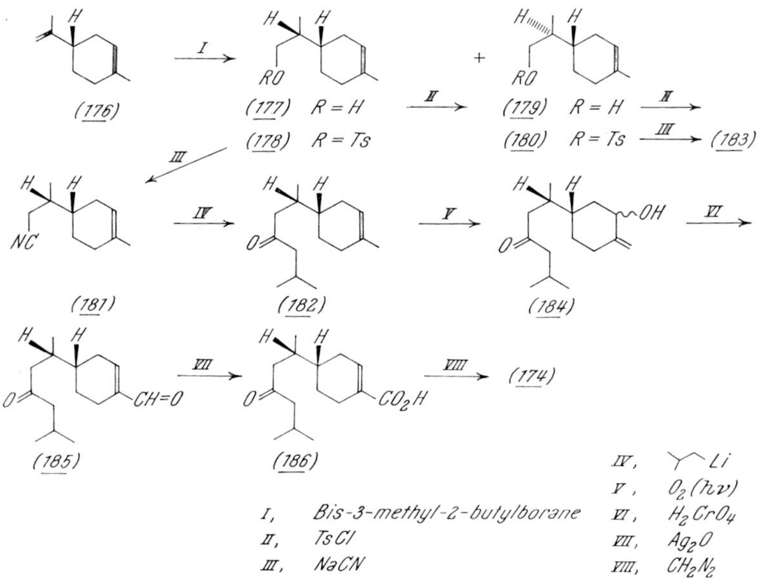

The syntheses of PAWSON *et al.* (Scheme 42) start from R-(+)-limonene on the one hand and S-(—)-limonene on the other; the absolute configuration of both limonenes had been reliably established earlier. Anti-Markovnikoff hydration of R-(+)-limonene (*176*) with bis-3-methyl-2-butylborane led to a mixture of diastereomeric alcohols which was separated by crystallisation of the corresponding 3,5-dinitrobenzoates into (4R, 8R)-p-menth-1-en-9-ol (*177*) and (4 R, 8 S)-p-menth-1-en-9-ol (*179*). The configuration of the former compound was determined by X-ray analysis of the p-iodobenzoate [21]. The p-toluenesulfonates (*178* and *180*) of these alcohols were converted by the action of an alkali metal cyanide to the appropriate nitrile (*181*) which, by the action of isobutyllithium, yielded the epimeric ketones (*182* and *183*). Photo-oxidation of the ketone (*182*) and subsequent reduction of the resulting hydroperoxides led to a mixture of alcohols containing the secondary alcohol (*184*). Chromic acid oxidation of the latter alcohol was accompanied by allylic rearrangement to give the aldehyde (*185*) which was converted by oxidation with silver oxide into todomatuic acid (*186*), identical in every respect with the naturally occurring material. Its esterification led to (+)-juvabione, which thus may be written as (4R)-(1-(S)-5-dimethyl-3-oxohexyl)-1-cyclohexene-1-carboxylic acid methyl ester (*174*).

(+)-Epijuvabione (*187*) was obtained analogously to (+)-juvabione from the ketone (*183*). Compound (*187*), methyl (4R)-(1-(R)-5-dimethyl-3-oxohexyl)-1-cyclohexene-1-carboxylate, possesses that absolute configuration which was erroneously ascribed to (+)-juvabione.

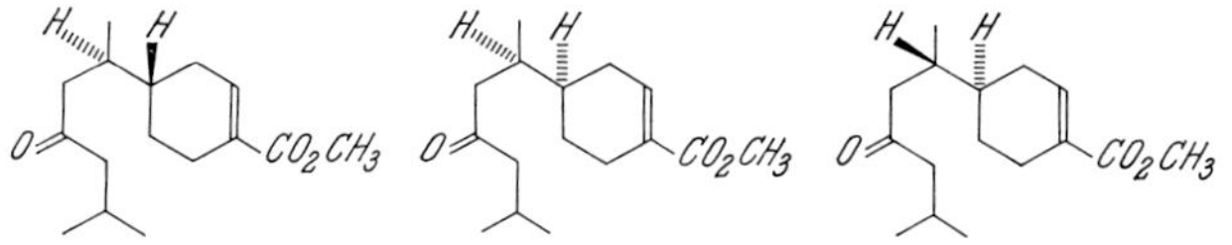

187 188 189

The enantiomeric (—)-juvabione (*188*) and (—)-epijuvabione (*189*) were prepared from (S)-(—)-limonene in an analogous manner to compounds (*174* and *187*). The configuration at the optically active centres of (—)-juvabione is S,R and that of (—)-epijuvabione is S,S.

As mentioned above, juvabione and dehydrojuvabione are biologically active exclusively on hemipterans of the family *Pyrrhocoridae*. It is noteworthy that the biological activities of the optical isomers of juvabione on *Pyrrhocoris apterus* and *Dysdercus cingulatus* differ only to a negligible extent, as observed by SLÁMA on specimens provided by PAWSON. This finding is at variance with the biological effects of peptidic analogues of juvabione derived from p-aminobenzoic acid; the biological activity of these analogues depends on the configuration of the incorporated amino acids (see below).

Similar to farnesol and its derivatives, juvabione served as a model substance in the synthesis of numerous compounds of potential insect juvenile hormone activity. It is noteworthy that many these compounds exhibited a considerably high and specific biological activity.

The first juvabione and dehydrojuvabione analogues of the aromatic series were prepared in Prague by SUCHÝ and ŠORM [315, 336] (cf. Scheme 43). p-Bromobenzyl alcohol, as the starting compound, was converted to the diol (*190*) by reaction of its lithium derivative and the corresponding methylheptenone. Oxidation of this diol gave the hydroxy acid (*191*), the methyl ester of which (*192*) was dehydrated to a mixture of dienes (*193* and *194*). The latter compounds exhibited a relatively high biological activity, but again exclusively on the hemipterans of the family *Pyrrhocoridae*. Further transformations led to compound (*195*), an aromatic analogue of bisdehydrojuvabione. Compounds (*196* and *197*) obtained by addition of hydrogen chloride or hydrogen bromide to the mixture of dienes (*193* and *194*) exhibited a high biological activity especially on the hemipteran *Dysdercus cingulatus*. Compounds (*196* and *197*) were prepared in connexion with the enormous increase of activity accompanying the conversion of methyl farnesoate into the corresponding dihydro dichloro derivative, as observed in the Prague

Scheme 43

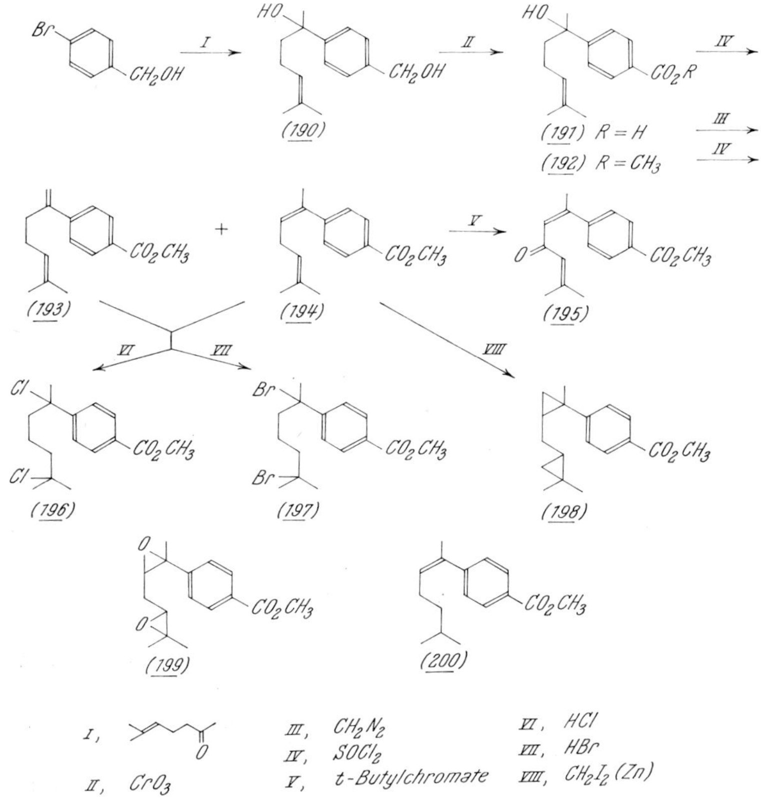

laboratories. Compound (*198*), prepared by the addition of carbene to the hydrocarbon (*194*), also exhibited a high biological activity, higher than the analogous diepoxide (*199*). Compound (*200*), obtained analogously to the diene (*194*) from p-bromobenzyl alcohol and the corresponding methylheptanone, exhibited a lower activity.

In the paper on the synthesis of racemic (+)-juvabione, Ayyar and Rao [6] also mention the preparation of an aromatic analogue of juvabione. The Grignard reaction of styryl isobutyl ketone (*201*) with methylmagnesium iodide gave 2-phenyl-6-methylheptan-4-one (*202*) which was chloromethylated in the *para* position. The resulting compound (*203*) was converted in the usual manner to the aldehyde (*204*), the oxidation of which gave the aromatic analogue of todomatuic acid (*205*). Esterification of compound (*205*) led to the aromatic analogue of juvabione (*206*), the biological activity of which was lower than that of the above-mentioned derivatives lacking the keto group.

Scheme 44

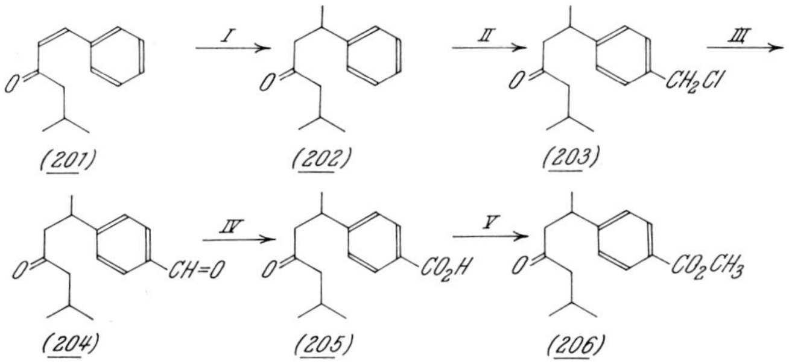

I, CH_3MgJ III, Hexamethylentetramin
II, $CH_3.O.CH_2Cl$ IV, $AgNO_3$–NaOH
V, CH_2N_2

Juvabione analogues lacking the methyl group in the side-chain were synthesized by Mori *et al.* [218] in a manner similar to the preparation of (±)-juvabione [214]. (±)-Demethyljuvabione (*207*) and demethyl-ar-juvabione (*208*) proved to be the most active substances.

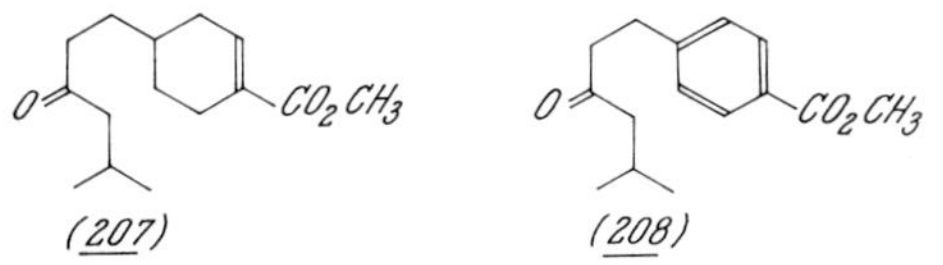

The same team [217] also synthesized some β-bisabolene derivatives starting from the known trans-4-hydroxycyclohexane-1-carboxylic acid (*209*). This acid was acetylated and the acetoxy derivative (*210*) converted to the acid chloride (*211*) which was treated with 4-methyl-3-pentenylmagnesium bromide to give the acetoxy ketone (*212*). Compound (*212*) was subjected to hydrolysis, treatment with dihydropyran, sodium borohydride reduction, acetylation, treatment with 5% aqueous-ethanolic hydrogen chloride, and oxidation with Jones reagent to give another acetoxy ketone (*213*). This compound was converted to the cyanohydrin (*214*) and then to the unsaturated nitrile (*215*). Saponification led to the unsaturated hydroxy acid (*216*) which was subjected to oxidation and then esterification to give the ester which was converted to chloro derivative (*218*). Compound (*217*) was transformed in several steps to (*219* and *220*). The juvenile hormone activity on *Pyrrhocoris apterus* of compound (*220*) was comparable with that of synthetic juvabione, while compounds (*216*, *218* and *219*) were less active.

Scheme 45

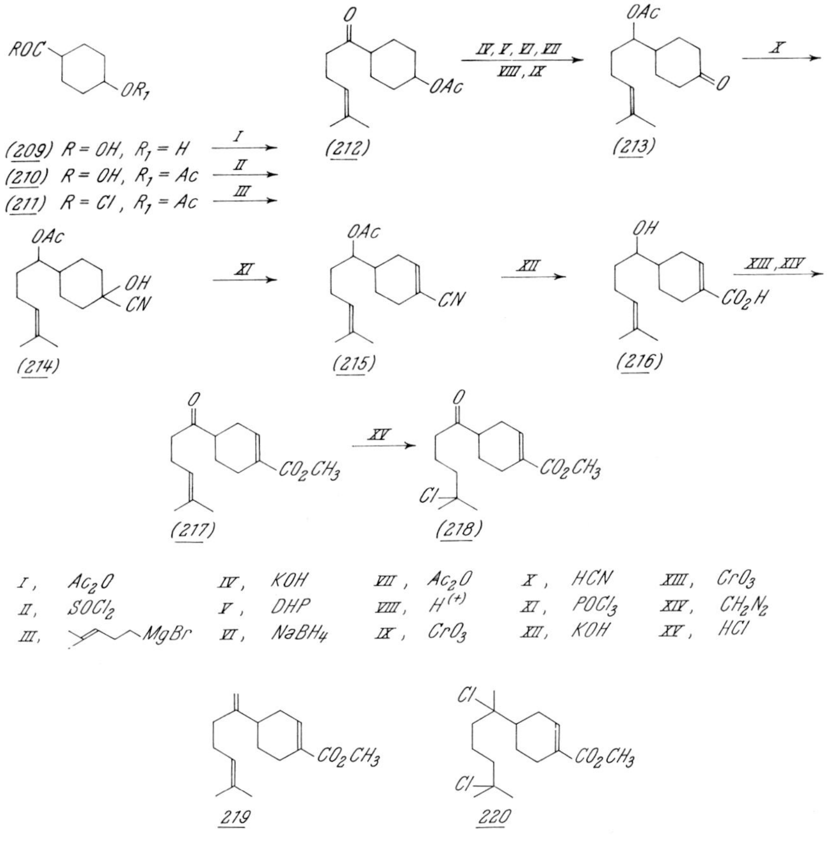

It is of interest that some compounds (e.g., 3-carbomethoxy-7-methyl-7-(4′-carboethoxy-3′-pentenyl)-bicyclo-[4 · 1 · 0]-hept-2-ene (*221*)) obtained by MORI and MATSUI [216] during the synthesis of racemic sirenin (*222*) (sperm attractant produced by the female gametes of the water mold Allomyces [189] show a juvenile hormone activity on the hemipteran *Pyrrhocoris apterus* analogous to that of juvabione, while sirenin itself is inactive.

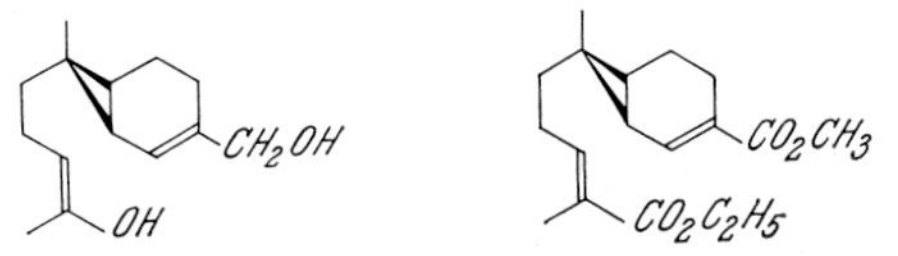

222 221

Derivatives of Benzoic Acid and Acetophenone

In connexion with investigations on aromatic methylenedioxy-derivatives, BOWERS [25] prepared the epoxide of geranyl phenyl ether (*222*) which showed some activity on the hemipteran *Oncopeltus fasciatus* [26]. The analogous ether (*223*) derived from p-hydroxyacetophenone also exhibited a low activity on this hemipteran and on *Tenebrio molitor*. On the other hand, a geranyl ether epoxide derived from methyl p-hydroxybenzoate, methyl p-(trans-6,7-epoxy-3,7-dimethyl-2-octenyloxy)-benzoate (*224*) showed a remarkable activity on *Oncopeltus fasciatus*. All these compounds were prepared by BOWERS by alkylation of hydroxylic groups of the corresponding phenols with geranyl bromide in dimethoxyethane and in the presence of powdered potassium hydroxide. Introduction of the epoxide grouping into position 6–7 of the geranyl residue was accomplished by the action of m-chloroperbenzoic acid.

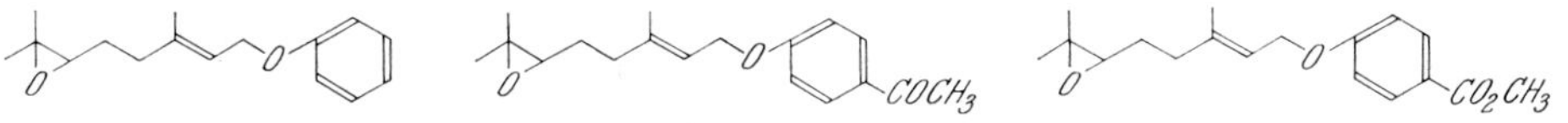

(222) (223) (224)

The ethers derived from p-hydroxybenzoic acid were also investigated in detail in Prague [272]. Some of them were found very active on the hemipteran *Graphosoma italicum*, which is closely related to the important grain pest *Eurygaster intergriceps* (family *Pentatomidae*). Both these hemipterans were repeatedly found to be slightly sensitive to previously known juvenile hormone analogues. Methyl p-(4,5-epoxy-1,5-dimethyl-hexyloxy)benzoate (*226*) was the most active compound; this methyl ester also showed a high activity on the hemipteran *Lygaeus equestris*. Compound (*226*) was prepared analogously to the related epoxides by a standard alkylation of methyl p-hydroxybenzoate with 6-bromo-2-methyl-2-heptene in the presence of methanolic sodium methoxide and the subsequent epoxidation of the resulting ether (*225*) with perphthalic acid.

Scheme 46

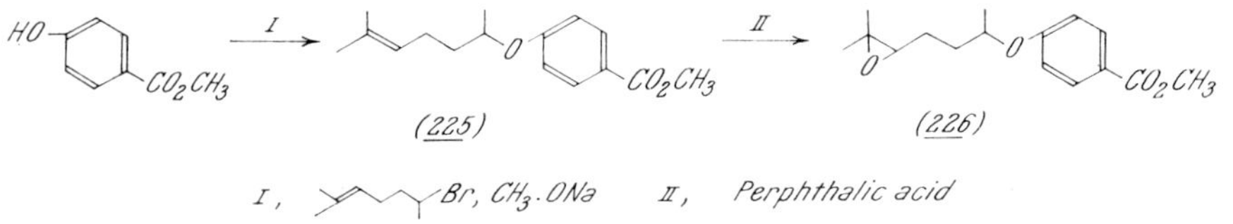

In this connexion, some aliphatic ethers derived from alkyl p-geranyloxybenzoates were prepared by so-called solvomercuration-demercuration in the presence of alcohols. Methyl p-(trans-7-ethoxy-3,7-dimethyl-2-octenyloxy)benzoate (*227*) was highly active on hemipterans of the family *Pyrrhocoridae* [15].

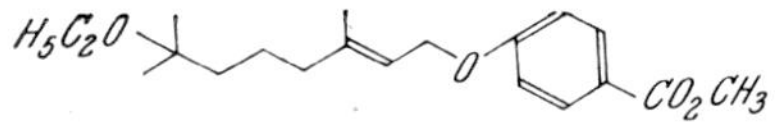

(227)

Numerous p-aminobenzoic acid derivatives [75, 76, 78] were prepared in Prague. Several compounds showed remarkable biological activity on some otherwise resistant hemipteran species. In these syntheses, esters of p-aminobenzoic acid were used as the starting material. Alkylation with geranyl bromide in the presence of pyridine gave, for example, a mixture of methyl N-geranyl-p-aminobenzoate (*228*) and methyl N,N-digeranyl-p-aminobenzoate (*229*) which were separated by chromatography. Addition of hydrogen chloride to these esters led to the corresponding chloro derivatives (*230* and *231*). Methyl p-(trans-6,7-epoxy-3,7-dimethyl-2-octenylamino)benzoate (*234*) was obtained by an indirect route. The N-trifluoroacetyl derivative of methyl N-geranyl-p-aminobenzoate (*232*) was converted by the action of perphthalic acid to the corresponding monoepoxide (*233*), the trifluoroacetyl group of which was then removed by mild alkaline hydrolysis with formation of the desired compound (*234*). Direct oxidation of methyl N,N-digeranyl-p-aminobenzoate (*229*) gave the diepoxide N-oxide (*235*). Methylation of methyl N-geranyl-p-aminobenzoate (*228*) with methyl iodide yielded methyl N-methyl-N-geranyl-p-aminobenzoate (*236*) which was converted by the action of perphthalic acid to the corresponding monoepoxide (*237*). All these compounds exhibited a high juvenile hormone activity on hemipterans of the family *Pyrrhocoridae*, but were practically inactive on other insect species. The highest biological activity was shown by the chloro derivatives (*230* and *231*) and the epoxy compounds (*234*). Compound (*236*) was rather active also on the hemipteran *Graphosoma italicum*, while compounds (*230* and *231*) also exhibited a remarkable activity on the hemipteran *Triatoma infestans*. It is of interest that both the free acid and the ethyl ester are quite inactive on larvae of this hemipteran, which obviously possesses esterases capable of saponifying the ethyl, but not methyl, esters of p-aminobenzoic acid derivatives. Alkylation of methyl p-aminobenzoate with citronellyl bromide led to methyl N-citronellyl-p-aminobenzoate (*238*) the biological activity of which was similar to that of the analogous N-geranyl derivative (*228*).

Scheme 47

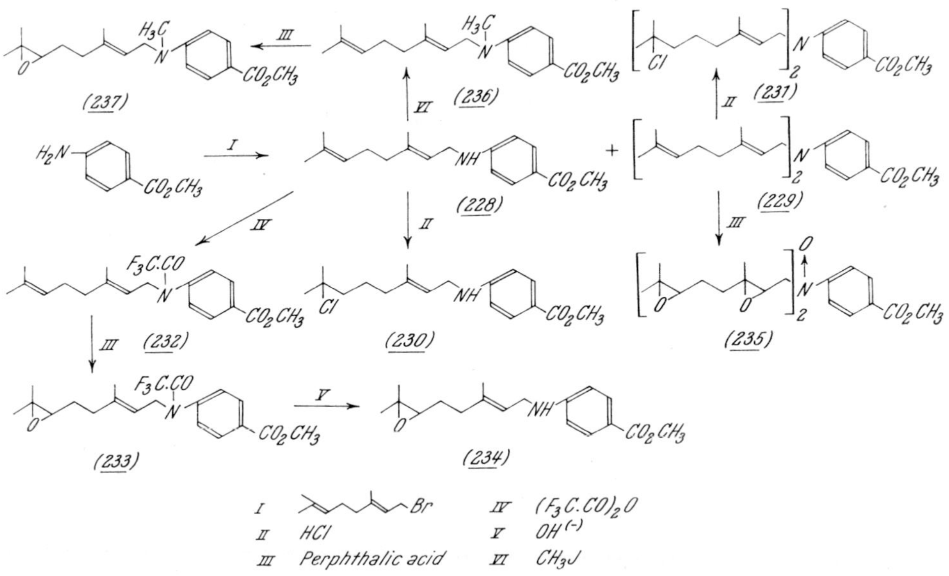

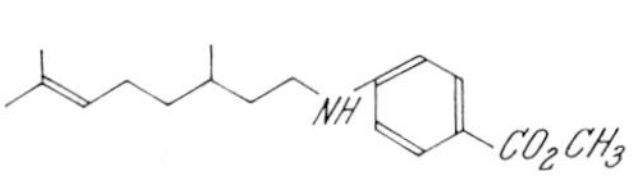

(238)

Some N-alkylated p-aminoacetophenone derivatives were also prepared [188] starting from p-aminoacetophenone, which was alkylated with citronellyl bromide or geranyl bromide. This reaction led to N-citronellyl-p-aminoacetophenone (*239*) or a mixture of N-geranyl- a

N,N-digeranyl-p-aminoacetophenone (*240* and *241*) which was separated by chromatography.

The preparation of p-(trans-6,7-epoxy-3,7-dimethyl-2-octenylamino)-acetophenone (*242*) was started from p-aminoacetophenone which was converted to N-trifluoroacetyl-p-aminoacetophenone. This was alkylated with geranyl bromide in the presence of methanolic sodium methoxide to yield N-trifluoroacetyl-N-geranyl-p-aminoacetophenone, which was converted with perphthalic acid to the monoepoxide. The protecting N-trifluoroacetyl group was removed by alkaline hydrolysis. The preparation of p-(6,7-epoxy-3,7-dimethyloctylamino)acetophenone (*243*) was carried out in a similar manner.

Scheme 48

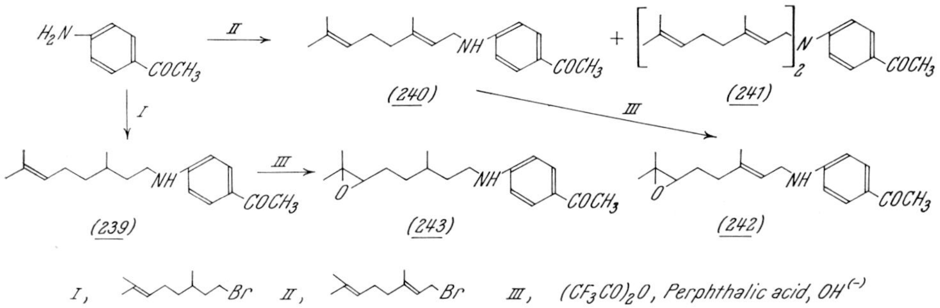

From all the above compounds, the epoxides were the most active juvenile-hormone-like substances. Thus, p-(trans-6,7-epoxy-3,7-dimethyl-2-octenylamino)acetophenone (*242*) was remarkably active on hemipterans of the family *Pyrrhocoridae*, as well as on *Graphosoma italicum*, and showed a moderate activity on *Tenebrio molitor*.

Phenyl Ethers and Aniline Derivatives

The syntheses of these compounds, some of which are highly active and specific juvenile hormone bioanalogues, were undertaken in connexion with preparations of aromatic juvabione analogues (see p. 174) and on the basis of observations of Bowers [25]. The latter concerned some natural and synthetic synergists of pyrethrum and carbamate insecticides, and their ability to mimic juvenile hormone in the case of the mealworm *Tenebrio molitor* and the milkweed bug *Oncopeltus fasciatus*. The activity of natural components of sesame oil, sesamine (*244*) and sesamoline (*245*), on these insects was low while synthetic compounds, e.g. sesoxane (*246*) or piperonyl butoxide (*247*) were somewhat more active.

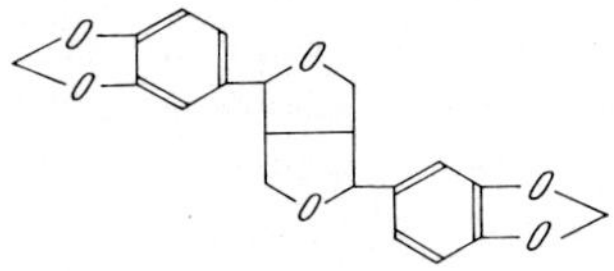

(244)

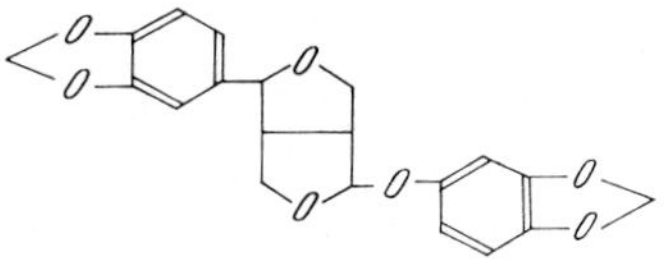

(245)

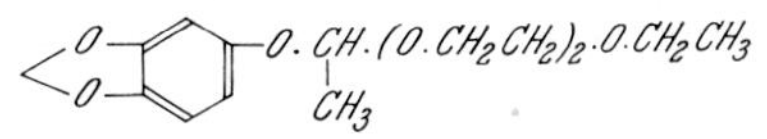

(246)

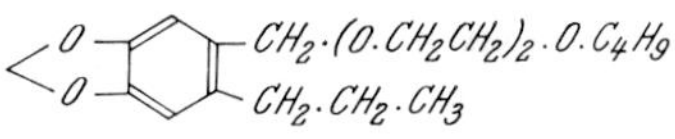

(247)

All these active compounds contain the characteristic methylenedioxy group attached to the aromatic nucleus; in other words, the active substances are derivatives of 1,3-benzodioxole.

As shown by later investigations, the presence of the methylenedioxy group markedly potentiates the juvenile hormone activity and has a great influence on the specificity of action. Bowers prepared synthetically from piperonyl alcohol and farnesyl chloride in the presence of dimethoxyethane and potassium tert-butoxide the piperonyl farnesyl ether (*248*). Treatment of this compound with m-chloroperbenzoic acid gave the corresponding monoepoxide (*249*). Particularly the latter compound was highly active on the milkweed bug (*Oncopeltus fasciatus*) [26].

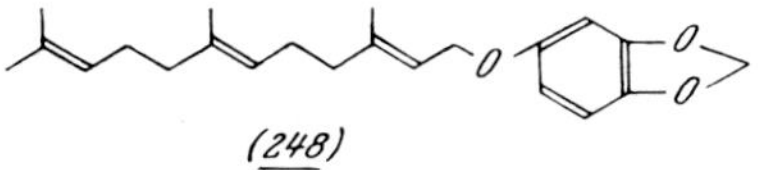

(248)

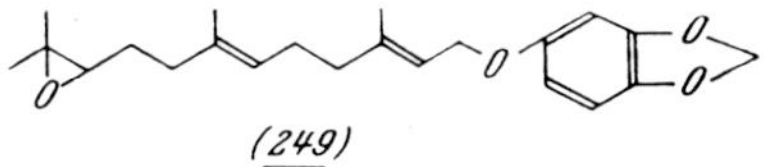

(249)

Bowers also reported [347] on some aromatic ethers derived from 3,4-methylenedioxyphenol, trans-3′,4′-methylenedioxyphenoxy-6,7-epoxy-3,7-dimethyl-2-octene (*250*), trans-3′,4′-methylenedioxyphenoxy-6,7-epoxy-3,7-dimethyl-2-nonene (*251*), and trans-3′,4′-methylenedioxyphenoxy-6,7-epoxy-3-ethyl-7-methyl-2-nonene (*252*). These compounds were prepared by standard procedures from 3,4-methylenedioxyphenol and aliphatic bromides and subsequent epoxidation of the resulting ethers.

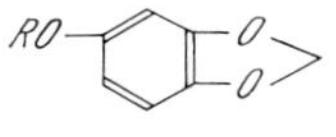

(250) R = 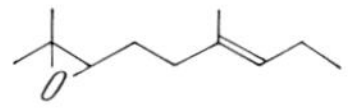

(251) R = 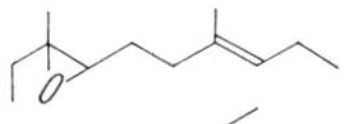

(252) R = 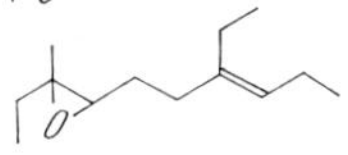

All these compounds exhibited morphogenetic properties as well as ovicidal activity on eggs of some *Coleoptera*, e.g. the Mexican bean beetle *Epilachna varivestis* and the cigarette beetle *Lasioderma serricorne*, and were also active as synergists of pyrethrum insecticides.

The methylenedioxy group is of great importance for the biological activity of these compounds since the analogous coumarone derivatives (recently synthesized in Prague) are considerably less active. Geranyloxycoumaranone (*253*) was prepared from resorcinol by the action of chloroacetyl chloride and alkylation with geranyl bromide.

Scheme 49

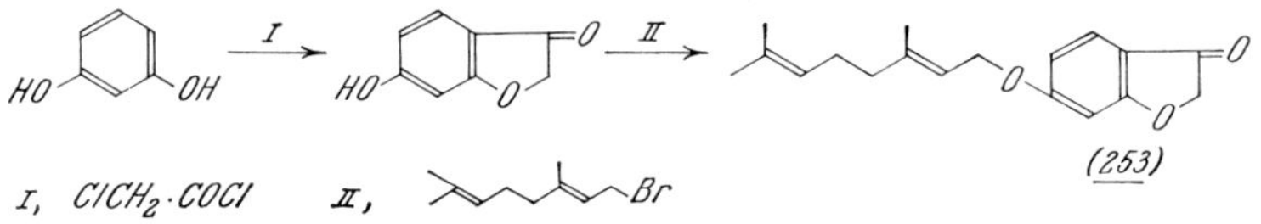

(253)

Some analogous compounds derived from 3,4-methylenedioxyaniline, in which the oxygen atom was replaced by nitrogen [75], were also synthesized in Prague. These compounds were prepared from the starting material by the action of geranyl bromide in the presence of dimethylformamide and potassium carbonate in the form of a mixture of the mono- and digeranyl derivatives (*254* and *255*). The latter compounds showed a moderate biological activity on the hemipterans *Pyrrhocoridae*, and were more active on the *Tenebrio molitor* and the South American hemipteran *Triatoma infestans*. It has been shown (see derivatives of p-aminobenzoic acid) that *T. infestans* is generally susceptible to aniline derivatives. Reduction of the side chain results in a compound with decreased activity on the *Triatoma* bug (*256*) while

Scheme 50

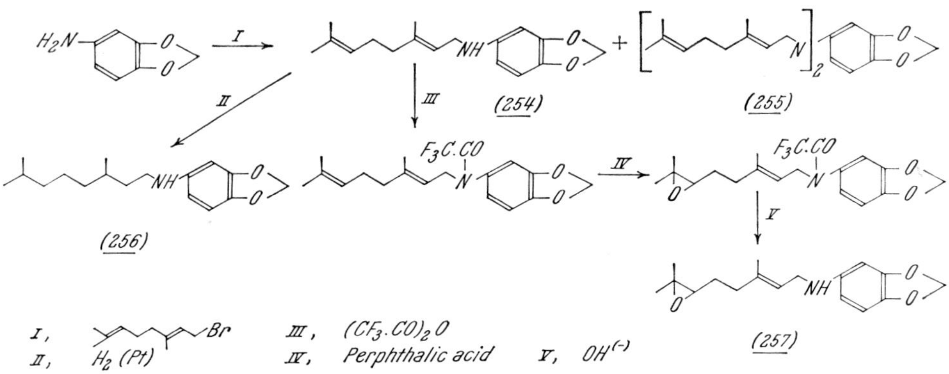

(257)

epoxidation of the 6,7-double bond of the geranyl residue does not change the activity of the parent compound. trans-3′,4′-Methylenedioxyphenylamino-6,7-epoxy-3,7-dimethyl-2-octene (*257*) was prepared from the parent compound by acylation with trifluoroacetic anhydride, oxidation with perphthalic acid and saponification with sodium hydroxide.

The derivatives of 4,5-benzo-1,3-dioxine are structurally closely related to the methylenedioxy compounds but, both N-monogeranylamino- and N,N-digeranylaminobenzodioxine (*258* and *259*) were poorly active even on *Pyrrhocoridae* [308]. Preparation of these compounds is shown in the following scheme:

Scheme 51

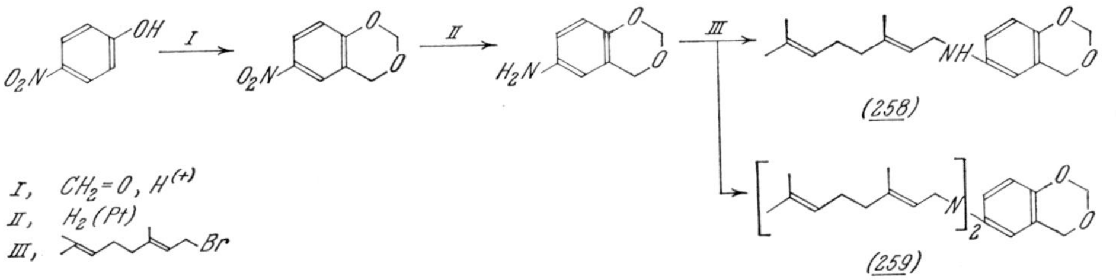

It is noteworthy in this connexion that ostruthin (*260*), the naturally occurring coumarin of the roots of *Imperatoria ostroruthium*, exhibits a weak juvenile hormone activity (shown by SLÁMA). On the other hand, aurapten (*261*) is biologically inactive in spite of a close structural relation to ostruthin.

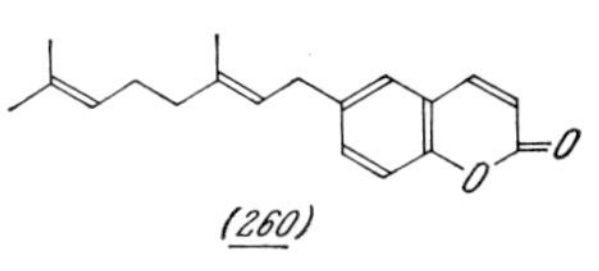

(*260*)

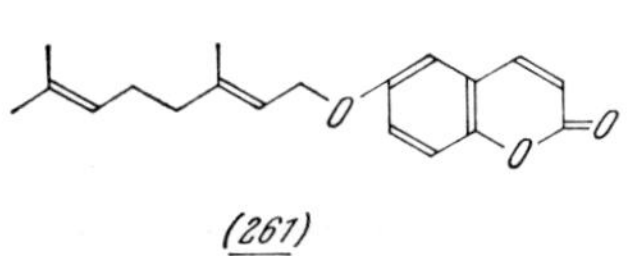

(*261*)

Attempts to discover to what extent the methylenedioxy group can be replaced by other structurally related groupings also led to the synthesis of some amino compounds derived from coumarone and dihydrocoumarone, with the use of 5-nitrocoumarone-2-carboxylic acid as the starting substance [158]. Decarboxylation of this acid yielded 5-nitrocoumarone, the nitro group and the heterocyclic double bond of which were reduced with Raney nickel. The resulting 5-aminodihydrocoumarone was alkylated with geranyl bromide in the presence of dimethylformamide and potassium carbonate to yield 5-N,N-digeranylaminodihydrocoumarone (*262*) which was poorly active on *Pyrrhocoridae*. Reduction of methyl 5-nitrocoumarone-2-carboxylate led to the corresponding amino compound, alkylation of which with geranyl bromide gave a mixture of methyl 5-aminocoumarone-2-carboxylate. N-geranyl and N,N-digeranyl derivatives (*263* and *264*). Both these compounds showed a very low biological activity.

Scheme 52

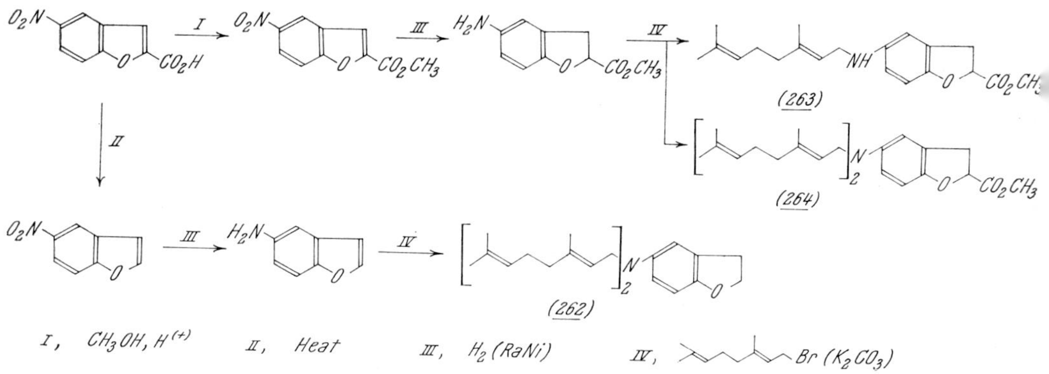

Thioethers, derived from 3,4-methylenedioxythiophenol [124] by direct alkylation of the corresponding sodium salt with geranyl bromide or citronellyl bromide in the presence of dimethylformamide [compounds (*265* and *266*)] were also prepared in Prague.

Alkylation of 3,4-methylenedioxythiophenol with 6,7-epoxygeranyl bromide led to the corresponding epoxide (*267*). Addition of hydrogen

Scheme 53

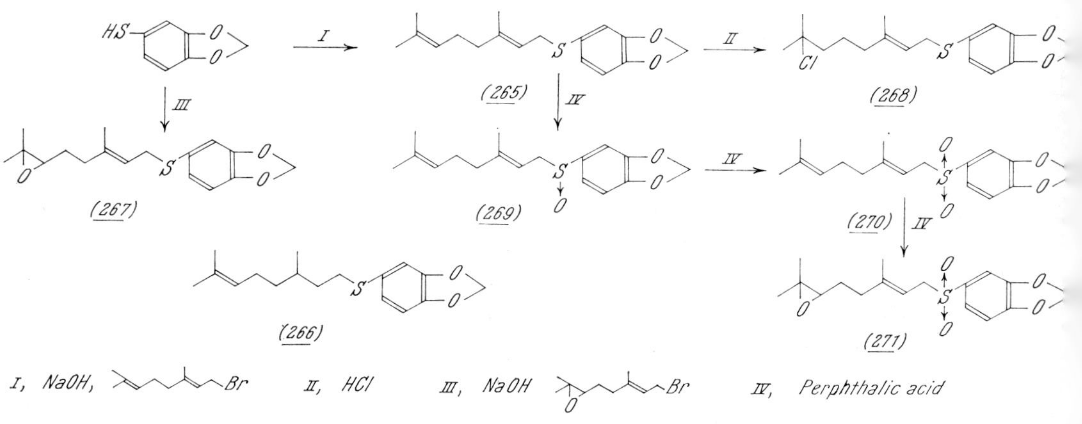

chloride to 3,4-methylenedioxyphenyl geranyl sulfide gave 4′-(trans-7-chloro-3,7-dimethyl-2-octenylthio)-1′,2′-methylenedioxybenzene (*268*). A mixture of the sulfoxide (*269*), the sulfone (*270*) and the epoxide of the sulfone (*271*) was obtained by oxidation of 3,4-methylenedioxyphenyl geranyl sulfide with perphthalic acid (the ratio of these compounds varied with the amount of the oxidising agent used). Only compound (*265*) was somewhat biologically active. Conversion of sulfides to sulfoxides or sulfones was accompanied by a complete loss of activity.

Replacement of the methylenedioxy group by an additional aromatic ring leads to biologically inactive compounds. Thus, β-naphthyl geranyl ether (*272*) as well as the monoethoxy and diethoxy derivatives (*273* and *274*) do not show any juvenile hormone activity on standard experimental insects. The latter compounds were prepared by a usual procedure from β-naphthol and geranyl bromide with the subsequent addition of ethanol [3].

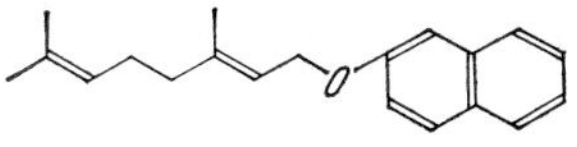

(272)

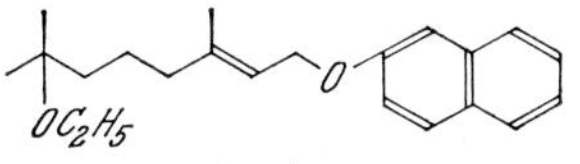

(273)

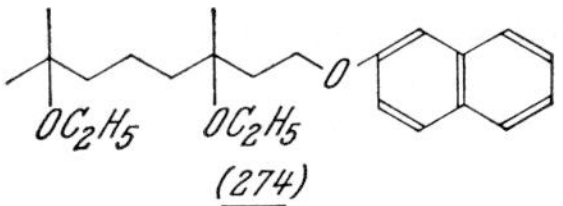

(274)

The geranyl ethers derived from 3,4-dimethylphenol and 4-tert-butylphenol are also very slightly active (*275* and *276*).

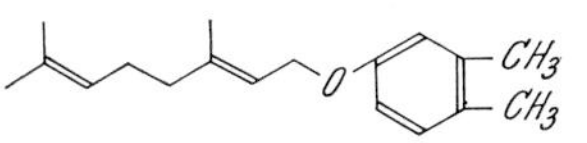

(275)

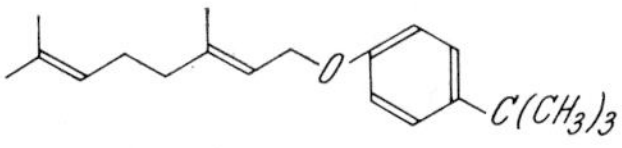

(276)

Derivatives of Nitrophenols, Halophenols, and Nitroanilines

The synthesis of p-nitrophenol derivatives in Prague was motivated by the high biological activity of some terpenoid derivatives of p-hydroxybenzoic acid and p-aminobenzoic acid [77]. The nitro group is known to closely resemble the carboxylic or alkoxycarbonyl group in its electronic structure and behaviour on the benzene nucleus. Treatment of sodium p-nitrophenoxide with geranyl bromide in dimethylformamide gave p-(geranyloxy)nitrobenzene (*277*) which showed a moderate activity on hemipterans and the bug *Tenebrio molitor*. Epoxidation of this compound with perphthalic acid led to the corresponding monoepoxide, p-(trans-6,7-epoxy-3,7-dimethyl-2-octenyloxy) nitrobenzene (*278*) which was far more active than the parent compound, especially on the hemipteran *Graphosoma italicum*. Compound (*278*) has also been recently prepared elsewhere [253]. Prolonged action of perphthalic acid led to the corresponding diepoxide (*279*) which, however, was only slightly active. Addition of alcohols to p-(geranyloxy)-nitrobenzene was carried out by application of alcohols in the presence of mercuric salts and subsequent treatment with sodium borohydride. We obtained a series of p-(7-alkoxy-3,7-dimethyl-2-octenyloxy)nitrobenzenes in which the alkyl was methyl, ethyl, propyl, isopropyl, butyl, and alkyls derived from more complex alcohols, e.g. tetrahydrofurfurylalcohol [323]. The ethoxy derivative (*280*) showed the highest activity on such species as *Tenebrio molitor*. The activity of the corresponding hydroxy compound (*281*) was considerably lower. The diethoxy derivative (*282*) was practically inactive. p-(7-Chloro-3,7-dimethyl-2-octenyloxy) nitrobenzene (*283*) as a crude product showed a high activity on *Graphosoma italicum* and *Coleoptera*.

Scheme 54

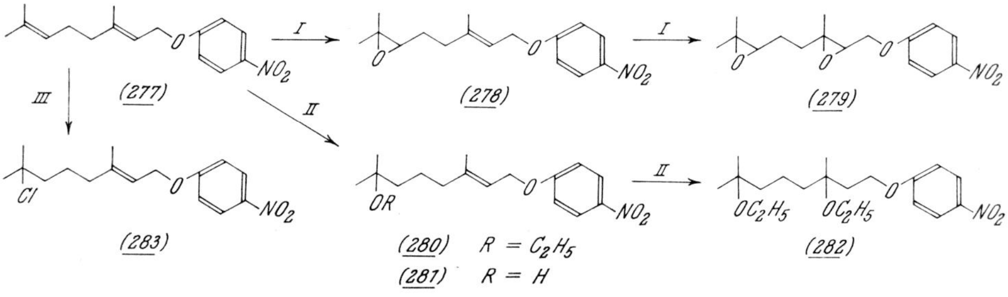

I, Perphthalic acid II, $C_2H_5OH, Hg(OAc)_2, NaBH_4$ III, HCl

Some o-nitrophenol and m-nitrophenol derivatives were synthesized analogously to para compounds. o-(Trans-6,7-oxido-3,7-dimethyl-2-octenyloxy)nitrobenzene (*284*) and its meta isomer (*285*) showed complete biological inactivity [3].

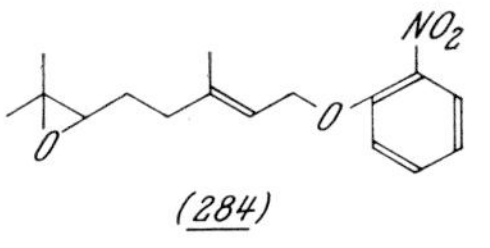

(284)

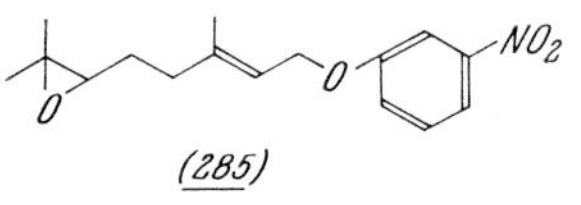

(285)

In connexion with the above mentioned relations between the biological activities of p-hydroxybenzoic acid and p-amino-benzoic acid derivatives, some terpenoid derivatives of p-nitroaniline were synthesized [237]. Alkylation of p-nitroaniline with geranyl bromide gave a mixture of p-N-geranylaminonitrobenzene (*268*) and p-N,N-digeranylaminonitrobenzene (*287*). Addition of hydrogen chloride led to p-N-(7-chloro-3,7-dimethyl-2-octenylamino)nitrobenzene (*288*) which showed in this series of compounds the highest juvenile hormone activity, but exclusively on hemipterans of the family *Pyrrhocoridae.* In contrast to the corresponding p-(geranyloxy)nitrobenzene derivative, the p-N-(7-ethoxy-3,7-dimethyl-2-octenylamino)nitrobenzene (*288*) was biologically inactive.

The earlier assumption, that the high juvenile hormone activity of some p-nitrophenol terpenoid derivatives depends on the electronic structure of the aromatic closely resembling that of benzoic acid, had to be corrected in view of the high biological activity of some p-halophenol terpenoid derivatives [3, 77]. The latter compounds were prepared by similar procedures as the p-nitrophenol derivatives. Condensation of sodium p-chlorophenoxide or p-bromophenoxide with citronellyl bromide or geranyl bromide gave the corresponding citronellyl or geranyl p-halophenyl ethers (*290*, *291*, *292*, and *293*). Subsequent treatment with perphthalic acid led to the corresponding epoxides (*294*, *295*, *296*, and *297*). Geranyl p-chlorophenyl ether and the analogous bromo compound were also used as starting compounds in the preparation of the diepoxides (*298* and *299*), 7-ethoxy derivatives (*300* and *301*), and 3,7-diethoxy derivatives (*302* and *303*). All these compounds (except for *302* and *303*) showed a moderate activity on hemipterans and *Tenebrio molitor*. The ethoxy derivatives (*300* and *301*) were highly active on *Graphosoma italicum* and *Coleoptera*. The monoepoxide (*296*) exhibited a relatively high activity on *Graphosoma italicum*.

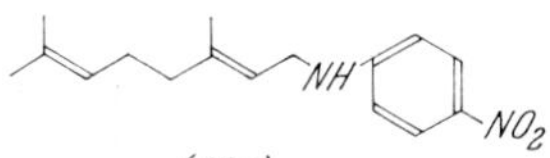

(286)

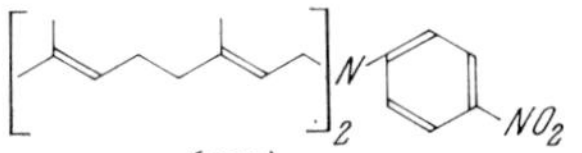

(287)

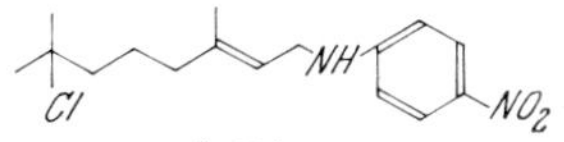

(288)

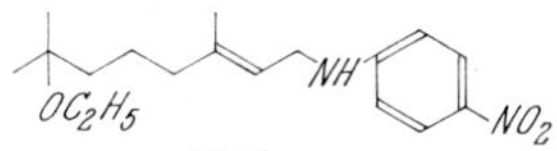

(289)

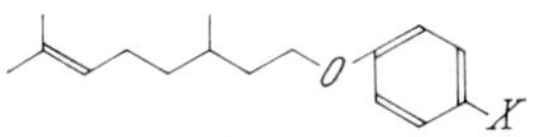

(290) X = Cl
(291) X = Br

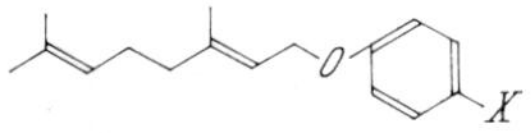

(292) X = Cl
(293) X = Br

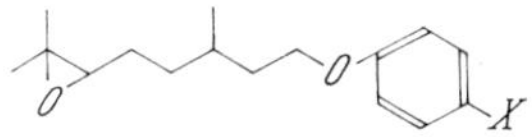

(294) X = Cl
(295) X = Br

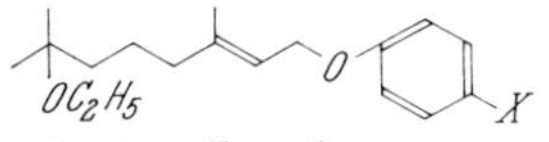

(300) X = Cl
(301) X = Br

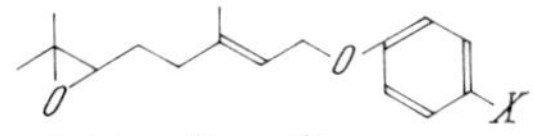

(296) X = Cl
(297) X = Br

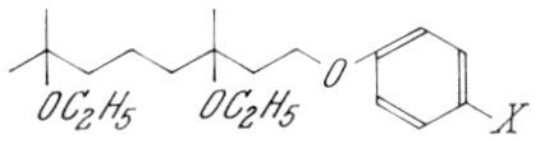

(302) X = Cl
(303) X = Br

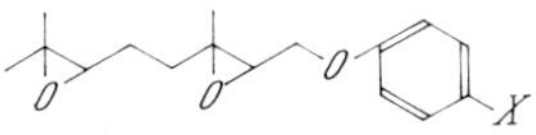

(298) X = Cl
(299) X = Br

Addition of some carbene compounds to the terminal double bond of geranyl p-chlorophenyl ether afforded the corresponding cyclopropane derivatives. The cyclopropane group mimics to a certain degree the epoxy group, which is known to increase strongly the juvenile hormone activity of parent compounds. Thus, addition of carbene [3], dichlorocarbene [4], and ethoxycarbonylcarbene [3] resp. to geranyl p-chlorophenyl ether yielded the corresponding derivatives *304*, *305*, and *306*. The activity of compounds *304* and *305* on the hemipterans *Pyrrhocoris apterus* and *Dysdercus cingulatus* was practically equal to that of the analogous epoxides, but very low on *Tenebrio molitor*. Compound (*306*) was biologically inactive. It is of interest that the analogous p-alkylphenol derivatives also show a relatively high activity on *Oncopeltus fasciatus* and *Tenebrio molitor*, e. g. p-(trans-6,7-epoxy-3,7-dimethyl-2-octenyloxy)-isopropylbenzene (*307*) [235].

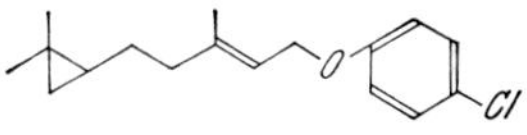

(*304*)

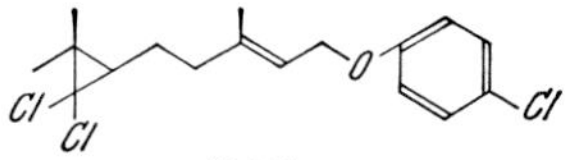

(*305*)

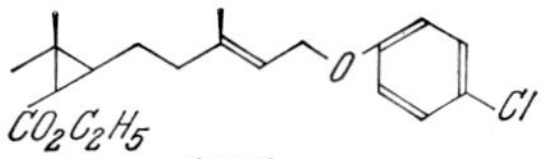

(*306*)

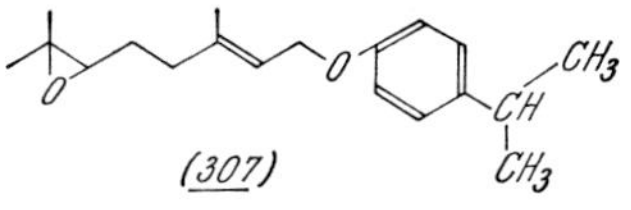

(*307*)

Derivatives of Benzenesulphonic Acid

In connexion with the high biological activity of terpenoid derivatives of p-hydroxybenzoic acid, p-aminobenzoic acid, and p-nitrophenol, it was of interest to prepare a series of p-substituted derivatives of benzenesulfonic acid and the analogous sulfoxides or sulfones. The synthesis of sulfanilic acid N,N-diethylamide derivatives was motivated by their potential activity on *Coleoptera* [157]. The starting acetanilide was converted by the action of chlorosulphonic acid into p-acetaminobenzenesulfonyl chloride (*308*) and then, on treatment with diethylamine, into the corresponding N,N-diethylamide (*309*). The acetyl group was removed by acidic hydrolysis (*310*) and the free amino group alkylated, according to a standard procedure with geranyl bromide in the presence of potassium carbonate, to yield a mixture of p-N-monogeranyl- and p-N,N-digeranylaminobenzenesulfonyl-N,N-diethylamide (*311* and *312*). Addition of hydrogen chloride to the monogeranyl derivative (*311*) gave the chloro derivative (*313*) while the ethoxy derivative (*314*) was prepared by addition of ethanol. None of these substances showed any activity on experimental insects.

Scheme 55

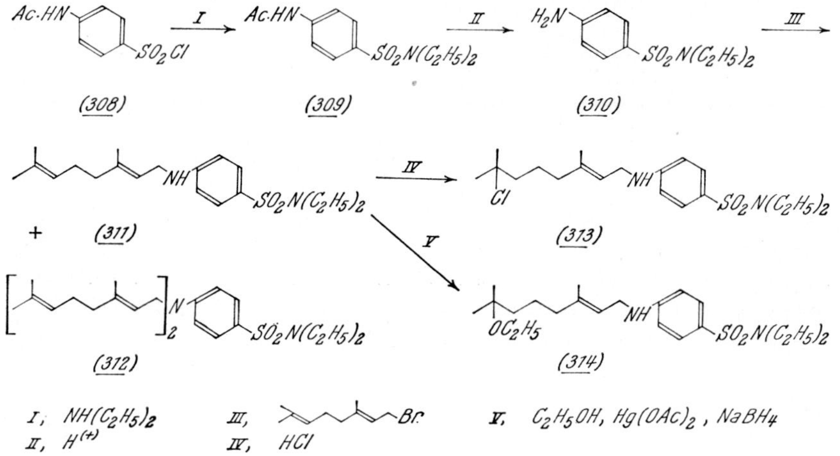

p-Acetaminobenzenesulphonyl chloride was also converted by the action of potassium fluoride into the fluoride, which was deacetylated and then alkylated with geranyl bromide in the presence of potassium carbonate to give p-N-geranylaminobenzenesulfonyl fluoride (*315*). Addition of hydrogen chloride to the latter fluoride led to compound (*316*). Compound (*315*), and particularly compound *316* showed high juvenile hormone activities on the hemipteran *Dysdercus cingulatus*. Methyl p-N-geranylaminobenzensulphonate (*317*), obtained from the fluoride (*315*) by the action of sodium methoxide, showed a moderate biological activity.

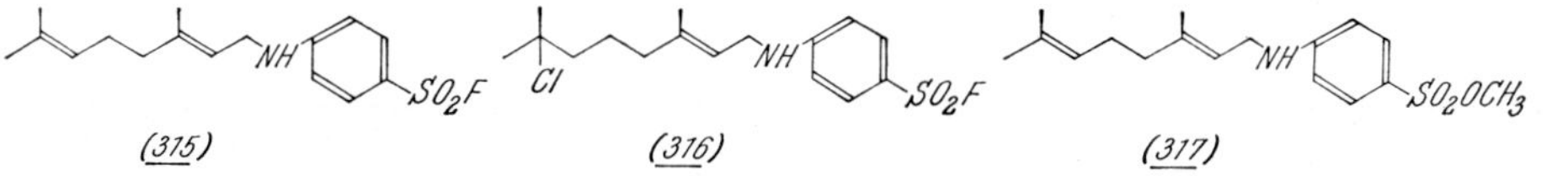

(315) (316) (317)

The N,N-diethylamide of p-aminobenzenesulphonic acid was also used as the starting material in the preparation of some derivatives of p-hydroxybenzoic acid. Thus, treatment with nitrous acid led to the N,N-diethylamide of p-hydroxybenzenesulphonic acid, which was alkylated with geranyl bromide in methanolic sodium methoxide to give the N,N-diethylamide of p-geranyloxybenzenesulphonic acid (*318*). The corresponding epoxide (*319*) was prepared by reaction of this compound with perphthalic acid. Addition of ethanol to compound (*318*) yielded the ethoxy derivative (*320*). None of these compounds exhibited any juvenile hormone activity.

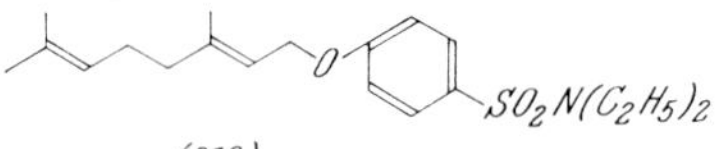

(318)

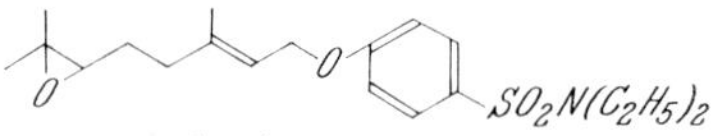

(319)

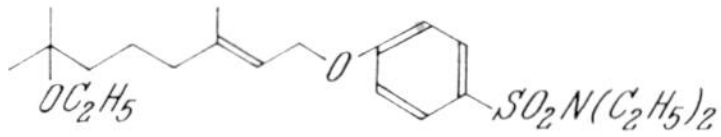

(320)

Alkylation of p-hydroxyphenyl methyl sulphide (*321*) with geranyl bromide in an alkaline medium led to p-geranyloxyphenyl methyl sulphide (*322*) which showed some activity on hemipterans. Oxidation of the sulphide (*321*) with hydrogen peroxide led to the sulphoxide (*323*) and sulphone (*324*), which were alkylated with geranyl bromide in the presence of potassium carbonate to give compounds (*325*) and (*326*), resp. Only p-geranyloxyphenyl methylsulphoxide (*325*) showed a low juvenile hormone activity on *Pyrrhocoridae*, while the sulphone (*326*) was inactive.

Scheme 56

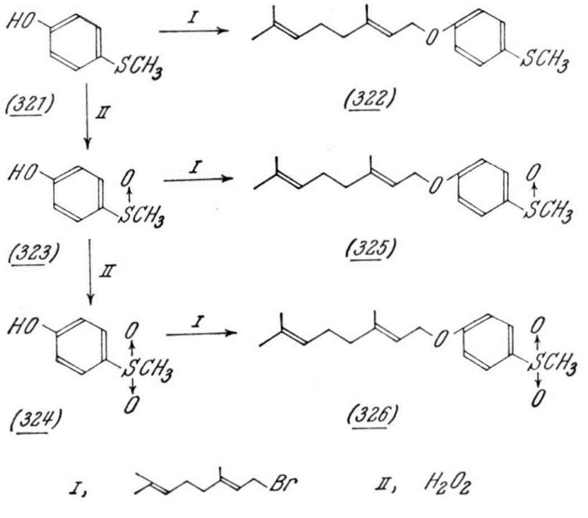

Peptide Derivatives

Some peptide derivatives of alkyl p-aminobenzoates show a remarkably high activity, but only on hemipterans of the family *Pyrrhocoridae*. The synthesis of these peptides was motivated by the high biological activity shown by terpenic derivatives of p-aminobenzoic acid. It was also of interest to what extent modification of aliphatic chains by introduction of hydrophilic groups might influence biological activity. The first peptidic juvenile-hormone-like compounds were prepared in

Prague by ZAORAL and SLÁMA [389]. Thus, ethyl L-isoleucyl-L-alanyl-p-aminobenzoate (*327*), a close analogue of ethyl N-geranyl-p-aminobenzoate, showed a relatively high activity on the hemipterans *Pyrrhocoris apterus* (0.5 μg per specimen) and *Dysdercus cingulatus* (0.5 μg). On the other hand, the peptide analogue (*328*) of farnesoic acid was entirely inactive. The derivative (*329*) of 3,4-methylenedioxyaniline was weakly active (about 100 μg).

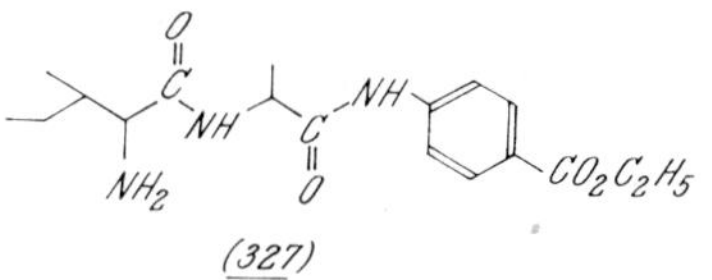

(327)

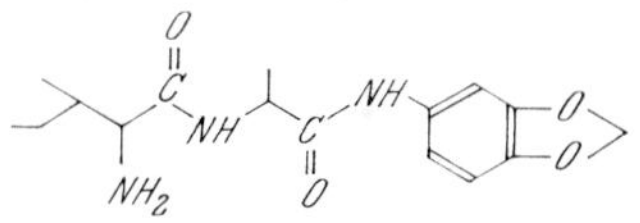

(329)

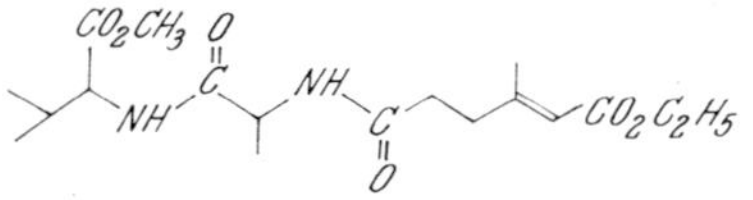

(328)

Later on, numerous peptide derivatives of alkyl p-aminobenzoates and of some other substituted anilines, especially 3,4-methylenedioxyaniline, were prepared in Prague by PODUŠKA *et al.* [246, 247, 248]. The amino groups of these peptide derivatives were substituted by an alkoxycarbonyl or acyl. The starting tert-butyloxycarbonyl (Boc) derivatives were obtained by reaction of the corresponding amino acids and tert-butyloxycarbonyl azide:

$$(BocN_3 + H_2N\cdot\overset{R}{\overset{|}{C}}H\cdot COOH \xrightarrow{OH^{(-)}} Boc\cdot NH\cdot\overset{R}{\overset{|}{C}}H\cdot COOH)$$

The ethoxycarbonyl derivatives were obtained by reaction of ethyl chloroformate and amino acids. Some other derivatives were prepared by the isocyanate method:

$$(R_1OH + OCN\cdot\overset{R}{\overset{|}{C}}H\cdot CO_2Et \longrightarrow R_1O\cdot CO\cdot NH\cdot\overset{R}{\overset{|}{C}}H\cdot CO_2Et \xrightarrow{OH^{(-)}} R_1O\cdot CO\cdot NH\cdot\overset{R}{\overset{|}{C}}H\cdot CO_2H)$$

The acyl derivatives of amino acids were obtained by reaction of acyl chlorides and amino acids.

The juvenile-hormone-like peptides were finally prepared from the above intermediate compounds by normal preparative methods, particularly by the N,N′-dicyclohexylcarbodiimide procedure using 1-hydroxybenzotriazole esters. The process may be illustrated by the following scheme.

Scheme 57

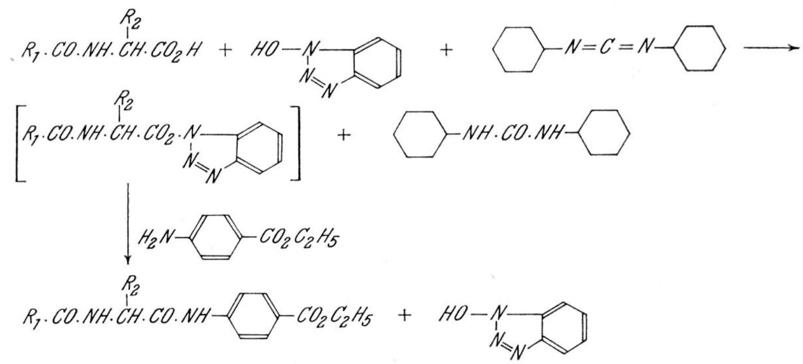

The alkoxycarbonyl derivatives were prepared with the use of phosphorus trichloride as follows:

Scheme 58

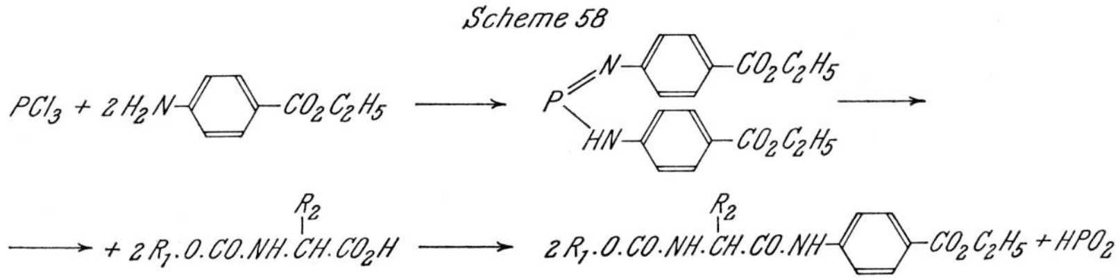

Compounds containing haloacyl groups were prepared by acylation of N-(aminoacyl)anilines with the use of the phosphorous oxychloride method:

Scheme 59

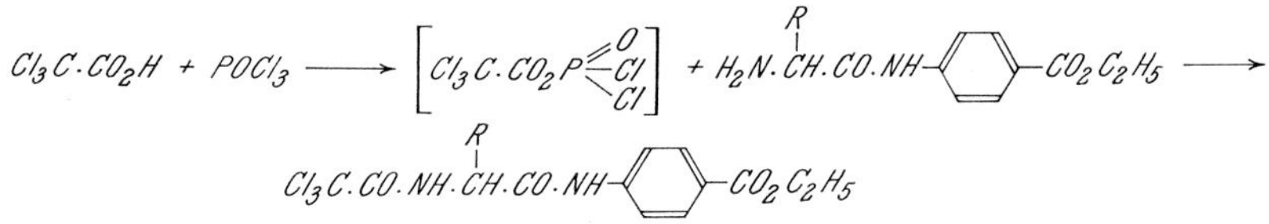

As mentioned above, the peptide derivatives of p-aminobenzoic acid are active only on some hemipterans species[1]. Introduction of a hydrophobic acyl group, instead of the hydrophilic amino groups, into the

[1] The biological activity of this type of compounds obviously depends on the presence of an esterified carboxylic group on the aromatic ring. The analogous peptide derivatives of 3,4-methylenedioxyaniline, p-nitroaniline, p-methoxyaniline, and p-toluidine are practically inactive.

peptide chain results in a large increase in biological activity of this type of compound. Thus, ethyl tert-butyloxycarbonyl-L-alanyl-p-aminobenzoate (*330*) is active on *Pyrrhocoris apterus* and *Dysdercus cingulatus* at a dose of 0.01 μg per specimen, while the activity of ethyl pivaloyl-L-alanyl-p-aminobenzoate (*331*) is even higher (0.00004 μg). Introduction of one chloro atom into the molecule of pivalic acid, i.e., replacement of this acid by α-chloroisobutyric acid in formation of the peptide derivative (*332*), leads to a further increase in activity (0.00001 μg). Further replacement of the L-alanine residue in compound (*332*) by L-valine (compound *333*) results in an extreme increase in activity to 0.000002 μg per specimen of the above-mentioned hemipterans. To our present knowledge, compound *333* represents the most active juvenile hormone bioanalogue. This finding may lead to considerations on the real nature of the juvenile hormone and to whether the structure of this hormone is not closely related to peptides. Noteworthy in this connexion is the high susceptibility of p-aminobenzoic acid peptide analogues to structural changes. Thus, only those derivatives which contain L-amino acids as the medium unit are active, while those containing a D-amino acid are entirely inactive, such as the D-analogue of compound (*330*). This observation is at variance with the known approximately equal biological activity of all optical isomers of juvabione (see p. 173). The presence of a short aliphatic residue on the middle unit of the peptide chain is also of a great importance for biological activity. Thus, replacement of this aliphatic residue by an aryl (with formation of compound *334*) leads to a complete loss of activity. Com-

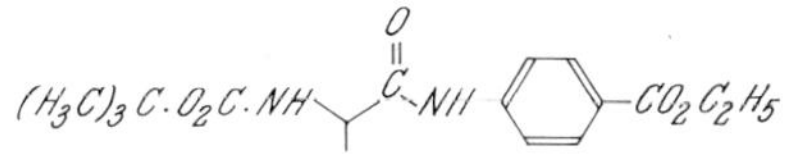

(330)

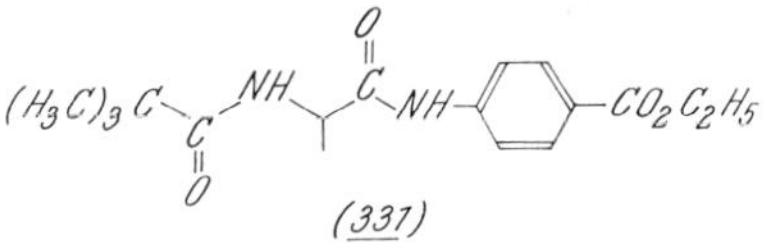

(331)

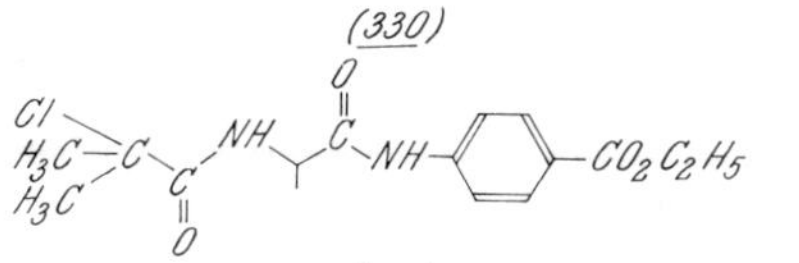

(332)

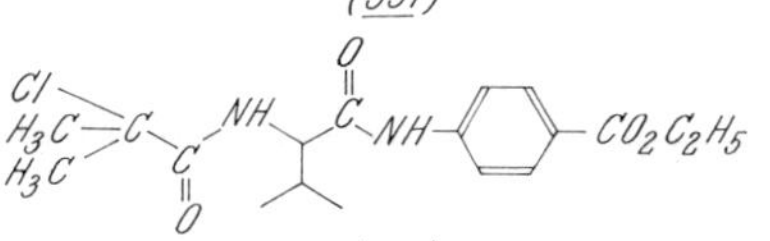

(333)

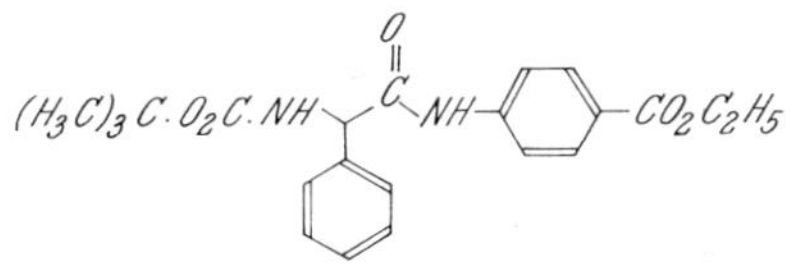

(334)

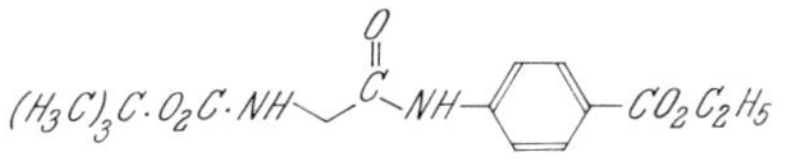

(335)

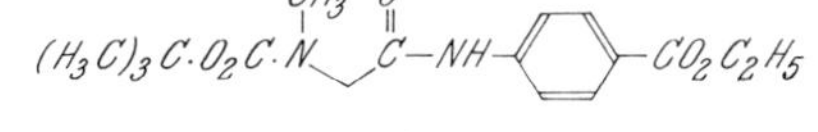

(336)

pounds lacking this aliphatic group are also inactive, e.g., ethyl tert-butyloxycarbonylglycyl-p-aminobenzoate (*335*) and its sarcosine analogue (*336*). On the other hand, substitution of the p-aminobenzoic acid amino group with a methyl group does not lower biological activity. On the basis of the above observations it may be assumed that the biological effectiveness of this type of p-aminobenzoic acid derivatives consists in a highly specific interaction with a precisely optically-oriented tissue receptor. This interaction depends on several important factors, such as the presence of an alkyl group on the middle L-unit of the peptide chain and the occurrence of a terminal hydrophobic acyl group, as well as of an aromatic residue.

D. Structure-Activity Relationships

In spite of the great number of juvenoid compounds, which have been reported in literature, relatively little attention has been given so far to the relationships between their chemical structure and biological activity. There exist countless difficulties when determining these relationships and those are of both objective and subjective nature. One of the objective difficulties is the structural heterogeneity of compounds which possess juvenile hormone activity. When compared with pharmaco-analogues of some other hormones, we discover that there exists enormous variability in chemical structures with respect to all the known compounds with juvenile hormone activity. A further obstacle includes wide diversity of species tested. For example, relationships found in one species may be similar to those found in another but may also be completely different [268]. A major subjective problem is presented by the lack of standard bioassay conditions. Even when single species is used in a bioassay, it is difficult to compare activities based on local epidermal effects with data on general morphological influence of the active compounds. In addition, various compounds may have different dose-response curves [274, 358], which makes it impossible to find a common denominator for data obtained with different techniques. There are still other factors which contribute to the problem, such as application methods, solvents and developmental stages employed.

In this section, we summarize the data on structure-activity correlations and special attention was paid to compounds prepared and tested on several selected insect species by our standardized methods. The conclusions we have made from our results are limited to a particular insect species or, to related groups of insects. The scope at this section does not allow us to discuss all the data presented in Appendix tables. We hope that the data may serve as a guide and source of inspiration to other authors for their own interpretations.

Comparative data from the literature are based mainly on observations obtained with a limited number of insect species. In general, those species which have been used extensively are pupae of *Tenebrio* and *Galleria* and larvae of hemipterans *Pyrrhocoris*, *Dysdercus*, *Graphosoma*, *Oncopeltus*, and *Rhodnius*. We are conscious of the fact that many substances, or even groups of substances, were overestimated or underestimated with respect to their probable potential ability to act on new insect species, including the economically important pests.

1. Physical Properties of Juvenoids

In addition to certain shape and size of the molecule, discussed in the next section, there exist clear correlations between water and lipid solubility and juvenile activity. Most of the highly active juvenoids are characterized by extremely low water solubility. On the other hand, the entirely nonpolar substances, such as hydrocarbons, are, as a rule, inactive [242, 358]. Thus, with respect to their solubility, the optimal juvenoids possess very low polarity, but are not absolutely nonpolar. One of the earlier observations was that substances with more polar groups generally possess less juvenile activity. This was documented in a comparison between the activity of farnesol and that of farnesyl methyl ether and farnesyl dimethylamine on *Tenebrio* [30, 280], *Rhodnius* [357], and *Galleria* [289]. Loss of activity was also observed with aromatic juvabione derivatives having free carboxylic or hydroxylic functions [315] when these substances were tested on the hemipterans *Pyrrhocoris* and *Dysdercus*. Further investigations on various insect species [312, 313, 358, 359] has led to the conclusion that the presence of polar groups such as hydroxyl, carboxyl or amino group result in decrease or complete loss of juvenile activity [242, 308]. Certain substituents or groups, such as chloro or bromo atoms, ethereal oxygen, and alkylated amidic nitrogen may in certain cases exert favourable effect on juvenile activity. However, the number of these groups must be in balance with an appropriate amount of —CH_2— groups in order not to affect the overall lipophilic character of the molecule. This principle seems to have general application for substances with terpenoid chains as well as for aromatic or even peptidic juvenoids. Supporting evidence for this conclusion can be found listed in the activity data contained in the Appendix tables.

Lipophilic character of the juvenoids increases the ability of the compound to penetrate the lipoid layer of the insect epicuticle. However, the lipophilic properties are also required when the test substances are applied by injection directly into the haemolymph. Moreover, some juvenoids have been reported to lose completely their juvenile hormone activity when injected as an aqueous emulsions [357].

2. Molecular Size

Dependence of juvenile activity on size of the molecule and length of the chain has been reported for various juvenoids tested on unrelated insect species. For example, dodecyl methyl ether was the most active while decyl methyl ether and tetradecyl methyl ether were almost inactive in *Galleria* assays [289]. We shall see that similar relationship to molecular weight, as shown for these unbranched acyclic alkanol derivatives, occurs for acyclic isoprenoids. Compounds with optimum activity are as a rule derived from a sesquiterpenoid skeleton (three isoprene units) and related structures of similar molecular size. Compounds of monoterpenoid character with only two isoprene units are almost exclusively inactive whereas relatively low activity is found in compounds of the diterpenoid character (four isoprene units).

According to Wakabayashi *et al.* [345] the aliphatic chain length of 15 carbon atoms is crucial for high juvenile hormone activity when assayed on *Tenebrio*, but this same relationship may not be so with respect to other species. For instance, juvenile activity of some para substituted aromatic esters (see e.g. T-156, T-246, etc.) was lower than that of the corresponding p-halo, p-nitro, p-ethyl, and 3,4-methylenedioxy derivatives (see T-181, T-188, T-215, T-289) on *Tenebrio*. This is in accordance with the above conclusions of Wakabayashi *et al.* [345]. However, the same compounds had lower but not higher activity in assays on various hemipterans. The limitation of the optimum chain length for activity to 15 carbon atoms thus does not exist in all insect species.

Correlations between the chain length and juvenile activity occurred also in sesamex-type compounds. Shortening or prolongation of the chain by a single or more carbon atoms led to decreased activity when the compounds were tested on *Tenebrio* [198, 252]. The chain-length effect was also exemplified on 7-alkoxygeranyl ethers derived from p-nitrophenol or p-chlorophenol (T-190, T-219, T-220, T-221). The presence of an ethoxy or propoxy group led to enormous increase in activity over the parent compound in *Tenebrio* (thousand times) while the presence of a butoxy group led to a drastic decrease in activity (hundred thousand times, see T-223). On the contrary, the presence of these alkoxy groups did not exert any larger effect on the activity of these juvenoids when assayed on several Hemipteran species (cf. T-220, T-223).

In certain species or groups of insects, such as *Pyrrhocorids,* we have observed a broad tolerance in activity with respect to differences in chain length of juvenoids. In *Pyrrhocoris,* such substances as juvabione or aromatic derivatives of juvabione having shorter side chains were approximately equal in activity as aromatic geranyl ethers having somewhat longer side chains or long-chain aromatic farnesyl ethers. The derivatives of N,N-digeranylaniline (T-260) or the esters of N,N-difarnesylaminobenzoic acid (T-256, molecular weight over 550), representing the largest juvenoid molecules known were still active on *Pyrrhocoris*.

The juvenile hormone activity of homologues of 3,7,11-trimethyl-2-dodecenic acid remained almost unchanged when the chain was shortened by one carbon atom between the original C-7 and C-11 atoms (T-115, T-119). A considerable decrease or even loss of activity as measured on representatives of six insect families was observed after simultaneous shortening (T-113) between both the C-7 and C-11 and also C-3 and C-7 atoms of the original carbon chain [268, 314].

These data indicated that the effect of chain length on the biological activity of juvenoids may be less important than specific change of distances between determined sensitive loci within the molecule. Equal shortening or prolongation performed at some part of the molecule resulted in completely different activity change than that performed on other part. We feel that it is not possible to understand relationships between the chain-length and biological activity of juvenoids until we actually know their exact space orientation in biological systems. At present we can generalize that a lower size limit for an active compound is restricted to at least 10 carbon atoms. Determination of an upper size limit is more difficult since there exist many compounds with chains approaching or exceeding twenty carbon atoms which retain juvenile hormone activity. Nevertheless, we can state that the majority of the most active compounds contain the basic skeletons which are equivalent in size to approximately 14 to 16 —CH_2— units [280, 308, 312, 314, 345, 358, 359]. Another factor which must be considered is the size of the molecule which is actually active within the target cells. For example, juvenoids with relatively long ester radicals may be hydrolysed in the insect by esterases, transesterified, and thus form the actual active compounds of a lower molecular weight. Similar explanation may perhaps be actual in the juvenile activity of farnesyl cholesteryl ether which has an exceptionally large molecular weight [34].

3. Shape of the Carbon Skeleton in Acyclic Juvenoids

As shown in the chemical section, most juvenoids contain at least a part of a branched carbon chain characteristic of isoprenoids. Each fourth carbon atom at the sequence, i.e. C-3, C-7, C-11, bears lower alkyl or other substituent as in the case of geraniol and farnesol. Such branching occurs in most natural juvenile hormone active substances, e.g. farnesol, juvabione, and the two Cecropia hormones. Some straight-chain compounds lacking any branching, such as some saturated or unsaturated fatty acids [303] or alcohols as 1-tetradecanol [304] were found to exhibit juvenile hormone activity. Other straight chain compounds such as 1-dodecanol and especially dodecyl methyl ether [30] proved active on *Tenebrio* and similar aliphatic unbranched ethers of various chain-lengths were active on *Galleria* [289]. The active substances containing just one branched C-atom in the side chain were also found among some insecticidal synergists [25]. Lack of branching at position C-3 in the chain of some sesquiterpenoid derivatives (i.e. absence of the methyl group at C-3) led to less active [250, 345] or highly active [213] juvenoids which may depend on additional structural features.

Comparison of juvenile activity of compounds derived from the parent 3,7,11-trimethyldodecane (farnesane) with related aliphatic juvenoids offers the following conclusions: a) it is not necessary that the branching alkyl be a methyl; b) it is not necessary that branching occur exclusively at the usual C-3, C-7, C-11 positions; and, c) the alkyl groups can be replaced by hydrogen or halogen atoms. For example, compounds of the homojuvenate and dihomojuvenate type (cf. T-151, T-155) were very active on *Tenebrio* [345], *Rhodnius* [358], and especially on Lepidoptera [265]. In homo- and dihomojuvenates and their geometrical isomers, the C-11 carbon atom bears a methyl and an ethyl group. Replacement of these substituents by the combination ethyl-ethyl (T-154) or n-propyl-methyl (T-157) led to the same or somewhat higher activity [213]. Some esters (T-159, T-160), the C-11 carbon atom of which is incorporated into a six-membered ring were also active on *Tenebrio* as shown by FINDLAY [94].

As already mentioned, changing the "classical" isoprenoid branching to C-3, C-7, C-10 did not influence juvenile hormone activity, whereas structural changes resulting in C-3, C-6, C-9 branching sequence resulted in the loss of activity [314]. Whether the loss of activity was due to the changed distance between the branching alkyl groups or due to the overall shortening of the chain remains to be determined.

Among the oxa-analogues of farnesoic acid (p. 165), the C-3, C-8, C-12 branched derivatives are, as a rule, less active than those branched at the usual 3–7–11 positions. Presence of an additional alkyl group at C-11 resulted in increased activity, especially on *Gryllus*. This has been demonstrated with 3,7,11,11-tetramethyl-2-dodecenoate and some homologous juvenoids [314] (see T-119, T-121). Introduction of additional alkyl substituents to positions other than 3-7-11 increased the activity on some insect species (*Graphosoma*, T-122, T-123) but in other species it did not lead to higher activity. Similarly, an additional substitution at C-7 atom resulted in increased activity on Pyrrhocoridae with simultaneous loss of activity on number of other insect species. Absence of the alkyl groups from both C-3 and C-7 was followed by inactivity in *Tenebrio* [346]. However, as shown by MORI [213], certain compounds with no alkyl substituent at C-3 (cf. T-131, T-139) may be active on *Tenebrio* or *Tribolium*. Similar juvenoid esters (T-137, T-141) with no substituents in the terminal part were almost inactive. It has also been observed that the methyl group at C-3 or C-7 can be replaced by chlorine atoms (cf. T-13, T-138) without significant changes in activity [308, 322]. Finally, from the data it appears that, sometimes, the branching at C-7 is not as essential for the activity as when it is present at C-3 (cf. T-118). Thus, all these data together suggest that branching at the terminal position of the carbon chain is more important than branching at other parts of juvenoid molecule.

Less attention has been focused, so far, on the role of alkyl groups in chains of aromatic terpenoids. The influence of alkyl groups at positions C-3 and C-7 of the geranyl residue on the activity appears to that of substituents at positions C-7 and C-11 of the farnesane type juvenoids [308]. Analogous relation in activity exist between geranyl and homogeranyl benzoates as exist between juvenates and homojuvenates [26, 213]. In some juvenoids with a wide activity spectrum, such as the terpenoid phenyl ethers, the distance between the benzene ring and the nearest side chain substituent is approximately the same as between the two alkyl substituents in an isoprenoid chain. When this distance was considerably shortened (juvabione and related substances) the juvenile activity (cf. T-172, T-173, T-239 to T-244) was limited to a few species or exclusively to Pyrrhocoridae [268]. With peptidic juvenoids, the activity was almost entirely dependent on the presence of an alkyl radical of the central amino acid [247].

4. Double Bonds and Geometrical Isomerism

The presence of ethylenic bonds has great significance in biological activity of juvenoids. With only a few exceptions, all the highly active substances contain at least one double bond. Juvenoids contain either isolated or conjugated double bonds. The latter are conjugated to each other or occur in α,β-position to various functional groups (particularly carbonyl) or aromatic systems. The occurrence of other types of C—C multiple bonds, namely of allenic ones, has not been reported in juvenoids so far.

The juvenoids derived from farnesoic acid esters are usually unsaturated at positions 2–3, 6–7, and 10–11. The absence of all three double bonds (T-12) resulted in a decrease or complete loss of juvenile hormone activity when tested on different insect species [134, 225, 268, 312, 345, 346]. Most authors agree that the $\Delta^{2,3}$ unsaturation is essential, probably because of its α,β-position to the functional group [308]. The absence of this $\Delta^{2,3}$ double bond results in the loss of activity of compounds tested on such different groups as Hemiptera [312], Orthoptera [225], Coleoptera [312], and Lepidoptera [134]. Thus, it may be assumed that the presence of this allylic, double bond is of general significance in these compounds. Relationships in the activity with respect to the $\Delta^{6,7}$ double bond are more complex. As exemplified on various insect species, i.e. *Rhodnius* [358], *Locusta* [225], *Galleria* [134], Hemipterans and *Tenebrio* [*312*], the presence of this "central" double bond was not as essential as the $\Delta^{2,3}$ double bond. Absence of this "central" double bond resulted in increased activity according to some reports, while other authors have reported that $\Delta^{6,7}$ unsaturated compounds were more active than the corresponding dihydroderivatives [345, 346, 351]. Although the discrepancy remains obscure, it most probably is due to differences in assay procedures. The positive activity data obtained with 6,7-dihydro compounds refer to ID-50 units while the negative data are based on localized epidermal effects. With *Tenebrio*, concentration-activity curves are very variable and the concentrations of some juvenoids based on localized effects may differ from those based on ID-50 units by several orders of magnitude [351]. The presence or absence of the $\Delta^{6,7}$ double bond in juvenoids produces compounds with different and often selective activity on different insect species. The occurrence of this double bond represents one of the multiple factors determining species specificity [312].

Most data indicate that the $\Delta^{10,11}$ double bond is not indispensable. Variations in the activity due to its presence or absence do not usually exceed, more than one order of magnitude change in concentration. Assays of similar compounds containing one to four double bonds in the molecule [312], have revealed the following structure-activity relationships: a) with increasing number of the double bonds no detectable change in activity occurs when compounds are tested on Pyrrhocoridae and Lygaeidae, but b) the activity increases in Pentatomidae, and c) it is considerably decreased in Tenebrionidae and Dermestidae. The changes in activity for these compounds when assayed on Acridiidae [225] and Pyralidae [134] were similar to those described in c). Similar structure-activity relationships with respect to the number of double bonds were observed with analogous oxa-esters (T-65, T-66, T-68). These examples document the fact that some insect species may be more susceptible to more unsaturated juvenoids whereas more saturated compounds may have better activity on other insects. It appears that the corresponding $\Delta^{6,7}$, $\Delta^{10,11}$ tetrahydro derivatives may be less specific, and act on diverse insect species [308]. The significance of the $\Delta^{2,3}$ double bond in the geranyl chain of aromatic terpenoid ethers or imines can be determined by comparing analogous substances with the geranyl terpenoid chain replaced by one containing the citronellyl group. As shown by the activity of these compounds in the Appendix tables, some insect species were more susceptible to geranyl-, others to citronellyl derivatives. Certain differences in the activity also occur between terpenoid aromatic ethers and terpenoid anilines. The corresponding geranyl and citronellyl derivatives of both alkyl and aryl carbamates act similarly on *Tenebrio* [292, 345]. These comparisons suggest that the presence of the allylic double bond in these aromatic terpenoids contribute to the juvenile hormone activity for certain species, but its presence is not essential. In this respect the presence of the isolated $\Delta^{6,7}$ double bond of geranyl or citronellyl side chains does not appear indispensable for juvenile hormone activity (cf. T-279, T-280).

The contribution of double bonds to the juvenile hormone activity is closely connected with that of geometrical isomerism. As early as 1962, YAMAMOTO and JACOBSON [388] pointed out differences in juvenile activity among four geometrical isomers of farnesol. At present we know that the orientation of substituents on the ethylenic carbon atoms is also an important factor in determining the activity of numerous unsaturated juvenoids. Furthermore, the shape of the molecule of the *cis* isomers considerably differs from that of the corresponding *trans* forms. The most detailed investigations have been performed on synthetic geometrical isomers of the racemic dihomojuvenate, where eight possible geometrical isomers have been tested on several insect species [242, 265, 274, 351, 358]. The activity data for these compounds are summarized in Table 11. The highest activity was obtained for the 2-*trans*-6-*trans*-10-*cis* isomer, the geometrical configuration of which is assumed to be identical with the naturally occurring hormone isolated from *Cecropia*. In contrast, one of the lowest activities reported was that of the isomer with just the opposite geometrical configuration. However, all-*trans* isomer was quite active while the activity of the all-*cis* isomer was low. Particularly important is the *trans* arrangement at $\Delta^{2,3}$ double bond (perhaps due to its neighbourhood with the functional groups). The *trans* configuration at $\Delta^{6,7}$ appears less essential, while almost unimportant when at $\Delta^{10,11}$. The esters of dichlorodihydrofarnesoic acid [313] or 6,7-dihydrofarnesoic acid [312] differ only by geometry at $\Delta^{2,3}$ double bond and their *trans* isomers (cf. T-30, to T-40) were more active than the corresponding *cis* ones [308]. The phenomenon that *trans* isomers possess more activity than *cis* isomers appears to have general application and should be considered in all cases when the juvenoid molecule contains an α,β-unsaturated system. This principle relates also to the corresponding α,β-γ,δ-unsaturated systems. In such dienoates (cf. T-46), the *trans*, *trans* isomer was the most active of the four possible geometrical isomers [395]. In addition, in aromatic

Table 11. *Morphogenetic Activity of Geometric Isomers of Methyl 10,11-Epoxy-7-ethyl-3,11-dimethyl-2,6-tridecadienoates in Various Insect Species*

Species:			*Tenebrio*	*Tenebrio*	*Tenebrio*	*Rhodnius*	*Blatella*	*Dysdercus*
Stage:			pupae	pupae	pupae	larvae	larvae	larvae
Application:			punctures	punctures	injections	topical	topical	topical
Activity Units:			ED-20[1]	ED-50[1]	T. U. patches	ID-50	ID-50	ID-50
References:								
Geometric Isomers:								
10,11	6,7	2,3						
trans	trans	trans	0.01	5	0.0005	1,8	0.5	5
cis	trans	trans	0.001	—	0.0002	0.5	0.5	0.1
cis	cis	trans	0.006	—	0.005	6.8	5	0.3
trans	cis	trans	0.04	80	0.007	30	50	0.3
trans	trans	cis	0.05	0.7	0.1	33	50	35
cis	trans	cis	0.006	—	0.1	15	10	300
trans	cis	cis	17	100	0.1	180	100	300
cis	cis	cis	—	—	0.1	230	500	5

[1] These data are not the actual ID-50; they reveal only relative differences in mg/ml of the testing solutions used for epidermal punctures.

juvenoids with geranyl side chains, the *trans* orientation of the allylic double bond may be essential for the juvenile activity [27].

5. Role of Hetero Atoms

Elucidation of the influence of hetero atoms on juvenile activity constitutes one of the most complex problems in determining structure-activity relationships. The hetero atoms may be arranged according to frequency of their occurrence in juvenoids as follows: oxygen, nitrogen, chlorine, and sulphur. These common hetero atoms occur as: a) substituents of hydrogen on the basic carbon chain; b) as links connecting parts of carbon chains; c) as substituents on aromatic rings; and d) as components of various functional groups. We believe that such a classification facilitates the description of structure-activity relationships, although there may be objections to this classification from the purely chemical point of view.

a) Hetero Atoms as Additional Substituents on Carbon Chains

Juvenoids related to farnesoic acid esters or 12,14-dihomofarnesoic acid esters resp., have been prepared very often with hetero atoms at positions C-3, C-7, and C-11. The data suggest that the presence of oxygen atom forming an oxirane ring was of great importance for juvenile hormone activity. Moreover, this 10,11-epoxy group occurs in homo- and dihomojuvenates isolated from insects (see p. 146). There exists ample evidence [308] that the presence of this epoxy group greatly enhanced (from 5 to 500 times) the juvenile hormone activity of juvenoids when assayed on a variety of unrelated insect groups or species. The already mentioned quaternization of the C-11 carbon atom, i.e., the presence of an additional alkyl group at C-11 (T-119) was not as effective as the 10,11-epoxy group (T-26) or chlorine atom (T-24) at C-11 with respect to juvenile activity [314]. Furthermore an oxygen atom present at C-11 in the form of an alkoxy group or a nitrogen atom in the form of an azirane ring were in many (but not in all) cases less effective than the original epoxy derivatives [260, 345, 395].

Quite a different activity relationships were found with respect to the presence or absence of a hetero atom at position C-7 of the farnesane type compounds [268, 308]. Such juvenoids containing hetero atoms at C-7 generally have lower juvenile hormone activity, especially on Endopterygote insects [134, 312, 313, 351] and also on *Rhodnius* [358, 359]. SLÁMA *et al.* [312] have shown that additional substituents at C-7 result in an enormous increase in juvenile hormone activity when tested on Pyrrhocoridae, while no change in activity was detected when the compounds were administered to representatives of other hemipteran family Pentatomidae. In *Tenebrio*, in contrast to Pyrrhocoridae these C-7 substituted acyclic juvenoids were almost inactive (cf. T-30 to T-32, T-23, T-24). From this it was concluded that the presence of any further substituents at the C-7 constitutes one factor determining species specificity and selectivity of action. Enormous differencies in juvenile activity between *Pyrrhocoris* and *Tenebrio* encountered with 7,11-dichloro derivatives were ascribed to the presence of the C-7 chlorine [268, 312, 313]. It is also known that the presence of the epoxide oxygen at C-3 or C-7, unlike that at C-11 remarkably suppressed the activity in *Tenebrio* [351] and *Rhodnius* [358].

In aromatic-terpenoid juvenoids with the geranyl side chain, hetero atoms at the terminal C-7 seem to affect the activity in a similar way as when present at the terminal C-11 of the farnesane type juvenoids. In these C-7 substituted geranyl-aromatic compounds (T-186, T-188, T-246, T-259, etc.) the juvenile hormone activity is considerably enhanced in various insect species [26, 27, 235, 293, 308, 322, 345). Compared with their parent compounds, some 7-alkoxygeranyl aromatic juvenoids (T-190, T-196, T-219 to 222, etc.) have especially high activity on *Tenebrio* and few other species. However, the presence of epoxy group or of the chlorine at C-3 of the geranyl side chain lowered the activity in *Tenebrio* and in many other species but mostly did not affect the activity in Pyrrhocoridae (T-189, T-195, T-216). These facts suggest that the role of the C-3 substituents on the geranyl side chains are equivalent to their effect at C-7 in farnesane type juvenoids.

b) Hetero Atoms as Links Connecting Parts of the Carbon Chains

In this section will be described certain structure-activity relationships for: i) acyclic juvenoids containing hetero atoms within their base chain; ii) aromatic juvenoids in which side chains are attached to the ring through a hetero atom; and, iii) peptidic juvenoids containing hetero atoms both in the chain and between the chain and the aromatic ring.

i) The first representative compounds of acyclic juvenoids with oxygen atom within the chain, such as farnesyl ethers or geranyl ethers derived from esters of β-hydroxy acids (cf. T-53, to T-56), had very low juvenile activity. More recent oxa-analogues resembling esters of farnesoic acid with one or more —CH_2— groups replaced by oxygen [322] sometimes have higher activity than the corresponding carbon analogues. When related to farnesane type compounds, the best juvenile activity has been found in oxa-analogues with C-5 replaced by oxygen, though other oxa- and dioxa-derivatives were also active. The original chain-length and the classical positions of the branching alkyl groups remained similar in these oxa-analogues as in the farnesane derivatives. Therefore, we expect that differencies found in activity were actually due to the presence of oxygen atom in the chain. In those cases (see: T-65 to T-73) in which oxygen atom did not replace any of the —CH_2— groups but was added to the basic chain the activity of the compounds was usually lower. This lower activity may have been due to the changed distance between the alkyl substituents as well as due to the overall prolongation of the basic chain. The 6,7-epoxygeranyl esters derived from β-ethoxyethanol or β-butoxyethanol have recently been found [27] to be active on *Tenebrio* and some other insect species (T-5, T-6).

Compounds containing nitrogen as part of the chain, namely the terpenoid carbamates (T-164 to T-171) reported by Schwarz *et al.* [292] and Wakabayashi *et al.* [345] belong also to this category of juvenoids. The activity of these compounds as assayed on *Tenebrio* depended particularly on the character of the terpenoid residue. The N-geranyl carbamates and N-citronellyl carbamates appeared to be more active than the corresponding oxa-analogues [345] while the citronellol and citronellylamine derivatives of the chemo- sterilants TEPA and HEMPA proved inactive in juvenile hormone assays [321].

ii) In a wide group of aromatic juvenoids, the hetero atoms form a bridge connecting the terpenoid chain with the benzene ring (see p. 177–190). Data concerning the relationships between the nature of such hetero atom and juvenile hormone activity are rather scarce. We have selected several series of juvenoids and have arranged them in Table 12 according to nature of the hetero atom, substituents on the geranyl side chains, and substituents on the aromatic ring. Further examples may be found in the Appendix tables. The activity data from Table 12 suggest that except for p-nitroaniline derivatives and few other compounds, there are no basic differences in activity between analogous compounds containing either oxygen or nitrogen or sulphur atom between terpenoid chain and aromatic ring. In some cases and in some particular species, the phenyl ethers or the phenylamines appeared somewhat more active. The sulphur containing analogues seemed to be generally less suitable juvenoids except for *Dysdercus*. In this species these aromatic terpenoid ethers, as well as thioethers, or amines derived from methylenedioxybenzene had almost equal activity (Table 12). Unfortunately, exact analogues to these compounds lacking hetero atom bridge at the above indicated position are not presently available for comparisons to be made. However, comparison of similar although not entirely analogous compounds (e.g. T-174, T-175) indicate that a C—C attachment of the terpenoid chain to the benzene ring is not as effective as the hetero atom-carbon linkage. Certain compounds of this type were

Table 12. *Comparisons of Juvenile Activity in Some Phenol, Thiophenol, and Aniline Derivatives*

Structural Units			ID-50 (topical appl.)				
Aromatic Component	Hetero-atom	Terpenic Moiety	DYS[1]	GRA[1]	TRI[1]	TEN[1]	No.[2]:
Phenol	O	Geranyl	100	△		△	T-176
Aniline	N	Geranyl	△	△		30	T-177
1,4-Dihydroxybenzene	O	Geranyl	△	△		△	T-199
p-Hydroxyaniline	N	Geranyl	△	△		△	T-200
Methyl p-Hydroxybenzoate	O	Geranyl	0.5	0.8		△	T-245
Ethyl p-Aminobenzoate	N	Geranyl	0.05	△	10	△	T-251
Ethyl p-Aminobenzoate	N	$(Geranyl)_2$	0.07	△	40	△	T-254
Methyl p-Hydroxybenzoate	O	6,7-Epoxygeranyl	0.05	0.01	0.08	100	T-246
Ethyl p-Aminobenzoate	N	6,7-Epoxygeranyl	0.008	0.5	0.03	△	T-262
3,4-Methylenedioxyphenol	O	6,7-Epoxygeranyl	4	5	0.0008	0.04	T-289
3,4-Methylenedioxyaniline	N	6,7-Epoxygeranyl	5	1	0.008	0.08	T-292
3,4-Methylenedioxythiophenol	S	6,7-Epoxygeranyl	3	10	0.8	5	T-293
p-Nitrophenol	O	Geranyl	4	0.8		30	T-210
p-Nitroaniline	N	Geranyl	7	△		△	T-211
p-Nitrophenol	O	7-Ethoxygeranyl	0.08	0.5	△	0.005	T-220
p-Nitroaniline	N	7-Ethoxygeranyl	△	△	△	△	T-227

[1] For abbreviations, see p. 390.
[2] See also Appendix, tables.

active on *Tenebrio*, as report by WAKABAYASHI *et al.* [345]. The location of the hetero atom in the side chain and its distance from the benzene ring appear to be important in determining the activity of substituted aromatic terpenoids. For example the phenoxy derivatives appeared more active than the corresponding benzyloxy derivatives as shown by REDFERN *et al.* [252].

Among series of nitrogen-containing juvenoids, the N,N-digeranyl derivatives were of interest since, they were equally or even more active than the corresponding N-monogeranyl compounds on Pyrrhocoridae. The N,N-digeranyl compounds were generally less active on other Hemipterans and only slightly active or entirely inactive when tested on other insect groups or species. In N-alkyl-N-geranyl anilines and N-acyl-N-geranyl anilines the juvenile activity remained intact when the alkyl was a small group such as methyl (T-257) or when the acyl was trifluoracetyl. When geranyl was replaced by geranoyl and attached to the aniline nitrogen atom forming thus an amidic linkage, the resulting compounds acted only on Pyrrhocoridae but not on other insects (T-305, T-346). Such selectivity was also found for phenyl geranoate-type compounds [270].

iii) As already indicated, the chain of peptide juvenoids contains nitrogen and in some cases also oxygen atoms. There are usually two nitrogen atoms representing amidic bonds in amino acids, or carbamic acid components. As seen in the Appendix tables (T-296 to T-353), compounds of this type acted only on Pyrrhocorid bugs. Replacement of the tert.-butoxycarbonyl group by the pivaloyl group which represented change of an urethane linkage to an amidic one led to an enormous increase in activity on the sensitive species. The presence of nitrogen atoms at the indicated positions of the chain and especially the character of the amidic linkage was, in our opinion, one of the principal reasons for selectivity of these compounds. When compared with remotely related N-geranyl-p-aminobenzoate (cf. T-250, T-251), the latter compounds had a much wider spectrum of juvenile hormone activity. The N-geranoyl-p-aminobenzoates (T-305) having one amidic bond common with the peptidic analogues were also completely specific to Pyrrhocorids. For additional structure-activity relationships in peptidic juvenoids the original papers should be consulted [246, 247].

Aromatic ethers derived from insecticidal synergists possess more than one etheric oxygen atom within their side chain [25, 27, 198, 252, 385]. They showed remarkable activity on *Tenebrio* and also a few other insects (cf. T-198, T-206, T-271, T-272, T-275, T-295). The fact that these ethers or the peptidic juvenoids or aromatic carbamates all possess juvenile hormone activity indicates that considerable variation is possible in side chain structures in aromatic juvenoids.

c) Hetero Atoms and Hetero Atom-Containing Groups as Substituents on Aromatic Rings

In the preceding sections we discussed structure-activity relationships with respect to the side chains of aromatic juvenoids. In this section the relationship between juvenile hormone activity and the nature and location of the hetero atom substituents on the benzene ring will be discussed. The first described alicyclic juvenoids such as juvabione [29], dehydrojuvabione [49], and bisabolene derivatives [217], the benzene ring containing compounds such as the aromatic analogues of juvabione [315, 335], the sesamex type substances [25] or the earlier geranyl phenyl ethers [26] all had the 1,4- or 1,3,4-substituted six-membered ring. The juvenile hormone activity of the corresponding meta substituted aromatic derivatives (e. g. T-187, T-202, T-248) was demonstrated by Redfern *et al.* [252]. However, the ortho substituted compounds (see T-208, T-247) were found to be poorly active or inactive. Compounds substituted at positions 3 and 4 by groups other than the usual methylenedioxy group were mostly active on *Tenebrio* (T-273, T-274, etc.) although less than similar para substituted compounds. The authors (McGovern *et al.* [198]) suggested that the presence of "electron donating" groups might yield higher activity than that of "electron withdrawing" groups, such as 3,4-dichlorophenyl. They [253] extended their juvenile hormone activity assays on the milkweed bug *Oncopeltus* using terpenoid aromatic ethers substituted at various positions of the benzene ring. From the large number of compounds tested, the 3,4-dichloro, 3,4-methylenedioxy, p-nitro, m-chloro, p-methyl, p-isopropyl, and p-tert.-butyl derivatives exhibited the highest activity. Compounds containing "electron donating" groups were more active on *Tenebrio* whereas those with "electron withdrawing" groups like p-nitro group were relatively more active on *Oncopeltus*. However, Schwarz *et al.* [293] noted that the presence of chlorine atoms in the benzene ring in juvenoids of terpenoid aniline type did not affect the juvenile activity despite suppresion of the "electron donating" ability by the chlorine. Finally, Wakabayashi *et al.* [345] studying relationships between juvenile hormone activity and electron resonance in the benzene ring in terpenoid phenyl carbamates, concluded that there were no direct correlations between these factors. In these studies the juvenile hormone activity was decreased by any substituent on the ring whether electron "donating" or "withdrawing". Bowers [27] observed that the para substituted terpenoid phenyl ethers were highly active, the meta substituted analogues were less active and other substituted ethers were inactive. These data are also in agreement with the juvenile activity for a number of aromatic juvenoids listed in Appendix tables. Despite the considerable number of compounds and activity data available for making comparisons, it is difficult to formulate a simple rule concerning ring substitution and juvenile hormone activity, except that the para substituted aromates were the most active juvenoids in all the species tested so far. Obviously, numerous additional factors are involved in this problem, such as general size of the whole molecule, mutual proportions of both substituents, electronic effects, chemical stability against enzymatic breakdown, polarity, etc.

It is also rather difficult to make any general statements concerning relationships between the activity and nature of particular substituents on the benzene ring since the data are restricted mostly to *Tenebrio* and to a few Hemipterans. Therefore, only a brief description of findings obtained with the para substituted terpenoid phenyl ethers will be presented. Absence of any para substituent in ethers with the geranyl side chains resulted in very low juvenile hormone activity or complete inactivity (T-176). Furthermore, when polar groups such as hydroxyl (T-199), amino group (T-230), free carboxyl (T-249), or sulphonylamido group were substituted at the para position, the phenyl ethers were almost inactive. However, phenyl ethers with alkoxycarbonyl groups substituted at para position were generally more active on Hemipterans and other Exopterygotes than on *Tenebrio* and other Endopterygotes. In contrast, Bowers [26] obtained just the opposite results with the corresponding methylenedioxyphenyl ethers [cf. 308]. Even the derivatives of p-hydroxyacetophenone were more active on *Tenebrio* than the corresponding p-hydroxybenzoate derivatives. When present in para position of the geranyl phenyl ethers a variety of atoms or groups yielded highly active juvenoids with a broad spectrum of activity when assayed on various species. The effective substituents were: halogen (T-185, T-193), nitro group (T-210), methylthio group (T-205), alkyl group (T-181) [235], or the already mentioned methylenedioxy group (T-289). The literature and our Appendix tables as well, list a wide variety of additional, very active compounds with variously substituted or otherwise modified terpenoid side chains. Further discussions on this point are left open and the reader is encouraged to investigate the data in the tables and make his own conclusions.

As shown in Table 12, relationships between the juvenile activity of phenyl geranyl ethers with respect to the kind of p-substituents were basically similar to those found in the corresponding aniline derivatives. Naturally, there are some exceptions mainly among compounds related to N-geranyl-p-nitroaniline (Table 12) or terpenoid phenyl carbamates which showed differences in activity when tested on *Tenebrio* [345]. The peptidic juvenoids which are in principle p-aminobenzoic acid derivatives, lost all activity on *Pyrrhocoris*, when, the p-alkoxycarbonyl group was replaced by para substituents of any other kind [246, 247].

Comparisons of the activity data suggested that presence of a hetero atom at the para position of the benzene ring was not a necessary condition for high juvenile activity of aromatic juvenoids. This conclusion is supported by observations that either straight or branched short-chain alkyls were equal or better para substituents than the halogen atoms, acyl groups, alkoxycarbonyl groups and other groups. This was true in activity assays on *Tenebrio* and many other insect species [235, 253]. In general, the para substituents of highly active juvenoids were relatively small groups or merely atoms. From earlier discussion concerning the critical length of the molecule, the loss in juvenile hormone activity might be expected for aromatic terpenoid compounds in which the para substituent was larger than a five-carbon alkyl group.

d) Hetero Atoms as Components of Functional Groups in Acyclic Juvenoids

Almost all acyclic juvenoids with definite juvenile activity possess functional groups containing oxygen or nitrogen atoms at the end of the chain. As demonstrated with a series of simple unbranched ethers derived from n-dodecylalcohol, the ethyl ether was the most active member; and the activity was found inversely proportional to the increasing length of the alkyl ether radical [289]. In rather large group of acyclic juvenoids containing a regular isoprenoid sequence, the functional groups occurred at the "head" of the molecule. Carboxylic functional group which was most common group formed an integrated unit with the allylic double bond. In the discussions on structure-activity relationships in these and similar juvenoids, it appears useful to consider such a conjugated system as a whole, wherever it is present.

It is well known [30, 265, 280, 292, 308, 357] that a free hydroxyl group like that in farnesol (T-7), is too polar for good juvenile hormone activity. Methoxy or ethoxy groups such as those in farnesyl ethers (T-8) are far better, especially when tested on *Tenebrio* [30, 280, 289]. The majority of natural as well as synthetic acyclic juvenoids which have been reported contained an ester functional group (alkoxycarbonyl group). The size and the shape of the ester radical appeared to be quite important in the juvenile hormone activity of the compound. The most efficacious group when these compounds were assayed on various species proved to be the ethyl ester group [134, 265, 268, 312, 313, 358]. In the series of alkyl esters of dichlorofarnesoic acid, activity was maximum with the ethyl ester and then gradually decreased as a function of increasing length of the aliphatic alkyl group [313]. A methyl keto analogue (T-163) of the methyl dihomojuvenate proved more active than the parent compound on *Tenebrio* but less active on flies [385].

14*

Table 13. *Comparison of Activities of Esters and Amides of Selected Juvenoids*

Substance: Formula:	R=	ID-50 (Morph.): PYR[1]	DYS[1]	GRA[1]	TEN(e)[1]
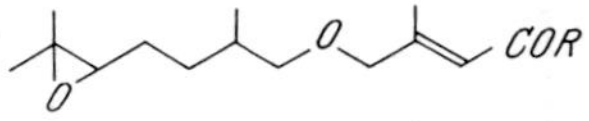	—OEt	0.01	0.005	0.1	50
	—NEt_2	8	4	△	0.005
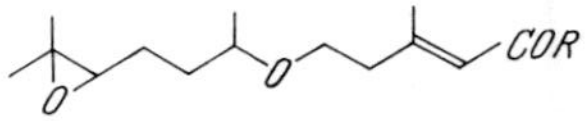	—OEt	0.08	0.05	4	100
	—NEt_2	100	100	△	0.01
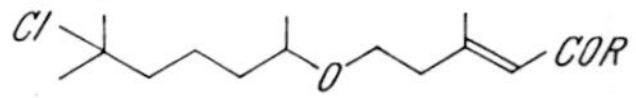	—OEt	—	0.5	10	△
	—NEt_2	—	100	△	0.1
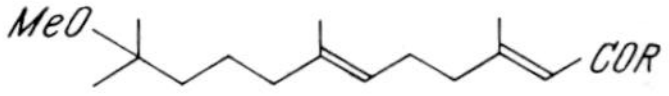	—OMe	1	0.5	10	100
	—NEt_2	1	—	△	0.07
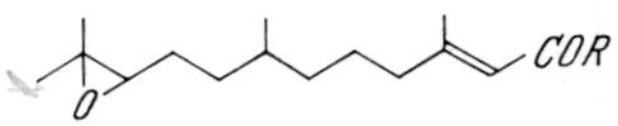	—OMe	0.5	0.05	0.5	10
	—NEt_2	4	—	△	0.05
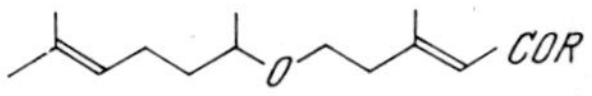	—OMe	—	5	100	100
	—NEt_2	10	10	△	0.5
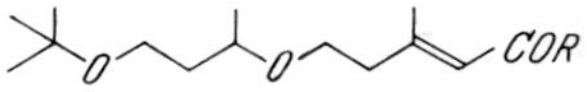	—OEt	0.005	—	100	△
	—NEt_2	100	50	△	1
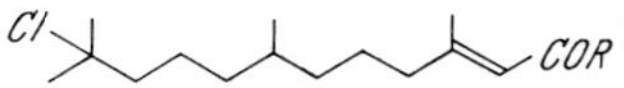	—OMe	0.05	—	5	1
	—NEt_2	0.7	0.4	△	0.04
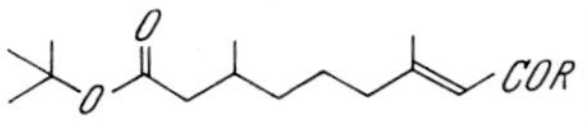	—OEt	0.001	0.007	100	△
	—NEt_2	1	1	△	10
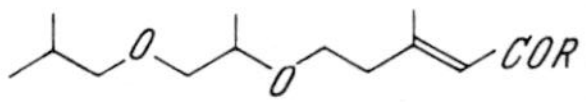	—OEt	0.08	0.1	△	△
	—NEt_2	—	△	△	100
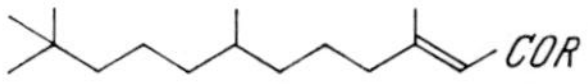	—OMe	1	0.8	50	50
	—NEt_2	—	5	△	8
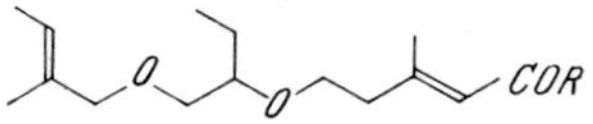	—OEt	5	—	△	100
	—NEt_2	—	△	△	500

[1] For abbreviations, see Appendix, p. 390.

The α,β-unsaturated carboxylic acid amides, especially dialkylamides were particularly active on *Tenebrio* (Table 13) [351]. The amines, namely dimethyl or diethylamines were more active than the corresponding hydroxy derivatives, i.e. alcohols [280, 289, 357] but less active than the mentioned N,N-dimethyl or N,N-diethylamides [308]. While the N,N-diethylamide was highly active on *Tenebrio* and showed little activity on Hemipterans, the N-monoethylamides were less specific having relatively high activity on both *Tenebrio* and the Hemipterans. The juvenile activity of amides derived from epoxyfarnesoic and dichlorodihydrofarnesoic acids was studied in more detail by Cruickshank [66]. Comparisons were made with activity of the corresponding alcohol, diethylamino derivative, methyl ether, methyl ester, and ethyl ester. In the case of N-monoalkylamides, the N-methyl- and especially the N-ethylamide were highly active, but activity then decreased with increasing chain length of the alkyl group. The change in activity with respect to size of the substituent was similar to that for ester radicals [313]. However, among the N,N-dialkylamides, very high activity was also shown for the N,N-di-n-butylamide [66].

The juvenile hormone activity of acyclic terpenoid carbamic acid esters was also influenced by nature of the ester alkyl group, as measured in assays on *Tenebrio* [292, 345].

An analysis of the chemical structures which have been reported reveals that simple and inactive geranyl or citronellyl chains *per se* (and other somewhat similar chains) were very often incorporated into active juvenile hormone substances by an attachement of various, sometimes very diverse structural groups of determined size and character. This was demonstrated for a large group of aromatic terpenoid juvenoids (p. 177), geranyl ethers derived from esters of aliphatic β-hydroxy acids (p. 142), geranyl ethers derived from ethylene glycol alkyl ethers [27], and terpenoid alkyl carbamates (p. 169), and various oxa-analogues [322] (p. 165). Importance of all these structural modifications in imparting juvenoid activity to aliphatic chains is understood if we realize that changes in the geranyl chain alone may increase the juvenile hormone activity by one to three orders of magnitude while alterations the attached structural groups may increase activity by as much as six orders of magnitude.

6. Juvenile Activity of Optical Isomers

Studies on relationships between biological activity and optical isomerism of juvenoids have been limited by the availability of experimental material. The synthetic (—)-enantiomer of the naturally occurring (+)-methyl dihomojuvenate (see chapter C, p. 157) was not

100 per cent pure so that it is not clear whether its activity was due to contamination with the native enantiomer. Our assays on (+)-juvabione and its enantiomer, (—)-juvabione, as well as on the corresponding diastereoisomers, (+)- and (—)-epijuvabione [240] obtained by the courtessy of Hofmann-LaRoche laboratories did not reveal any large differences in the activity of the optical isomers [322]. On the other hand, the activity of peptidic juvenoids was strictly dependent on the L-configuration of the central amino acid. The presence of any amino acid of the D-series in this central position led to the loss of juvenile hormone activity [246, 247, 308, 322]. The spatial arrangement of the peptidic chain and the considerable rigidity of the whole molecule appear to constitute one of the most important factors in determining the activity of peptidic juvenoids. These requirements cannot be, however, applied to juvenoids of other types, e. g. juvabione and other terpenoid compounds.

7. Chemical Structure and Species Specificity

In the preceeding sections of this chapter the effects of specific structural changes in juvenoid molecules on juvenile activity were described. It remains to identify which types of juvenoids preferentially affect certain insect groups or species. Our conclusions should be viewed as a starting point for further discussions for they are based on limited data obtained on a limited number of laboratory species.

The results of juvenile activity assays indicated that some insect species have different sensitivity to certain types of juvenoids. Minor structural changes in the substance led to a general increase or decrease in activity in all species tested. On the other hand, a minor structural change sometimes yielded compounds whose activity was completely different for one species than for others. A "classical" example of this is the difference in activity of farnesoates (T-17, T-18) following conversion to dichlorodihydrofarnesoates (T-30 to T-32). This conversion led to tremendous increase in the juvenile hormone activity for Pyrrhocorid bugs but a drastic loss of activity was observed when the compound was tested on *Tenebrio* [268, 313].

With regard to insect taxonomy, the group of Apterygota have not been discussed because of the lack of information and smaller economic importance. In the Pterygote insects the juvenoids act on representatives of both the Exopterygotes and Endopterygotes. It has already been pointed out that there are series of acyclic and aromatic juvenoids which were equally active on various orders of Exopterygote as well as Endopterygotes. A major difficulty in determining the species specificity of juvenoids and their action on higher taxonomic levels is that the available data are mostly derived from comparisons made between such unrelated insects as *Pyrrhocoris* (Exopterygota, Hemiptera, Pyrrhocoridae) and *Tenebrio* (Endopterygota, Coleoptera, Tenebrionidae).

Suchý *et al.* [335] concluded that only a relatively slight change in the specificity of juvenoids occurred at the species level, but a somewhat greater took place at the genus level whereas considerably different specificity occurred at the family or at higher taxonomic levels. Although these conclusions were based on experiments with aromatic juvabione-type compounds (cf. T-239 to T-242) in Hemiptera, they seem to have wide applicability [308]. The idea of similarity in juvenoid action on representatives of the same family [335] is supported by the work of Sláma *et al.* [312] obtained on representative species of three genera of the Pentatomid family, two genera of Pyrrhocorid family, and one representative species of Lygaeidae. There are juvenoids affecting representatives of more families with more or less equal intensity while there are also such compounds which are selectively acting on families belonging to the same higher taxonomic unit. This was found in studies performed on representatives of six different families of insects [314]. Comparison of our data on *Triatoma* (cf. Appendix tables) with those presented by Wigglesworth [358] for *Rhodnius* which belongs to the same family (Reduviidae) suggest similar sensitivity to the juvenoids.

These data indicate that insects with respect to their taxonomic classification share certain similarity in their sensitivity to juvenoids. A question which is often raised is, whether it would be possible to predict which juvenoids would be effective on new insect species. From our experience we know that using taxonomical position as a criterion such prediction has been possible in many cases. However, it has to be noted that for some unknown reasons, representatives of two taxonomically unrelated groups had the same responses to various juvenoids while representatives of some closely related groups were quite different in their response to the juvenoids. Therefore, the taxonomical criterion for judging the effectiveness of juvenoids is not universal.

One insect family with an unusually high susceptibility to juvenoids are the Pyrrhocoridae. Its representative species have been sensitive to almost all kinds of juvenoids. Moreover, many types of juvenoids act exclusively on insects belonging to this family. These compounds include for example all juvabione type compounds, aromatic derivatives of juvabione, and peptidic juvenoids. Some peptidic juvenoids (cf. T-335 to T-337, T-340) were active on the larvae of Pyrrhocoridae in picogram amounts whereas no activity was observed on any other insect species tested including other Hemipterans. In addition to those juvenoids which showed exclusive selectivity on Pyrrhocoridae, there were compounds that possessed a thousand times more activity on Pyrrhocoridae than on any other species tested. Among these compounds with high activity were the N-geranyl and N,N-digeranyl derivatives of alkyl p-aminobenzoates (cf. T-250, T-253, T-258, T-263), numerous acyclic juvenoids of the farnesoic acid type with additional substituents at C-7 such as chlorine, epoxy group, alkyl or alkoxyl, various types of oxa-analogues, and other compounds; for more information see the Appendix tables.

In contrast to the effect on Pyrrhocoridae, the hydrochlorination of farnesoates did not lead to higher activity on the Hemipterans of the family Pentatomidae. Furthermore, Pentatomids were quite resistant to numerous types of juvenoids with sometimes enormous activity on Pyrrhocoridae. In *Triatoma* (Reduviidae), the N-geranyl and N,N-digeranyl derivatives of alkyl p-aminobenzoates (T-246, T-250, T-253), were also highly active as they were in Pyrrhocoridae. With this exception, there were enormous differences between both families in their susceptibility to juvenoid types (cf. T-31, T-123, T-178, T-220). Another peculiarity of the family Reduviidae was that aromatic juvenoids containing methyl esters showed higher activity than ethyl esters when tested on species of this family (T-250, T-251, T-253, T-254). In other insects these relations were reversed. In another Hemipteran family Lygaeidae (*Lygaeus*) the aromatic juvenoids with a shortened side chain (T-244) were relatively more active than the compounds carrying the geranyl chain (T-246) which is not the case in other Hemipteran families [268].

The most intensive studies of structure-activity relationships in juvenoids were performed on *Tenebrio*, an insect easy to breed in the laboratory. However, comparisons of the activities are complicated by the fact that the literature data were expressed in various arbitrary units and thus it was difficult to convert them into standard ID-50 values. Some of the juvenoids, which are more or less specific for *Tenebrio* were N-alkyl or N,N-dialkyl amides of some acyclic acids. Replacement of the ester functional group by the amidic group—in somecases—caused an increase in activity by as much as four orders of magnitude when tested on *Tenebrio* while loss of activity can occur for many other insect groups [268]. A few examples of this change in activity due to replacement of a functional group are shown in Table 13. Acyclic amides containing additional substituents at the branched terminal part ("tail") of the carbon skeleton exhibit exceptionally high activity on *Tenebrio*. Geranyl phenyl ethers with ethoxy or propoxy groups at the C-7 of the geranyl chain (T-190, T-219 to T-221) were more or less specific aromatic juvenoids for *Tenebrio*. With butoxy (T-223) or larger groups instead of the propoxy one there was a sharp fall in activity when assayed on *Tenebrio*. In contrast to their high activity on Hemipterans, aromatic juvenoid esters were relatively little active or inactive on *Tenebrio* (cf. T-244, T-254, T-257, etc.).

Mori [213] has presented structure-activity data for compounds active on *Tenebrio* and also *Tribolium* which belongs to the same family, Tenebrionidae. The data obtained on both species further support the assumption of similarities in sensitivity to juvenoids within one insect family. Similar structure-activity relationships with Tenebrionidae were obtained in another family of Coleoptera-Dermestidae. However, unlike Tenebrionidae, this family was less sensitive to amides and more sensitive to dichlorodihydrofarnesoates. Except for data on Tenebrionidae and Dermestidae, little is known about other families of Coleoptera. Our preliminary data with some juvenoids suggest that the structure-activity relationships in Chrysomelidae and in some other Coleopterans whose larvae are feeding on fresh leaves may be considerably different from the situation as found in *Tenebrio*.

Lepidopteran insects were sensitive to substances related to homojuvenates and also to the esters of 6,7-dihydrofarnesoic acid (T-16) and their 11-chloroderivatives (T-23, T-24). However, the activity of the latter compounds was far below the activity of homojuvenates (cf. T-151, T-155). By comparing the activity of compounds on *Galleria* (Pyralidae) with that on *Tenebrio*, it can be shown that the amides of acyclic juvenoids were much less active on this Lepidopteran. On the other hand, both *Tenebrio* and *Galleria* responded in a similar manner to 6,7-dihydrofarnesoates (T-16), or 6,7—10,11-tetrahydrofarnesoates [134] (cf. T-13), as well as to 3,4-methylenedioxyphenyl geranyl ethers (T-289), but were resistant to dihydrodichlorofarnesoates. Other similarities in activities may be found by comparing the data in Appendix tables. Since the activity of homojuvenates (T-151, T-155) on *Galleria* and other Lepidopterans [241] was up to three orders of magnitude higher than that of other juvenoids, it appears reasonable to assume that these juvenoids were at least partly specific for the Lepidopterans [294, 308]. However, as the data indicate the homojuvenates also acted on many other unrelated insects.

In Orthopteran insects such as families Blattidae [242], Acridiidae [225], and Gryllidae [314] the homojuvenates also have relatively good activity. Additional compounds which were active on these insects were terpenoid phenyl ethers, N-geranyl or N,N-digeranyl p-aminobenzoic acid esters, and other aromatic juvenoids, but their activity was not as high as that of some acyclic juvenoids [314]. This was true for *Gryllus* as shown in the Appendix (T-121 etc.), and also for *Locusta* and *Schistocerca* [225].

From all the available literature data and that compiled in the Appendix one could speculate about many other possible structure-activity relationships between the juvenoids and the sensitivity of particular insect species or groups. We feel that this is premature since complementary data on additional insect species are lacking. In conclusion, we may point out that there are certain types of juvenoids with a broad activity spectrum on diverse insect species or groups. An example of these compounds are dihomojuvenate derivatives, farnesoates and similar acyclic juvenoids, and 3,4-methylenedioxyphenyl ethers among the aromatic ones. Others, like variously substituted aromatic terpenoid juvenoids act on a limited number of insects. Finally, juvabione or peptidic type juvenoids are strictly specific for a single insect family, the Pyrrhocoridae. The final properties and utilization value of new synthetic materials with insect hormone activity can only be estimated from activity data obtained on representatives of few insect families, because the family of insects is the lowest taxonomical unit where large divergence to juvenoid responses takes place.

E. Physiological and Biochemical Effects of Juvenoids

The compounds known as juvenile hormone analogues or juvenoids, comprise a rather wide spectrum of acyclic or aromatic structures. With the exception of compounds, such as homo- and dihomojuvenate isolated from Cecropia, and some compounds of plant origin, most of the juvenoids were prepared synthetically. They represent hormono-mimetic agents which mimic the physiological effects of the endogenous CAH, but it still remains to be resolved whether there exist common structural patterns between some of these synthetic juvenoids and CAH. Such a hormonomimetic action has been known in several other hormones. We may recall the situation with oestradiol, a steroid sexual hormone of vertebrates, and diethylstilbestrol which has some structural features in common with oestradiol but is not even a steroid compound.

The growth hormones induce sequences of biochemical, physiological, and morphological changes in the body. Our problem with juvenoids is that we still do not know exactly whether they stimulate the above sequence of developmental events from the very beginning, as does the CAH, or from some intermediate point, since most of our knowledge is based on analysis of the end products of the hormonal reaction. However, as it is documented later in the text, similarities between the effects caused by juvenoids and those stimulated by the endogenous hormone agree in specific biochemical and physiological details. This may indicate that juvenoids influence the developmental processes at a very low subcellular level, if not from the same receptor site which has been originally tuned for the CAH molecule.

In this chapter we have discussed physiological conditions for juvenoid action on insect development using comparative criteria obtained from our studies with juvenoids of different types, and with insects belonging to different taxonomic groups. The basic information on physiological effects of juvenoids can be found in previous review articles by BERKOFF [16], EL-IBRASHY [80], HIGHNAM and HILL [118], RÖLLER and DAHM [265], SLÁMA [308], WIGGLESWORTH [361], WILLIAMS [374], and WYATT [386].

1. Developmental Periods Sensitive to Juvenoids

The first apparent similarity in physiological action between juvenoids and CAH is manifested by the fact that both are effective at certain strictly delimited developmental periods, referred to as sensitive periods. It is assumed that physiological and biochemical conditions which are associated with these sensitive periods may play a dominant role in understanding the mode of action of juvenoids. One of such conditions is represented by a long known observation that CAH action is always correlated with the prothoracic gland hormone activity, in all immature insect stages [118, 227, 243, 361]. RÖLLER and DAHM [265] confirmed the existence of the relationships using combined injections of dihomojuvenate and ecdysone into isolated parts of the body in *Galleria*. They came to a conclusion that juvenile hormone should be present before ecdysone in order to obtain maximum juvenile effects. At present it becomes quite obvious that the effects of juvenoids on all larval and pupal stages are indeed realized only in presence of prothoracic gland hormone or in presence of exogenous supply of ecdysoids. The exact nature of these interactions is not clearly understood. There is a possibility of direct chemical interaction between both the hormonal agents but this is not likely to occur for reasons discussed in further sections. More probably, the prothoracic gland hormone or ecdysoids induce specific developmental changes which, once being initiated, become—for a certain time—a subject of simultaneous juvenoid regulations. This assumption is supported by the fact that the sensitive period to juvenoids occurs at different times in various epidermal cells [173] or internal organs, and this is correlated with developmental rate rather than with prothoracic gland hormone titre. Furthermore, the sensitive periods to juvenoids occur in and are correlated with certain stages of embryonic development where the prothoracic gland hormone is absent. Thus it appears that the prothoracic gland hormone or ecdysoids stimulate development in various larval or pupal cells, whereas the CAH or juvenoids determine the degree of qualitative morphogenetical changes in the initial phases of development. In short, the former hormones decide whether development will proceed, while the latter decide on how it will proceed.

During the course of developmental cycles in an insect cell, whether this is stimulated by ecdysone or not, there occur certain critical moments after which CAH or juvenoids are unable to further change the developmental programme. For example, juvenoids were unable to affect epidermal cells when applied at the end of the moult cycle, e.g. at the time when the cells started to secrete a new cuticle [173]. Moreover, it appears that this critical moment for juvenoid action is over as soon as the cells start to divide [308]. We may thus conclude that the sensitive period to juvenoids begins when the cells are in the interphase state, and that it is extended until the period of caryokinesis. This is the time of DNA replication which coincides very well with a period when prothoracic gland hormone brings the cells from dormancy into an "activated" state. In general, these initial stages of cell development, sensitive to juvenoids, represent a dynamic state which can be influenced not only by juvenoids but also by many other physiologically active compounds. In embryonic development the blastoderm stage is the most sensitive to juvenoids. In this stage the embryos own genome is turned on and a derepression of specific genes, operons, and polyoperons takes place [373].

In connexion with the bioassays for juvenoids (p. 97), we have already described locations of the sensitive periods in the development of several insect species. We have also indicated that the usual position and duration of these periods can be influenced by some physiological factors or by alterations of the usual developmental conditions.

For example, epidermal injury or induced regeneration can prolong duration of the sensitive period, or possibly reinduce the sensitive period in the regenerating cells. In this instance the sensitive period has been influenced on localized parts of the body. However, of special interest are such factors which delay or prolong the sensitive period in the whole body. These factors are characterized by two general features: suppression of the developmental rate (low temperature) and inhibition of the prothoracic gland hormone secretion (starvation).

The rates of development or morphogenesis are profoundly inhibited or completely arrested with temperatures below 5 °C in most insect species. As shown in Fig. 17, the critical period for ID-50 effect is prolonged in dependence on decreasing temperature. Furthermore, it is evident that this prolongation is directly proportional to the length of the whole intermoult period. Also in many other insect species the sensitive period can be prolonged or postponed by low temperature. This has practical significance for bioassays on rapidly developing stages with relatively short sensitive periods. Slightly lowered temperature merely prolongs the duration of the sensitive period, whereas much lower temperatures (around 5 °C) can cause its complete delay, provided that the experimental animals survive. On the other hand, starvation does not prolong the sensitive period but usually causes its delay. This is connected with the fact that in most insect larvae secretion of the prothoracic gland hormone is dependent on feeding [118, 361]. The necessary feeding period is commonly 24 to 36 hours but in some cases, such as in Dipteran larvae [391] it is less than 10 hrs. When larvae were starved since the moment of ecdysis, the hormone was not secreted, the development was suspended, and the sensitive period to juvenoids was delayed until

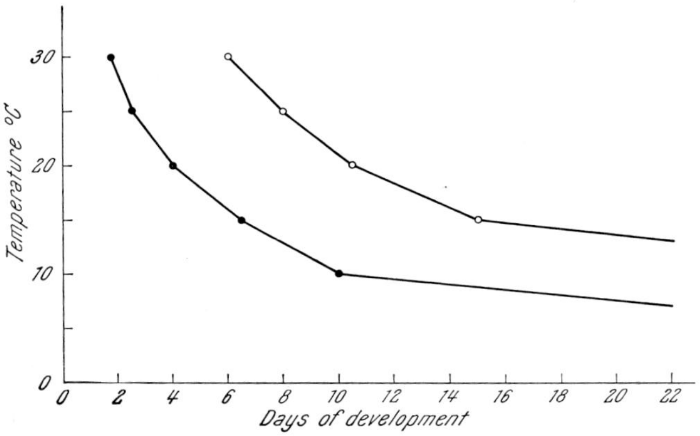

Fig. 17. Developmental periods in last instar larvae of *Pyrrhocoris* when excessive amounts of juvenoids cause just 50 per cent morphological change at different breeding temperatures. Open circles indicate the total length of the intermoult period. Note that the sensitive period to juvenoids as well as the length of the intermoult period are very prolonged at temperatures below 10 °C. The exact values cannot be determined due to high mortality

feeding and development were resumed. It may be assumed that other factors which inhibit development in general may also prolong or postpone this sensitive period to juvenoids.

An analysis of various morphological and physiological effects caused by juvenoids has revealed the existence of considerable tissue or organ differences with respect to the critical moments after which the cells become insensitive to juvenoids. In most species the epidermal cells of imaginal discs in Endopterygotes or the epidermal cells of the wing lobes, thoracic and head appendages, and external genitalia lose their sensitivity to juvenoids sooner than, for instance, the epidermal cells of the abdomen [308]. Such a shortened sensitive period is usually found in epidermal structures associated with most profound metamorphosis transformations. The described differences in sensitivity of the epidermal cells to juvenoids are commonly seen on intermediates produced in all bioassays with metamorphosis stages. Furthermore, some internal organs such as the nervous system, ovaries, or prothoracic glands [390] may loose their sensitivity to juvenoids sooner than do most of the epidermal cells. This was documented by appearance of special developmental deviations, as for instance extra-larvae which were unable to moult, almost perfect extra-larvae which contained ripe eggs in the ovaries, and larvae with adult sexual instincts [390]. It appears that the differences in position of the critical moments of sensitivity to juvenoids in various tissues or cells are most probably associated with different velocities of growth and developmental processes. Thus, mitoses are initiated sooner in tissues with shorter duration of the sensitive period to juvenoids, and vice versa.

According to the above statements, juvenoids can cause formation of tissue intermediates which possess a mixture of structural patterns of two or even three developmental stages. However, at the cellular level it has not been clearly resolved whether juvenoids act according to an all or nothing rule or whether they can cause appearance of some intermediate cell types. LAWRENCE [173] found mosaics of larval and adult patterns or intermediate types in an epidermal cell of *Oncopeltus*. In contrast, the cellular pigment pattern was influenced according to all or no effect. We have analysed a large number of adultoids with mosaics of larval and adult abdominal epidermal patterns in *Pyrrhocoris*. There were only larval or adult cell types with no positive evidence for the intermediate types, although the cuticle was intermediate at the margins of larval and adult cell islands, and some cells had atypical shape.

2. Physiological Conditions for Induction of Supernumerary Instars and Intermediate Forms

It is believed that one of the main physiological functions of the CAH is to mobilize products obtained from assimilated food, and to stimulate utilization of these products for growth of the larval tissues or eventually, for growth of the ovaries. Indeed, there is hardly a period in insect development where CAH would be secreted in a nonfeeding stage. In larvae, the number of feeding instars is proportional to the time in which the CA secrete the hormone. And, during this time there occurs a tremendous increase of the larval body size. One might speculate about production of some real giant forms of insects as a result of prolongation of the feeding period by means of juvenoid treatments. However, there is no experimental proof so far available which would confirm such theoretical expectations. In all instances the rate of growth of the larval body decreases rapidly or stops completely with each of the supernumerary larval instar. Such an inability of supernumerary larvae produced by juvenoid treatment to keep growing is associated with a whole complex of physiological factors. First of all, growth of the larval body seems to be adjusted by evolutionary adaptations to certain optimum size which is limited by physical factors, such as solidity and permeability of internal membranes, diffusion of gases, or strength of the integument. Furthermore, the body contains surface structures such as wing lobes, cylindrical structures as are legs or antennae, and also spherical structures as is head and sometimes abdomen. The biometrical proportions between these structures change progressively from one instar to another. In the supernumerary larval instars the bio-

Table 14. *The Fate of Extra-Larval Instars of Dysdercus cingulatus. Indirect Topical Application with Compound No. T-326 on the Penultimate Larval Instar. Calculated Dose Approx. 1 µg per Spec. Only those Extra-Larvae which Successfully Ecdysed Were Used Out of Some 2,000 Spec. Treated*

Developmental fate of the extra-larvae after next moult	No. of spec.	Per cent
Died during development	180	21.2
Unable to moult at all	125	14.7
Ecdysed successfully into 2nd extra-larval instar	5	0.6
Failed to shedd off the old exuviae	543	63.5
Total	853	100

metrical differences become critical and result frequently in appearance of ecdysial failures followed by death. Finally, after treatment with juvenoids there appear disproportions in function of other components of the neuroendocrine system, which usually leads to disturbed synchronization of the moult cycles. Thus, using juvenoid applications on insect larvae, one can obtain mostly one supernumerary larval instar, rarely two, and quite exceptionally more than two. Dimensions of such supernumerary larvae are considerably larger when compared with a normal last instar larvae (see Figs. 9–11), but further growth of these extra-larvae is limited, even when additional supernumerary moults do occur.

Unless special methods of application have been used, it is rather uncommon that the supernumerary larvae would metamorphose into large adults. Under normal conditions of application this usually happens in less than 1 per cent of the animals. All the rest of them are unable to moult again or die during an unsuccessful attempt to ecdyse into adultoids, as seen in Table 14. In pupae of some insects, such as *Dermestes*, it is possible to obtain up to four successive extra-pupal instars (see Figs. 23 and 25). Since these are nonfeeding stages, the size of their body is progressively diminished and all of them, including also adultoids, are unable to escape from the old cuticle.

With regard to comparative aspects, it appears that the transformation of larvae into adults in Exopterygote insects, as well as the transformation of pupae into adults in Endopterygote insects can be easily affected by juvenoids. The supernumerary larval or pupal instars and all types of adultoids have already been induced in most orders of both Exopterygota and Endopterygota. In contrast, the larval-pupal transformation in many Endopterygote insects is controled by special hormonal conditions which prevent juvenoids from causing supernumerary instars. Thus, there are many species of Diptera-Cyclorrhapha, Hymenoptera, and Coleoptera, where treatment by any amount of juvenoid at any time of the last larval instar is never followed by a supernumerary larval moult, although the same juvenoid compound might be extremely active on the pupal-adult transformation in the same species. The usual effect of juvenoids on last larval instars of these insects is characterized by a delay of pupation or a delay of puparium formation respectively. Since all these insects have rapid development in the young larval period, it was assumed [308, 311] that there probably exist some special, endogenous, anti-juvenile mechanisms. These mechanisms may prevent occurrence of spontaneous lethal metathetelies which might result from incomplete breakdown or excretion of the endogenous CAH present in the earlier young larval stages. Due to such a physiological adaption, development in the last larval instars appears to be suspended for as long as the juvenile hormone remains in the body. This is confirmed by the fact that the delay of pupation in these species is correlated with the dose of the juvenoid applied.

These mechanisms protecting certain Endopterygote larvae against spontaneous disturbances of metamorphosis, do not occur in all Endopterygotes. For example, they may not occur in *Galleria*, or at least, their presence is limited here to a short transitory period at the beginning of the last larval instar. Thus, in *Galleria* it was possible to cause supernumerary larval instars or larval-pupal intermediates by treatment of 5-day-old last instar larvae. Treatments performed at the beginning of the last instar resulted in the formation of either perfect extra-larvae or pupae without intermediate forms [295]. We have a feeling that the effect of juvenoids on the larval-pupal transformation is a complex problem which requires further experimental analysis. Therefore, we have concentrated further discussions on the effects of juvenoids to the larval-adult and pupal-adult transformations.

In contrast to the above discussed situation, juvenoids do not inhibit development in the last instar larvae of Exopterygotes and in the pupal stage. In this case physiological conditions required for induction of supernumerary instars, e.g. extra-larval or extra-pupal instars, as well as those for obtaining intermediate forms, e.g. larval or pupal adultoids, are relatively simple. Thus, in order to cause a perfect supernumerary instar, the concentration of juvenoid in the body must reach certain effective threshold level and, in addition, it must keep beyond this critical concentration throughout the whole course of the sensitive period. On the other hand, in order to cause intermediates (adultoids), the concentration of juvenoid must traverse either up or down the critical effective level within the course of the sensitive period. There are three main parameters which determine the morphological and physiological character of an adultoid: a) moment in the sensitive period when the

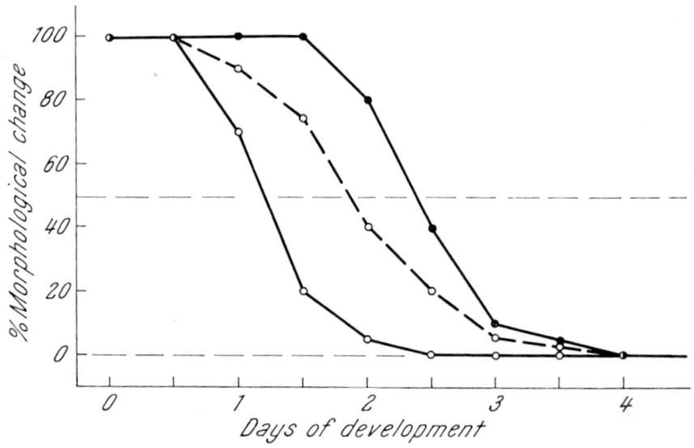

Fig. 18. Effect of excessive doses of juvenoids (1000-fold ID-50 applied in successive 12 hours intervals to tarsal segments) on differential retention of larval epidermal patterns (per cent morphological change) in the abdominal epidermal cells (filled circles), wing epidermis (open circles), and in the whole body (broken line) of *Pyrrhocoris* larvae at 27 °C

critical effective concentration was traversed; b) direction of the concentration change, e.g. whether above or below the critical level, and c) species specific differences in the sensitive periods between separate cells or tissues.

In the foregoing section we have noted that the epidermal cells which construct the head and thoracic appendages may pass the critical period to juvenoids sooner than the epidermal cells of the abdomen. This is slightly different in different unrelated groups of insects, but the general relationships in this respect may be similar to a situation which is illustrated in Fig. 18. Due to enormous diversity of insect species we can distinguish many types of adultoids according to specific patterns. However, there are two kinds of adultoids of common occurrence which deserve increased attention since their formation is conditioned by the above described differential responses by abdominal and thoracic epidermal cells. Thus, we can distinguish an "abdominal" type of adultoids and a "thoracic" type. The former, "abdominal" type is represented by intermediates which retain larval or pupal epidermal patterns on the abdomen while the head and thoracic appendages contain more or less developed adult epidermal patterns. This type of adultoids can be produced experimentally when juvenoids are applied in the middle of the sensitive period. The second, "thoracic" type of adultoids is represented by intermediates with larval or pupal patterns on thoracic appendages while abdomen shows more or less adult morphological patterns. This is the most common type of adultoid caused by ID-50 doses in the assays on Exopterygote larvae, but it is rather uncommon in the assays on pupae. Both these types of adultoids have completely different morphological appearance, although the average degree of morphological change related to the whole body may be the same in both. We did not want to complicate the evaluation procedures by introducing these additional criteria into the standard assays. However, the distinction between the "thoracic" and "abdominal" types of adultoids may appear extremely useful for judging on some properties of juvenoids and for estimation of physiological conditions connected with the sensitive periods. For example, appearance of the "thoracic" type of adultoids may indicate a rapid inactivation of juvenoids or, may suggest relatively long duration of the sensitive period. On the other hand, appearance of the "abdominal" type of adultoids suggests slow transport of juvenoids, relatively short sensitive period or late application in the course of the sensitive period.

Most of the work with juvenoids has been performed on stages with inactive CA in the body. One might argue that the effects of juvenoids could be perhaps caused by their activation of the CA. We have discussed some effects of juvenoids on functions of the CA in one of the next sections (p. 265), however, it has been well established that juvenoids act directly on the peripheral tissue without mediation of their effects through CA. This was demonstrated with juvabione on allatectomized larvae of *Pyrrhocoris* [318, 379] and with the Cecropia extracts on allatectomized larvae of *Galleria* [295]. Further evidence on direct action of juvenoids on the peripheral tissue can be derived from a general occurrence of local epidermal effects, from their action at the early embryonic stage, and, especially, from stimulation of ovarian growth in allatectomized females. Furthermore, the localized epidermal effects reveal that the action of juvenoids would not be mediated by any other centralized insect tissue.

3. Effect of Juvenoids on Growth and Duration of the Intermoult Cycles

Further similarity between juvenoids and CAH is represented by the inability of exogenous juvenoid supply to influence growth and development of the young larval instars. In general, this problem attracts less attention because the young larval instars contain actively secreting CA. Just for academic reasons, we performed repeated applications of 10,000 times increased ID-50 doses on second larval instars of several Exopterygote larvae. The results revealed that even such extremely large doses of juvenoids had no effect on development of the young larval instars. Thus, the normal growth rates and feeding periods were unaltered, ecdyses occurred in regular intervals and there was no remarkable change in morphology of the epidermal structures, until the last larval instar. Evidently, juvenoids did not interfere with the processes regulating the cyclicity of development in the young larval stage.

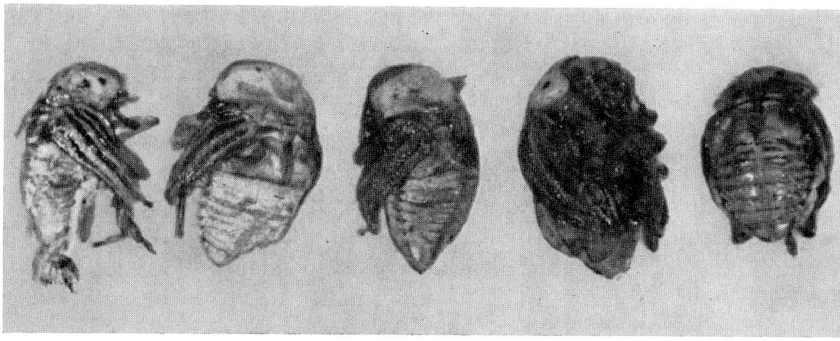

Fig. 19. Adultoids and deformed adults of *Leptinotarsa* caused by treatments of last instar larvae or pupae with juvenoids

Duration of the intermoult period, i.e. the period between two successive ecdyses, appears to be more or less constant under constant nutritional and environmental conditions. As a rule, duration of the period between the two larval ecdyses is shorter than that of the last larval instar, i.e. between larval and pupal or larval and adult ecdyses. For example, in most Hemipterans the period between two larval ecdyses is about 5 days whereas the larval-adult ecdysis occurs after 7 days. The difference is far more apparent in Endopterygote insects. Thus, for instance, in higher Diptera the larval-larval ecdyses occur in 1 or 2-day intervals, whereas pupariation takes place after more than 6 days. In *Dermestes vulpinus* each intermoult period between two larval ecdyses lasts 2–3 days, while pupation occurs after 10 days. In *Galleria* this difference is 5 to 8 days, etc. The prolonged duration of the last larval instars is undoubtedly associated with the occurrence of metamorphosis and this is in turn affected by juvenoids.

In the previous section we have mentioned that in rather exceptional cases juvenoids can delay the time of pupation or puparium formation in last instar larvae of some Endopterygote insects. Such a prolongation of the last larval instar is dependent on special hormonal conditions. In some of these insects, such as in *Leptinotarsa*, the delay of pupation caused by juvenoids is counterbalanced by shortening of the pupal-adult intermoult period. The pupal adultoids which are commonly produced in this species after treatment of the last instar larvae (Fig. 19) secrete the new cuticle earlier than do the adults.

In the majority of other cases, however, juvenoids cause an apparent shortening of the intermoult period. It is commonly observed that the

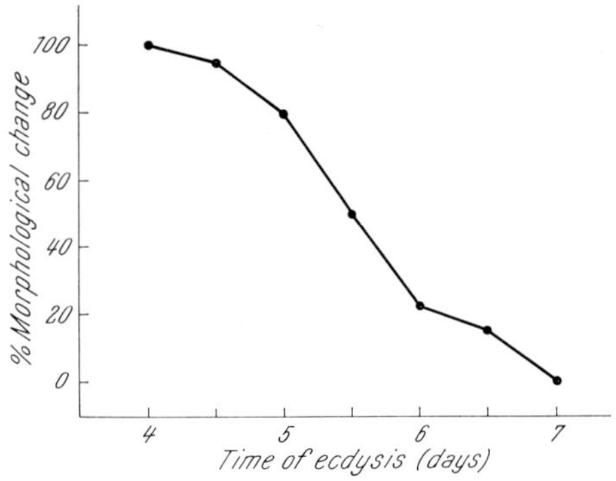

Fig. 20. Relationship between the degree of morphological change induced by juvenoids and the length of the intermoult period in last instar larvae of *Pyrrhocoris* at 27 °C

extra-pupae induced by juvenoid treatments are ready for ecdysis sooner than are normal adults. In addition, such a shortening of the intermoult period appeared to be correlated with the degree of morphological change induced by juvenoids [173, 358, 379]. This is also documented by our investigations on *Pyrrhocoris*, the results of which are presented in Fig. 20. Here the morphologically perfect extra-larvae ecdyse first, normal adults last, and in between there are adultoids of different types. However, the curve in Fig. 20 which reflects relationships between duration of the intermoult period and the degree of morphological influence, is not linear. This suggests that some physiological processes associated but not identical with the actual morphological changes are responsible for the effect. Moreover, in this insect it appears that the duration of the intermoult period in an instar developing into an extra-larva is the same as that of the previous larval instars which develop in the presence of CAH. It thus occurs that the effect of juvenoids on the duration of the intermoult period is quite similar to that produced by endogenous CAH.

These comparisons allow us to conclude that the larval-larval ecdyses on one hand and the larval-adult metamorphosis on the other hand, require different but relatively constant time. Further comparisons of the main developmental events between larval development and metamorphosis show little differences in the rate of feeding or growth of the body but large differences in the extent of cell differentiation. This occurs predominantly in the second half-period before secretion of the new cuticle. We believe, therefore, that inhibition of the cell differentiation process by juvenoids is followed by an earlier secretion of the new cuticle and that this in turn results in an earlier ecdysis. It was observed that certain adult patterns, the formation of which requires extensive differentiation, may be commonly missing on the body of adultoids [173]. Finally, it has to be added that the cuticle of supernumerary instars or adultoids is quite normal and continuous all over the body. Even in the case of supernumerary pupal instars produced by juvenoids, the new second pupal cuticle is released sooner than the adult cuticle, although these stages are never able to perform successful ecdysis.

4. The Nature of Ecdysial Failures

Inability of insects to cast off the old cuticle during ecdysis represents a most common side effect associated with juvenoid action. The effect occurs also after CA implantations, and with juvenoids it is caused by a wide spectrum of chemically unrelated compounds. Therefore, an induction of these ecdysial failures is not correlated with presence of some specific structural features in the juvenoid molecules.

According to their physiological nature we may distinguish several types of ecdysial failures caused by juvenoids. These are mainly: a) relatively slight ecdysial failures when ecdysis proceeds almost normally but the old cuticle remains fixed to the body at the tips of the legs, at the tips of the abdomen or at the distal part of the wings. This type of ecdysial failures is not lethal provided that the specimens are able to move and feed. It occurs commonly in extra-larvae or in relatively little affected adultoids of Exopterygotes. The second type (b) is also characterized by presence of ecdysial movements and other physiological changes associated with ecdysis. The new cuticle has been properly formed, the old one properly digested and split at the ecdysial line. However, these specimens are unable to shed the old exuviae and they quickly dessicate and die. This very common type of ecdysial failure occurs in adultoids of both the Exopterygote and Endopterygote insects. Finally, the third type (c) of ecdysial failures caused by juvenoids is characterized by a complete absence of ecdysial behaviours. The new cuticle may be well formed, the old properly digested, the ecdysial line may eventually rupture, but these specimens remain completely enclosed within the old cuticle. This type occurs mainly in extra-pupal instars or in pupal adultoids.

It appears that the a) and b) types of the ecdysial failures are not consequences of disturbed regulation and timing of the ecdysial process. More probably they result as a consequence of disturbed biometrical proportions of separate parts of the body. Their occurrence is not directly proportional to dose of the applied juvenoids but there exist certain correlations between their occurence and the degree of morphological change induced. Namely, these ecdysial failures occur

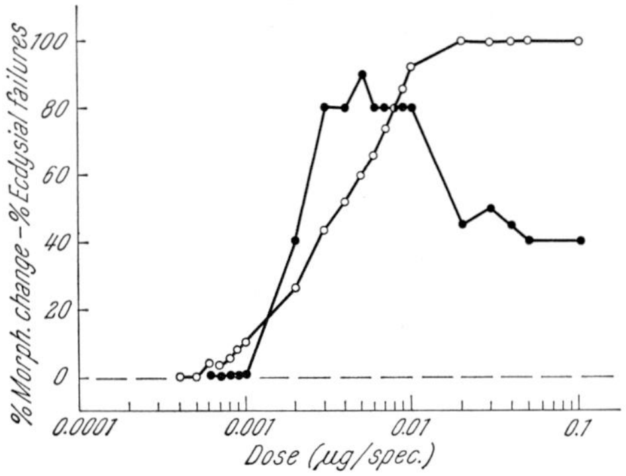

Fig. 21. Relationship between the degree of morphological change induced by juvenoids (open circles) and the occurrence of ecdysial failures (filled circles) in last instar larvae of *Pyrrhocoris* (topical treatment with methyl dihydrodichlorofarnesoate)

most frequently in adultoids with about 50 per cent morphological influence, e.g. they occur in intermediates with maximum morphological deviations from both the old and new developmental stages. These relationships are graphically illustrated in Fig. 21, which shows dependence of ecdysial failures on the degree of morphological influence obtained on a large number of *Pyrrhocoris* larvae. It is evident that the frequency of the ecdysial failures increases from 0 to about 50 per cent morphological change and then decreases again with increasing proportions of extra-larval patterns. Most of these ecdysial failures were of the b) type, as seen in Fig. 22.

Further observations on a large number of successfully ecdysed extra-larval instars of *Dysdercus intermedius* revealed that ecdysial failures were far more frequent at the time of the next extra-larval ecdysis (Table 14). Thus, only less than 1 per cent of the extra larvae out of those which attempted to ecdyse were able to form a free second extra-larval instars. In contrast to this, however, we have never observed ecdysial failures caused by juvenoids in young larval instars. Therefore we conclude that juvenoids have no direct effect on the occurrence of ecdysial failures of the a) and b) types. These can be considered as special side effects resulting from developmental abnormalities, such as are disturbed biometric proportions in the shape of the body.

Relationships between the degree of morphological influence and the occurrence of ecdysial failures of the c) type are quite different. In

Fig. 22. Adultoids and supernumerary larvae of *Pyrrhocoris* produced by contact with American paper containing-juvabione. Note ecdysial failures of different kind

this case only adultoids with less than about 30 per cent morphological change are able to undergo ecdysis, depending on species. All the intermediates with higher percentage of morphological change, including all extra-pupal instars, remain enclosed within the old cuticle. This shows that the c) type of ecdysial failures would not depend on disproportions of the biometrical values. More probably, it is caused by disturbances in regulation of the ecdysial process because some typical ecdysial symptoms such as specific ecdysial movements or increase of the internal pressure of the coelomic fluid are lacking. Furthermore, this effect occurs mainly in pupal stages where adult emergence is brought about by contractions of specific somatic muscles. Development of these adult muscles is most probably inhibited together with the inhibited histogenesis of many other adult tissues by juvenoids. It was found that in some insects specific groups of intersegmental muscles develop in each instar [361]. These muscles were specifically used for rhythmical contractions of the abdomen at ecdysis. According to this we may expect that insect pupae treated with juvenoids may lose the muscles which brought about the larval-pupal ecdysis but no equivalent muscle system for the extra-pupal ecdysis has developed.

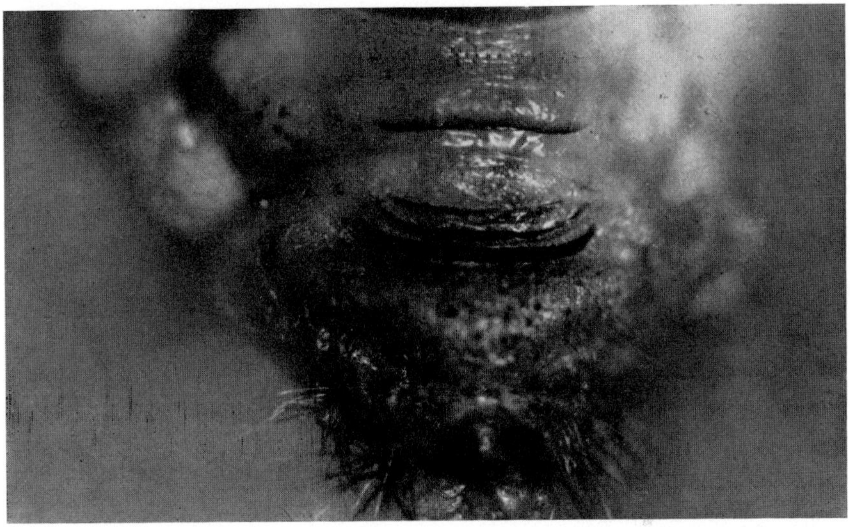

Fig. 23. Dorsal view of the distal end of *Dermestes vulpinus* pupa which underwent two extra-pupal moults after treatment with juvenoids during the prepupal period. The presence of three telescopically situated pupal cuticles is documented on three parallel layers of the tergal gintraps visible through the upper, transparent cuticular layers

Some insects like *Dermestes vulpinus* have tremendous ability to survive within the old cuticle. Treatment with juvenoids leads in this species to formation of several telescopically interlocated extra-pupal instars, all being enclosed in the first pupal cuticle, as seen in Figs. 23 and 25. A real ecdysis never occurs. Moreover, some insects with long wings in the adult stage, such as Locusts or aphids, are very sensitive to the a) type ecdysial failures. Finally, we may add that the described inability of insects to ecdyse properly after juvenoid treatment is a side effect of considerable practical significance because most or sometimes all the specimens which fail to ecdyse are ruled out from further reproduction.

5. The Nature of Juvenoid Effects on Differentiation

One of the most conspicuous effects of juvenoids on insect metamorphosis is often characterized as a maintenance of developmental *status quo*. This has been derived from a generally suppressed or delayed formation of new, qualitatively different structures, which normally originate by a process of cytodifferentiation. Consequently, in numerous papers and review articles we can find traditional statements that juvenoids act through an inhibition of cellular differentiation [16, 80, 118, 173, 295, 308, 361, 374, 386].

Recent studies with rapidly penetrating and highly active juvenoids have revealed a surprising recognition; namely that the process of cytodifferentiation occurs in insect development long after the tissues have lost all their sensitivity to juvenoids. In other words, juvenoids appear completely ineffective at the time of intensive cell divisions followed by cellular differentiation. This phenomenon is quite apparent in metamorphosis of Exopterygote insects, although a brief duration of the sensitive period in pupae suggests similar relationships. Because of such a lack of coincidence between the timing of juvenoid action and cytodifferentiation, an inhibition of the latter process by juvenoids must be regarded as an indirect effect. Thus, in order to be effective a juvenoid must act long before the process of cellular differentiation is set into motion, probably at a time when the gene sets are being reprogrammed for a new developmental cycle.

Further evidence which favours an indirect effect of juvenoids on cellular differentiation is based on the finding that there exist special forms of "larval" differentiation which are not inhibited but are, in contrast, stimulated by juvenoids. An example of such "larval" differentiation is represented by development of some small epidermal organs in larvae. In *Rhodnius* [358] these structures differentiate in each larval-larval moult and this happens also during development to extra-larvae. Another example is given by special "pupal" differentiation which takes place during development of extra pupal instars induced by juvenoids in *Dermestes* [311]. As seen in Figs. 24 and 25, there exist special epidermal tubercles on thoracic tergites of pupae. With each of the successive extra-pupal moults induced by juvenoids more (about twice as much) of these exclusively pupal structures differentiate. Hence, we observe that juvenoids indirectly inhibit differentiation which is a prerequisite for the adult morphogenesis but simultaneously stimulate the pre-existing type of "larval" or "pupal" differentiation.

The extent of differentiation among the tissues is largely dependent on the state, amounts, and inherited developmental programme of special stemm cells (germ cells). Under influence of juvenoids the

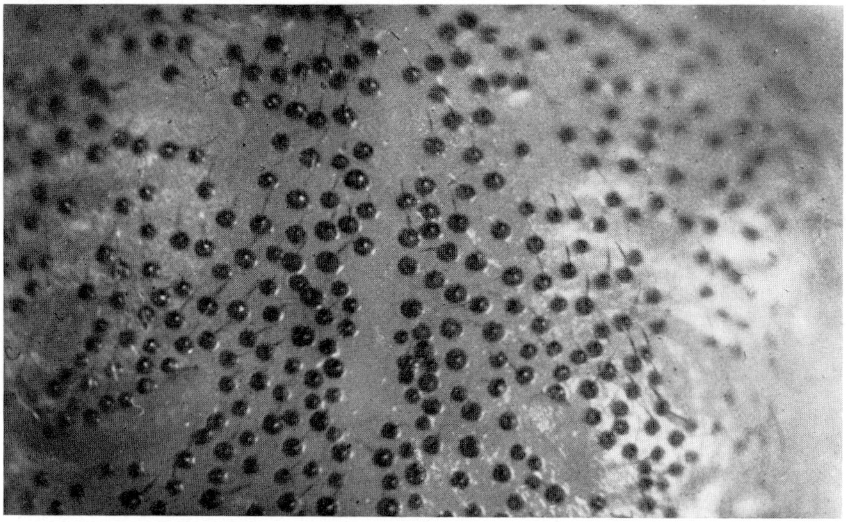

Fig. 24. Epidermal structures in the form of tubercles with hairs on the thoracic tergites of *Dermestes vulpinus* pupae. These structures represent specific pupal formations which are preserved and multiplied after treatment with juvenoids

Fig. 25. Normal pupa of *Dermestes vulpinus* (left), the cuticle of the second pupal instar containing the third pupal instar inside (middle), and the third pupal instar induced by juvenoid treatment. Note that there are no signs of "larvalization" of the integument

undifferentiated stemm cells are maintained in their existing immature form while all their capabilities for further growth and differentiation remain preserved. Moreover, in each developmental cycle which takes place in the presence of juvenoids, the number of these stemm cells increases according to a genetically programmed schedule. Later on, during metamorphosis, all these mother stemm cells differentiate and construct correspondingly large or more numerous adult structures. Actually, these observations were originally made with extirpations and transplantations of the CA in *Pyrrhocoris* [310] but we have confirmed that juvenoids cause the same effects as CAH in this respect. A highly schematic outline of the above relationships is shown in Fig. 26.

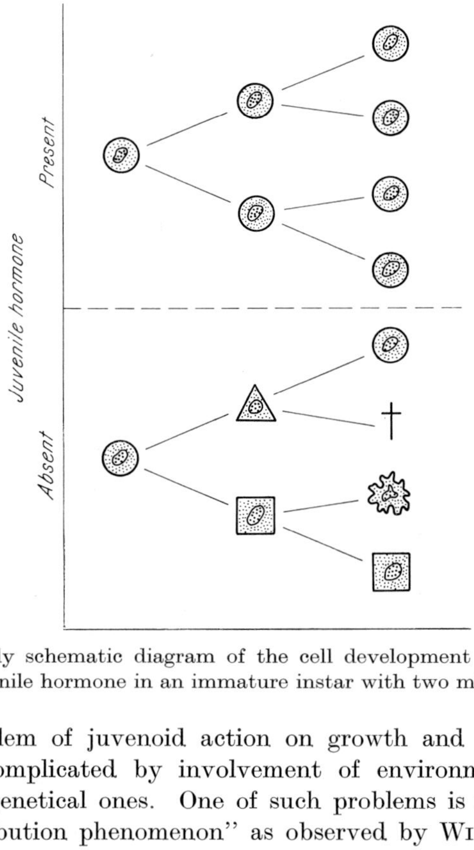

Fig. 26. Highly schematic diagram of the cell development in presence and in absence of juvenile hormone in an immature instar with two mitotic waves

The problem of juvenoid action on growth and differentiation is sometimes complicated by involvement of environmental factors in addition to genetical ones. One of such problems is represented by a "space distribution phenomenon" as observed by WIGGLESWORTH, see LAWRENCE [173]. In this system which occurs usually in feeding stages such as in larvae of *Rhodnius*, the extent of differentiation of new structures is qualitatively dependent on hormones, although quantitatively it is dependent on certain feed-back mechanisms dependent on the extent of feeding or on the degree of stretching of the epidermis. In this case juvenoids are conditioning differentiation of the "larval" epidermal organs but the actual number of cells which undergo differentiation depends on the degree of stretching of the abdomen. However, the above described tubercles on thoracic tergites of *Dermestes* pupae (Figs. 24 and 25) show a completely different distribution patterns. Their number is doubled on the same area with each extra-pupal moult induced by juvenoids. This shows that an extent of differentiation stimulated by juvenoids in pupae of *Dermestes* is either genetically determined or regulated by other than nutritional factors.

When making comparisons between metamorphosis development on one hand and extra-larval or extra-pupal development on the other hand we observe certain differences in timing and extent of mitoses among various tissues of one and the same developmental instar (cf. p. 34). For example, some organs or tissues forming specific adult structures, such as imaginal discs of appendages or wing lobe cells, show, as a rule, earlier appearance and more extensive mitotic activity during metamorphosis, which is inhibited by juvenoids. Other tissues as for instance abdominal epidermal cells show relatively less pronounced differences in mitotic activity between metamorphosis and larval development. Finally, specific larval tissues such as abdominal pseudopods are able to undergo mitotic activity only in the presence of CAH or juvenoids. This shows that the developmental responses of separate larval or pupal tissues to juvenoids are not universal. Another fact we have to realize is that there is no evidence so far for any direct stimulation of mitotic divisions by juvenoids in the absence of active prothoracic glands or ecdysone in any larval or pupal tissue. For instance, in *Aeschna* there was one wave of mitoses stimulated by ecdysone; juvenoids did not stimulate mitoses but affected the determination of differentiation. In tissues where mitoses were independent of ecdysone, mitoses could only occur in the absence of juvenile hormone activity [220]. It is quite probable that the differences in mitotic activity between extra-larval and larval-adult moult cycles are not a direct cause of juvenoid action as one may assume; they are rather a consequence of an altered developmental programme.

At the beginning of metamorphosis some cells or whole tissues associated with specific larval functions undego rapid disintegration. Their breakdown is manifested by formation of chromatic droplets as has been discussed in Part I, B. This process of "chromatolysis" requires ecdysone and its intensity differs from tissue to tissue. It was observed [392] that Cecropia extracts inhibited mitoses in some tissues as well as they inhibited the chromatolysis process in other tissues. Furthermore, it appeared that certain tissues having no alternative developmental programme but differentiation into adult structures, such as the rectal pads in Saturniid pupae, showed an overall growth inhibition in response to juvenoids but no disintegration [155]. Numerous examples of extra-larvae and extra-pupae produced by treatment with juvenoids (these stages retain larval or pupal tissues which otherwise disintegrate) suggest that prevention of disintegration of the old tissues by juvenoids may be as common consequence of juvenoid action as is inhibition of adult histogenesis in imaginal discs. It seems that the action of juvenoids on the chromatolysis process may thus be similar to their action on mitoses; e.g. an inhibition of mitoses in cells which realize a new "adult" programme on one hand, and an inhibition of chromatolysis and disintegration in cells which realize an old "larval" or "pupal" programme on the other hand, both being consequences of altered developmental instructions.

Juvenoid influence on differentiation among epidermal cells can be recognized according to structural changes in the newly secreted cuticle. Zlotkin and Levinson [393] considered the newly formed cuticle secreted by *Tenebrio* pupae after juvenoid treatment as abnormal. Suggestions were made on a possible "larvalization" of the pupal cuticle. Also Caveney [46] assumed a possibility of reversal of metamorphosis caused by juvenoid treatments; in agreement with some other suggestions made in this respect [173]. However, due to enormous diversity of insect species and developmental habits the problem of reversal of metamorphosis or "larvalization" remains still opened for further discussions. For example, we have a good argument to show that in case of repeated extra-pupal moults induced by juvenoid treatments in *Dermestes* (cf. Fig. 25), neither the slightest signs of "larvalization" of the pupal cuticle nor any other signs of reversal of metamorphosis have ever appeared. In addition, according to our experience with extra-larvae or extra-pupae produced by juvenoids in various unrelated insects, we believe that these abnormal developmental stages possess a structurally and physiologically perfect larval or pupal cuticle.

6. Effects on Reproduction and Adult Diapause

The fact that a single juvenoid compound can simultaneously influence insect metamorphosis as well as reproduction documents a common background for juvenoid action on both these developmental processes. Retrospectively, this fact has pointed out that there is no real necessity for larval and adult CA to secrete different hormones, as it was disputed several years ago. Before going into details with descriptions of juvenoid effects on reproduction we should like to attract attention to discussions on the effects of CAH on reproduction in Part I. Also, we shall concentrate here on problems of developmental physiology whereas problems associated with metabolism and biochemical aspects of juvenoid effects on reproduction have been discussed in further sections.

LARSEN and BODENSTEIN [169] were the first to obtain stimulation of ovarian development in mosquitoes by Cecropia extracts. There arose some doubts concerning specificity of the effect [cf. 306], but LARSEN [168] demonstrated later its dependence on juvenile hormone activity. After isolation and identification of farnesol, WIGGLESWORTH [356] succeeded to stimulate ovarian development in *Rhodnius* with this compound. Similar results with farnesol were obtained in some other species, such as *Cimex* [72] and *Euthystira* [221]. However, in no case has farnesol produced effects completely equivalent to CA implantations. Moreover, in certain species farnesol appeared almost inactive in assays on adult females, as e.g. in *Musca, Calliphora* and *Schistocerca* [119, 120, 348], or was completely ineffective in *Pyrrhocoris* [306] or in *Bombyx* [98].

The findings by WIGGLESWORTH [356] were soon supported by considerably stronger gonadotropic effects found with some farnesol derivatives, namely with farnesylmethylether in *Rhodnius* [357], and 10,11-epoxyfarnesoate in *Periplaneta* [31]. This stimulated more extensive study of gonadotropic effects of juvenoids. A suitable material in this respect appeared to be various species of cockroaches. So, the experiments performed on *Periplaneta* [11, 263, 265], *Leucophaea* [35, 36, 88] and *Byrsotria* [12, 84] confirmed the effectiveness of farnesylethers and the hormones isolated from Cecropia extracts. The gonadotropic effects of juvenoids were further demonstrated in adult females of *Locusta* [152, 154, 208] and *Pyrrhocoris* [190], and various aspects of juvenoid influence on ovarian development were also studied in Ichneumonids [32], in *Pterostichus* beetles [86], *Tenebrio* [171], and in some other insects [89, 308].

The most apparent effects of juvenoids on insect reproduction are those on growth and development of the ovaries. Depending on the developmental state of the ovaries the effects of juvenoids on this tissue can be separated into two groups. These are effects on immature, nondifferentiated ovaries, and on the fully differentiated adult ovarian tissue. The immature ovaries respond to juvenoids in the same way as do most of the other larval or pupal tissues, e.g. juvenoids inhibit metamorphosis associated with differentiation of the ovaries. On the other hand, ovaries which have undergone metamorphosis and have attained a differentiated and functionally adapted stage respond to juvenoids in a different way. In this case of gonadotropic action, juvenoids stimulate growth and development of the oocytes which results in formation of ripe eggs. The ability to mimic the gonadotropic effects of CAH represents one of the most specific actions of juvenoids. In principle, the differentiated adult ovaries do not develop beyond the point when deposition of yolk within the oocytes is initiated. An exogenous supply of juvenoids or presence of endogenous CAH, are associated with profound morphological changes, especially among the ovarian follicle cells. Only after such a structural rearrangement of the follicular epithelium can the ovaries perform their main physiological function of vitellogenesis.

With respect to species specificity the above mentioned dependence of vitellogenesis on CAH or juvenoids is not universal, although it seems to occur in the majority of insect species. In some insects, such as in some species of Lepidoptera and Hymenoptera, ovarian metamorphosis is immediately followed by vitellogenesis in the pupal stage. Here formation of the ripe eggs proceeds as a part of metamorphosis process in the absence of CAH. Consequently, juvenoids do not stimulate but in contrast they do inhibit such a pupal vitellogenesis hand in hand with a general inhibition of metamorphosis. A characteristic feature in these insects is a lack of special regulations of female reproduction. However, certain species of Lepidoptera do require special regulation of reproduction, such as synchronization of oocyte growth with nutrition and seasonal variations. And, in this case vitellogenesis which occurs in the adult stage is stimulated by juvenoids. This was documented on *Pieris* [14], *Manduca* [328], *Danaus* [236], see also *Malacosoma* [350]. As a rule it appears that gonadotropic effects can be induced by a proper juvenoid compound in all females where ovarian growth is positively under control of CAH.

Acting on ovaries of susceptible adult females, juvenoids induce whole series of structural and functional changes such as stimulation of growth and divisions in the follicle cells, stimulation of nucleic acid and protein synthesis, induction of new enzymes, and other biochemical changes which characterize tissue growth, as has been discussed in further sections. Most important on these growth changes which are usually stimulated by prothoracic gland hormone or ecdysones in all larval or pupal stages, is that they proceed in complete absence of the mentioned hormones. Thus, we come to rather a conflicting conclusion that juvenoids cause exactly the same developmental changes in the ovaries as do ecdysoids in larval or pupal tissues. Moreover, as has been already mentioned, the prothoracic gland hormone or ecdysoids can stimulate vitellogenesis in pupae of some insects. According to these relationships, both juvenoids and ecdysoids can be used for stimulation of the same developmental process. Their recognition by tissues is genetically determined.

For several years we have systematically studied the effects of juvenoids on reproduction in adult females of *Pyrrhocoris*. Results of this study are briefly described in the following paragraphs.

The telotrophic ovaries of *Pyrrhocoris* females are composed of 7 ovarioles on each side. The oocytes develop in distinctive gonotrophic or reproduction cycles. In each cycle 3 to 6 terminal oocytes develop simultaneously in one ovariole and the ripe eggs are deposited at once. The normal active females deposit the eggs in regular intervals. The first batch of eggs is laid after 5 to 6 days since adult emergence, the next ovipositions occur in about 4-day intervals [306]. Treatment of normal females with large doses of juvenoids (10,000 times of ID-50 Morph.) just after emergence had no profound influence on the course of the reproduction cycles. The treated females deposited regular batches of morphologically perfect eggs. The feeding periods were normal as well as their ability to mate with normal males. Only occasionally there appeared a slight acceleration of ovipositions. However, the presence of juvenoids was manifested later in the egg stage as an inhibition of embryonic development [191, 192]. Thus, the effects of juvenoids on reproduction cycles in normal adult females are rather small. They are analogous to the effects of juvenoids on the moult cycles in young larval instars.

Females which were starved since adult emergence did not oviposit. When treated with large doses of juvenoids they did oviposit, although they produced only about 20 eggs per female whereas the usual number is 50 to 70 eggs. They were rarely able to oviposit a second time but deposited only about 2 to 5 eggs per female. Normally fed females lay as many as 10 or more successive batches of eggs. Thus, juvenoids induced formation of ripe eggs even in the starved females despite almost complete exhaustion of their fat bodies, as revealed by dissections.

The females of *Pyrrhocoris* which were allatectomized soon after emergence increased their body size, accumulated reserve products in the body but never produced ripe eggs in their ovaries [306]. When treated with large doses of juvenoids they started immediately to produce and oviposit the eggs. Oviposition occurred in regular time intervals similarly like in normal females. The first oviposition took place after 5 days following the treatment. Small doses (about 10 times ID-50 Morph.) induced only one reproduction cycle, two cycles were stimulated by 5 to 10 times larger doses whereas 100 times more juvenoid was required for induction of more than three cycles. Without respect to the amount of reproduction cycles or number of eggs produced, all eggs deposited were sterile. Sterility occurred despite the fact that there was apparently normal copulation between these females and normal males. So far, we have never obtained any hatchable egg by juvenoid treatment of the allatectomized females, although the sterility was not irreversible. Thus, after induction of one or two reproduction cycles in response to juvenoids the allatectomized females were able to produce healthy eggs following transplantations of active CA. In connexion with bioassays (p. 117) we have indicated that the percentage of allatectomized females which start to oviposit has been correlated with dose the of juvenoid applied. With larger doses a certain correlation between the dose of juvenoid and the number of successive ovipositions also appears.

The diapausing females of *Pyrrhocoris* responded to juvenoid treatments in exactly the same way as did the allatectomized ones. However, there was no correlation between the dose of juvenoid and the number of consecutive ovipositions. Once a diapausing female was induced to form ripe eggs and oviposit, even with relatively low doses, it has always continued to reproduce further for a longer or shorter period of time. Initially, the diapausing females treated with juvenoids deposited sterile eggs as did the allatectomized ones. Then, sooner or later, depending on dose they continued to produce apparently normal hatchable eggs. As revealed by dissections, their previously inactive diapausing corpora allata were reactivated in the course of the first reproductive cycle induced by juvenoids. It has thus been assumed that the insect's own neuroendocrine system has taken over the control of reproduction from the time when the first hatchable eggs appeared. These results also indicate that at least with respect to ovarian growth the synthetic juvenoids are fully competitive with physiological functions of the neuroendocrine system.

According to our results those juvenoids which had higher activity on metamorphosis also had higher activity on adult females. However, perhaps due to their larger body size, the ID-50 Diap. doses per females were, as a rule, about 10 times larger than ID-50 Morph. obtained with larvae. Also, there was quite an expressed proportionality between the effective doses required for specific effects on females of different physiological state. Thus, doses required for induction of oviposition in 50 per cent of the allatectomized or diapausing females as well as the doses which caused 50 per cent ovicidal effects in normal females were almost identical, as has been illustrated in Fig. 27. In contrast, we were unable to detect direct relationships between the number of eggs produced in one reproduction cycle and the dose of juvenoid applied. On the other hand, there existed some correlations between the dose and the total sum of eggs produced per female, as e.g. in allatectomized females. It is expected that the number of oocytes which develop simultaneously in each ovariole is regulated by feed-back reactions which are synchronized somehow with the available amount of nutrients in the body.

The above results on *Pyrrhocoris* females may be quite instructive but one has to realize that they were made on a very sensitive laboratory model insect. In other species, such as in females of Diptera and many Coleoptera we were unable to affect reproduction even with juvenoids that had high activity on metamorphosis in the same species. Furthermore, in some females juvenoids stimulated ovarian growth, e.g. vitellogenesis but the eggs were not oviposited [32, 356, 357], probably because of a lack of synchronization between ovarian growth and other functions of the neuroendocrine system. The data obtained on *Pyrrhocoris* were reached with doses far below the toxic level. We may recall that ovarian growth represents one of the most sensitive metabolic process which is easily inhibited by all kinds of physiologically active or toxic chemicals. We observed that large doses of farnesol, farnesylmethylether, or especially terpenoid aromatic ethers derived from p-nitro, p-chloro, or 3,4-methylenedioxyphenol actually inhibited ovarian growth in normal females, whereas low doses of these latter compounds had pronounced gonadotropic activity.

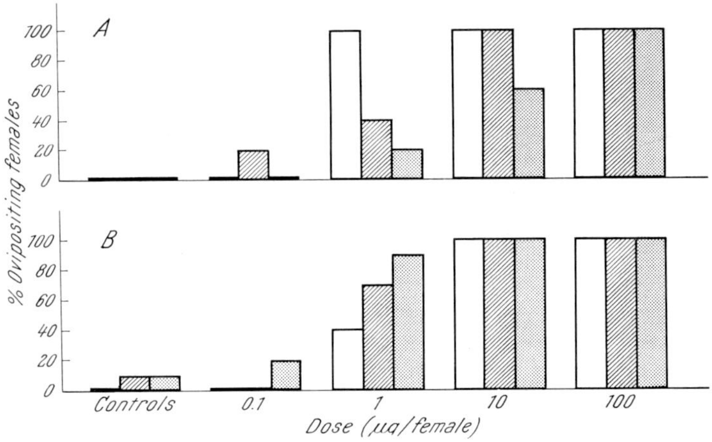

Fig. 27. Effect of dihydrodichlorofarnesoate on oviposition of allatectomized (A) and diapausing (B) females of *Pyrrhocoris*. Allatectomy was made immediately after emergence. Diapausing females were obtained by rearing the larvae and adults in short (8 hours) photoperiod at 25 °C. All the females were treated the 20th day after emergence. Only part of the eggs deposited by diapausing females later than 7 days after treatment with 1 μg of the juvenoid were actually able to hatch into morphologically perfect larvae. All remaining eggs showed inhibition of embryonic development and did not hatch. Open columns indicate oviposition in the first 7 days after treatment, lined columns that between 7 to 10 days, and dotted columns 10 to 15 days respectively

In addition to ovarian growth, juvenoids may also affect various other processes associated with reproduction, such as prevention of oocyte resorption [120], changes in behaviour (p. 270), changes in haemolymph protein composition (cf. p. 258), or functions of the fat body (p. 258).

Another adult tissue which shows similar dependence on CAH and juvenoids appears to be accessory sexual glands of both sexes but especially those of the females. Physiological functions of these glands are closely correlated with ovarian growth. Their excretory products are essential for oviposition. The product gives special properties to the egg chorion. In insects with determined reproduction cycles the content of the product in the accessory sexual glands accumulated progressively in the course of oocyte growth is entirely used at the time of oviposition. In allatectomized females the accessory sexual glands contain very little or no excretory product. In *Periplaneta* for instance, the product consisted of proteins and glucosides of protocatechuic acid. Synthesis of the product was stimulated by juvenoids [296]. Also in some other insect species juvenoids appeared to stimulate synthesis of the accessory sexual gland product in females [23, 89, 381, 386]. However, its chemical nature might be completely different from that found in *Periplaneta*. For example, in females of *Pyrrhocoris* the product consists of yellow oily lipid. In general, it seems that juvenoids would affect the female accessory sexual gland functions in all females where they also stimulate ovarian growth. It was suggested [23] that changes in the accessory sexual glands could be used as a sensitive bioassay for juvenoids.

Except some results concerning the effects on male sexual behaviour (see p. 271) very little is known so far about the effects of juvenoids on development and function of the male gonads. In most species the testes contain some mature spermatozoids already at the time of adult emergence. The spermatogenesis process occurs partially or completely during metamorphosis. It is generally expected that juvenoids may affect the process of spermiogenesis in a similar way as they affect morphogenesis in other adult tissues during metamorphosis. However, as far as the adult stage is concerned, the male specimens tolerate considerably large doses of juvenoids without remarkable changes in their ability to copulate and inseminate the females. So, it was observed in *Pyrrhocoris* [192, 193, 378] that the males treated with large doses of juvenoids can transfer the compound to the females by mating. The females which once copulated with these treated males became sterile due to ovicidal action of juvenoids on embryonic development. The effect did not depend on transmission of defective sperm because the eggs showed initiation of embryonic development and, in addition, the effect was not decreased by mating of the females with normal untreated males. Studies on the effects of juvenoids on the male sterility are actually complicated by the described possibility of transfer of the compounds from males to females. There does not exist any adequate method so far which would allow us to distinguish the effects caused merely by a transfer of juvenoids during copulation from a real male sterility.

Sometimes it appears that treatment of metamorphosis stages by juvenoids may result in curious effects on development of gonads and on sexual behaviour. For example, Žďárek and Sláma [390] obtained extra-larvae of *Pyrrhocoris* which had ripe eggs in the ovaries or mature spermatozoids in the sperm ducts respectively. Curiously enough, these extra larvae exhibited active sexual behaviour which was normally present only in adult stage. An earlier application of juvenoids caused extra-larvae with incompletely differentiated gonads in both sexes as is the usual case in most adultoids, especially in Endopterygote insects such as *Dermestes*, *Tenebrio*, *Galleria* and others.

In connexion with bioassays (p. 117) we have noted already that one of the main features of adult diapause in insects is an arrest of reproduction. It appears that unlike larval or pupal diapause which is little, if at all, influenced by juvenoids [127], the adult diapause can easily be broken by properly selected juvenoid compounds. As a typical example we can take the above described effect of juvenoids on adult diapause in *Pyrrhocoris* females. In a similar way we succeeded to terminate adult diapause in various other Hemipteran species, such as *Eurygaster intergriceps* and *Graphosoma italicum* (Pentatomidae) or in *Lygaeus equestris* (Lygaeidae). The most effective juvenoid compounds to break diapause in these species were those with the best effects on metamorphosis in the respective species. This relationship seems to have a widespread applicability, at least among Exopterygote insects.

Bowers and Blickenstaff [28] terminated adult diapause in alfalfa weevil by treatment with 10,11-epoxyfarnesoate. They observed relationships between dose of the juvenoid applied and number of eggs oviposited per female. Reactivation of the neuroendocrine system was not likely to occur in this case. Some parasites which were present in the body of the diapausing beetles emerged prematurely suggesting their synchronization with development of the host. Connin *et al.* [56] obtained similar effects in diapausing adults of *Oulema* beetles. The treated females laid hatchable eggs and the hatched larvae developed normally. Thus, no ovicidal effects were recorded in this study. Later, Bowers [cf. 378] reported successful termination of adult diapause in several other insect species.

The artificial termination of adult diapause by juvenoids is considered to have a great importance for insect control. Without respect to possible reactivation of the neurosecretory system and formation of hatchable eggs, an insect with an artificially terminated diapause is faced with the problem of reproducing under highly unfavourable environmental conditions, in absence of food, in low temperatures, etc., which normally leads to lethality.

7. Action of Juvenoids on Embryonic Development

According to current developmental theories [227], morphogenesis of an insect embryo is finished at the moment when newly formed CA start to secrete CAH. This may occur sooner in Endopterygotes than in Exopterygotes and therefore the larvae of the latter group hatch at more advanced morphogenetic stage. It appears that embryonic development and metamorphosis are characterized by many common patterns. Together they represent two main periods in insect development where most of the morphogenetical changes have taken place. For practical reasons we may consider metamorphosis as continuation of the embryonic development at a larger body size level. Despite accumulated evidence for juvenoid action on metamorphosis, there were no results to show their action in embryonic development until 1965 when SLÁMA and WILLIAMS discovered the effect of juvabione on the eggs of *Pyrrhocoris* [317]. They found a real inhibition of embryonic morphogenesis from early stages of germ band formation to more advanced forms of the embryos. The observed phenomenon was similar to extra-larvae and adultoids caused by juvenoids in metamorphosis. However, it appeared that there was a sensitive period at the beginning of embryonic development just as in metamorphosis. The juvenoid preparation acted through the female's body and there was the same selectivity of the effects on embryogenesis as found in metamorphosis. For example, the juvenoid material used was inactive on eggs of those species which were insensitive to it during metamorphosis. These original observations [317] were fully confirmed by RIDDIFORD and WILLIAMS [261] on Lepidoptera eggs with a different juvenoid preparation. Further results obtained in human lice [342], in Orthoptera [229, 230], Coleoptera [347] and in some other insects [80, 89, 308, 386] suggested that the effect of juvenoids on embryonic development of insects may be a widespread phenomenon.

Earlier in the text (p. 236) we mentioned that some tissues which have no alternative developmental programme but differentiation into adult structure show an overall inhibition of growth in response to juvenoids. In analogy with this, the early embryonic tissues have most probably no alternative developmental programme but differentiation into larval structures. This may be responsible, among other factors, for a common inhibition of embryogenesis by juvenoids. Thus, in presence of juvenoids the fertilized egg cell is unable to develop at all or more commonly, it develops only to a certain critical stage. Furthermore, some embryonic cells in the presence of juvenoids deviate from the process of genetically programmed differentiation which leads to drastic developmental abnormalities and embryonic malformations.

In the original observations [317] juvenoids did not seem to interfere with embryonic development until the stage of germ band formation. Later it appeared, however, that a very early application of juvenoids affected development of embryos in Saturniids already at the blastoderm level [261]. This conclusion was supported by further studies on

Pyrrhocoris [90, 194, 258, 259], and *Schistocerca* [230]. With respect to different time of juvenoid application it was possible to cause abnormal development at the blastoderm stage, germ band formation, blastokinesis, and also at the final period of larval differentiation [258]. Unequal penetration of juvenoids through the egg chorion to different parts of the egg at different time of the sensitive period is associated with relatively large variability of morphological effects. There appear miniature embryos containing excess of nonutilized yolk, nonsegmented or only partially segmented embryonic creatures, quite assymmetric embryos or, some structures like antennae or legs appear in differentiated larval form whereas the rest of the body remains in completely undifferentiated embryonic stage. When applied towards the end of the

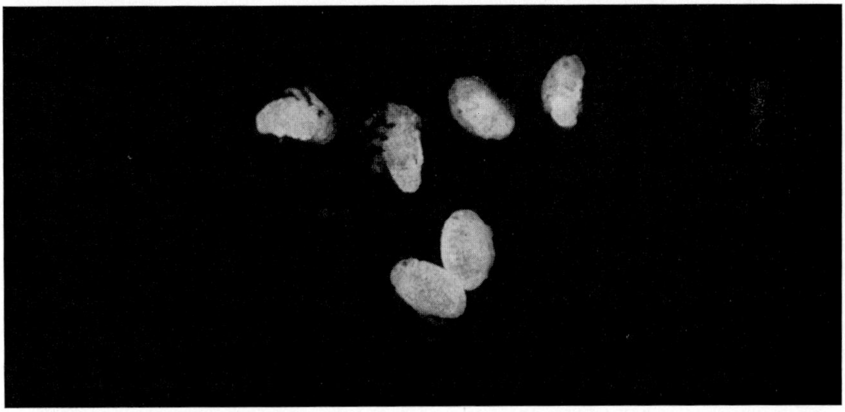

Fig. 28. Embryos of *Pyrrhocoris* removed from the egg shell showing deformities caused by juvenoid treatment of the eggs

sensitive period juvenoids may cause formation of more or less morphologically perfect larvae which appear to have incompletely differentiated and malfunctioning internal organs. These are usually immobile and unable to hatch. If they do hatch they die soon afterwards [194, 230, 317]. In the eggs of Hemipterans this effect is connected with secretion of larval cuticle by incompletely differentiated larval epidermal tissues. This results in appearance of embryonic creatures reminiscent of miniature Lepidopteran pupae [317]. Finally, we may add that under humid conditions these malformed embryos obtained after treatments of insect eggs with juvenoids may surive without further morphological changes for a long time after the controls have hatched. Certain types of malformed embryos of *Pyrrhocoris* are shown in Fig. 28. All these data favour the view that action of juvenoids at cellular level is indeed identical in both embryogenesis and metamorphosis.

A continuous presence of juvenoids since the time of oocyte growth does not seem to intervene with the process of egg fertilization. Neither does it intervene with the initial cleavage divisions, at least in certain species. Embryonic development usually proceeds until the stage of blastoderm formation when the embryonic cells begin to realize their own genetic programme. And, similarly as in metamorphosis, since this time development of an embryo becomes a subject of "status quo" maintenance by juvenoids [258, 259, 261, 373]. Both natural compounds isolated from Cecropia extracts had similar effect on embryonic development as other juvenoids [258]. However, due to unsolved technical problems, it is not known whether CA would cause the same effect in embryonic development as do synthetic juvenoids. WALKER and BOWERS [347] observed pronounced ovicidal effects on eggs exposed to juvenoid vapours, suggesting that extremely small amounts were effective with this method of application. We observed proportionality between the effects of juvenoids on metamorphosis and embryonic development in a few Hemipteran species. However, the eggs of certain Endopterygote insects were completely insensitive to selected juvenoids with pronounced effects on metamorphosis despite the use of all possible combinations of application methods.

It was found earlier that eggs of *Hyalophora* [261] or *Pyrrhocoris* [191] treated with juvenoids at the end of the sensitive period gave rise to larvae which suffered from high mortality during the young larval period. The reason for such a high larval mortality may depend on the mentioned malfunctions of some internal organs. RIDDIFORD [256, 257, 258, 259], observed quite specific effects of juvenoids at the late embryonic and in the postembryonic development of insects. Essentially, the eggs were treated at the end of embryogenesis when the critical period to juvenoids was passed. The larvae hatched and developed normally until the last larval instar. Then, however, instead of metamorphosing into normal adults, they underwent extra-larval moults. Since there was approximately one month interval which separated the time of application from the observed effects on metamorphosis, these results were interpreted as due to possible delayed, irreversible changes in a mechanism determining the future developmental programme [256, 258].

It is well known [308] that juvenoids can cause delayed effects between different developmental stages. Such effects are due either to persistence of the compounds from one stage to the next, or to developmental disturbances in differentiation of the gonads. Common examples for this phenomenon are pupal adultoids which appear after treatment of last instar larvae in certain Endopterygote insects, appearance of female sterility after treatment of last instar larvae of Exopterygotes, or female sterility after treatment of pupal stages. However, persistence or recontamination were not considered as possible reasons for the above effects observed by RIDDIFORD. It was argued by WILLIS [380] that juvenoids might then intervene with physiological mechanisms which are switching off functions of CA at the end of larval stage. Such a possibility was further investigated experimentally by RIDDIFORD [257] using transplantations of CA taken from last instar larvae of *Pyrrhocoris* which either developed from the treated or untreated eggs. Results of these studies [257, 259] suggested that the delayed effects might indeed depend on a juvenoid interference with developmental programming of secretory activity of the CA. Thus, according to these assumptions, treatment with juvenoids at the late embryonic period would prevent the CA from being turned off at the beginning of metamorphosis, subjecting an insect to "permanent" larval development.

When fully confirmed, this effect might indeed have a far reaching theoretical and practical implications. However, there remain still some uncertain points in its interpretation. For example, WILLIS and LAWRENCE [382] performed mutual epidermal transplantations between larvae of *Oncopeltus* which were either treated at the end of embryogenesis by juvenoids or developed normally. Their results suggested that persistence or re-contamination was involved. Furthermore, it was found in *Pyrrhocoris* [232] and in some other insects as well [243] that even the CA taken from last instar larvae became active when transplanted as hosts into other last instar larvae. Finally, we also tried to reproduce these effects on *Pyrrhocoris* but did not succed even with extremely larger doses than ID-50 Morph., provided that special care was paid to avoid recontamination, as seen in Table 7.

8. Physiological Conditions for Female Sterility

Whereas inhibition of embryonic development by juvenoids has been called ovicidal effects [317], the production of nonhatchable and nonviable eggs by females treated with juvenoids can be regarded as a special case of female sterility. Basic conditions required for induction of such a sterility in adult females of *Pyrrhocoris* were partly described by MASNER *et al.* [191], and some remarks on this problem have been made in the last sections above. One common feature of this type of female sterility is represented by the fact that the sterility is not caused by suppression of the egg production which is rather a common case when using "classical" chemosterilants. Instead, female fecundity is normal or may temporarily be increased beyond the normal rate. Perhaps the most essential feature is that the treated females deposit defective eggs of very low or zero hatchability. In addition, intensity and duration of these sterilizing effects are clearly dependent on dose of the juvenoid applied. Excessive doses can induce permanent sterility in a single application as illustrated in Fig. 29. Occasionally it appears that longevity of the treated females is shortened with increased egg production. However, this relationship occurs also in normal females so that the effects of juvenoids on longevity of females can be considered as physiological rather than pathological. Although, it sometimes happens that females treated with large doses of juvenoids discontinue for some reason egg laying sooner than the control females. Their fecundity is then smaller than that of the controls. In certain species, such as in females

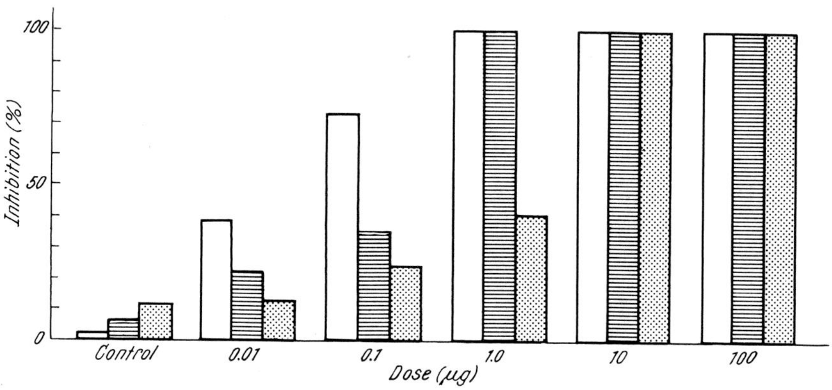

Fig. 29. Inhibition of embryonic development (larval hatching) by dihydrodichlorofarnesoate which was applied topically to freshly emerged adult females of *Pyrrhocoris*. Open columns indicate the effect on eggs of the first reproductive cycle, lined areas the eggs of the 2nd to 4th reproductive cycle, and the dotted areas indicate the effect on eggs of the 5th and later cycles

of higher Diptera, various juvenoids can decrease the rate of oviposition. Physiological explanation of this effect is not known but according to our knowledge nobody has succeeded to induce typical ovicidal effects by juvenoids in these insects.

In Hemipterans the sterilization of females can be achieved by application of juvenoids at any moment of the reproduction cycle. This includes also stages when the eggs were already chorionated [191]. Also in females with concurrent oviposition the effect appears on eggs deposited few hours after the treatment. However, in certain Endopterygote insects where oocytes develop before adult emergence juvenoids may cause more pronounced sterility effects when applied at a certain determined stage of oocyte development [261, 258].

We have noted already (p. 242) that studies on possible male sterilization by juvenoids are complicated by the fact that females which have once mated with the treated males may become sterile due to their contamination by juvenoids transferred from the males [192, 193]. However, the treated males can inseminate normal females and these can produce hatchable eggs, provided that mating has taken place after prolonged periods of time since the treatment. This fact indicates that if the sperm of the treated males were defective shortly after the treatment, the effect should not be at least irreversible.

9. Relationships in the Effects on Metamorphosis, Embryogenesis and Reproduction

In chapter D on structure-activity relationships we have described changes in the effectiveness of various juvenoids in the metamorphosis assays on various insect species. It has been further suggested that a juvenoid may exhibit about the same effectiveness on adult females of different physiological state, such as on the normal active females, allatectomized, or diapausing females (see Fig. 27). However, there still remains a problem whether juvenoids also exhibit similar or different effectiveness at different levels of ontogenetic development, e.g. on eggs, metamorphosis stages and adults of one species.

There exist many suggestions in literature regarding a common nature of both the effects of juvenoids on metamorphosis and embryonic development [cf. 259, 261, 308, 317, 380, 382]. Similarities between these effects can be derived from comparisons of juvenoid action in *Pyrrhocoris* [191, 317] and *Hyalophora* [257, 261], where an inhibition of morphogenesis is a common physiological principle. Moreover, even the effects of juvenoids on ovarian development have revealed at least similar structure-activity relationships with the effects on metamorphosis. This comes out from studies on *Rhodnius* [357, 358] and similar conclusions can be drawn from results on other species [cf. 28, 31, 35, 56, 265, 308, 342, 347, 390]. We have a feeling that the action of juvenoids on eggs, metamorphosis, and ovarian development can only be understood if we realize that the respective target tissues of different ontogenetic developmental stages have different characters. At different stages of differentiation the tissues acquire new developmental programmes and it is not uncommon that their developmental responses to the same hormone may get significantly altered in the course of ontogeny. For example, embryonic tissues are characterized by the fact that they do not require hormones of the main endocrine glands for their development, unlike larval or pupal tissues which occur at more advanced ontogenetic stage. In both instances, whether stimulated by hormones or not, juvenoids inhibit further ontogenetic progress. However, the adult target tissues of juvenoids, such as are ovaries, occur at the terminal ontogenetic stage. This tissue does not require other hor-

Table 15. *Comparison of ID-50 Values in µg per g of Fresh Weight of Selected Juvenoid Esters on Larvae, Adult Females and Eggs of Pyrrhocoris apterus*[1]

Standard Unit/g Application to: Egg batches taken (days)	ID-50 Morph. last inst. lar.	 0–7	ID-50 Ovic. on females females 2-day-old 7–11	 11–15	 15–23	ID-50 Ovic. on eggs eggs 0–20 hrs
Compound number in Appendix tables:						
T-16	47.0	6.3	12.5	63.0	310.0	100.0
T-151	33.0	50.0	137.0	500.0	1000	25
T-30	0.13	0.25	2.5	12.5	50	0.5
T-60	0.33	0.25	0.63	7.5	38	0.25
T-62	0.66	0.12	5.1	19	87	0.25
T-240	66	50	110	1100	—	500
T-172	66	125	400	1250	—	250
T-246	3.3	63	140	1250	—	25
T-259	0.33	0.01	0.75	2.5	10	12.5

[1] Respective weights of eggs, larvae, and females were 0.4, 15 and 80 mg. For each concentration 50 eggs were treated each with 0.2 µl of acetone solution; 5 females were used for each concentration. They were treated individually only once with 1 µl of acetone solution; for ovicidal evaluation after treatment of females 200 eggs were evaluated in respective time intervals. All applications were topical.

mones but CAH or juvenoids which simultaneously serve as a messenger for the one and ultimate function of this tissue, which is production of ripe eggs. It is reasonable to assume that all the effects of juvenoids at different ontogenetic levels may be based on a common, still unknown, mode of action. All the known physiological and biochemical features described in this book are most probably only consequences but not a real cause of juvenoid action. In the whole group of insects the ontogenctic differences in responses of the target tissues to juvenoids are multiplied by taxonomic differences which finally result in enormous diversification of the effects in various species.

Nevertheless, despite the above discouraging statements on diversity of the target tissues, we made an attempt to compare relative units of biological activity of several selected juvenoid compounds on eggs, larvae and adult females of *Pyrrhocoris*. The results were expressed in ID-50 units per gram of fresh weight and used for construction of Table 15. It is evident from these data that several juvenoids, especially the acyclic ones, had surprisingly consistent effectiveness on all the developmental stages. The aromatic juvenoids were relatively more effective on larvae than on eggs, whereas the last compound which is a derivative of p-aminobenzoic acid had especially pronounced ability to sterilize females and cause ovicidal effects. These data are most probably valid only for the species studied, but at any rate they indicate that in general there do exist correlations in the effectiveness of juvenoids at various stages of insect ontogenetic development. With practical aspects in mind, this suggests that juvenoids with pronounced effects on metamorphosis in a given species should preferentially be used in search for new female sterilants and ovicides.

10. Synergism and Potentiation of Juvenoid Effects

In studies on insecticides we can frequently encounter examples that some compounds, being inactive alone, can considerably increase toxicity when used in combination with insecticides. These insecticide synergists are assumed to inhibit metabolic degradation, oxidation, or excretion of the toxic compounds. So, also in hormonal studies we find that some compounds, such as cyclic AMP potentiate biological activity of the hormones by intermediating hormonal stimuli. We have discussed in special sections of chapter B and later in this chapter that juvenoids undergo also metabolic degradation, biological inactivation or excretion. Obviously, compounds which would increase chemical or biological stability of juvenoids might possibly synergize their effects in analogy with insecticides.

Bowers [25] was the first to explore possible potentiation of juvenoid effects by insecticide synergists. However, he discovered that certain synergists had juvenile activity *per se*. Later he suggested [27] occurrence of some slight potentiation of juvenoid effects by insecticide synergists, although this was considered of no practical value. More recently FALES *et al.* [91] approached these studies differently. They used juvenoids with some common structural features in the molecule with insecticide synergists and tested them for synergistic action along with some insecticides. They recorded potentiation of toxicity of pyrethrins against flies but none of the juvenoid-synergist combinations was as good as the commonly used insecticide synergist piperonyl butoxide. Nevertheless, these observations have shown that insecticide synergists can slightly potentiate juvenoid activity and, that juvenoids with partly synergistic structural patterns can in turn potentiate the action of normal insecticides.

Further support for synergism of juvenoid action was given by CRUICKSHANK [66] who found that the juvenile activity of 10,11-epoxy-farnesenic acid amides was potentiated by additions of some insecticide synergists. Still greater potentiation of juvenile activity occurred after combination of two juvenoids, the above epoxy-ethylamide with structurally related 7,11-dichloro ethylamide of the farnesenic acid [67]. Mixture of both these juvenoids was up to 1000-fold more active than any one of them applied alone. However, juvenile activity of the indicated dichloroamide was also potentiated after its application in squalene. This indicates that the use of isoprenoid solvents or carriers may enhance the biological activity of isoprenoid juvenoids. Probable explanation may be that an inactive isoprenoid like squalene did not compete with juvenoids for specific cellular receptor sites while it was fully competitive with regard to metabolic degradation or excretion of these isoprenoid compounds.

COREY [57] and RIDDIFORD *et al.* [260] found potentiation of juvenile activity of methyl dihomojuvenate by its synthetic 10,11-iminoanalogue. The effect was explained as being due to inhibition of biochemical mechanisms which might cause opening of the oxirane rings forming thus more polar and biologically inactive glycols. These conclusions were based on analogous stabilization of the epoxy-group by corresponding imino-analogues in other biological systems [57]. However, SIDDAL *et al.* [300] did not confirm such a potentiation of juvenile activity of epoxy-juvenoids by their imino-analogues.

11. Effects of Mitotic and Metabolic Inhibitors

In the early studies with lipid extracts from Cecropia, WILLIAMS [370] assumed that juvenile hormone might inhibit mitoses among the epidermal cells with simultaneous nuclear enlargement. They tested whole series of antimitotic agents but obtained no signs of real juvenile hormone activity. Moreover, BLAUSTEIN and SCHNEIDERMAN [20] found prevention of scale formation in metamorphosis of Lepidoptera after treatment with some antimetabolites and toxic compounds, but did not observe typical juvenile hormone effects.

In addition, SLÁMA [303, 304] obtained some nonspecific juvenile hormone effects with colchicine and, especially, with some nonesterified fatty acids and some long chain alcohols which had local toxicity to epidermal cells. Induction of these "pseudo-juvenilizing" effects showed apparent correlations with cell divisions and differentiation.

Later ILAN *et al.* [129] reported formation of intermediates containing adult head and thorax while having pupal abdomen after injection of actinomycine D into pupae of *Tenebrio*. Indeed, this antibiotic inhibiting RNA synthesis caused more or less similar effects on metamorphosis as did juvenoids, e.g. it induced retention of pupal cuticular patterns on the body of an adult. These effects of actinomycine D were confirmed by CHASE [50] who recorded additionally an existence of correlations among the degree of the effect, the dose, and the time of application. Furthermore, EMMERICH [83] supported the assumptions on similarity of the above effects of actinomycine D with the effects of juvenoids. He also found that such similarity may exist also with respect to development of intestinal epithelial cells. Later it appeared that these juvenile hormone like effects in *Tenebrio* pupae could be also caused by mitomycine C and ouabaine, whereas colchicine was inactive [51].

It is known that a highly accelerated development which is caused by ecdysone may lead to developmental abnormalities called prothetely [227]. These prothetelic forms, though originating by different physiological mechanisms, have many morphological features in common with the intermediates produced by juvenoids. Recently SOCHA and SEHNAL [320] analysed differencies in juvenile hormone effects produced by antibiotics, ecdysoids, and juvenoids. They uncovered certain physiological differences in action of these chemical agents, which all induced juvenile-like morphological effects in *Tenebrio* pupae. For example, juvenoids never inhibited development and apolysis of the old cuticle; they shortened the period between ecdyses but there was always a complete resorption of the old cuticle. Ecdysterone did not inhibit development either, but it induced precocious ecdysis associated with incomplete resorption of the old cuticle. Finally, in distinction to the hormones antibiotics such as actinomycine D, mitomycine, and some others induced usually total inhibition of the development or caused mortality, never accelerated ecdysis, nor produced generalized effects manifested by appearance of perfect extra-pupal instars [320].

The pseudo-juvenile effects of antibiotics seem to confirm the earlier conclusions [304] on the possibility of obtaining juvenile-like local effects by nonspecific agents. We have already explained that the mode of action of real specific juvenoids does not depend on direct inhibition of the cell differentiation process, although such inhibition is most important for a combination of morphological patterns according to which we evaluate the juvenile hormone activity. The original stimulus which juvenoids give to the target cells is followed by a chain of dynamic developmental events before the final morphological form is established. Somewhere in between the "activated" cell and the resulting structural forms there is intercalated a whole series of physiological changes, such as those involving membrane permeabilities, ionic balance, synthesis of regulatory and constitutional macromolecules, construction of internal cellular microstructures, exchange of substrates and waste products, and many still unknown intermediary steps. It may be anticipated that endogenous hormones would act at the very beginning of this chain of physiological events because of their decisive regulatory functions. Juvenoids may act at rather similar, if not identical levels, whereas antibiotics and other of the mentioned nonspecific chemicals may reach more or less similar final results by acting from some other, perhaps more advanced point of this chain. The old hypothesis by SLÁMA [304] that virtually any compound which would directly or indirectly inhibit the differentiation process without simultaneous inhibition of the moult cycle could theoretically cause juvenile or "pseudo-juvenile" effects in insect metamorphosis, remains still to be considered after more than one decade. At present we know already that these nonspecific juvenile hormone mimics may be free fatty acids which inhibit insect growth [197] and oxidative phosphorylation [304]; ouabaine which influences Na/K-dependent ATP systems [51]; colchicine which causes repetitive c-mitoses and inhibits cell divisions [51, 173, 320]; actinomycine D which inhibits DNA dependent RNA syntheses [51, 83, 129, 320]; protein denaturants such as chromium trioxide, formaldehyde, mercuric chloride, picric acid, osmium tetroxide, etc. [175]; and, under certain circumstances even precisely timed epidermal injury. Diversity of pharmaco-biological properties of all these compounds points out the probability of their action at different loci of the above described developmental chain of events. And, as a matter of fact, according to selective influence on determined epidermal structures LAWRENCE [173] concluded that mitomycine C came into action later than juvenoids.

After analysis of all these nonspecific juvenile hormone-like effects we have reached the conclusion that virtually none of the above listed nonspecific compounds can be equalized in their action with juvenoids listed in the Appendix tables, both qualitatively or quantitatively. In addition, these nonspecific effects are most frequently found in *Tenebrio* pupae the cuticle of which lacks "specific" small epidermal organs very common in other insects. This makes distinction between the real juvenile and nonspecific effects rather difficult. It is quite possible that extensive screenings for new juvenoids will soon reveal more of such nonspecific compounds. We recommend that in these instances compounds with equivocal juvenile or nonspecific properties be subjected to ovarian assays in order to confirm their real juvenoid character. Hence, in contrast to metamorphosis assays, we do not know a single case of positive responses in ovarian assays by any of the mentioned antibiotics or other antimitotic or antimetabolic materials.

12. Effects of Juvenoids on Respiratory Metabolism

The first attempts to study the influence of Cecropia extracts on respiratory metabolism were performed with homogenates prepared from adult *Leptinotarsa*. There occurred considerable stimulation of homogenate respiration after addition of the extracts [333, 362, 368]. This work was made, however, with crude extracts containing minimum amounts of the active compound. Further, GILBERT and SCHNEIDERMAN [107] and STEEN [332] performed *in vivo* measurements of O_2 consumption in Lepidopteran pupae injected with Cecropia extracts. They found certain deviations from a common U-shaped course of the respiration curves which normally occur in metamorphosis of all Endopterygotes. For example, there occurred a slight acceleration in the rise of respiration corresponding to the right hand part of the U-shaped curve, but no basic differences were found. Quite similar results were obtained by SCHMIALEK and DREWS [281] and by GEYER *et al.* [100] after treatment of *Tenebrio* pupae with farnesylmethylether. Here the U-shaped decline of the respiratory curves did not reach the bottom values registered in the control group. It was concluded that possible similarities between these effects and those found by SLÁMA after the CA implantations in metamorphosis of Hemiptera might exist.

Farnesol and some other juvenoids were found ineffective on oxygen consumption of adult females of *Locusta* [211] or *Pyrrhocoris* [306]. This was most probably due to a generally low biological activity or inactivity of these juvenoids. Further experiments with some highly active juvenoids revealed apparent similarity between the effects of juvenoids and the effects of CAH on respiratory metabolism in general. This is documented for instance in Fig. 30 which shows both the effects of juvenoids and CAH on oxygen consumption in the last larval instar and in adult females of *Pyrrhocoris*. It is quite apparent that juvenoids had an almost identical effect on respiratory metabolism as had the implanted CA. Moreover, these effects of juvenoids were consistent with CAH in metamorphosis as well as during reproduction. As seen in Fig. 30, the results obtained with last larval instars of *Pyrrhocoris* are also consistent with SCHMIALEK and DREWS's [281] findings on *Tenebrio* pupae which

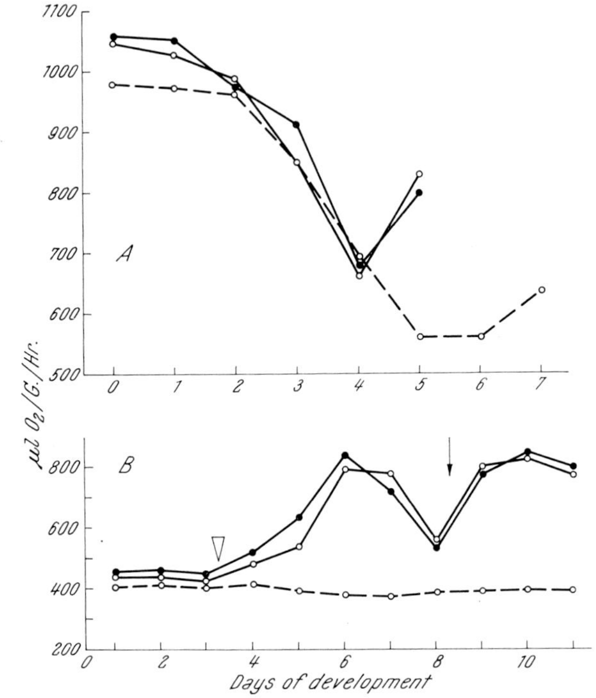

Fig. 30. A. The course of oxygen consumption curves in the last instar larvae of *Pyrrhocoris* which developed into: i) normal adults (broken line), ii) extra-larvae after treatment with juvenoids (filled circles), and iii) extra-larvae after implantations of CA (open circles). The larval-adult ecdysis occurred after 7 days, the larval-larval one after 5 days. B. The same in allatectomized females which were: i) treated with 50 μg of methyl dihydrodichlorofarnesoate (filled circles), ii) implanted with active corpus allatum (open circles), and iii) shame-operated as controls (broken line). The females were 15 days after allatectomy and emergence at the point indicated by triangle, which also indicates the treatment with juvenoids and implantation. The females treated with juvenoid deposited 43 eggs, the implanted ones 54 eggs per female at the point indicated by arrow. The control allatectomized females did not produce eggs

is in favour of their conclusions. On page 40, 42 we have discussed a possibility that CAH may affect the rate of metabolism in an indirect way, i.e. with respect to particular tissues and in connexion with induced morphogenetical changes. These relationships between the rate of respiratory metabolism and the degree of growth and morphological changes stimulated or inhibited by hormones apply very well to juvenoids. In this respect juvenoids would influence respiratory metabolism only in particular target tissues, in particular species, under determined conditions, and according to a genetically determined developmental programme. This was confirmed by measurements of O_2 consumption in allatectomized adult males of *Pyrrhocoris*. These, in contrast to the allatectomized females did not change their respiratory rate after application of any amounts of juvenoids. In addition, the CA implantation was here also ineffective [305].

13. The Effect of Juvenoids on Enzymatic Activity

Our knowledge on the effects of juvenoids on enzymatic activity in general is rather scarce. Any evidence that the mode of action of juvenoids would depend on induction of some specific enzyme systems is lacking. However, with respect to indirect metabolic role of CAH and juvenoids we may anticipate that the effects of juvenoids in induction or breakdown of some enzyme systems will be dependent on specificity of the separate target tissues, their ontogenetic stage, and on the general character of the developmental change to be realized.

In addition to some experimental results which are described below, some information on the effect of juvenoids on enzymatic activity can be derived from their effects on tissue growth and morphogenesis. For example, in response to stimulated ovarian growth by juvenoids, the ovarian follicle cells increase enormously their activity of mitochondrial cytochrome enzymes and also the activity of other enzyme systems associated with cellular growth and divisions. In contrast, juvenoid action on metamorphosis is followed by drastic inhibition of synthesis of all these enzyme systems, for instance in indirect flight musculature, as a consequence of inhibited histogenesis of this tissue by juvenoids. Moreover, induced by juvenoids retention and survival of specific larval tissues may suggest that an appearance of hydrolytic enzymes such as acid phosphatase, ribonucleases, cathepsin, etc. (which always accompany disintegration of these cells) was prevented as a consequence of juvenoid action. In addition, this prolonged survival of specific larval tissues may be associated also with retention of the old larval enzyme systems. We have mentioned these examples just for illustration that juvenoids can in fact induce synthesis of some enzyme systems in some target tissue whereas their action can be followed by oposite effects on the same enzyme systems in a different target tissue.

As found in *Tenebrio* pupae, farnesylmethylether did not affect the activity of malate dehydrogenase or lactate dehydrogenase whereas it stimulated activities of some other enzymes such as glucoso-6-phosphate dehydrogenase, fructoso-1,6-diphosphate aldolase and glutamate-oxalacetate transaminase [87]. In *Pyrrhocoris* the activity of transaminase enzymes was not affected by juvenoids, although it was indirectly influenced by changes in total metabolism, provided that these had taken place [226]. In connexion with induced extra-pupal moults by dihomojuvenate in *Antheraea* there appeared qualitative changes in esterase enzymes [22]. In the accessory sexual glands of adult females of some species juvenoids induced synthesis of specific enzymes associated with formation of specific secretory products [12, 23, 296].

14. The Effect on Carbohydrate, Lipid, and Protein Metabolism

It appears that stimulation or inhibition of growth and development by juvenoids is often reflected by changes in transfer and content of the reserve materials. This is less significant in metamorphosis where one type of growth is merely replaced by other, but it is quite apparent in reproduction due to intensive growth of ovaries. Thus, the body composition of *Dysdercus* larvae was little influenced when metamorphosis was delayed with farnesoate dihydrochloride. The body content of water, fatty acid esters and glycogen was somewhat smaller in extra-larvae than in adults. However, the total content of protein was the same [133]. Farnesyl methyl ether had no remarkable influence on haemolymph glucose and trehalose content, nor on proteosynthetic intensity in *Tenebrio* pupae [100]. On the other hand, experiments with dihomojuvenate in *Antheraea* suggested appearance of new proteins with esterase activity [22].

In adult females, enormous amount of sugar or aminosugar, lipid and phospholipid, protein and other nitrogen containing compounds are transferred from the fat body into the growing oocytes. A special role in this transport which is also associated with the effects of juvenoids on reproduction is ascribed to egg yolk proteins or vitellogenic proteins. Their synthesis in the fat body of *Leucophaea* was stimulated by juvenoids [36]. Similar effects of juvenoids on synthesis of vitellogenic proteins were observed in other cockroaches [12], and also in adult females of Endopterygote insects such as in *Leptinotarsa* [185], *Manduca* [328], or *Danaus* [236]. With respect to haemolymph proteins it appeared that some of the protein fractions may considerably change in quality and quantity, after juvenoid application [187, 387], see also a review [386] by Wyatt. Most of these changes in protein patterns after juvenoid application on adult females were dependent on presence of the growing ovaries in the body. However, in certain species juvenoids did influence metabolic changes in some other tissues, especially in the fat body, even in the absence of the growing ovaries [40, 184].

15. The Effect on Metabolism of Nucleic Acids and on Puffing Patterns

In all the previous sections we have collected enough evidence to show that juvenoids influence growth and morphogenetic development. Essentially, any living cell which was stimulated to grow, divide, or perform some specific physiological functions revealed also certain nonspecific changes in metabolism of nucleic acids. These changes are induced by all hormones affecting morphogenesis and tissue growth in different animals. With insects, KRISHNAKUMARAN *et al.* [164] observed that virtually each cell synthesized DNA during the larval-larval or larval-pupal moult cycles in Saturniids. The synthesis was abolished during diapause in absence of hormones whereas it reappeared again as one of the earliest signs of post diapause development. Further they observed that after absolving a moult cycle in absence of juvenile homone, most of the tissues lost the ability for future DNA synthesis. In the absence of ecdysone Cecropia extracts had no effects on DNA synthesis in any tissue of these insects. Comparisons of DNA synthesis patterns between the two types of development, e. g. the one leading to extra-pupal moults after treatments with Cecropia extracts and the other leading to formation of normal adults, did not reveal any significant differences. These results indicate that the effects of juvenoids on metamorphosis do not depend on general control of DNA synthesis and cell replication, as was presumed by some theories [231].

As far as the effects of juvenoids on some nonspecific changes in RNA synthesis is concerned, OBERLANDER and SCHNEIDERMAN [234] detected slightly enhanced rate of RNA synthesis in pupae which developed into extra-pupal instars. Such a slightly increased incorporation of nucleic acid and protein precursors was also found in pupae of *Tenebrio* after application of farnesylmethylether [100]. At present it is rather difficult to appreciate how important these relatively small differences in the total RNA synthesis may be for both the types of development. At any rate, the above authors [234] demonstrated in experiments with isolated abdominal fragments that Cecropia extracts had no effect on RNA synthesis in the absence of active prothoracic glands or ecdysone. And, this is in good agreement with all the morphological observations concerning the effect of juvenoids on tissue growth.

Somewhat different conditions with respect to juvenoid effects on nucleic acid metabolism seem to exist in adult female insects. Here the effects of juvenoids are associated with an enormous increase of both DNA and RNA synthesis rates especially in the ovarian follicle cells, without dependence on prothoracic gland hormone or ecdysone. In addition, some other tissues in the body of adult females may remain unaffected in this respect [125]. However, the increased rate of nucleic acid synthesis in the ovarian tissue is not an exclusive feature caused by juvenoids. It is obvious that this happens also during metamorphosis of certain Endopterygote species without any dependence on CAH or juvenoids, suggesting that genetical factors may be involved.

Lezzi and Frigg [178] have pointed out that any effects of hormones on RNA synthesis are often thought to be the evidence for gene activation, while these may be quite general responses in the target cells. They mention two methods for determination of specific gene activation, e.g. RNA-DNA hybridization and induction of puffs on the polytene chromosomes. Since both the methods were already used for studies of juvenoid action we give a brief account on the results obtained in the following paragraphs.

The RNA-DNA hybridization method was used by Karlson *et al.* [159] for studies on the effect of dihomojuvenate on RNA metabolism in isolated nuclei of the fat body cells of *Calliphora*. It was observed that not only ecdysone but also dihomojuvenate stimulated RNA synthesis in this system [54, 55]. Both hormones applied simultaneously had smaller effect. Hybridization techniques [159] further revealed that ecdysone as well as dihomojuvenate caused appearance of new RNA molecules. In addition, at least part of the genes activated by ecdysone appeared also to be activated by dihomojuvenate. Smaller effect of both the hormones used simultaneously was interpreted so that certain repressors could bind with both hormones. When this would happen, derepression and transcription of that particular locus was not realized, which might also indicate a competitive effect of both hormones at subcellular level.

The method based on puffing patterns was employed by Laufer and Holt [170] for studies on the effect of farnesoate hydrochlorination products on the polytene chromosomes in salivary glands of *Chironomus thummi*. They observed that an inhibition of pupation which was caused by the above mixture of juvenoids was associated with specific inhibition of certain puffs. Moreover, one of the puffs, e.g. the Balbiani ring IV b was enlarged. They assumed that juvenoids might regulate transcription of certain gene loci. Specific changes in RNA synthesis might then lead to changes in metamorphosis. Lezzi and Gilbert [179] working with *Chironomus tentans* found inhibition of certain ecdysone-sensitive Balbiani ring by dihomojuvenate. In addition, the dihomojuvenate stimulated formation of other puffs which were in turn inhibited by ecdysone. In other words, both these hormones worked antagonistically with respect to puff formation. Assuming that juvenoids might increase intracellular transport of sodium ions, Lezzi and Frigg [178] created a theory which interpreted possible antagonistic action of both the hormones on the basis of selective influence on some Na-sensitive or K-sensitive parts of the chromosome. However, there still remains a problem whether these effects on puffing patterns were really specific to juvenoid action. It may be argued that at the time when the salivary glands were used for the experimentation the cells were already committed unconditionally to metamorphosis and their critical period to juvenoids was passed over. We shall discuss this problem more in details in connexion with ecdysoids in part III.

The results obtained with antibiotics and other inhibitors of nucleic acid metabolism (p. 253) have pointed out that there exist close relations between the effects of juvenoids and changes in nucleic acid metabolism [320]. But with regard to juvenoid action at various ontogenetic and taxonomic levels, it still remains undecided what is a cause, what is a consequence, and what is merely a coincidence of juvenoid action at the molecular level. Nevertheless, there are many theories which try to explain the mode of action of juvenoids on basis of their effects on nucleic acid metabolism. For instance, some theories assume that juvenoids may act through: i) stimulation of DNA synthesis and replication [231]; ii) formation of mRNA [83] or other regulatory RNA molecules [159]; regulation of translation through tRNA molecules [128]; iv) control of transcription of the regulatory genes [375]; and, v) induction or suppression of gene activity as revealed by puffing patterns [170, 178]. Some of these theories are relevant to juvenoid action on metamorphosis, others to their action on ovarian growth but none of them gives universal explanation for all the juvenoid effects.

16. The Mode of Action of Juvenoids

We are going to discuss here the mode of action in the strict sense of the term, e.g. the primary effect of juvenoids on the target cells. It seems reasonable to assume that such a primary effect of juvenoids will be identical in various stages as well as in various insect species, while a sequence of developmental events that follow may be completely different with respect to stage and species. We must confess honestly that the primary effect of juvenoids has not been unequivocally identified so far. However, there are some theories in this respect which will be briefly discussed and offered to reader's own considerations. For reasons mentioned formerly in the text, we do not discuss here theories which explain the mode of action of juvenoids on basis of their stimulation of some general metabolic processes. In analogy with other vertebrate or invertebrate hormones, the current theories on the mode of action of juvenoids are mainly recruited from three biochemical disciplines. These are as follows: a) specific influence on nucleic acid metabolism and gene activation, b) intracellular balance of potassium and sodium ions, and c) effects on a membrane permeability.

a) In the previous section we have quoted some theories explaining the mode of action of juvenoids by specific changes in nucleic acid metabolism and gene activation. We restrict our discussions to this point here to the theory by WILLIAMS and KAFATOS [375] which represents an application of the Jacob-Monod's theory on gene activation in micro-organisms to insect development. According to this theory the genome of each cell is divided into three main sets of the genes, e.g. larval, pupal, and adult. Each of them is under control of certain "master regulatory gene" containing a promoter which is recognized by RNA polymerase present in all stages, and an operator which is subjected to inhibition by a special repressor. The repressors are activated or inactivated with respect to concentration of the juvenile hormone, and this is followed by realization of larval, pupal, or adult developmental programmes. However, this theory is not universal too. It is relevant to metamorphosis but does not apply on ovarian growth. Moreover, results obtained with juvenoids suggest that juvenile hormone comes into action already at the blastoderm stage. Since then it can stop the morphogenetic progress practically at any stage. Thus each cell should perhaps contain not three but impossibly large number of the above complete gene sets with a complex system of promoters, operons, and specific repressors for each step of the sequential polymorphism of insect development.

b) Based on some biochemical studies concerning juvenoid action in *Locusta*, MINKS [211] concluded that according to a considerable importance of sodium/potassium balance across the mitochondrial membranes, juvenoids could influence the Na/K "pump" which may then result in a decrease of ATP-ase activity. The following scheme of action was suggested: juvenoid reaches the cell membrane—stimulates the Na/K pump—this results in a decrease of intracellular Na and in an increase of intracellular K ratio—ATP-ase activity decreases which leads to stimulation of phosphorylating ability. It occurs that similar mechanisms depending on changes of intracellular Na concentration were used earlier by KROEGER and LEZZI for explanation of the mode of action of ecdysone (see p. 365). A further theory by LEZZI and FRIGG [178] is based on entirely different asumptions than that of MINKS. It does not expect a decrease but in contrast it claims that an increase of intracellular Na concentration occurs after juvenoid action, e.g. that juvenoids create a reciprocal ionic balance situation to that caused by ecdysone.

c) It was assumed by WIGGLESWORTH [358], that an important factor in the mechanism of juvenoid action might be their influence on internal lipoprotein membranes, irrespectively whether the receptor sites would occur in the nucleus or in the cytoplasm. Such action of juvenoids on the permeability relations within the cells was thought to be associated somehow with activation of the genetically controled enzymatic systems responsible for formation of the larval patterns. According to BAUMANN [8] certain juvenoids indeed cause a depolarization of the cell membrane in salivary glands of *Galleria*, or influence conductivity of bimolecular lipid-protein membranes [9]. The effect was explained by possible interactions of juvenoids with the membrane lipids, although interactions with the translocator system used could not be excluded.

We have seen in the list of juvenoid structures contained in Appendix tables that the range of their activity is limited by certain size and shape of the molecule, by presence of specific functional groups and substituents, and by some physico-chemical features such as the low polarity. It seems, therefore, that similarly like with other hormones, the primary effects of juvenoids may depend on their ability to bind with certain yet unidentified intracellular receptor sites which were specifically tuned to fit and recognize molecules of this size, shape, type, and properties. Isoprenoid compounds are known ligands to proteins [111], such as are the nonpolar steroids [342] which form nondialyzable complexes. Some isoprenoids resembling remotely juvenoids, like retinene for instance [111] are able to form lamellar lipoprotein complexes. The above hypothetical receptor sites for juvenoids can thus well be on protein or nucleoprotein molecules. After testing a large number of active and inactive compounds we have finally adopted this hypothesis of juvenoid binding with specific cellular receptor sites. Moreover, by combining all the available structure-activity data we have calculated approximative figures and dimensions of these receptor sites. But still we have no idea where in a cell the receptors might be situated. In agreement with some theories on juvenoid action [159, 358, 375] the receptor loci should be closely linked with genetical factors which determine tissue, stage, and species specificity of the biological responses to juvenoids.

17. Possibility of Anti-Juvenile or Anti-Juvenoid Effects

It is a general belief that a discovery of an anti-juvenile activity might be extremely important from theoretical as well as from practical aspects. We have indicated already (p. 223) that some physiological anti-juvenile mechanisms may be effective at the end of larval development in some Endopterygote insects. Studies on these natural anti-juvenile mechanisms in *Dermestes vulpinus* [311] revealed that the anti-juvenile effect was caused by low concentrations of ecdysone or other ecdysoids.

Accepting the above hypothesis that juvenoids act by binding with some intracellular receptor sites we can deduce two possibilities with respect to anti-juvenile action. Developmental responses may be realized either: a) by a simple contact of juvenoid molecule with the receptor site or, b) when juvenoid combines with a receptor but in addition it exhibits some further biochemical interactions. Being it the former eventuality, a compound may either bind or not and, therefore, it would be active or inactive, but no anti-hormonal activity can be expected. However, being it the second eventuality, a compound may bind with the receptor but due to a lack of further chemical interactions the developmental responses will not be affected. In other words, such compounds may compete with the active molecules for the receptor sites, they may "block" the receptor and in this way cause anti-hormonal effects. In order to investigate the possibility of anti-juvenile activity of this kind, we have tested a large number (about 200) different compounds which were structurally related to juvenoids but were inactive due to factors such as wrong stereochemical or optical isomerism, content of wrong functional groups, altered positions of the branching radicals on the chains, and, because of other structural reasons. Some of the compounds tested had very small juvenile activity or were active on other species. The assays were performed on freshly moulted penultimate instar larvae of *Pyrrhocoris*. An expected positive anti-juvenile response was precocious metamorphosis, e.g. appearance of adult patterns on the body of last instar larvae. However, in not a single case we obtained any positive anti-juvenile result in these assays. We do not present this information in order to discourage everybody from doing further anti-juvenile assays, but just to show that this way has already been explored and is not as easy as one might think.

18. Inactivation of Juvenoids in the Body

Because of their regulatory role the hormones must be inactivated or excreted soon after their stimulus to the target tissue was given. For example, excretion of the isoprenoid hormones such as are sexual hormones of vertebrates is a common phenomenon. With insects it is known that isoprenoids are contained in the excrements since SCHMIALEK [161] isolated farnesol from the excrements of *Tenebrio*. In *Pyrrhocoris*, certain juvenoids were present in the excrements in biologically active form (p. 130). GILBERT and SCHNEIDERMAN [105, 284] observed inactivation of Cecropia extracts in the body. They found that temperature and various application procedures had considerable effect on this process.

SCHMIALEK and his co-workers studied some biosynthetic features of isoprenoids in adult Saturniids. They observed conversion of labelled mevalonate into farnesol, farnesal and nerolidol among other compounds [279]. Farnesol was oxidized to farnesoic acid in *Tenebrio*; in addition, there occurred a potent alcohol dehydrogenase with higher affinity to farnesol than to some other alcohols [85]. They also suggested a possibility of allylic transformation of farnesol into inactive nerolidol, hydrogenation of the double bonds, and some other biochemical changes with respect to a higher biological stability of farnesylmethylether over farnesol. Finally, they discussed a possibility of faster metabolic degradation of isoprenoid juvenoids when injected in water emulsions [108].

It is expected that one of the main factors of inactivation of the dihomojuvenates may be an opening of the oxirane ring, which is usually followed by formation of the corresponding inactive diols. Further inactivation of these compounds may occur by ester hydrolysis, yielding rather polar and biologically inactive acid, as has been pointed out by SIDDALL and his co-workers [299, 300], and by STAAL [330]. The ester hydrolysis problem applies to large part of juvenoids which are the esters of aliphatic or aromatic acids. It is generally known that insect haemolymph and some tissues such as intestine, Malpighian tubes, and pericardial cells contain very potent nonspecific esterases which have especially high affinity to split the aromatic esters. As seen in the Appendix tables, the terpenoid phenyl-ethers are very active juvenoids whereas the phenyl esters of the corresponding terpenoid acids are almost inactive. This is most probably due to enzymatic ester hydrolysis. However, it is not quite clear whether a generally higher juvenile activity of ethyl esters when related to corresponding compounds with other ester radicals would be associated with slower hydrolysis of the ethyl esters or whether this would be due to other factors.

METZLER *et al.* [203] extensively studied possible biosynthetic mechanisms for dihomojuvenate in adult Cecropia moths. According to an incorporation of the labelled precursors they concluded that L-methionine provided the ester methyl group but did not contribute to formation of the carbon skeleton. Farnesol, farnesylpyrophosphate, and propionate were not utilized. Mevalonate was extensively incorporated into *trans, trans* farnesol, but not into the sesquiterpene-like hormone. This result indicated that the compound was not synthesized *via* mevalonate in the adult moths. The label of 2-^{14}C-acetate was, however, recovered in the carbon skeleton.

19. The Effect of Juvenoids on Prothoracic Glands and on Corpora Allata

It is a common feature in comparative endocrinology that hormones exhibit either positive or negative feedback effects on their parental or related endocrine glands. These feedback effects are realized as a direct response to hormonal concentration in the blood stream or, indirectly trough certain physiological changes which were induced by the hormone. The first type of these feedback effects may prevent an overproduction of the hormone and may enable a more flexible regulation. The second is important for mutual synchronization of several hormone dependent physiological functions.

In insects, WILLIAMS [369] and GILBERT and SCHNEIDERMAN [104] observed activation of prothoracic glands after application of Cecropia extracts to Saturniid pupae. Further it appeared that the extracts could prevent degeneration of the prothoracic glands [101]. Moreover, KRISHNAKUMARAN and SCHNEIDERMAN [165, 166] demonstrated that the observed prothoracotropic effects of the Cecropia extracts were indeed associated with juvenile activity, since positive responses were also obtained with series of synthetic juvenoids. Recently, SIEW and GILBERT [302] reported an increased rate of RNA synthesis in the prothoracic glands and in corpora allata after treatment with dihomojuvenate. They supposed possible occurrence of a positive feedback between the hormone and the endocrine glands. They considered an increased rate of RNA synthesis to be a criterion for the endocrine gland activity. However, BERRY *et al.* [18] found an increased rate of RNA synthesis in most tissues of these insects following a simple mechanical injury.

Ability of juvenoids to prevent degeneration and breakdown of the prothoracic glands is a widespread phenomenon commonly associated with the occurrence of extra-larval or extra-pupal moults. This effect of juvenoids on survival of the prothoracic glands is in principle quite consistent with the effects of CAH. In addition, similarly like with many other specifically "larval" tissues, juvenoids affect retention and survival of these glands only at the determined sensitive periods which take place long before the glands actually start to degenerate.

A more complicated problem than survival is concerned with the above discussed stimulation of the prothoracic gland activity by juvenoids. According to our knowledge, these prothoracotropic effects were observed in diapausing pupae of Lepidoptera. In all other systems, such as in starved larvae of Exopterygotes in general, or in diapausing larvae or pupae of other than Lepidopteran insects, we found that juvenoids did not stimulate the activity of prothoracic glands and, consequently, never induced moult cycles *per se*. However, juvenoids caused a considerable delay or even complete inhibition of pupation in last instar larvae, which has been discussed in connexion with juvenoid effects on metamorphosis. We believe that juvenoids in this case inhibited the secretory activity of the prothoracic glands. When analysing these phenomena we came to a problem whether such a negative feedback effect of juvenoids or CAH on prothoracic glands would be of any selective advantage for development in these insects. And, it occurred to us that it would. For example, such an adaptation might prevent initiation of metamorphosis before all residues of the endogenous CAH left from previous stages were extinguished. Moreover, this might prevent spontaneous appearance of intermediate forms and lethal metathetelies which are extremely rare under normal developmental conditions.

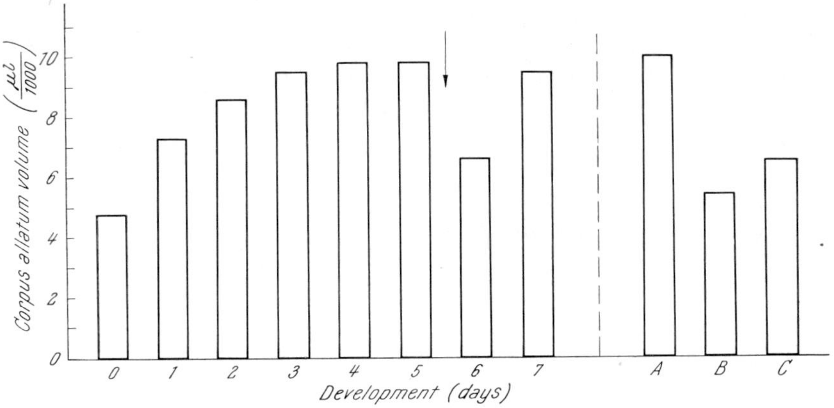

Fig. 31. Growth of the CA volume during the initial 7 days of development in adult female *Pyrrhocoris* and a compensatory effect of dihydrodichlorofarnesoate (100 μg per female) on growth of the CA. A = CA volume of 4-day-old control females, B = CA volume of 4-day-old females treated with the juvenoid, C = CA volume of 4-day-old females implanted with one active CA. The juvenoid treatment and implantation of CA were performed just after adult emergence. Arrow indicates the moment of oviposition

Further relationships between the effects of juvenoids and the insect neuroendocrine system are characterized by reactivation of the hormonal activity in diapausing adult females (cf. p. 240). This is most probably a result of indirect feedback mechanisms which synchronize hormonal activity with the course of reproductive cycles. In *Pyrrhocoris* a necessary condition for reactivation of the neuroendocrine system is that juvenoids stimulate at least one complete reproduction cycle including oviposition of the ripe eggs. In contrast to this species, reactivation of the neuroendocrine system by juvenoids is rather limited in adults of *Graphosoma*. The effect is not universal and depends on species.

As far as the interactions between juvenoids and functions of the corpora allata is concerned, our informations on this subject are limited to the observed stimulation of RNA synthesis in the CA by juvenoids [302] and, to assumptions on the delayed effects of juvenoids on later programming of the CA functions [259]. Our approach to this problem was to study the effects of juvenoids on one of the most specific interactions between CA and their hormone. This interaction was represented by the so called compensatory effect (see p. 30). Results of this study are illustrated in Fig. 31. They show that a regular growth of the CA which occurs after emergence of adult females is inhibited by implantation of active CA taken from an old adult female. This confirms the presence of the compensatory effect in this system. Furthermore, the CA of females treated with the juvenoid compound show the same growth inhibition as in the former case with CA implantations. The affected CA remain small but active as indicated by their yellow colour (in *Pyrrhocoris* females the active CA are always yellow whereas the inactive ones are colourless). A considerably small size of the affected CA suggests a decreased synthesis of the hormone. Additional experiments performed with several other juvenoid compounds revealed that the occurrence of compensatory effects was closely correlated with doses near to ID-50 Ovic. per female, irrespective whether highly or little active compounds were used. The compensatory effect is clearly caused by humoral way since it reciprocally occurs also on young CA which have been implanted into females with large CA. Physiological functions of the compensatory mechanism probably depend on maintenance of certain steady titre of the hormone in the haemolymph, or at least it prevents an overproduction of the hormone. Such ability of synthetic juvenoids to copy even the very specific compensatory effects of CAH illustrates clearly that the similarities go into rather specific physiological details.

20. Interactions between Plants and Insects by Juvenoids

It is generally known that plants contain various secondary substances of no nutritional value. These can influence various physiological and developmental functions or can be toxic to phytophagous insects [10]. The insects are in turn able to identify and discriminate between plants that are not suitable for their development, which evolved as a result of evolutionary adaptations [290]. Herout [115] pointed out that a large number of sesquiterpenoid compounds present in plants had pronounced physiological and biochemical effects on animals. Some of the compounds listed had similar structural patterns with certain juvenoids. In a review on insect-plant interactions through isoprenoids [116] he mentioned several possibilities of hormonal interactions. For example, juvenoids had some structural features in common with a plant sesquiterpenoid hormone—abscissic acid. Its functions in a plant system are to some extent analogous with that of juvenoids in insects, e.g. inhibition of morphogenesis.

Findings on presence of a juvenoid in the American paper products and in the wood of the Canadian balsam fir trees by SLÁMA and WILLIAMS [316, 318, 379] suggested a real possibility of interactions between plants and insects by these hormonal agents. However, in contrast to a common occurrence of terpenoid compounds in conifers, the hormonally active juvenoid compounds occurred only in certain evergreen trees, as suggested by STAAL [329]. General information on these insect-plant interactions by juvenoids are contained in special review articles by WILLIAMS [374], SLÁMA [307], and in other papers [116, 329]. We have, therefore, restricted our discussions to only a few comments.

It appeared that a contact of sensitive insects with paper, wood, and other parts of a plant, or treatment of these insects with raw plant extracts [42, 43, 244, 276] was enough to produce serious developmental disturbances. This confirmed that these interactions between plants and insects could practically well exist in nature. We may add that juvabione which is the active principle in the balsam fir wood is specifically active on insects of the family Pyrrhocoridae. However, more recently it has been reported that plants may contain less selective juvenoids which are active on some other insects such as on *Tenebrio* [378] or *Galleria* [195]. A plant *Echinacea angustifolia* is reported to contain some potent juvenoids [132]. These may be present also in other plants [378].

These studies on insect-plant interactions through juvenoids may not be as simple as they were thought at the time of "paper factor" discovery. The problem is associated with a lack of logistical reasons for selection of the proper plants and proper insect species for testing the activity. For example, various plants may contain different bisabolene derivatives which are active on *Pyrrhocorid* bugs but are virtually inactive on any other insect species. Hence, such juvenoids will never be discovered unless tested right on these susceptible species. Moreover, some active plant products may turn to be rather trivial sesquiterpenoid compounds which have been known from synthetic studies on juvenoids. As suggested by SLÁMA [307], the active juvenoid compounds may be contained only in certain parts of the plant sytem and their amount may additionally change with respect to the vegetation period, especially in the annual plants. We have a feeling that a random screening of plants for possible juvenoid presence may occasionally give positive responses but it is quite inefficient. A more efficient and qualitatively different attitude to these studies is to investigate possible involvement of juvenoids in specific ecological situations where certain insect species or pests carefully avoid certain plants.

21. Other Effects of Juvenoids, Action on Different Groups of Insects

In addition to the effects of juvenoids on morphogenesis and ovarian growth, there appears also an influence on some other physiological processes, such as on the formation of epidermal pigments or on the insect behaviour. We are going to discuss these effects here along with the effects of juvenoids on other organisms than insects, and, finally, we mention also differences in responsiveness to juvenoids in insects of different taxonomic groups.

During insect metamorphosis there commonly occurs a change of epidermal pigments. Similarly like in the case of CA implantation (cf. p. 46), the induction of extra-larval instars by juvenoids is associated with retention of the old larval pigmentation, whereas formation of the new pupa- or adult-specific pigments is inhibited. (This can be seen in Figs. 8–12.) Such changes in pigmentation are dependent on the degree of morphological change. Under certain circumstances (see p. 136), however, juvenoids may cause a change of pigments without direct dependence on the morphological changes. Thus, by special treatments at the end or after the sensitive period to juvenoids, it was possible to obtain morphologically perfect adults with juvenile coloration. These effects were most frequent in grasshoppers and in other Orthopterans after treatments with Cecropia extracts, farnesol derivatives, or other juvenoids [99, 152, 153, 154, 225]. This phenomenon commonly occurs in assays for biological activity of juvenoids in various Exopterygote insects. It is recognized by the fact that extra-larvae or adultoids caused by excessive doses have different pigmentation than adultoids produced by doses around ID-50 Morph. For example, in *Graphosoma* large doses of juvenoids produce yellowish-green specimens, whereas adultoids produced by small amounts of juvenoids contain extensive areas with dark pigmentation. Further it appears that juvenoids may influence pigmentation also in Endopterygote insects. So, Hitze [121] observed pronounced pigmentation changes in *Cerura* catterpillars after application of farnesoate dihydrochloride. She also noted [122] that the effect of juvenoids on pigmentation was antagonized by ecdysone. The latter accelerated pupation which was connected with specific changes in the pigment patterns. On the other hand, juvenoids delayed pupation and inhibited also the associated pigmentation changes. In *Pieris* farnesol did not mimic formation of green patches which readily occurred after CA implantations [117], but this was probably due to a generally low biological activity of farnesol in this species.

Further effects of juvenoids involve some changes in insect behaviour. In consent with the earlier Piepho's observations concerning the effect of CA implantations on behaviour in *Galleria*, Hintze-Podufal [122] observed that juvenoids inhibited appearance of behavioural patterns characteristic for the onset of metamorphosis in the last instar larvae of *Cerura*. In addition to metamorphosis, insect behaviour associated with reproduction is also influenced by juvenoids. This is perhaps most apparent on changes in the sexual behaviour. We have already mentioned (p. 243) curious observations by ŽĎÁREK and SLÁMA [390] on extra-larvae of *Pyrrhocoris* which exhibited both male and female sexual behaviour, e.g. the male larvae were attracted to female larvae and both exhibited mating instincts. Needless to say, copulation was never realized because these extra-larvae of both sexes had completely undeveloped external genital system. In adult Diptera, epoxidated or hydrochlorinated farnesoates induced receptivity in allatectomized females or shortened the refractory period. This was found in *Musca* [1] and in *Aedes* [110]. These effects of juvenoids on sexual behaviour are usually connected with simultaneous stimulation of ovarian growth. Similarly like in the case of CA implantations the effects are more apparent on females than on the males. It seems that shortening of the refractory period after mating is a common effect. In some insects, such as in Orthopteran *Euthystira*, this may occur even with farnesol which generally has very small activity [221]. It was also suggested that juvenoids may influence insect behaviour by stimulating production of sexual pheromenes, as e.g. in *Byrsotria* [84], *Tenebrio* [201], or *Ips* [24].

With respect to taxonomic criteria it appears that representatives of some insect groups show considerable deviation from the common type of developmental responses to juvenoids. These differences are common between higher taxonomic units but they reach down to the family level. We do not have in mind differences in relative activity of juvenoids which have been described in section D on structure-activity relationships (p. 194), but we are going to discuss here some special differences in response to all kinds of juvenoids which may result from special developmental adaptions in a given group of insects. Thus, we have already seen that there exist certain "reliable" insect groups such as Exopterygote insects where juvenoids induce rather consistent and reproducible responses in various orders and families. In addition to this there exist certain "difficult" insect groups represented mostly by various orders and families of the Endopterygote insects where juvenoids cause either atypical responses or are inactive at all.

One of such "difficult" groups of insects are Diptera. In mosquitoes, juvenoids cause disturbances of pupal-adult transformation which result in inability of adults or adultoids to emerge. This was observed at first by SPIELMAN and WILLIAMS [325] and later it was confirmed by others [176, 222, 324]. Several juvenoids appeared to cause sterilization or ovicidal effects in *Aedes* [239]. In addition, treatment of mosquito larvae by juvenoids may cause inhibition of pupation which is frequently associated with high larval mortality. This effect is of considerable practical importance for calculations on a possible mosquito control by juvenoids.

Perhaps the most difficult problem in juvenoid research was associated with limited action on higher Diptera-Cyclorrhapha. Thus, farnesol derivatives failed to induce positive juvenile responses in these insects, and for a long time it was not known whether this was caused by i) use of wrong chemical compounds; ii) inability of farnesol derivatives to penetrate within the tissues, or, iii) completely changed sensitive periods to juvenoids. Finally, SRIVASTAVA and GILBERT [326, 327] succeeded to produce pupal-adult intermediates with definite morphological juvenile patterns. They were unable to induce also extra-larval moults by juvenoids, but were able to cause a prolongation of the last larval instar [cf. also 73], e.g. they obtained delayed puparium formation. These original observations on *Sarcophaga* were soon confirmed in *Drosophila* by ASHBURNER [5], in *Stomoxys* by WRIGHT [384] in *Calliphora* by PILAN [245] and others [385]. Therefore, the effect of juvenoids on metamorphosis of higher Diptera showed certain deviations but was firmly established. However, it still remains to determine whether juvenoids could also induce female sterility and ovicidal effects in this group of insects. In *Drosophila*, BRYANT and SANG [38, 39] observed that farnesol derivatives and some other juvenoids increased the frequency occurrence of melanotic tumours. Also farnesol increased the number of facets in the Bar-eye of the Bar-eyed mutant of *Drosophila* [73]. However, dichlorofarnesoate had no mutagenic activity in this species [13].

Among Lepidoptera there are species like *Galleria* where extra-larval instars can easily be induced by various juvenoids [134, 265]. In some other species juvenoids only delayed pupation when applied in the last larval instar [122]. Species specific differences occurred also with respect to ovarian growth. Thus, in some species such as in *Pieris* [14] or in the monarch butterfly [236] juvenoids stimulated ovarian development whereas in species like *Malacosoma* [275, 350], or *Heliothis* [109], farnesol or farnesyl methylether inhibited ovarian growth and decreased female fecundity. Moreover, ovicidal effects were easily produced by juvenoids in Cecropia (p. 244), and also in Tortricids [254, 255].

We have a personal feeling that some of the results obtained in Lepidoptera with farnesol and farnesyl methyl ether represent rather side effects than real physiological effects of these poorly active juvenoids because they are quite inconsistent with other findings and theoretical expectations. For example, EL-IBRASHY and MANSOUR [81] concluded that farnesol induced precocious pupation of small larvae in *Agrotis* which is exactly opposite from what all juvenoids do in other insects. In contrast, they found induction of extra-larvae in this species by "Alar" which is succinic acid 2,2-dimethyl hydrazide- an apparent juvenile hormone effect in a compound completely unrelated to any known juvenoids. Quite usual side effects of high dosages of farnesol or farnesyl methyl ether are high mortality associated with ecdysial failures and other nonspecific developmental abnormalities as found in *Malacosoma* by WELLINGTON [349]. Furthermore, FUKAYA [97] observed also high mortality and delay of pupation with farnesol in larvae of the rice stem borer. In addition he found retention of diapause by farnesol which was in consent with the author's assumptions on regulation of diapause in this species by CAH. Some effects of juvenoids on larval development in Lepidoptera were described by STAAL [331].

Among Coleopterans we can find both "reliable" and "difficult" families with respect to juvenoid action. In a large number of Coleopteran species juvenoids cause inhibition of pupal-adult tranformation as exemplified on pupae of *Tenebrio* or *Dermestes*. However, similarly like in other Endopterygotes, it is extremely difficult here to produce extra-larval instars by juvenoids. A typical effect of juvenoids on last instar larvae is prolongation of the instar, or the delay of pupation. Finally, when pupation occurs, there usually appear slight juvenile patterns on the legs or other appendages and these pupae almost invariably moult into adultoids. This was found for instance in *Leptinotarsa* [200], *Dermestes* [311] or *Caryedon* [202], but it seems to be a phenomenon of common occurrence. Exceptionally, the duration of the pupal instar was also prolonged by juvenoids [65, 202], whereas in most species it was rather abbreviated. In some Dermestid beetles, such as *Trogoderma*, there occurred spontaneous extra-larval instars. After treatment with juvenoids the number of additional extra-larval instars was increased [202]. The pupal adultoids which were finally obtained as a result of previous treatments of Coleopteran larvae with juvenoids, contained a mixture of larval, pupal, and adult epidermal patterns [202]. In contrast to Diptera, ovicidal effects of juvenoids do exist in Coleoptera. They were obtained by juvenoid application to fresh eggs [347]. Finally, we may point out that some higher Coleoptera such as Curculionidae and Scolytidae are extremely resistant to juvenoids.

Little information concerning the action of juvenoids is available for Hymenoptera. In contrast to other Endopterygote groups where the effects of juvenoids on pupae were more pronounced than those on larvae, we did not find any activity of juvenoids when applied to pupae of sawflies. However, when applied at the end of the larval stage or at the beginning of prepupal instar, juvenoids were able to inhibit metamorphosis. Similar inhibition with metamorphosis of juvenoids was earlier observed in the alkali bee *Nomia melanderi* by HSIAO and HSIAO [127].

Despite relative uniformity of juvenoid action on Exopterygote group, there were some old problems with aphids. An exact evaluation of the juvenoid effects on metamorphosis was in aphids complicated by relatively small differences between the larval and the adult epidermal patterns. Convenient epidermal structures such as are the wings can be used as a criterion only in populations with predominating winged forms. And, it appeared that farnesol affected the wing development in *Doralis* [74]. Further results concerning the effects of juvenoids in aphids were described by WHITE [353, 354) in the cabbage aphid. Recently it is clear that aphids are the regular subject of juvenoid influence. They are no exception and it is possible to induce here formation of perfect extralarvae and larval adultoids of different types. The adultoids may contain dead or malformed embryos inside the female's body. Hence, the effects of juvenoids on metamorphosis are here combined with ovicidal action.

For a long time it was known that CAH had an important role in regulation of the caste differentiation in Termites. Under normal conditions the density of individual castes in the colony was precisely regulated. LÜSCHER [186] discovered that injections or feeding of larvae and pseudergates with juvenoids induced soldier development. The juvenoid treated colonies contained therefore considerably increased number of soldier caste.

Finally, there remains a question whether juvenoids may also affect other Arthropods or other forms of organisms than insects. Problems concerned with possible toxicity of juvenoids to mammals have been discussed in the next chapter. The function of juvenile hormone is so specific for insect development that one may rather expect a possibility of some nonspecific effects than a real similarity in action on other organisms.

Nevertheless, extracts with juvenile hormone activity possessed antimicrobial effects [319]. Furthermore, farnesol and synthetic dihomojuvenates promoted growth in pea stem sections [334]. Or, a hydrochlorinated mixture of farnesoate inhibited growth of a protozoan *Crithidia* in culture [130]. In addition, juvenoids also prolonged larval life in *Trichinella* [123] and delayed moults in other Nematodes [71]. However, it is very difficult to comprehend how general and specific these findings may be with respect to so many different types and structures of juvenoids that we already know.

According to some theories juvenoids seem to stabilize in immature insects the existing situation on the genome preventing derepression and realization of new information. In other words they are maintaining the genetical flux. This may explain the increased interest in juvenoids by the oncological laboratories where many compounds are being tested against tumor growth.

F. Juvenoids as Potential Pesticides

The idea of using materials with insect juvenile hormone activity as pesticides was known since 1956. It was offered by WILLIAMS [368] in his initial studies on Cecropia extracts. However, the first possibilities of its realization came into consideration only one decade later when SLÁMA and WILLIAMS [316, 317, 318, 379] reported complete inability of certain insect species to reproduce in contact with a juvenoid or "paper factor" present in American paper products. Then, since 1966 there occurred a rapid progress in the research of juvenoids which was supported by both academic and industrial institutions.

Actually, the general principles concerned with practical use of juvenoids in insect control occur to be quite simple. Their basic idea is that developmental disturbances caused by juvenoids at different ontogenetic stages may lead to mortality of the affected specimens or, at least, may prevent insects from reproduction. Since a real potency of insect pest gradiations is primarily associated with their tremendous capacity to reproduce, it is easy to realize how serious consequences for development of insect populations such an arrest of reproduction might have. With juvenoids, the economic entomologists get an access to chemical growth regulators which have no direct toxicity to insects and as far as we know they are not toxic to other forms of life either. Thus, like antibiotics, certain selective plant-control agents or hormones used for birth control, juvenoids form a new class of chemicals which can be characterized as specific regulators of insect growth and development.

General aspects of using juvenoids as insect pesticides have been frequently discussed in literature. There are WILLIAMS's reviews on "Third generation pesticides" [371] and on "Juvenile hormone insecticides" [372]. Further we may mention articles by SCHNEIDERMAN and his associates on strategy of controlling insect pests with growth regulators [282, 288], discussions to this point by DE WILDE [363, 364], ELLIS [82], and EL-IBRASHY [80]. Various aspects of this problem have been touched in series of other papers [16, 79, 102, 131, 163, 308, 338, 374, 378]. Quite a realistic account of practical use of juvenoids in insect control has recently been expressed by STAAL [330] who is going to prepare a review article on the field use of juvenoids (personal communication)*.

It may be observed that many data presented in other sections of this book have a direct relevance to practical use of juvenoids, as e.g.

* After completion of the manuscript a collection of papers on the practical use of juvenoids as pesticides appeared in the book: MENN, J. J., and M. BEROZA, eds.: Insect Juvenile Hormones: Chemistry and Action. New York-London: Academic Press. 1972.

methods of application and evaluation of the field experiments, structure-activity relationships, selectivity, survey of activity in Appendix tables, etc. However, when the manuscript of this text was prepared there were almost no data in literature concerning actual results of the field experiments with juvenoids. For this reason, we have given here only a brief discussion of the basic problems pertinent to field effectiveness of juvenoids, using experience obtained from our field experiments.

There are two categories of juvenoids with respect to their suitability for insect control. First of them is represented by compounds with realtively low selectivity which act on many unrelated insect species. Biological activity of these more or less "universal" juvenoids expressed in ID-50 Morph. per spec. is equal or may surpass LD-50 of the best insecticides. However, the activity of juvenoids of the second category, e.g. compounds with selective action on some insect species, is, as a rule, much higher. Despite this, the tasks to produce juvenoids for commercial use are in favour of the "universal" juvenoids. This is due to the fact that it may be easier to succeed in laborious and costly procedures required for authorized registration of a new pesticide with one "universal" juvenoid than with several "selective" compounds.

Further problems in practical use of juvenoids are associated with difficulties in selection of a proper juvenoid candidate compound. There are many laboratories which have hundreds of juvenoid compounds at hand without realization which of the insect pests the compounds may be good for. Originally we thought that the best criterion for selection of juvenoid compounds for field use was their highest biological activity in the laboratory assays. Then, after several unsuccessful field experiments we came to a conclusion that equally or even more important were the factors associated with chemical stability of juvenoids or persistence of their activity under field conditions. For example, 6,7-epoxygeranyl-p-nitrophenylether had the best biological activity on some species in laboratory but it appeared to be completely ineffective in the field due to its decomposition in less than 24 hours. On the other hand, far less active but more stable juvenoids than the above compound appeared to be effective under field conditions. Thus, the situation with persistence of juvenoids is quite different from the situation with insecticides. The tendency is to use nonpersistent insecticides in order to minimize contamination of the fields and agricultural products by their toxic residues. However, insecticides cause an almost immediate kill of insects at almost any developmental stage. Not so with juvenoids. These affect development of insects only at certain determined sensitive periods and in order to get a maximum control juvenoid must persist in active state or concentration as long as all the specimens have crossed over their period of sensitivity. And, as a rule, this may take several days or even weeks depending on the degree of developmental heterogeneity in a given species. Therefore, further development of juvenoids suitable for field control should be directed towards chemically stable compounds with relatively long persistence, or towards new methods of formulation which would increase persistence of juvenoids under field conditions. Unlike insecticides, persistent juvenoids will not represent a hazard in nature, provided that juvenoid residues are nontoxic and harmless. Apparently, juvabione which retains its juvenile hormone activity in the balsam fir wood for a hundred years is a good natural example.

It is quite evident that there is no juvenoid compound which could be universally used to control all insect pests. There are certain harmful species of insects, for instance Scolytid and Curculionid beetles, higher Diptera, some sawflies, and perhaps others which may resist any treatment with so far known juvenoids. We have already discussed that there are particular types of juvenoids which act preferentially on particular insect species. However, for practical reasons it appears more important to know whether there exist insect pests which may be generally more convenient for juvenoid control and, in addition, what criteria may serve for their identification. As pointed out by Staal [330], one such criterion may be relatively fast developmental rate. Insect pests which seem to be suitable for juvenoid control are, for instance, aphids, mosquitoes, and stored product pests. Moreover, equally suitable for juvenoid control appear to be insect pests with seasonally synchronized development where sensitive stages occur in limited time interval. This usually occurs during the first generations in polyvoltine species and generally in insects where occurrence of certain developmental stages is synchronized with determined stages of the food plant development. Less convenient for juvenoid control appear insect pests with rather slow and heterogeneous development because it enormously increases requirements for persistence of the activity.

We have already seen (p. 104) that evaluation of the effectiveness of juvenoids is not based on direct toxicity. It is based on a complex of integrated physiological effects on the whole population. Using persistent compounds on insects with asynchronous development we can observe appearance of all kinds of developmental disturbances, from female sterility, ovicidal effects, ecdysial or adult emergence failures, inability to feed or move, defects in structure and function of some organs, to a final inability of mating and insemination. Sometimes, one of these effects may be dominant as for instance ecdysial failures which eventually occur in almost 100 per cent of the specimens and are fully responsible for successful juvenoid control measures. Another of such dominant control measures under field conditions may be suppressed feeding. As shown by Varjas [341], the effect of juvenoids appeared considerably more effective when evaluated according to the food consumption, whereas it was less effective when evaluated according to morphological criteria. This shows that additional factors which were insignificant in laboratory assays may occasionally act in favour of juvenoids in the field. Unfortunately, reciprocal relationships also exist; this is exemplified by the highly active but unstable juvenoid compound as has been described in one of the previous paragraphs. Nevertheless, all these problems concerned with evaluation of juvenoid effectiveness in the field point out to a necessity to involve trained specialists in insect population dynamics for these studies, at least at a stage before juvenoids reach routine application. This may help to sieve the preliminary informations which show very often quite contradicting results, e.g. conspicuously positive results with apparently wrongly selected juvenoids, as well as failures with well selected compounds. It is essential that these scientists moderate too much enthusiasm and too much skepticism as it comes out from the preliminary results.

In order to prove a real usefulness of juvenoids for insect control, we performed a model experiment with very active peptidic juvenoid (T-335) against the Indian red cotton staining bug *Dysdercus cingulatus* in an artificial cotton plantation. The compound was extremely selective, highly persistent, and non-toxic. A full account of the results will be published elsewhere but we just want to mention here that this compound seems to be an excellent control agent for this species. It was found that amounts as small as 0.1 g per Hectare exhibited 100 per cent control with persistence for several weeks. Other examples of such suitable juvenoid-insect systems are known from literature. For example, the dienoic juvenoids of Zoecon Comp. were acting against mosquitoes in 0.0001 p. p. m. in water [395]. Other examples were juvenoids specifically acting on Tenebrionid stored product pests [66, 235], sesamyl geranylethers and their epoxides acting down to nanogram levels [27], and some other aromatic terpenoid ethers [235, 385]. It may be expected that the usefulness of juvenoids in insect control will be amplified when this is combined with other methods of the integrated pest control programme. Especially promising in this respect is the fact that most of the insect parasites belong to Hymenopteran or Dipteran groups which are relatively most resistant to juvenoid action.

Practical use of juvenoids is still dependent on toxicological and pharmacological results. It occurs that the insect juvenile activity in itself does not represent any danger to men or warmblooded animals. We have pointed out (p. 90) that compounds with juvenile hormone activity on insects were found in human organs such as adrenal cortex, thymus, and placenta. In addition, these compounds are actually present in our ordinary food. They were found in milk or cream, they occur also in various vegetable and plant food products, and eventually also in bran or bread which commonly contain farnesol and some of its derivatives. It is quite reasonable to expect however, that some synthetic juvenoids may possess certain structural features which will not be fully compatible with all the physiological functions in plants or animal systems. Recent studies performed with some selected acyclic and aromatic juvenoids suggest no acute oral toxicity [301]. In mice, many juvenoids appeared completely nontoxic in peroral applications of 5–10 g per kg of the body weight [66, 235, 301, 364]. Despite such large doses there were no deviations from regular growth of the body weight, consumption of food, and water intake in the treated mice. In addition, there were no significant differences in the content of glucose, blood urea nitrogen, and sodium, nor in alkaline phosphatase activities, nor in the haemoglobine haematocrit, total white blood cell or differential cell counts between the animals treated with juvenoids and the untreated ones. [301]. However, the whole problem does not depend only on determination of the toxicity of juvenoids. It is equally or perhaps even more important (though more expensive and labourious) to determine whether juvenoids would not cause some additional pharmacological effects, such as terratologic, carcinogenic, allergetic, and the like.

Ellis [82] pointed out that premature use of juvenoids may represent a real hazard when used without exact toxicological data. She suggested that juvenoids related to juvabione be inspected first of all for possible carcinogenic activity. However, this precaution seems to be derived from a misinterpretation of the larval character of epidermal patches in *Pyrrhocoris* which were confused by Carlisle *et al.* [44] for tumorous growth. We have calculated that some peptidic juvenoids act in a range of 1 μg per metric ton of insect weight. This is enormous activity which might really represent a hazard to human health provided that any parallelity in action between insects and warmblooded animals would occur. Yet the same compound has no toxic influence on insects even when applied in million-times larger doses than ID-50 Morph. [308].

Before the synthetic juvenoids were known, Williams [368] expressed an assumption that insects could hardly develop resistance against their juvenile hormone. And, as a matter of fact, juvenile hormone activity is absolutely essential to any insect in order to develop into reproducible adults. However, recent situation is somewhat different. The juvenile hormone activity is now exhibited by hundreds of synthetic juvenoids which possess different structural patterns as well as different properties. We have shown that some juvenoids are inactivated or excreted from the body faster than others, some are easily transported through the haemolymph whereas others are not. Moreover, some juvenoids contained in plants are active only on certain species of insects whereas other species remain unaffected. This suggests that insects might have evolved resistance in contact with naturally occurring juvenoids during millions of years of insect evolution. Probably, this process is also associated with an apparent diversification of insect sensitivity to juvenoids [318]. Essentially, we do not expect development of resistance against hormonal activity as such, but it is likely that resistance may develop against particular properties of a particular juvenoid compound. It has been pointed out by SCHNEIDERMAN [282, 288] that any insecticide, hormonally based or not, is bound to act as a powerful sieve for concentrating resistant mutants that are present in low frequencies in the original populations. Resistant forms may appear when the effects of juvenoids will be used as selection factors, e.g. when specimens which survived juvenoid treatments in reproducible state had a chance to produce offsprings subjected to the same treatment. Actually, this is never realized in laboratory screening procedures, but it may easily happen in the treated fields. According to STAAL [330] speculations about development of insect resistance to juvenoids is premature since this phenomenon can be studied only as it develops under the field conditions. It may be argued, however, that laboratory experiments programmed for selection of resistant strains to juvenoids can be well designed according to analogous experiments used for studies on resistance to insecticides. In this respect it might be quite important to know whether resistance to juvenoids would depend on the same or different selection factors than that to insecticides. One has to agree with SCHNEIDERMAN [282] that we have to identify mechanisms of resistance to juvenoids before it actually occurs in the field and learn how to minimize it in advance.

Finally, it remains to discuss what are the actual advantages and disadvantages of juvenoid pesticides versus insecticides. Traditionally it is thought that a main advantage of juvenoids may depend on their low or practically zero toxicity which also applies to their breakdown products. Moreover, whole series of juvenoids are selectively acting on certain species of insect pests which may enable to apply selective control with minimum disturbances of natural zoocenoses, including normal development of the pest's natural enemies such as parasites and predators. Further advantage is enormously high biological activity of certain juvenoids which make it possible to use considerably smaller amounts of the active compounds and so to minimize contamination of the environment. Because of a lack of toxicity to insects juvenoids enable to use qualitatively new methods in insect control such as sexual distribution of female sterility.

Disadvantage of juvenoids over insecticides are dependent mainly on the fact that their use must be cautiously prepared and programmed according to specific developmental conditions in each species. Further, insecticides can destroy harmful stages almost immediately with less attention to developmental stage. They can be used in emergency situations whereas juvenoid action is less immediate and most efficient in preventing gradations and reproduction of the pest populations. Juvenoids must also persist for considerably longer periods than insecticides in order to achieve a complete control.

These general calculations are superficially reflecting a recent situation which is, of course, a subject of future alterations. From the discussions given in this section it is clear that juvenoids will never fully substitute for insecticides. The main reason is that they are not universal. However, it will be quite satisfactory if juvenoids can replace insecticides in few of the main insect pests. This is still dependent on identification of the best juvenoid compounds for a given pest species and on finding of the best conditions for their use in the field. We have indicated that very promising progress on the way to such selective juvenoid pesticides have already been made. A lack of universality warns against exaggerated optimism but it is not a good reason for premature skepticism about potential usefulness of juvenoids for insect control. The situation may be illustrated on an example of antibiotics, which may be less universal than sulfonamides but with respect to specific diseases their utility is unmatched.

It is beyond the scope of this publication to go into more details with discussions on further problems concerning the field experiments with juvenoids. This applied discipline belongs to the field of economic entomology. We are reasonably certain that due to a rapid progress in this discipline a lot of new results will be published in the near future.

References

[1] Adams, T. S., and A. M. Hintz: Relationship of age, ovarian development, and the corpus allatum to mating in the house fly *Musca domestica*. J. Insect Physiol. *15*, 201–215 (1969).

[2] Anderson, R. I., C. A. Henrick, and J. B. Siddall: Stereoselective synthesis of olefins. Reaction of dialkylcopper-lithium reagents with allylic acetates. J. Amer. chem. Soc. *92*, 735–737 (1970).

[3] Arnold, Z., J. Kahovcová, M. Pánková, M. Svoboda, M. Tichý, and F. Šorm: Natural and synthetic materials with insect hormone activity XIV. Synthesis of substituted phenylgeranylethers and related compounds. Coll. Czech. Chem. Commun. *38*, 261–269 (1973).

[4] Arnold, Z., and F. Šorm: Unpublished results.

[5] Ashburner, M.: Effects of juvenile hormone on adult differentiation of *Drosophila melanogaster*. Nature *227*, 187–189 (1970).

[6] Ayyar, K. S., and G. S. Krishna Rao: Studies in terpenoids IV. Synthetic studies in juvabiones and analogues. Conversion of ar(+)-turmerone to ar(+)-juvabione. Canad. J. Chem. *46*, 1467–1472 (1968).

[7] Babu, T. H., and K. Sláma: Systemic activity of a juvenile hormone analog. Science *175*, 78–79 (1972).
[8] Baumann, G.: Zur Wirkung des Juvenilhormons: elektrophysiologische Messungen an der Zellmembran der Speicheldrüse von *Galleria mellonella*. J. Insect Physiol. *14*, 1459–1476 (1968).
[9] Baumann, G.: Juvenile hormone: effect on bimolecular lipid membranes. Nature *223*, 316–317 (1969).
[10] Beck, S. D.: Resistance of plants to insects. Ann. Rev. Entomol. *10*, 207–232 (1965).
[11] Bell, W. J.: Dual role of juvenile hormone in the control of yolk formation in *Periplaneta americana*. J. Insect Physiol. *15*, 1279–1290 (1969).
[12] Bell, W. J., and R. H. Barth, Jr.: Quantitative effects of juvenile hormone on reproduction in the cockroach *Byrsotria fumigata*. J. Insect Physiol. *16*, 2303–2313 (1970).
[13] Beneš, V., and R. Šram: Mutagenic activity of some pesticides in *Drosophila melanogaster*. Ind. Med. Surg. *38*, 442–444 (1969).
[14] Benz, G.: The stimulation of oogenesis in *Pieris brassicae* by the juvenile hormone derivative farnesenic acid ethyl ester. Experientia *26*, 1012 (1970).
[15] Beran, P., and F. Šorm: Unpublished results.
[16] Berkoff, C. E.: The chemistry and biochemistry of insect hormones. Quart. Rev. *23*, 372–391 (1969).
[17] Berkoff, C. E.: Synthetic juvenile hormone and "Synthetic juvenile hormone". Science *168*, 1607 (1970).
[18] Berry, S. J., A. Krishnakumaran, H. Oberlander, and H. A. Schneiderman: Effects of hormones and injury on RNA synthesis in *Saturniid* moths. J. Insect Physiol. *13*, 1511–1537 (1967).
[19] Birch, A. J., P. L. Macdonald, and V. H. Powell: A stereoselective synthesis of (±)-juvabione. Tetrahedron Letters *1969*, 351–354.
[20] Blaustein, M. P., and H. A. Schneiderman: A brief survey of the effects of potential antimetabolites and enzymes on the development of giant silkmoths. J. Insect Physiol. *5*, 143–159 (1960).
[21] Blount, J. F., B. A. Pawson, G. Saucy: X-Ray structure determination of 4 R, 8R-p-menth-1-en-9-ol p-iodobenzoate. Revision of the absolute stereochemistry of natural (+)-juvabione. Chem. Communications *1969*, 715, 1016.
[22] Blumenfeld, M., and H. A. Schneiderman: Effect of juvenile hormone on the synthesis of specific proteins in silkworm pupae. Am. Zoologist *6*, 97 (1966).
[23] Bodenstein, D., and E. Shaaya: The function of the accessory sex glands in *Periplaneta americana* (L.) I. A quantitative bioassay for juvenile hormone. Proc. nat. Acad. Sci. (Amer.) *59*, 1279–1290 (1968).
[24] Borden, J. H., K. K. Nair, and C. E. Slater: Synthetic juvenile hormone: induction of sex pheromone production in *Ips confusus*. Science *166*, 1626–1627 (1969).
[25] Bowers, W. S.: Juvenile hormone: activity of natural and synthetic synergists. Science *161*, 895–897 (1968).
[26] Bowers, W. S.: Juvenile hormone: activity of aromatic terpenoid ethers. Science *164*, 323–325 (1969).
[27] Bowers, W. S.: Chemistry and biological activity of morphogenetic agents. Mitt. Schweiz. Ent. Ges. *44*, 115–130 (1971).

[28] Bowers, W. S., and C. C. Blickenstaff: Hormonal termination of diapause in the alfalfa weevil. Science *154*, 1673–1674 (1966).

[29] Bowers, W. S., H. M. Fales, M. J. Thompson, and E. C. Uebel: Juvenile hormone: identification of an active compound from Balsam fir. Science *154*, 1020–1021 (1966).

[30] Bowers, W. S., and M. J. Thompson: Juvenile hormone activity: effects of isoprenoid and straight chain alcohols on insects. Science *142*, 1469–1470 (1963).

[31] Bowers, W. S., M. J. Thompson, and E. Uebel: Juvenile and gonadotropic hormone activity of 10,11-epoxyfarnesoic acid methyl ester. Life Sciences *4*, 2323–2331 (1965).

[32] Bracken, G. K., and K. K. Nair: Stimulation of yolk deposition in an *Ichneumonid* parasitoid by feeding synthetic juvenile hormone. Nature *216*, 483–484 (1967).

[33] Brady, S. F., M. A. Ilton, and W. S. Johnson: A highly stereoselective synthesis of trans-trisubstituted olefinic bonds. J. Amer. chem. Soc. *90*, 2882–2889 (1968).

[34] Brieger, G.: Juvenile hormone mimics: structure-activity relationships for *Oncopeltus fasciatus*. J. Insect Physiol. *17*, 2085–2093 (1971).

[35] Brookes, V. J.: The maturation of the oöcytes in the isolated abdomen of *Leucophaea maderae*. J. Insect Physiol. *15*, 621–631 (1969).

[36] Brookes, V. J.: The induction of yolk protein synthesis in the fat body of an insect *Leucophaea maderae*, by an analog of the juvenile hormone. Dev. Biol. *120*, 459–471 (1969).

[37] Brown, H. C., and Min-hon Rei: The solvomercuration-demercuration of representative olefins in the presence of alcohols. Convenient procedures for the synthesis of ethers. J. Amer. chem. Soc. *91*, 5646–5647 (1969).

[38] Bryant, P. J., and J. H. Sang: *Drosophila:* lethal derangements of metamorphosis and modifications of gene expression caused by juvenile hormone mimics. Nature (London) *220*, 393–394 (1968).

[39] Bryant, P. J., and J. H. Sang: Physiological genetics of melanotic tumors in *Drosophila melanogaster*. VI. The tumorigenic effects of juvenile hormone-like substances. Genetics *62*, 321–336 (1969).

[40] Butterworth, F. M., and D. Bodenstein: Adipose tissue of *Drosophila melanogaster*. IV. The effect of the corpus allatum and synthetic juvenile hormone on the tissue of the adult male. Gen. Comp. Endocrinol. *13*, 68–74 (1969).

[41] Caliezi, A., and H. Schintz: Zur Kenntnis der Seskviterpene und Azulene 90. Die Cyclisation der Farnesylsäure. Helv. Chim. Acta *32*, 2556–2560 (1949).

[42] Carayon, J., and M. Thouvenin: Emploi d'une substance mimétique de l'hormone juvénile pour la lutte contre les *Dysdercus*, Hémiptéres nuisibles au cotonier. C. R. Acad. Agric. France *52*, 340-346 (1966).

[43] Carlisle, D. B., and P. E. Ellis: Abnormalities of growth and metamorphosis in some Pyrrhocorid bugs: the paper factor. Bull. Ent. Res. *57*, 405–417 (1967).

[44] Carlisle, D. B., P. E. Ellis, Z. Brettschneiderová, and V. J. A. Novák: Growth disorders of the bug *Pyrrhocoris apterus* and the "juvenile hormone" effect of certain papers. J. Endocrinol. *35*, 211–212 (1966).

[45] CARROLL, M. F.: Addition of β,γ-unsaturated alcohols to the active methylene group. Part II. The action of ethyl acetoacetate on cinnamylalcohol and phenylvinylcarbinol. J. Chem. Soc. *1940*, 1266–1268.

[46] CAVENEY, S.: Juvenile hormone and wound modelling of *Tenebrio* cuticle architecture. J. Insect Physiol. *16*, 1087–1107 (1970).

[47] CAVILL, G. W. K., D. G. LAING, and P. J. WILLIAMS: A synthesis of the juvenile hormone, methyl cis-10,11-epoxy-7-ethyl-3,11-dimethyl-trideca-trans-2-trans-6-dienoate. Austr. J. Chem. *22*, 2145–2160 (1969).

[48] CAVILL, G. W. K., and P. J. WILLIAMS: A synthesis of methyl 10,11-epoxy-3,7,11-trimethyl-2,6-dodecadienoate. Austr. J. Chem. *22*, 1737–1744 (1969).

[49] ČERNÝ, V., L. DOLEJŠ, L. LÁBLER, F. ŠORM, and K. SLÁMA: Dehydrojuvabione — a new compound with juvenile hormone activity from balsam fir. Coll. Czech. Chem. Commun. *32*, 3926–3933 (1967); Tetrahedron Letters *1967*, 1053–1057.

[50] CHASE, A. M.: Effects of actinomycin D on development in pupae of *Tenebrio molitor*. Nature *215*, 1516–1517 (1967).

[51] CHASE, A. M.: Effects of antibiotics on epidermal metamorphosis and nucleic acid synthesis in *Tenebrio molitor*. J. Insect Physiol. *16*, 865–884 (1970).

[52] CHEN, D. H., W. E. ROBBINS, and R. E. MONROE: The gonadotropic action of cecropia extracts in allatectomized American cockroach. Experientia *18*, 577 (1962).

[53] CIZIN, YU. S., and A. A. DRABKINA: Yuvenilnyi gormon nasekomykh i ego analogi (Juvenile hormone and its analogues). Uspechi Chimii *39*, 1074–1094 (1970).

[54] CONGOTE, L. F., C. E. SEKERIS, and P. KARLSON: On the mechanism of hormone action. XIII. Stimulating effects of ecdysone, juvenile hormone and ions on RNA synthesis in fat body cell nuclei from *Calliphora erythrocephala* isolated by a filtration technique. Exp. Cell Res. *56*, 338–346 (1969).

[55] CONGOTE, L. F., C. E. SEKERIS, and P. KARLSON: On the mechanism of hormone action. XVIII. Alterations of the nature of RNA synthesized in isolated fat body cell nuclei as a result of ecdysone and juvenile hormone action. Z. Naturforsch. *25 b*, 279–284 (1970).

[56] CONNIN, R. V., O. K. JANTZ, and W. S. BOWERS: Termination of diapause in the cereal leaf beetle by hormones. J. Econ. Entomol. *60*, 1752 1753 (1967).

[57] COREY, E. J.: Stereospecific chemical synthesis of juvenile hormones. Mitt. Schweiz. Ent. Ges. *44*, 87–96 (1971).

[58] COREY, E. J., and K. ACHIWA: A method for deoxygenation of allylic and benzylic alcohols. J. Org. Chem. *34*, 3667–3668 (1969).

[59] COREY, E. J., N. W. GILMAN, and B. E. GANEM: New methods for the oxidation of aldehydes to carboxylic acids and esters. J. Amer. chem. Soc. *90*, 5616–5617 (1968).

[60] COREY, E. J., J. A. KATZENELLENBOGEN, N. W. GILMAN, S. A. ROMAN, and B. W. ERICKSON: Stereospecific total synthesis of the dl-C_{18} Cecropia juvenile hormone. J. Amer. chem. Soc. *90*, 5618–5620 (1968).

[61] COREY, E. J., J. A. KATZENELLENBOGEN, and G. H. POSNER: A new stereospecific synthesis of trisubstituted olefins. Stereospecific synthesis of farnesol. J. Amer. chem. Soc. *89*, 4245–4247 (1967).

[62] Corey, E. J., W. L. Mock, and D. J. Pasto: Chemistry of diimide. Some new systems for the hydrogenation of multiple bonds. Tetrahedron Lett. *1961*, 347–352.

[63] Corey, E. J., and H. Yamamoto: Simple stereospecific synthesis of C_{17} and C_{18} Cecropia juvenile hormones (racemic) from a common intermediate. J. Amer. chem. Soc. *92*, 6636–6637 (1970).

[64] Corey, E. J., and H. Yamamoto: New stereospecific synthetic routes to farnesol and its derivatives, including a biologically active position isomer of C_{17} Cecropia juvenile hormone. J. Amer. chem. Soc. *92*, 6637–6638 (1970).

[65] Critchley, B. R., and D. G. Campion: Effects of synthetic juvenile hormone and a juvenile hormone analogue, methyl farnesoate dihydrochloride on pupal development of the yellow mealworm *Tenebrio molitor* L. Bull. ent. Res. *61*, 293–297 (1971).

[66] Cruickshank, P. A.: Some juvenile hormone analogs.–A critical appraisal. Mitt. Schweiz. Ent. Ges. *44*, 97–114 (1971).

[67] Cruickshank, P. A., and R. M. Palmere: Terpenoid amides as insect juvenile hormones. Nature (London) *233*, 488–489 (1971).

[68] Dahm, K. H., and H. Roeller: The juvenile hormone of the giant silk moth *Hyalophora gloveri* (Strecker). Life Sciences *9*, 1397–1400 (1970).

[69] Dahm, K. H., H. Roeller, and B. M. Trost: The juvenile hormone. IV. Stereochemistry of juvenile hormone and biological activity of some of its isomers and related compounds. Life Sciences *7*, 129–137 (1968).

[70] Dahm, K. H., B. M. Trost, and H. Roeller: The juvenile hormone. V. Synthesis of the racemic juvenile hormone. J. Amer. chem. Soc. *89*, 5292–5294 (1967).

[71] Davey, K. G.: Molting in a parasitic nematode, *Phocanema decipiens*. IV. The mode of action of insect juvenile hormone and farnesyl methyl ether. Int. J. Parasitol *1*, 61–66 (1971).

[72] Davis, N. T.: Studies of the reproductive physiology of *Cimicidae* (Hemiptera). I. Fecundation and egg maturation. J. Insect Physiol. *10*, 947–963 (1964).

[73] Dearden, M.: Experiments on the effect of farnesol on the development of normal and bareyed *Drosophila*. J. Insect Physiol. *10*, 195–210 (1964).

[74] Dehn, M. von: Hemmung der Flügelbildung durch Farnesol bei der schwarzen Bohnenlaus, *Doralis fabae* Scop. Naturwissenschaften *50*, 578–579 (1963).

[75] Dolejš, L., P. Beran, J. Kahovcová, Z. Machková, K. Sláma, and F. Šorm: Czech. Pat. Appl. *PV* 8513 (1969).

[76] Dolejš, L., P. Beran, K. Sláma, and F. Šorm: Czech. Pat. Appl. *PV* 3122 (1970).

[77] Dolejš, L., J. Kahovcová, K. Sláma, and F. Šorm: Czech. Pat. Appl. *PV* 652 (1970).

[78] Dolejš, L., Z. Machková, P. Beran, J. Kahovcová, K. Sláma, and F. Šorm: Czech. Pat. Appl. *PV* 3123 (1970).

[79] Dyte, C. E.: Evolutionary aspects of insecticide selectivity. Proc. 5th Br. Insectic. Fungic. Conf., 393–397 (1969).

[80] El-Ibrashy, M. T.: Insect hormones and analogues: chemistry, biology and insecticidal potencies. Z. angew. Ent. *66*, 113–144 (1970).

[81] El-Ibrashy, M. T., and M. H. Mansour: Hormonal action of certain biologically active compounds in *Agrotis ypsilon* larvae. Experientia *26*, 1095–1096 (1970).

[82] Ellis, P.: Can insect hormones and their mimics be used to control pests. Pest Articles and News Summaries (A) *14*, 329–342 (1968).

[83] Emmerich, H.: Beeinflussung der Imaginalentwicklung von *Tenebrio molitor* durch Farnesylmethyläther und Actinomycin. Verh. deutsch. Zool. Ges., Innsbruck 1968, pp. 519–526. Leipzig: Geest & Portig K.-G. 1968.

[84] Emmerich, H., and R. H. Barth, Jr.: Effect of farnesyl methyl ether on reproductive physiology in the cockroach *Byrsotria fumigata* (Guérin) 1. Z. Naturforsch. *23 b*, 1019–1020 (1968).

[85] Emmerich, H., G. Drews, K. Trautmann, and P. Schmialek: Über den Stoffwechsel des Farnesols in *Tenebrio molitor*-Larven. Z. Naturforsch. *20 b*, 211–213 (1965).

[86] Emmerich, H., and H. V. Thiele: Wirkung von Farnesylmethyläther auf die Ovarienreifung von *Pterostichus nigrita* F. Naturwissenschaften *56*, 641 (1969).

[87] Emmerich, H., A. Zahn, and P. Schmialek: Über die Aktivität einiger Dehydrogenasen und Transaminasen bei *Tenebrio molitor* unter dem Einfluß von Farnesylmethyläther. J. Insect Physiol. *11*, 1161–1168 (1965).

[88] Engelmann, F.: Female-specific protein: biosynthesis controlled by corpus allatum in *Leucophaea maderae*. Science *165*, 407–409 (1969).

[89] Engelmann, F.: (see reference No. 70 on p. 71).

[90] Enslee, E., and L. M. Riddiford: Morphological effects of juvenile hormone analogues on embryos of *Pyrrhocoris apterus*. Amer. Zool. *10*, 527 (1970).

[91] Fales, J. H., O. F. Bodenstein, and W. S. Bowers: Seven juvenile hormone analogues as synergists for pyrethrins against house flies. J. Econ. Entomol. *63*, 1379–1380 (1970).

[92] Faulkner, D. J., and M. R. Petersen: A synthesis of trans-trisubstituted olefins using the Claisen rearrangement. Tetrahedron Letters *1969*, 3243–3246.

[93] Faulkner, D. J., and M. R. Petersen: Synthesis of C_{18} Cecropia juvenile hormone to obtain optically active forms of known absolute configuration. J. Amer. chem. Soc. *93*, 3766–3767 (1971).

[94] Findlay, J. A.: Synthesis of Cecropia juvenile hormones and related compounds. Mitt. Schweiz. Ent. Ges. *44*, 65–72 (1971).

[95] Findlay, J. A., and W. D. MacKay: A convenient synthetic route to insect juvenile hormone. Chem. Communications *1969*, 733–734.

[96] Fisher, F. M., Jr., and R. C. Sanborn: Production of insect juvenile hormone by the microsporidian parasite *Nosema*. Nature *194*, 1193 (1962).

[97] Fukaya, M.: The inhibitory action of farnesol on the development of the rice stem borer in post diapause. Jap. J. Appl. Ent. Zool. *6*, 298 (1962).

[98] Fukuda, S.: Effect of farnesol on pupae of the silkworm *Bombyx mori*. Annot. Zool. Jap. *37*, 86–93 (1964).

[99] Fuzeau-Braesch, S.: Action de l'hormone juvénile de synthèse dans la morphogenèse et la pigmentogenèse de *Gryllus bimaculatus* (Orthoptéres). C. R. Séan. Soc. Biol. *162*, 1086–1090 (1969).

[100] GEYER, A., G. HERDA, and P. SCHMIALEK: Beeinflussung des Stoffwechsels von *Tenebrio molitor* (Coleoptera) durch Farnesylmethyläther während des Puppenstadiums. Acta ent. bohemoslov. *65*, 253–262 (1968).

[101] GILBERT, L. I.: Maintenance of the prothoracic gland by the juvenile hormone in insects. Nature (London) *193*, 1205–1207 (1962).

[102] GILBERT, L. I.: Invertebrate hormones. I. The chemistry of insect hormones. Proc. III Int. Congr. Endocrin., Mexico, 340–346 (1968).

[103] GILBERT, L. I., and H. A. SCHNEIDERMAN: Occurrence of substances with juvenile hormone activity in adrenal cortex of vertebrates. Science *128*, 844 (1958).

[104] GILBERT, L. I., and H. A. SCHNEIDERMAN: Prothoracic gland stimulation by juvenile hormone extracts of insects. Nature (London) *184*, 171–173 (1959).

[105] GILBERT, L. I., and H. A. SCHNEIDERMAN: The development of a bioassay for the juvenile hormone of insects. Trans. Amer. micr. Soc. *79*, 38–67 (1960).

[106] GILBERT, L. I., and H. A. SCHNEIDERMAN: The content of juvenile hormone and lipid in Lepidoptera: sexual differences and developmental changes. Gen. Comp. Endocrinol. *1*, 453–472 (1961).

[107] GILBERT, L. I., and H. A. SCHNEIDERMAN: Some aspects of insect metamorphosis. Am. Zoologist *1*, 11–51 (1961).

[108] GRÄSSMANN, A., A. GEYER, G. HERDA, and P. SCHMIALEK: Die Inaktivierung von Juvenilhormonwirksamen Substanzen in Insekten. Acta ent. bohemoslov. *65*, 92–99 (1968).

[109] GUERRA, A. A.: Effect of biologically active substances in the diet on development and reproduction of *Heliothis* spp. J. Econ. Entomol. *63*, 1518–1521 (1970).

[110] GWADZ, R. W., L. P. LOUNIBOS, and G. B. CRAIG: Precocious sexual receptivity induced by juvenile hormone analog in female of the yellow mosquito, *Aedes aegypti*. Gen. Comp. Endocrinol. *16*, 47–51 (1971).

[111] HANAHAN, D. J., F. R. N. GURD, and I. ZABIN: Lipide chemistry, 300 pp. New York: J. Wiley. 1960.

[112] HEJNO, K., V. JAROLÍM, K. SLÁMA, and F. ŠORM: Czech. Pat. Appl. *PV* 5534 (1968).

[113] HEJNO, K., V. JAROLÍM, K. SLÁMA, and F. ŠORM: Czech. Pat. Appl. *PV* 6793 (1968).

[114] HEJNO, K., V. JAROLÍM, K. SLÁMA, and F. ŠORM: Czech. Pat. 138448 (1970).

[115] HEROUT, V.: Biochemistry of sesquiterpenoids. In: GOODWIN, T. W., ed.: Aspects of terpenoid chemistry and biochemistry, pp. 53–94. (Proc. Phytochem. Soc. Symp.) Academic Press. 1971.

[116] HEROUT, V.: Some relations between plants, insects and their isoprenoids. Progr. Phytochem., Vol. II, pp. 143–202. Interscience Publ. 1970.

[117] HIDAKA, T., and T. OHTAKI: Effet de l'hormone juvénile et du farnésol sur la coloration tégumentaire de la nymphe de *Pieris rapae crucivora* Boisd. C. R. Séanc. Soc. Biol. *157*, 928–930 (1963).

[118] HIGHNAM, K. C., and L. HILL: (see reference No. 126 on p. 74).

[119] HIGHNAM, K. C., O. LUSIS, and L. HILL: Factors affecting oocyte resorption in the desert locust, *Schistocerca gregaria* (Forskal). J. Insect Physiol. *9*, 587–596 (1963).

[120] Highnam, K. C., O. Lusis, and L. Hill: The role of the corpora allata during oocyte growth in the desert locust, *Schistocerca gregaria* (Forskal). J. Insect Physiol. *9*, 827–837 (1963).

[121] Hintze, Ch.: Die Wirkung von Methylfarnesoatdihydrochlorid auf die Umfärbung und das Verpuppungsverhalten der Raupe von *Cerura vinula* L. (Lepidoptera). Biol. Zbl. *88*, 77–83 (1969).

[122] Hintze-Podufal, Ch.: The juvenile hormone-like effect of farnesol derivatives on the pupal development of *Cerura vinula* L. Z. Naturforsch. *26 b*, 154–157 (1971).

[123] Hitcho, P. J., and R. E. Thorson: Possible molting and maturation controls in *Trichinella spiralis*. J. Parasitol. *57*, 787–793 (1971).

[124] Hlaváček, J., L. Dolejš, and F. Šorm: Natural and syntethic materials with insect hormone activity. XIII. Aromatic terpenic thioethers. Coll. Czech. Chem. Commun. *37*, 3905–3908 (1972).

[125] Hodková, M.: The effect of juvenile hormone on the nucleic acid content in various organs during the reproduction cycle of *Pyrrhocoris* females. Acta Soc. Zool. Czechoslov. (In press.)

[126] Hoffmann, W. H., H. Pasedach, and H. Pommer: Reaktionen von Allylalkoholen mit aktiven Methin- und Methylenverbindungen. Ann. *729*, 52–63 (1969).

[127] Hsiao, C., and T. H. Hsiao: Insect hormones: their effects on diapause and development of Hymenoptera. Life Sciences *8*, 767–774 (1969).

[128] Ilan, J., J. Ilan, and N. Patel: Mechanism of gene expression in *Tenebrio molitor*—juvenile hormone determination of translational control through transfer ribonucleic acid and enzyme. J. Biol. Chem. *245*, 1275–1281 (1970).

[129] Ilan, J., J. Ilan, and J. H. Quastel: Effects of actinomycin D on nucleic acid metabolism and protein biosynthesis during metamorphosis of *Tenebrio molitor* L. Biochem. J. *100*, 441–447 (1966).

[130] Ilan, J., J. Ilan, and S. Ricklis: Inhibition by juvenile hormone of growth of *Crithidia fasciculata* in culture. Nature *224*, 179–180 (1969).

[131] Irving, G. W., Jr.: Agricultural pest control and the environment. Science *168*, 1419–1424 (1970).

[132] Jacobson, M.: Chemistry of natural products with juvenile hormone activity. Mitt. Schweiz. Ent. Ges. *44*, 73–77 (1971).

[133] Janda, V., Jr.: Einfluß einer juvenilhormonwirksamen Substanz auf den Protein-, Fett- und Glykogenmetabolismus der Larven von *Dysdercus cingulatus* (Fabr.). Zool. Jb. Physiol. *75*, 361–369 (1970).

[134] Jarolím, V., K. Hejno, F. Sehnal, and F. Šorm: Natural and synthetic materials with insect hormone activity. 8. Juvenile activity of the farnesane-type compounds on *Galleria mellonella*. Life Sciences *8*, 831–841 (1969).

[135] Jarolím, V., K. Hejno, K. Sláma, and F. Šorm: Czech. Pat. 138444 (1970).

[136] Jarolím, V., K. Hejno, K. Sláma, and F. Šorm: Czech. Pat. 142094 (1971).

[137] Jarolím, V., K. Hejno, K. Sláma, and F. Šorm: Czech. Pat. 144091 (1971).

[138] Jarolím, V., K. Hejno, K. Sláma, and F. Šorm: Czech. Pat. 142098 (1971).

[139] Jarolím, V., K. Hejno, K. Sláma, and F. Šorm: Czech. Pat. 142096 (1971).

[140] Jarolím, V., K. Hejno, K. Sláma, and F. Šorm: Czech. Pat. 142095 (1971).

[141] Jarolím, V., K. Sláma, and F. Šorm: Czech. Pat. Appl. *PV* 2371 (1970).

[142] Jarolím, V., K. Sláma, and F. Šorm: Czech. Pat. Appl. *PV* 3119 (1970).

[143] Jarolím, V., K. Sláma, and F. Šorm: Czech. Pat. Appl. *PV* 7674 (1970).

[144] Jarolím, V., K. Sláma, and F. Šorm: Czech. Pat. Appl. *PV* 7868 (1970).

[145] Jarolím, V., K. Sláma, and F. Šorm: Czech. Pat. Appl. *PV* 1872 (1971).

[146] Jarolím, V., K. Sláma, and F. Šorm: Czech. Pat. Appl. *PV* 3941 (1971).

[147] Jarolím, V., K. Sláma, and F. Šorm: Czech. Pat. Appl. *PV* 7002 (1971).

[148] Johnson, W. S., T. J. Brocksom, P. Loew, D. H. Rich, L. Werthemann, R. A. Arnold, T. Li, and J. D. Faulkner: Olefinic ketal Claisen rearrangement. A facile route to juvenile hormone. J. Amer. chem. Soc. *92*, 4463–4464 (1970).

[149] Johnson, W. S., S. F. Campbell, A. Krishnakumaran, and A. S. Meyer: Total synthesis of the racemic form of the second juvenile hormone (Methyl 12-homojuvenate) from the Cecropia silk moth. Proc. nat. Acad. Sci. *62*, 1005–1009 (1969).

[150] Johnson, W. S., Tsung-Tee Li, D. J. Faulkner, and S. F. Campbell: A highly stereoselective synthesis of the racemic juvenile hormone. J. Amer. chem. Soc. *90*, 6225–6226 (1968).

[151] Johnson, W. S., L. Werthemann, W. R. Bartlett, T. J. Brockson, I. Li, D. J. Faulkner, and M. R. Petersen: A simple stereoselective version of the Claisen rearrangement leading to trans-trisubstituted olefinic bonds. Synthesis of squalene. J. Amer. chem. Soc. *92*, 741–743 (1970).

[152] Joly, P.: Réactions de *Locusta migratoria* dans substances juvénilisantes. Gen. Comp. Endocrinol. *5*, 53 (1965).

[153] Joly, P.: Résultats d'injections de fortes doses d'hormone juvénile à *Locusta migratoria* en phase grégaire. C. R. Hebd. Séanc. Acad. Sci. (Paris) *D 268*, 1634–1635 (1969).

[154] Joly, P., and A. S. Meyer: Action de l'hormone juvénile sur *Locusta migratoria* (Orthoptère) en phase grégaire. Arch. Zool. Exp. Gén. *111*, 51–63 (1970).

[155] Judy, K. J., and L. I. Gilbert: Effects of juvenile hormone and molting hormone on rectal pad development in *Hyalophora cecropia*. J. Morph. *131*, 301–314 (1970).

[156] Julia, M., S. Julia, and R. Guégan: Préparation de composés terpéniques et apparentes, à partir de méthyl cyclopropyl cétone. Bull. Soc. Chim. (France) 1960, 1072–1079.

[157] Kahovcová, J., Z. Arnold, and F. Šorm: Natural and synthetic materials with insect hormone activity. XV. Coll. Czech. Chem. Commun. *38*, 765–769 (1973).

[158] Kahovcová, J., Z. Arnold, and F. Šorm: Natural and synthetic materials with insect hormone activity. XVI. Coll. Czech. Chem. Commun. *38*, 1165–1167 (1973).

[159] Karlson, P., L. F. Congote, and C. E. Sekeris: Stimulation of RNA synthesis by juvenile hormone in isolated fat body nuclei from *Calliphora erythrocephala*. Mitt. Schweiz. Ent. Ges. *44*, 171–176 (1971).

[160] Karlson, P., and M. Nachtigall: Ein biologischer Test zur quantitativen Bestimmung der Juvenilhormon-Aktivität von Insektenextrakten. J. Insect Physiol. *7*, 210–215 (1961).

[161] Karlson, P., and P. Schmialek: Nachweis der Exkretion von Juvenilhormonen. Z. Naturforsch. *14 b*, 821 (1959).

[162] Killian, D. B., G. T. Hennion, and J. A. Nieuwland: Addition of methanol to alkenyl acetylenes. J. Amer. chem. Soc. *58*, 892–894 (1936).

[163] Knipling, E. F.: Alternate methods of controlling insect pests. FDA Papers *3*, 16–24 (1969).

[164] Krishnakumaran, A., S. J. Berry, H. Oberlander, and H. A. Schneiderman: see ref. [174] on page 77.

[165] Krishnakumaran, A., and H. A. Schneiderman: Brain hormone activity of certain terpene derivatives in insects. Amer. Zool. *3*, 532 (1963).

[166] Krishnakumaran, A., and H. A. Schneiderman: Prothoracotrophic activity of compounds that mimic juvenile hormone. J. Insect Physiol. *11*, 1517–1532 (1965).

[167] Kroeger, H., and M. Lezzi: Regulation of gene action in insect development. Ann. Rev. Entom. *11*, 1–22 (1966).

[168] Larsen, J. R.: Inability of metabolizable oils to mimic corpus allatum hormone stimulation of ovarian development in mosquitoes (*Aedes aegypti*, *Culex pipiens*). Nature *206*, 428 (1965).

[169] Larsen, J. R., and D. Bodenstein: The humoral control of egg maturation in the mosquito. J. Exp. Zool. *140*, 343–382 (1959).

[170] Laufer, H., and T. K. H. Holt: Juvenile hormone effects on chromosomal puffing and development in *Chironomus thummi*. J. Exp. Zool. *173*, 341–352 (1970).

[171] Laverdure, A. M.: Action de l'ecdysone et de l'ester méthylique du farnesol sur l'ovaire nymphal de *Tenebrio molitor* (Coleoptére) cultivé *in vitro*. Ann. Endocrin. *31*, 516–524 (1970).

[172] Law, J. H., C. Yuan, and C. M. Williams: Synthesis of a material with high juvenile hormone activity. Proc. nat. Acad. Sci. (Amer.) *55*, 576–578 (1966).

[173] Lawrence, P. A.: Cellular differentiation and pattern formation during metamorphosis of the milkweed bug *Oncopeltus*. Dev. Biol. *19*, 12–40 (1969).

[174] Levinson, H. Z.: Studies on the juvenile hormone (neotenin) activity of various hormonomimetic materials. Riv. Parasitol. *27*, 47–63 (1966).

[175] Levinson, H. Z., and E. Zlotkin: The pseudojuvenilizing effect of protein denaturants on the integument of *Tenebrio molitor*. J. Insect Physiol. *18*, 511–519 (1972).

[176] Lewallen, L. L.: Effects of farnesol and ziram on mosquito larvae. Mosquito News *24*, 43–45 (1964).

[177] Lewis, C. T.: Diffusion of oil films over insects. Nature *193*, 904 (1962).

[178] Lezzi, M., and M. Frigg: Specific effects of juvenile hormone on chromosome function. Mitt. Schweiz. Ent. Ges. *44*, 163–170 (1971).

[179] Lezzi, M., and L. I. Gilbert: Control of gene activities in the polytene chromosomes of *Chironomus tentans* by ecdysone and juvenile hormone. Proc. nat. Acad. Sci. (Amer.) *64*, 498–503 (1969).

[180] Linstead, R. P., J. C. Lunt, B. C. L. Weedon, and B. R. Shephard: Anodic syntheses. Part VI. Chain extension of fatty acids by an isoprene unit. J. Chem. Soc. *1952*, 3621–3624.

[181] Loew, P., and W. S. Johnson: The synthesis of the optically active form of the C-18 Cecropia juvenile hormone. J. Amer. chem. Soc. *93*, 3765–3766 (1971).

[182] Loew, P., J. B. Siddall, V. L. Spain, and L. Werthemann: A short stereoselective synthesis of Cecropia juvenile hormone. Proc. nat. Acad. Sci. (Amer.) *67*, 1462–1464 (1970).

[183] Loew, P., J. B. Siddall, V. L. Spain, and L. Werthemann: A short stereoselective synthesis of Cecropia juvenile hormone. Experimental details. Proc. nat. Acad. Sci. (Amer.) *67*, 1824–1826 (1970).

[184] Loof A. de, and A. Lagasse: Juvenile hormone and the ultrastructural properties of the fat body of the adult Colorado beetle, *Leptinotarsa decemlineata* Say. Z. Zellforsch. *106*, 439–450 (1970).

[185] Loof, A. de, and J. de Wilde: Hormonal control of synthesis of vitellogenic female protein in the Colorado beetle, *Leptinotarsa decemlineata*. J. Insect Physiol. *16*, 1455–1466 (1970).

[186] Lüscher, M.: Die Bedeutung des Juvenilhormons für die Differenzierung der Soldaten bei der Termite *Kalotermes flavicollis*. Proc. VI Congr. IUSSI, Bern, pp. 165–170 (1969).

[187] Lüscher, M., G. Mühlmann, and M. Wyss-Huber: Juvenile hormone and protein synthesis in adult female cockroaches. Mitt. Schweiz. Ent. Ges. *44*, 197–206 (1971).

[188] Machková, Z., L. Dolejš, and F. Šorm: Derivatives of p-aminoacetophenone with juvenile hormone activity. Coll. Czech. Chem. Commun. *38*, 595–598 (1973).

[189] Machlis, L., W. H. Nutting, M. W. Williams, and H. Rapoport: Production, isolation, and characterisation of sirenin. Biochemistry *5*, 2147–2152 (1966).

[190] Masner, P.: The application of juvenile hormone-like substances for the study of morphogenesis of the gonads of *Pyrrhocoris apterus* (Insecta, Heteroptera). Gen. Comp. Endocrinol. *9*, 472 (1967).

[191] Masner, P., K. Sláma, and V. Landa: Natural and synthetic materials with insect hormone activity. 4. Specific female sterility effects produced by a juvenile hormone analogue. J. Embryol. Exp. Morph. *20*, 25–31 (1968).

[192] Masner, P., K. Sláma, and V. Landa: Sexually spread insect sterility induced by the analogue of juvenile hormone. Nature *219*, 395–396 (1968).

[193] Masner, P., K. Sláma, J. Žďárek, and V. Landa: Natural and synthetic materials with insect hormone activity. 10. A method of sexually spread insect sterility. J. Econ. Entomol. *63*, 706–710 (1970).

[194] Matolín, S.: Effects of a juvenile hormone analogue on embryogenesis in *Pyrrhocoris apterus* L. Acta ent. bohemoslov. *67*, 9–12 (1970).

[195] Mansingh, A., T. S. Sahota, and D. A. Shaw: Juvenile hormone activity in the wood and bark extracts of some forest trees. Canad. Entomol. *102*, 49–53 (1970).

[196] Mayo, P. de: Mono- and sesquiterpenoids. New York: Interscience Publ. 1959.

[197] McFarlane, J. E.: Fatty acids, methyl esters and insect growth. Comp. Biochem. Physiol. *24*, 377–384 (1968).

[198] McGovern, T. P., R. E. Redfern, and M. Beroza: Juvenile hormone activity of acetals applied topically and as a vapor to the yellow mealworm. J. Econ. Entomol. *64*, 238–241 (1971).

[199] Melor, J. M., and S. Munavali: Synthesis of sesquiterpenes. Quart. Revs. *18*, 270–294 (1964).

[200] Meltzer, J., and S. Heijnsbergen: Experiments with neotenine and isoprenoids on the Colorado potato beetle. Meded. Landbouwhogesch. Opzoeking Staat Gent *30*, 1599–1608 (1966).

[201] Menon, M.: Hormone-pheromone relationship in the beetle *Tenebrio molitor*. J. Insect Physiol. *16*, 1123–1139 (1970).

[202] Metwally, M. M., and F. Sehnal: Effects of juvenile hormone analogues on the metamorphosis of beetles *Trogoderma granarium* (Dermestidae) and *Caryedon gonagra* (Bruchidae). Biol. Bull. *144*, 368–383 (1973).

[203] Metzler, M., K. H. Dahm, D. Meyer, and H. Röller: On the biosynthesis of juvenile hormone in the adult Cecropia moth. Z. Naturforsch. *26 b*, 1270–1276 (1971).

[204] Metzler, M., D. Meyer, K. H. Dahm, H. Röller, and J. B. Siddall: Biosynthesis of juvenile hormone from 10-epoxy-7-ethyl-3,11-dimethyl-2,6-tridecadienoic acid in the adult Cecropia moth. Z. Naturforsch. *27 b*, 321–322 (1972).

[205] Meyer, A. S.: The juvenile hormones of the Cecropia silk moth. A chronicle. Mitt. Schweiz. Ent. Ges. *44*, 37–63 (1971).

[206] Meyer, A. S., and H. A. Ax: Liquid-solid chromatographic procedures for juvenile hormone purification and characterisation. J. Insect Physiol. *11*, 695–702 (1965).

[207] Meyer, A. S., and E. Hanzmann: The optical activity of the Cecropia juvenile hormones. Biochem. biophys. Res. Commun. *41*, 891–893 (1970).

[208] Meyer, A. S., E. Hanzmann, H. A. Schneiderman, L. I. Gilbert, and M. Boyette: The isolation and identification of the two juvenile hormones from the Cecropia silk moth. Arch. Biochem. Biophys. *137*, 190–213 (1970).

[209] Meyer, A. S., H. A. Schneiderman, and L. I. Gilbert: A highly purified preparation of juvenile hormone from the silk moth *Hyalophora cecropia*. Nature *206*, 272–275 (1965).

[210] Meyer, A. S., H. A. Schneiderman, E. Hanzmann, and J. H. Ko: The two juvenile hormones from the Cecropia silk moth. Proc. nat. Acad. Sci. (Amer.) *60*, 853–860 (1968).

[211] Minks, A. K.: Biochemical aspects of juvenile hormone action in the adult *Locusta migratoria*. Arch. Néerl. Zool. *17*, 175–258 (1967).

[212] Monro, J., and P. T. Bailey: Influence of radiation on ovarian maturation and histolysis of pupal fat body in Diptera. Nature *207*, 437–438 (1965).

[213] Mori, K.: Synthesis of compounds with juvenile hormone activity. Mitt. Schweiz. Ent. Ges. *44*, 17–35 (1971).

[214] Mori, K., and M. Matsui: Synthesis of (±)-juvabione (methyl (±)-todomatuate), a sesquiterpene ester with juvenile hormone activity. Tetrahedron Letters *1967*, 2515–2518.

[215] Mori, K., and M. Matsui: Synthesis of compounds with juvenile hormone activity I. (±)-Juvabione (methyl (±)-todomatuate). Tetrahedron *24*, 3127–3138 (1968).

[216] Mori, K., and M. Matsui: Synthesis of mono- and sesquiterpenoids II. (±)-Sesquicarene and a mixture of (±)-sirenin and its C-7-epimer. Tetrahedron *26*, 2801–2814 (1970).

[217] Mori, K., and M. Matsui: Synthesis of compounds with juvenile hormone activity. Part V. Some β-bisabolene derivatives. Agr. Biol. Chem. *34*, 115–121 (1970).

[218] Mori, K., T. Miyake, I. Yoshimura, and M. Matsui: Synthesis of compounds with juvenile hormone activity. Part III. (±)-Demethyl-juvabione and demethyl-ar-juvabione. Agr. Biol. Chem. *33*, 1745–1750 (1969).

[219] Mori, K., B. Stalla-Bourdillon, M. Ohki, M. Matsui, and W. S. Bowers: Synthese de composes d'activité hormonale juvenile —II. Syntheses du melange des stereoisomeres de hormone juvenile de *Hyalophora cecropia* et de ses analogues. Tetrahedron *25*, 1667–1677 (1969).

[220] Mouse, M. M.: Rôle de l'hormone juvénile dans la métamorphose oculaire de larves *d'Aeschna cyanea* Müll. (Insecte Odonate). C. R. Acad. Sci. (Paris) *D 273*, 2316–2319 (1971).

[221] Müller, H. P.: Zur Frage der Steuerung des Paarungsverhaltens und der Eireifung bei der Feldheuschrecke *Euthystira brachyptera* Oesk. unter besonderer Berücksichtigung der Rolle der Corpora allata. Z. Vergl. Physiol. *50*, 447–497 (1965).

[222] Nair, K. K.: Susceptibility of mosquito larvae to a synthetic juvenile hormone, farnesyldiethylamine. Naturwissenschaften *54*, 494–495 (1967).

[223] Nakanishi, K., D. A. Schooley, M. Koreeda, and J. Dillon: Absolute configuration of the C_{18}-juvenile hormone: application of a new circular dichroism method using tris(dipivaloylmethanato) praseodymium. Chem. Communications *1971*, 1235–1236.

[224] Nakazaki, M., S. Isoe: The structure of todomatuic acid. Synthesis of (±)-dihydro-desoxo-todomatuic acid. Bull. Chem. Soc. Jap. *34*, 741–742 (1961); *36*, 1198–1204 (1963).

[225] Němec, V., V. Jarolím, K. Hejno, and F. Šorm: Natural and synthetic materials with insect hormone activity. 7. Juvenile activity of the farnesane-type compounds on *Locusta migratoria* L. and *Schistocerca gregaria* (Forskal.). Life Sciences *9*, 821–831 (1970).

[226] Nohel, P., and K. Sláma: 1972 (see reference No. 216 on p. 79).

[227] Novák, V. J. A.: (see reference No. 218 on p. 79).

[228] Novák, V. J. A.: The juvenile hormone and the problem of animal morphogenesis. In: Beament, J. W. L., and J. E. Treherne, eds.: Insects and Physiology, pp. 119–132. Edinburgh-London: Oliver & Boyd. 1967.

[229] Novák, V. J. A.: Insect embryology and its hormonal influencing. Proc. XIII Int. Congr. Entomol. (Moscow) *1*, 423–424 (1968).

[230] Novák, V. J. A.: Morphological analysis of the effects of juvenile hormone analogues and other morphogenetically active substances on embryos of *Schistocerca gregaria* (Forskal.). J. Embryol. Exp. Morph. *21*, 1–21 (1969).

[231] Novák, V. J. A.: Juvenile hormone action as a special type of DNA-derepression. Insect endocrines, Academia Praha, 139–148 (1971).

[232] Novák, V. J. A., and E. Červenková: The function of corpus allatum in the last larval instar of metabolic insects. The Ontogeny of Insects, Nakladatelství ČSAV, Praha, pp. 152–156 (1960).

[233] Newman, M. S., and P. F. Beal: A new synthesis of α-alkoxy ketones. J. Amer. chem. Soc. *72*, 5161–5163 (1950).

[234] Oberlander, H., and H. A. Schneiderman: Juvenile hormone and RNA synthesis in pupal tissues of saturniid moths. J. Insect Physiol. *12*, 37–41 (1966).
[235] Pallos, F. M., J. J. Menn, P. E. Letchworth, and J. B. Miaulis: Synthetic mimics of insect juvenile hormone. Nature *232*, 486–487 (1971).
[236] Pan, M. L., and G. R. Wyatt: Juvenile hormone induced vitellogenin synthesis in the monarch butterfly. Science *174*, 503–505 (1971).
[237] Pánková, M., and F. Šorm: Unpublished results.
[238] Pattenden, G., and B. C. L. Weedon: Carotenoids and related compounds. Part XVIII. Synthesis of cis- and di-cis-polyenes by reaction of Wittig type. J. Chem. Soc. *1968*, 1984–1986.
[239] Patterson, J. W.: Critical sensitivity of the ovary of *Aedes aegypti* adults to sterilization by juvenile hormone mimics. Nature (London) *233*, 176–177 (1971).
[240] Pawson, B. A., H. C. Cheung, S. Gurbaxani, and G. Saucy: Syntheses of natural (+)-juvabione, its enantiomer (—)-juvabione, and their diastereoisomers (+)- and (—)-epijuvabione. J. Amer. chem. Soc. *92*, 336–343 (1970).
[241] Peyer, B., and W. Vogel: Correlations in the spectrum of activity of juvenile hormone analogues. Mitt. Schweiz. Ent. Ges. *44*, 187–192 (1971).
[242] Pfiffner, A.: Juvenile hormones. In: Goodwin, T. W., ed.: Aspects of terpenoid chemistry and biochemistry, pp. 95–133. London-New York: Academic Press. 1971.
[243] Pflugfelder, O.: (see reference No. 273 on p. 80).
[244] Pierrard, G.: Effet juvénile de la substance SW sur *Dysdercus völkeri* Schmidt et ses possibilités d'utilisation dans la lutte contre ce déprédateur. Med. Rijksfac. Landbouw. Wetenschap. Gent *32*, 623–632 (1967).
[245] Pilan, J. C., and A. M. Bautz: Utilisation des mimétiques de l'hormone juvénile comme agents tératogénes de la morphogenése tégumentaire chez *Calliphora erythrocephala* Meigen (Insecte Diptère). C. R. Hebd. Séanc. Acad. Sci. (Paris) *D 270*, 2973–2976 (1970).
[246] Poduška, K., J. Hlaváček, F. Šorm, and K. Sláma: Simple peptide derivatives with insect juvenile hormone activity. In: Nesvadba, H., ed.: Peptides, pp. 286–289. Amsterdam: North-Holland Publishing Comp. 1973.
[247] Poduška, K., F. Šorm, and K. Sláma: Natural and synthetic materials with insect hormone activity. 9. Structure-juvenile hormone activity relationships in simple peptides. Z. Naturforsch. *26 b*, 719–722 (1971).
[248] Poduška, K., M. Zaoral, K. Sláma, and F. Šorm: Czech. Pat. Appl. *PV* 6240 (1970).
[249] Ratuský, J.: Czech. Pat. Appl. *PV* 3415 (1969).
[250] Ratuský, J., K. Sláma, and F. Šorm: Natural and synthetic materials with insect hormone activity VI. Juvenile hormone effects of some alkyl-ethers derived from aliphatic β-hydroxy acids. J. Stored Prod. Res. *5*, 111–117 (1969).
[251] Ratuský, J., and F. Šorm: Czech. Pat. 132932 (1969).
[252] Redfern, R. E., T. P. McGovern, and M. Beroza: Juvenile hormone activity of sesamex and related compounds in tests on the yellow mealworm. J. Econ. Entomol. *63*, 540–545 (1970).

[253] Redfern, R. E., T. P. McGovern, R. Sarmiento, and M. Beroza: Juvenile hormone activity of mixed ethers containing a phenyl and a terpenoid moiety applied topically to the large milkweed bug and the yellow mealworm. J. Econ. Entomol. *64*, 374–376 (1971).

[254] Retnakaran, A.: Blocking of embryonic development in the spruce budworm, *Choristoneura fumiferana* (Lepidoptera: Tortricidae), by some compounds with juvenile hormone activity. Canad. Entomol. *102*, 1592–1596 (1970).

[255] Retnakaran, A., and D. Grisdale: The ovicidal effect of the hydrochlorination reaction mixture on the spruce budworm, *Choristoneura fumiferana* Clemens. J. Econ. Entomol. *63*, 907–909 (1970).

[256] Riddiford, L. M.: Juvenile hormone application to Hemipteran eggs: delayed effects on postembryonic development. Amer. Zool. *9*, 1120 (1969).

[257] Riddiford, L. M.: Prevention of metamorphosis by exposure of insect eggs to juvenile hormone analogues. Science *167*, 287–288 (1970).

[258] Riddiford, L. M.: Effects of juvenile hormone on the programming of postembryonic development in eggs of the silkworm, *Hyalophora cecropia*. Develop. Biol. *22*, 249–263 (1970).

[259] Riddiford, L. M.: Juvenile hormone and insect embryogenesis. Mitt. Schweiz. Entomol. Ges. *44*, 177–186 (1971).

[260] Riddiford, L. M., A. M. Ajami, E. J. Corey, H. Yamamoto, and J. E. Anderson: Synthetic imino analogs of Cecropia juvenile hormones as potentiators of juvenile hormone activity. J. Amer. chem. Soc. *93*, 1815–1816 (1971).

[261] Riddiford, L. M., and C. M. Williams: The effect of juvenile hormone analogues on the embryonic development of silkworms. Proc. nat. Acad. Sci. (Amer.) *57*, 595–601 (1967).

[262] Röller, H., and J. S. Bjerke: Purification and isolation of juvenile hormone and its action in Lepidopteran larvae. Life Sciences *4*, 1617–1624 (1965).

[263] Röller, H., J. S. Bjerke, L. M. Holthaus, D. W. Norgard, and W. H. McShan: Isolation and biological properties of the juvenile hormone. J. Insect Physiol. *15*, 379–389 (1969).

[264] Röller, H., J. S. Bjerke, and W. H. McShan: The juvenile hormone. I. Methods of purification and isolation. J. Insect Physiol. *11*, 1185–1197 (1965).

[265] Röller, H., and K. H. Dahm: The chemistry and biology of juvenile hormone. Rec. Progr. Horm. Res. *24*, 651–680 (1968).

[266] Röller, H., and K. H. Dahm: The identity of juvenile hormone produced by corpora allata *in vitro*. Naturwissenschaften *57*, 454–455 (1970).

[267] Röller, H., K. H. Dahm, C. C. Sweeley, and B. M. Trost: Die Struktur des Juvenilhormons. Angew. Chem. *79*, 190–191 (1967).

[268] Romaňuk, M.: Sélectivité de l'effet de certains analogues de l'hormone juvénile. Arch. Zool. Exp. Gén. *112*, 553–563 (1971).

[269] Romaňuk, M.: Coll. Czech. Chem. Commun. (in press).

[270] Romaňuk, M., and K. Sláma: Unpublished results.

[271] Romaňuk, M., K. Sláma, and F. Šorm: Constitution of a compound with pronounced juvenile hormone activity. Proc. nat. Acad. Sci. (Amer.) *57*, 349–352 (1967).

[272] ROMAŇUK, M., L. STREINZ, F. ŠORM, and K. SLÁMA: Czech. Pat. Appl. *PV* 7824 (1969).

[273] ROMAŇUK, M., L. STREINZ, and F. ŠORM: Natural and synthetic materials with insect hormone activity. XII. Synthesis of methyl 3,7,11,11-tetramethyl-2-dodecenoate and some of its homologues. Coll. Czech. Chem. Commun. *37*, 1755–1761 (1972).

[274] ROSE, M., J. WESTERMANN, H. TRAUTMANN, P. SCHMIALEK, and J. KLAUSKE: Juvenilhormonwirksame Verbindungen. I. Juvenilhormonwirkungen bei *Tenebrio molitor* L. in Abhängigkeit von der Konzentration der hormonalen Substanz. Z. Naturforsch. *23 b*, 1245–1248 (1968).

[275] SAHOTA, T. S.: Hormonal control of ovarian development and metamorphosis in *Malacosoma pluviale*. Canad. J. Zool. *47*, 917–920 (1969).

[276] SAXENA, K. N., and C. M. WILLIAMS: Paper factor as an inhibitor of the metamorphosis of the red cotton bug, *Dysdercus koenigii* F. Nature (London) *210*, 441–442 (1966).

[277] SCHMIALEK, P.: Die Identifizierung zweier in Tenebriokot und in Hefe vorkommender Substanzen mit Juvenilhormonwirkung. Z. Naturforsch. *16 b*, 461–464 (1961).

[278] SCHMIALEK, P.: Metamorphosehemmung von *Tenebrio molitor* durch Farnesylmethyläther. Z. Naturforsch. *18 b*, 513–515 (1963).

[279] SCHMIALEK, P.: Über die Bildung von Juvenilhormonen in Wildseidenspinnern. Z. Naturforsch. *18 b*, 462–465 (1963).

[280] SCHMIALEK, P.: Über Verbindungen mit Juvenilhormonwirkung. Z. Naturforsch. *18 b*, 516–519 (1963).

[281] SCHMIALEK, P., and G. DREWS: Wirkung von Farnesylmethyläther auf die Atmung von *Tenebrio molitor* Puppen. Z. Naturforsch. *20 b*, 214–215 (1965).

[282] SCHNEIDERMAN, H. A.: The strategy of controlling insect pests with growth regulators. Bull. Soc. Ent. Suisse *44*, 141–149 (1971).

[283] SCHNEIDERMAN, H. A., and L. I. GILBERT: Substances with juvenile hormone activity in Crustacea and other invertebrates. Biol. Bull. *115*, 530–535 (1958).

[284] SCHNEIDERMAN, H. A., and L. I. GILBERT: The inactivation of juvenile hormone extracts by pupal silkworms. Anat. Rec. *131*, 557–558 (1958).

[285] SCHNEIDERMAN, H. A., and L. I. GILBERT: The chemistry and physiology of insect growth hormones. In: RUDNICK, D., ed.: Cell, Organism and Milieu, pp. 157–187. Ronald Press Comp. 1959.

[286] SCHNEIDERMAN, H. A., and L. I. GILBERT: Control of growth and development in insects. Science *143*, 325–333 (1964).

[287] SCHNEIDERMAN, H. A., L. I. GILBERT, and M. J. WEINSTEIN: Juvenile hormone activity in microorganisms and plants. Nature *188*, 1041–1042 (1960).

[288] SCHNEIDERMAN, H. A., A. KRISHNAKUMARAN, P. J. BRYANT, and F. SEHNAL: Endocrinological and genetic strategies in insect control. In: Proc. Symp. Potent. Crop Protect., N. Y. State Agric. Exp. Sta., Geneva, pp. 14–25 (1969).

[289] SCHNEIDERMAN, H. A., A. KRISHNAKUMARAN, V. G. KULKARNI, and L. FRIEDMAN: Juvenile hormone activity of structurally unrelated compounds. J. Insect Physiol. *11*, 1641–1649 (1965).

[290] SCHOONHOVEN, L. M.: Chemosensory bases of host plant selection. Ann. Rev. Entomol. *13*, 115–136 (1968).

[291] Schulz, H., and I. Sprung: Synthese des dl-Juvenilhormons des Riesenseidenspinners *Hyalophora cecropia*. Angew. Chem. *81*, 258 (1969).

[292] Schwarz, M., P. E. Sonnet, and N. Wakabayashi: Insect juvenile hormone: activity of selected terpenoid compounds. Science *167*, 191–192 (1970).

[293] Schwarz, M., N. Wakabayashi, P. E. Sonnet, and R. E. Redfern: Compounds related to juvenile hormone. VII. Activity of selected nitrogen-containing terpenoid compounds on the yellow mealworm. J. Econ. Entomol. *63*, 1858–1860 (1970).

[294] Sehnal, F.: (personal communications).

[295] Sehnal, F., and A. S. Meyer: Larval-pupal transformation: control by juvenile hormone. Science *159*, 981–984 (1968).

[296] Shaaya, E., and D. Bodenstein: The function of the accessory sex glands in *Periplaneta americana* (L.) II. The role of the juvenile hormone in the synthesis of protein and protocatechuic acid glucoside. J. Exp. Zool. *170*, 281–292 (1969).

[297] Shepard, H. H.: Methods of testing chemicals on insects. Burgess Publ. Co., Minneapolis, I, 0–356 (1958).

[298] Shepard, H. H.: Methods of testing chemicals on insects. Burgess Publ. Co., Minneapolis, II, 0–248 (1960).

[299] Siddall, J. B.: (personal communication).

[300] Siddall, J. B., R. J. Anderson, and C. A. Henrick: Studies on insect hormones. Proc. XXIIIrd Int. Congr. Pure & Appl. Chem., pp. 17–25. London: Butterworths. 1971.

[301] Siddall, J. B., and M. Slade: Absence of acute oral toxicity of *Hyalophora cecropia* juvenile hormone in mice. Nature New Biol. *229*, 158 (1971).

[302] Siew, Y. C., and L. I. Gilbert: Effects of moulting hormone and juvenile hormone on insect endocrine gland activity. J. Insect Physiol. *17*, 2095–2104 (1971).

[303] Sláma, K.: Pseudojuvenilizing effect in insects. Acta Soc. Ent. Čechoslov. *58*, 117–120 (1961).

[304] Sláma, K.: The juvenile hormone like effect of fatty acids, fatty alcohols, and some other compounds in insect metamorphosis. Acta Soc. Ent. Čechoslov. *59*, 323–340 (1962).

[305] Sláma, K.: (see reference No. 313 on p. 84).

[306] Sláma, K.: The effect of hormone mimetic substances on the ovarian development and oxygen consumption in allatectomized adult females of *Pyrrhocoris apterus* L. (Hemiptera). J. Insect Physiol. *11*, 1121–1129 (1965).

[307] Sláma, K.: Plants as a source of materials with insect hormone activity. Ent. Exp. & Appl. *12*, 721–728 (1969).

[308] Sláma, K.: Insect juvenile hormone analogues. Ann. Rev. Biochem. *40*, 1079–1102 (1971).

[309] Sláma, K.: Some aspects of insect hormone research. In: Novák, V. J. A., and K. Sláma, eds.: Insect Endocrines, pp. 149–156. Academia Praha. 1971.

[310] Sláma, K.: (see reference No. 310 on p. 84).

[311] Sláma, K.: Insect hormones: the developmental story of *Dermestes*. Biol. Bull. (In press.)

[312] Sláma, K., K. Hejno, V. Jarolím, and F. Šorm: Natural and synthetic materials with insect hormone activity. 5. Specific juvenile hormone effects in aliphatic sesquiterpenes. Biol. Bull. *139*, 222–228 (1970).

[313] Sláma, K., M. Romaňuk, and F. Šorm: Natural and synthetic materials with insect hormone activity. 2. Juvenile hormone activity of some derivatives of farnesenic acid. Biol. Bull. *136*, 91-95 (1969).

[314] Sláma, K., M. Romaňuk, and F. Šorm: Natural and synthetic materials with insect hormone activity. 11. Juvenile activity of methyl 3,7,11,11-tetramethyl-2-dodecenoate and its homologues. J. Insect Physiol. *18*, 19–24 (1972).

[315] Sláma, K., M. Suchý, and F. Šorm: Natural and synthetic materials with insect hormone activity. 3. Juvenile hormone activity of derivatives of p-(1,5-dimethylhexyl) benzoic acid. Biol. Bull. *134*, 154–159 (1968).

[316] Sláma, K., and C. M. Williams: Juvenile hormone activity for the bug *Pyrrhocoris apterus*. Proc. nat. Acad. Sci. (Amer.) *54*, 411–414 (1965).

[317] Sláma, K., and C. M. Williams: „Paper factor" as an inhibitor of the embryonic development of the European bug, *Pyrrhocoris apterus*. Nature *210*, 329–330 (1966).

[318] Sláma, K., and C. M. Williams: The juvenile hormone. V. The sensitivity of the bug, *Pyrrhocoris apterus*, to a hormonally active factor in American paper-pulp. Biol. Bull. *130*, 235–246 (1966).

[319] Slepecky, R. A., and L. I. Gilbert: The antimicrobial activity of some insect extracts possessing juvenile hormone activity. Experientia *18*, 1–5 (1962).

[320] Socha, R., and F. Sehnal: Inhibition of adult development in *Tenebrio* molitor by insect hormones and antibiotics. J. Insect Physiol. *18*, 317–337 (1972).

[321] Sonnet, P. E., M. Schwarz, and N. Wakabayashi: Compounds related to insect juvenile hormone. V. Derivatives of citronellol and citronellylamine. J. Med. Chem. *13*, 1247–1248 (1970).

[322] Šorm, F.: Some juvenile hormone analogues. Mitt. Schweiz. Ent. Ges. *44*, 7–16 (1971).

[323] Šorm, F., M. Svoboda, J. Závada, K. Sláma, and Z. Arnold: Czech. Pat. Appl. *PV* 3600 (1970).

[324] Spielman, A., and V. Skaff: Inhibition of metamorphosis and of ecdysis in mosquitoes. J. Insect Physiol. *13*, 1087–1095 (1967).

[325] Spielman, A., and C. M. Williams: Lethal effects of synthetic juvenile hormone on larvae of the yellow fever mosquito, *Aedes aegypti*. Science *154*, 1043–1044 (1966).

[326] Srivastava, U. S., and L. I. Gilbert: Juvenile hormone: effects on a higher dipteran. Science *161*, 61–62 (1968).

[327] Srivastava, U. S., and L. I. Gilbert: The influence of juvenile hormone on the metamorphosis of *Sarcophaga bullata*. J. Insect. Physiol. *15*, 177–189 (1969).

[328] Sroka, P., and L. I. Gilbert: Studies on the endocrine control of postemergence ovarian maturation in *Manduca sexta*. J. Insect Physiol. *17*, 2409–2419 (1971).

[329] Staal, G. B.: Insect hormones in plants. Med. Rijksfac. Landbouw. Wetenschappen Gent *32*, 393–400 (1967).

[330] Staal, G. B.: Practical aspects of insect control by juvenile hormone. Bull. Wld. Hlth. Org. *44*, 391–394 (1971).

[331] Staal, G. B.: The role of juvenile hormone in the morphogenetical development of larval instar in *Lepidoptera*. Endocrinol. exp. *5*, 35–38 (1971).

[332] Steen, J. B.: The effect of juvenile hormone on the respiratory metabolism of silkworm pupae, as recorded with a new semi-micro device. Acta Physiol. Scand. *51*, 275–282 (1961).

[333] Stegwee, D.: Metabolic effect of a corpus allatum hormone in diapausing *Leptinotarsa decemlineata* Say. Verh. XI. Int. Kongr. Entom. (Wien) *B 3*, 218–221 (1960).

[334] Stowe, B. B., and V. W. Hudson: Growth promotion in pea stem sections. III. By alkyl nitriles, alkyl acetylenes and insect juvenile hormones. Plant Physiol. *44*, 1051–1057 (1969).

[335] Suchý, M., K. Sláma, and F. Šorm: Insect hormone activity of p-(1,5-dimethylhexyl) benzoic acid derivatives in *Dysdercus* species. Science *162*, 582–583 (1968).

[336] Suchý, M., and F. Šorm: Czech. Patents No. 131632 (1969), 131635 (1969), 134459 (1969).

[337] Tamelen, E. E. van, and J. P. McCormick: Synthesis of Cecropia juvenile hormone from trans,trans-Farnesol. J. Amer. chem. Soc. *92*, 737–738 (1970).

[338] Thomas, P. J., and P. L. Bhatnagar-Thomas: Use of a juvenile hormone analogue as insecticide for pests of stored grain. Nature (London) *219*, 949 (1968).

[339] Treherne, J. E.: The diffusion of non-electrolytes through the isolated cuticle of *Schistocerca gregaria*. J. Insect Physiol. *1*, 178–186 (1957).

[340] Trost, B. M.: The juvenile hormone of *Hyalophora cecropia*. Accounts Chem. Res. *3*, 120–130 (1970).

[341] Varjas, L.: The effective dosis of juvenile hormone analogues from the point of view of insect control. Acta Phytopath. Acad. Sci. Hung. *6*, 219–227 (1971).

[342] Vinson, J. W., and C. M. Williams: Lethal effects of synthetic juvenile hormone on the human body louse. Proc. nat. Acad. Sci. (Amer.) *58*, 294–297 (1967).

[343] Wadsworth, W. S., and W. D. Emmons: The utility of phosphonate carbanions in olefin synthesis. J. Amer. chem. Soc. *83*, 1733–1738 (1961).

[344] Wakabayashi, N.: Compounds related to insect juvenile hormone. J. Med. Chem. *12*, 191–192 (1969).

[345] Wakabayashi, N., M. Schwarz, P. E. Sonnet, R. M. Waters, R. E. Redfern, and M. Jacobson: Compounds related to insect juvenile hormones. Mitt. Schweiz. Ent. Ges. *44*, 131–140 (1971).

[346] Wakabayashi, N., P. E. Sonnet, and M. W. Law: Compounds related to insect juvenile hormone IV. J. Med. Chem. *12*, 911–913 (1969).

[347] Walker, W. F., and W. S. Bowers: Synthetic juvenile hormones as potential coleopteran ovicides. J. Econ. Entomol. *63*, 1231–1233 (1970).

[348] Weirich, G.: Zur Frage der hormonalen Regulation der Eireifung bei Insekten. Inaug. Diss. Naturwiss. Fak. Ludwig-Maxmilians Univ., München (1963).

[349] WELLINGTON, W. G.: Effects of three hormonal mimics on mortality, metamorphosis, and reproduction of the western tent caterpillar, *Malacosoma californicum pluviale*. Canad. Entomol. *101*, 1163–1172 (1969).

[350] WELLINGTON, W. G., and D. A. MAELZER: Effects of farnesyl methyl ether on the reproduction of western tent caterpillar, *Malacosoma pluviale*. Some physiological, ecological and practical implications. Canad. Entomol. *99*, 249–263 (1967).

[351] WESTERMANN, J., M. ROSE, H. TRAUTMANN, P. SCHMIALEK, and J. KLAUSKE: Juvenilhormon-wirksame Verbindungen. II. Dosis-Wirkungsbeziehungen juvenilhormonaktiver Substanzen. Z. Naturforsch. *24 b*, 378–380 (1969).

[352] WESTPHAL, V., and B. D. ASHLEY: Steroid-protein interactions. VI. Stereochemical aspects of interaction between Δ^4–3–ketosteroids and human serum albumin. J. Biol. Chem. *234*, 2847–2851 (1959).

[353] WHITE, D. F.: Postnatal treatment of the cabbage aphid with a synthetic juvenile hormone. J. Insect Physiol. *14*, 901–912 (1968).

[354] WHITE, D. F., and K. P. LAMB: Effect of a synthetic juvenile hormone on adult cabbage aphids and their progeny. J. Insect Physiol. *14*, 395–402 (1968).

[355] WIGGLESWORTH, V. B.: Some methods for assaying extracts of the juvenile hormone in insects. J. Insect Physiol. *2*, 73–84 (1958).

[356] WIGGLESWORTH, V. B.: Some observations on the juvenile hormone effect of farnesol in *Rhodnius prolixus* Stal (Hemiptera). J. Insect Physiol. *7*, 73–78 (1961).

[357] WIGGLESWORTH, V. B.: The juvenile hormone effect of farnesol and some related compounds: quantitative experiments. J. Insect Physiol. *9*, 105–119 (1963).

[358] WIGGLESWORTH, V. B.: Chemical structure and juvenile hormone activity: Comparative tests on *Rhodnius prolixus*. J. Insect Physiol. *15*, 73–94 (1969).

[359] WIGGLESWORTH, V. B.: Chemical structure and juvenile hormone activity. Nature *221*, 190–191 (1969).

[360] WIGGLESWORTH, V. B.: Juvenile hormone assays in insects, with special reference to metabolic breakdown. Gen. Comp. Endocrinol. *13*, 163 (1969).

[361] WIGGLESWORTH, V. B.: (see reference No. 379 on p. 88).

[362] WILDE, J. DE: Diapause in the Colorado beetle *Leptinotarsa decemlineata* Say as an endocrine deficiency syndrome of the corpora allata. Acta Symp. Evol. Ins., Prague, pp. 226–230 (1959).

[363] WILDE, J. DE: New elements for integration in pest control schedules. Meded. Rijksfak. Landbouw-Wettenschap. Gent *33*, 597–604 (1968).

[364] WILDE, J. DE: The present status of hormonal insect control. EPPO Bull. *1*, 17–23 (1971).

[365] WILDE, J. DE, C. A. D. DE KORT, and A. DE LOOF: The significance of juvenile hormone titers. Mitt. Schweiz. Ent. Ges. *44*, 79–86 (1971).

[366] WILDE, J. DE, G. B. STAAL, C. A. D. DE KORT, A. DE LOOF, and G. BAARD: Juvenile hormone titer in the haemolymph as a function of photoperiodic treatment in the adult Colorado beetle, *Leptinotarsa decemlineata* Say. Koninkl. Nederl. Akad. Wettenschap. *71*, 321–326 (1968).

[367] WILDE, J. DE, and D. STEGWEE: Two major effects of the corpus allatum in the adult Colorado beetle. Arch. Néerl. Zool. *13*, 277–289 (1958).

[368] WILLIAMS, C. M.: The juvenile hormone of insects. Nature *178*, 212–213 (1956).

[369] WILLIAMS, C. M.: The juvenile hormone. I. Endocrine activity of the corpora allata of the adult Cecropia silkworm. Biol. Bull., Woods Hole *116*, 323–338 (1959).

[370] WILLIAMS, C. M.: The juvenile hormone. Proc. 1st Intern. Congr. Endocrin. (Symp. X), pp. 189–191. Copenhagen, 1960.

[371] WILLIAMS, C. M.: Third generation pesticides. Scientific American *217*, 13–17 (1967).

[372] WILLIAMS, C. M.: Juvenile hormone insecticides. Accademia Nazionale dei Lincei, Roma, Quaderno *128*, 79-87 (1969).

[373] WILLIAMS, C. M.: Nervous and hormonal communication in insect development. Develop. Biol. Suppl. *3*, 133–150 (1969).

[374] WILLIAMS, C. M.: Hormonal interactions between plants and insects. Chem. Ecol. *1970*, 103–132.

[375] WILLIAMS, C. M., and F. C. KAFATOS: Theoretical aspects of the action of juvenile hormone. Mitt. Schweiz. Ent. Ges. *44*, 151–162 (1971).

[376] WILLIAMS, C. M., and J. H. LAW: The juvenile hormone. IV. Its extraction, assay, and purification. J. Insect Physiol. *11*, 569–580 (1965).

[377] WILLIAMS, C. M., L. V. MOORHEAD, and J. F. PULIS: Juvenile hormone in thymus, human placenta and other mammalian organs. Nature *183*, 405 (1959).

[378] WILLIAMS, C. M., and W. E. ROBBINS: Conference on insect-plant interactions. Bio Science *18*, 791–799 (1968).

[379] WILLIAMS, C. M., and K. SLÁMA: The juvenile hormone. VI. Effects of the "paper factor" on the growth and metamorphosis of the bug, *Pyrrhocoris apterus*. Biol. Bull. *130*, 247–253 (1966).

[380] WILLIS, J. H.: The programming of the differentiation and its control by juvenile hormone in saturniids. J. Embryol. Exp. Morph. *22*, 27–44 (1969).

[381] WILLIS, J. H., and P. C. J. BRUNET: Hormonal control of colleterial gland secretion. J. Exp. Biol. *44*, 363–378 (1966).

[382] WILLIS, J. H., and P. A. LAWRENCE: Deferred action of juvenile hormone. Nature *225*, 81–83 (1970).

[383] WITTIG, G., and W. HAAG: Über Triphenyl-phosphinmethylen als olefinbildende Reagenzien (II. Mitt.). Ber. *88*, 1654–1666 (1955).

[384] WRIGHT, J. E.: Hormones for control of livestock arthropodes. Development of an assay to select candidate compounds with juvenile hormone activity in the stable fly. J. Econ. Entomol. *63*,878–883 (1970).

[385] WRIGHT, J. E., and G. E. SPATES: Biological evaluation of juvenile hormone compounds against pupae of the stable fly. J. Agr. Food Chem. *19*, 289–290 (1971).

[386] WYATT, G. R.: Insect hormones. In: Biochemical actions of hormone *2*, 385–490 (1972).

[387] WYSS-HUBER, M., and M. LÜSCHER: Influence of juvenile hormone on ovarian protein synthesis in *Leucophaea maderae*. Gen. Comp. Endocrinol. *13*, 166 (1969).

[388] Yamamoto, R. T., and M. Jacobson: Juvenile hormone activity of isomers of farnesol. Nature *196*, 908–909 (1962).
[389] Zaoral, M., and K. Sláma: Peptides with juvenile hormone activity. Science *170*, 92–93 (1970).
[390] Žďárek, J., and K. Sláma: Mating activity in adultoids or supernumerary larvae induced by agents with high juvenile hormone activity. J. Insect Physiol. *14*, 563–567 (1968).
[391] Žďárek, J., and K. Sláma: Supernumerary larval instars in Cyclorhaphous Diptera. Biol. Bull. *142*, 350–357 (1972).
[392] Zlotkin, E., and H. Z. Levinson: Influence of Cecropia oil on the epidermis of *Tenebrio molitor*. J. Insect Physiol. *14*, 1719–1723 (1968).
[393] Zlotkin, E., and H. Z. Levinson: Influence of Cecropia oil on the cuticle of *Tenebrio molitor*. J. Insect Physiol. *15*, 105–110 (1969).
[394] Zurflüh, R., E. N. Wall, J. B. Siddall, and J. A. Edwards: Synthetic studies on insect hormones. VII. An approach to stereospecific synthesis of juvenile hormones. J. Amer. chem. Soc. *90*, 6224–6225 (1968).
[395] Anonymous: Insect hormone mimics near market. Chem. & Eng. News *49*, 33–34 (1971).

III. Chemistry and Physiology of Ecdysoids

A. The Chemistry of Moulting Hormones (Ecdysones — Ecdysoids)

The history of the discovery of insect moulting hormone began with papers by the Polish entomologist KOPEČ [172, 173] which demonstrated for the first time that the moulting process was controlled by a humoral factor. Further significant contributions to the physiology of the moulting process came from WIGGLESWORTH [323]. Of great importance for the chemistry of moulting hormones was the observation that the moulting process is controlled by a factor present in the prothoracic glands (FUKUDA [68–70]). PLAGGE and BECKER [14, 234] discovered that moulting hormone could be extracted from blowfly (*Calliphora erythrocephala*) larvae with boiling ethanol.

An effective and reliable biological assay (the Calliphora test) for the moulting hormone was devised by FRAENKEL [62, 63].

Ecdysone (α-Ecdysone)

BUTENANDT and KARLSON [28] isolated pure moulting hormone in 1954. A complex processing of 500 kg of dried silkworm (*Bombyx mori*) pupae yielded 25 mg of a crystalline substance which was named ecdysone. The yield was about 50% of the total hormone present. The silkworm pupae were extracted with methanol, the extract concentrated and the concentrate diluted with butanol. After removal of acidic and alkaline components, the solution was evaporated and the residue repeatedly chromatographed on alumina, the fractions being checked by the Calliphora test. The final pure substance was obtained by crystallisation from ethyl acetate and methanol. The activity of this substance was 0.0075 μg per one unit of the Calliphora test. In 1959, STAMM [287] isolated from 10 kg of the adult locust *Dociostaurus maroccanus Thurnberg* 24 mg of ecdysone and a small amount of ecdysterone (β-ecdysone), the main moulting hormone of crustacea. KAPLANIS *et al.* [141] isolated crystalline ecdysone and ecdysterone from 7 day-old pupae of the tobacco hornworm *Manduca sexta*. HORN *et al.* [109] isolated ecdysone from the oak silk moth, *Antherea pernyi*.

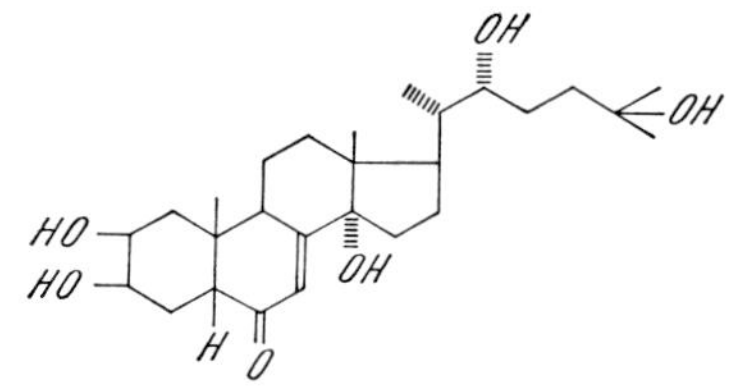

(1)

The correct formula of ecdysone $C_{27}H_{44}O_6$ was determined by KARLSON *et al.* [153] in 1965 with dehydrogenation on palladium and the formation of methylcyclopentenophenanthrene. Isolation of this hydrocarbon indicated that ecdysone was a steroid. Additional chemical and physical evidence, especially the X-ray studies [106, 111] obtained mainly by KARLSON and his school [105, 154] established the 2β,3β, 14α,22α(R),25-pentahydroxy-5β-cholest-7-en-6-one (*1*) structure of ecdysone. This structure was confirmed by syntheses published independently, and almost simultaneously, by scientific teams from Syntex and Schering-Hoffmann-LaRoche. Both syntheses start from 23,24-bis-norcholenic acid derivatives.

SIDDALL *et al.* [277, 279] (see Scheme 1) from Syntex started from methyl 3β-acetoxy-23,24-bisnorchol-5-en-22-oate which was converted by the action of performic acid and subsequent hydrolysis to the 3β,5α, 6β-triol (*2*). The hydroxyl group at position 6 is selectively oxidised by the action of N-bromosuccinimide to a keto group, and the hydroxyl at position 3 is eliminated via the tosyl derivative with formation of the olefinic derivative (*3*). cis-Dihydroxylation of compound (*3*) with silver acetate and iodine in moist acetic acid lead to the triol (*4*) which was converted to the diacetate. This compound was brominated at position 7 and the product dehydrobrominated to give the unsaturated keto derivative (*5*). The hydroxyl function at position 5 was acetylated, position C-14 was oxidised with selenium dioxide with formation of the required 14α-hydroxy derivative. The acetoxy group at position 5 was reduced with chromous chloride and the acetyl groups at position 2 and 3 were saponified. The cyclic moiety of the resulting steroid derivative (*6*) then contained all the functional groups of ecdysone. The cis-diol configuration in ring A is protected by the formation of the isopropylidene derivative which is equilibrated in the presence of potassium carbonate to yield almost exclusively compound (*7*) with cis-fused A/B rings as required.

Compounds (*7*) was reduced with tri-tert-butoxylithium aluminium hydride to a mixture of 6α- and 6β-hydroxy derivatives (*8*). This mixture was converted, on treatment with the lithium salt of 2-methyl-4-phenyl-sulphinylbutan-2-ol tetrahydropyranyl ether (*9*), to an analogous mixture

Scheme 1

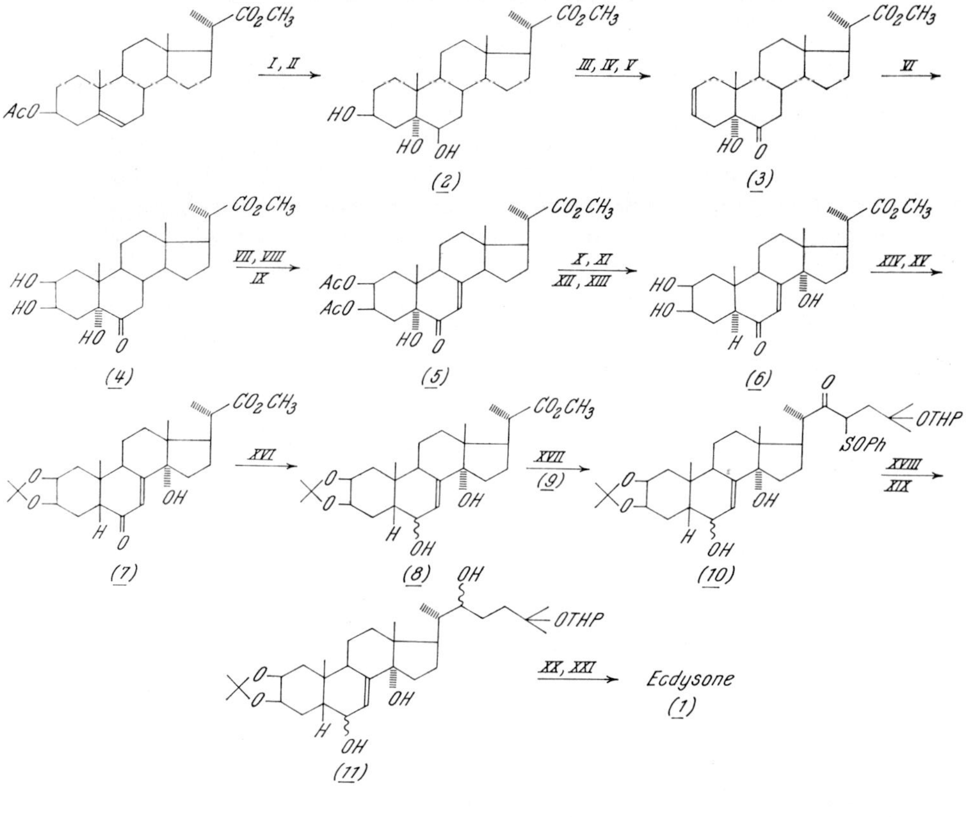

I, $HCOOH, H_2O_2$	VII, Ac_2O	XII, $CrCl_2$	XVII, $9 = Li{\cdot}CH{\cdot}CH_2{\cdot}C(CH_3)_2$ with SOPh on CH and OTHP on C
II, NaOH	VIII, Br_2	XIII, H_2O	XVIII, AlHg
III, Br-succinimide	IX, Li_2CO_3	XIV, $(CH_3)_2CO$	XIX, $LiAlH_4$
IV, TsCl	X, Ac_2O	XV, K_2CO_3	XX, MnO_2
V, Li_2CO_3	XI, SeO_2	XVI, $(t\text{-}BuO)_3LiAlH$	XXI, HCl
VI, AgOAc, J_2			

of ketones (*10*). These ketones contain the fundamental carbon skeleton of ecdysone with corresponding oxygen functional groups at appropriate positions. The phenylsulphinyl residue was removed by the action of aluminium amalgam and the resulting 22-ketone was reduced with lithium aluminium hydride to give mixture of epimeric 22-carbinols (*11*). The hydroxylic function at position 6 was selectively oxidised with manganese dioxide to the keto group and the protecting groups were removed by acidic hydrolysis. The resulting mixture of epimeric 22-carbinols was separated by chromatography to give ecdysone (*1*), m. p. 237.0–239.5 °C, in a 12% yield (referred to the mixture of epimers). The same authors later reported another variant of ecdysone synthesis [87].

The team [73, 163, 164, 322] from Schering and Hoffmann-LaRoche started their synthesis of ecdysone (Scheme 2) from methyl 3,3-ethylenedioxy-23,24-bisnorchol-5-en-22-oate (*12*) which was epoxidised and the resulting epoxide converted in acidic media to the stable 3,6-dioxo derivative (*13*). Compound (*13*) was brominated and the resulting 2-bromo derivative (*14*) reduced with tri-tert-butoxylithium aluminium hydride to give the bromohydrin (*15*). Compound (*15*) was acetylated and the acetate (*16*) treated with silver acetate in moist acetic acid to afford the 5-isomeric diacetates (*17*) and (*18*) which were separated by fractional crystallisation. Bromination of both diacetates lead to the stable 7α-bromo-5α-derivative (*19*), which was converted by dehydrobromination to the unsaturated ketone (*20*). Halolysis of the ester with lithium iodide in lutidine, and simultaneous partial epimerisation at C-5 and C-20, gave the free acid (*21*). The required (20S)-isomer was obtained in pure condition by chromatography. The acid (*21*) was reduced as imidazolide with tri-tert-butoxylithium aluminium hydride. The resulting 20-formyl derivative (*22*) was treated with 3-methyl-3-tetrahydropyranyloxy-1-butynylmagnesium bromide (*23*) to afford a mixture of 22-epimeric carbinols, which was separated by chromatography. Hydrogenation of the triple bond in the (22R)-epimer (*24*), oxidation of this compound with selenium dioxide to (*25*), and removal of protecting groups by alkaline and acidic hydrolyses resulted in ecdysone (*1*), m. p. 232–233 °C.

The Schering and Hoffmann-LaRoche team [71, 72] also accomplished the synthesis of ecdysone from ergosterol as starting material. This synthesis was considerably shorter (14 reaction steps) than the earlier procedures. Japanese chemists [212, 213] synthesized ecdysone from stigmasterol. Recently, the synthesis of Barton *et al.* [12], starting from 10 g of ergosterol, gave 7 mg of ecdysone and 3 mg of 5α-ecdysone.

Ecdysone is a crystalline substance which is readily soluble in polar organic solvents. The melting point data vary in the range 232–242 °C, depending on the solvents used for crystallisation and on the synthetic procedures.

Scheme 2

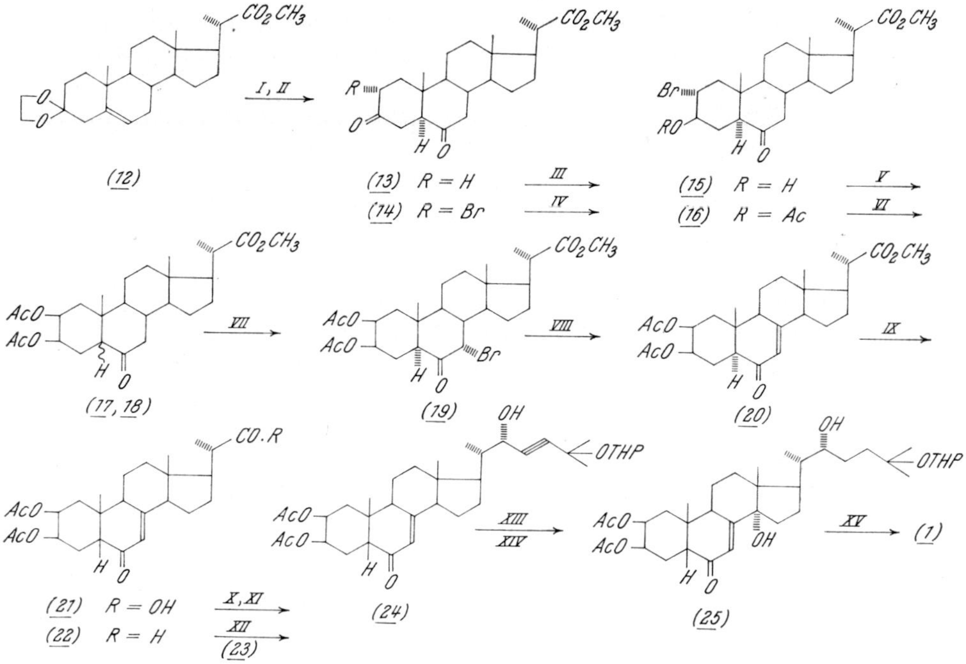

I,	AcO_2H	VI,	AcOAg	XI,	$(t\text{-}BuO)_3LiAlH$
II,	$HClO_4$	VII,	Br_2, AcOH	XII,	23 = Mg.Br.C≡C.C$(CH_3)_2$ OTHP
III,	Br_2	VIII,	Li_2CO_3–LiBr	XIII,	H_2(Pt)
IV,	$(t\text{-}BuO)_3LiAlH$	IX,	Lutidine–LiJ	XIV,	SeO_2
V,	Ac_2O	X,	Carbonyldiimidazol	XV,	$H^{\cdot}$

Ecdysterone (20β-Hydroxyecdysone, β-Ecdysone, Crustecdysone, Iso-inokosterone, Polypodine A)

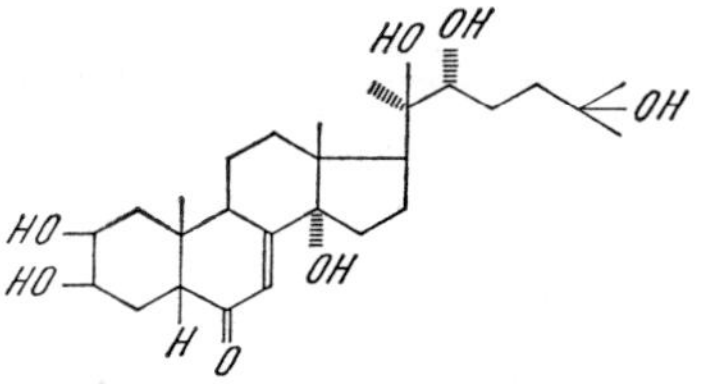

(26)

Ecdysterone is a specific hormone which controls ecdysis of crustacea. Horn *et al.* [85] isolated 2 mg of pure but amorphous ecdysterone (activity of 0.03 μg per 1 unit of the Calliphora bioassay) from 1 ton of frozen crayfish (*Jasus lalandei*) waste consisting chiefly of ventral thoraxes. The structure of 20β-hydroxyecdysone (*26*) was ascribed to ecdysterone on the basis of chemical evidence (formula $C_{27}H_{44}O_7$) and in particular spectral data [57, 108]. The same authors also isolated ecdysterone (200 μg) from pupae of the oak silk moth, *Antherea pernyi*, and established its identity with the crayfish component. Another isolation of ecdysterone was carried out from an extract of the female marine crab *Callinectes sapidus* just before moulting [57]. The general occurrence of this hormone in insect organisms was confirmed by its isolation [102] from the silkworm *Bombyx mori* pupae (along with the predominant ecdysone); 2.8 tons of dried pupae afforded 48 mg of ecdysterone and 206 mg of ecdysone. Ecdysterone was obtained also from adults of the locust *Dociostaurus maroccanus* [287] and from caterpillars of the tobacco hornworm *Manduca sexta* [141].

Ecdysterone is a crystalline substance, m. p. 240–242 °C. The structure of ecdysterone was confirmed by several syntheses. The first synthesis was performed by the Syntex team [113] (see Scheme 3). The starting material, 2β,3β-isopropylidenedioxy-20-formyl-5α-pregn-7-en-6-one (*27*), contains the majority of functional groups located at appropriate positions of the steroid skeleton. The C-20 hydroxylic function was introduced via the Δ^{20}-enol acetate (*28*) and epoxidation. The mixture of epoxides was converted to a mixture of (20R)- and (20S)-

Scheme 3

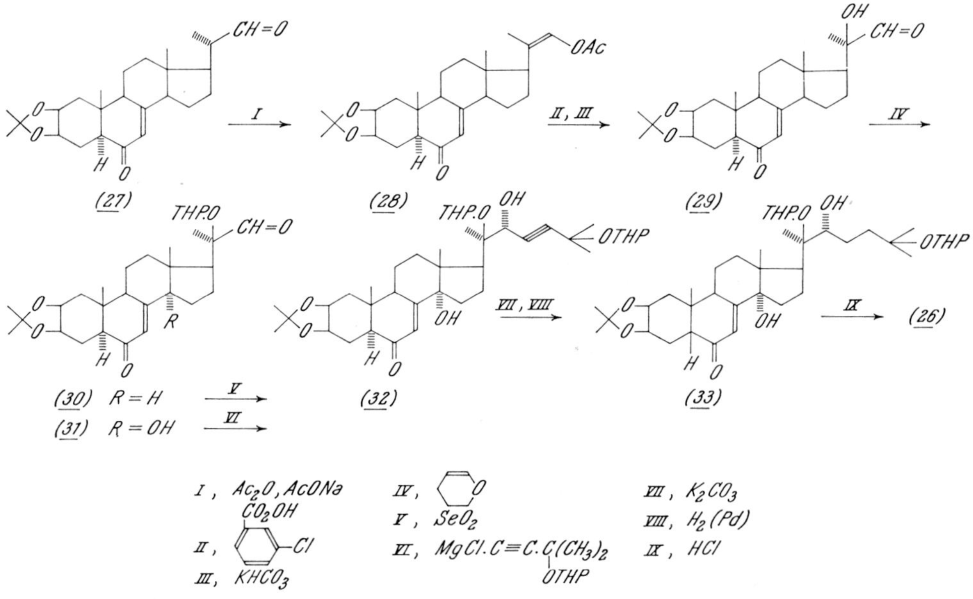

carbinols, which were separated. The free hydroxylic function of the required (20S)-isomer (*29*) was protected by the formation of the tetrahydropyranyl ether (*30*) and this substance was hydroxylated at C-14 α with selenium dioxide. The aldehyde group of the resulting 14 α-carbinol (*31*) was selectively alkylated with the chloromagnesium derivative of 3-methyl-3-(tetrahydropyran-2-yloxy)-butyne to give compound (*32*) which possesses all the required oxygen-containing functions at appropriate positions. At this stage of the synthesis, an inversion of configuration at C-5 was effected by the action of potassium carbonate in aqueous methanol, and the required 5β-isomer was isolated. Selective hydrogenation of the sidechain triple bond on palladium-charcoal resulted in (*33*), and removal of protecting groups by acidic hydrolysis gave the final crystalline ecdysterone (*26*).

Another synthesis of ecdysterone was carried out independently by the Schering-Hoffmann-LaRoche team [165] using compound (*36*) as the starting material (see Scheme 4), prepared by a series of reactions from the nitro derivative of the hydroxy ketone (*34*) via (*35*) . Selective treatment of compound (*36*) with the bromomagnesium derivative of 4-methyl-4-(tetrahydropyran-2-yloxy)-pentyne resulted in compound (*37*) which was hydrated by the action of $Hg(OAc)_2$ and $BF_3 \cdot Et_2O$ to afford the corresponding 22-oxo derivative. This compound was hydroxylated at C-14 with the formation of the triol (*38*). Alkaline hydrolysis

Scheme 4

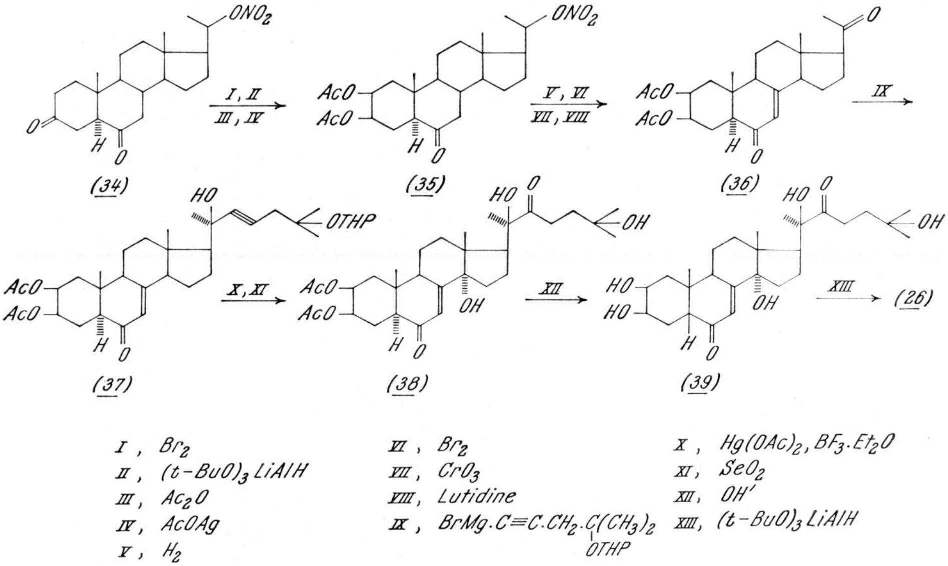

and simultaneous inversion at C-5 gave the 5β-pentol (*39*). The last step of the synthesis consisted in selective reduction of the 22-oxo group with tri-tert-butoxylithium aluminium hydride.

A further synthesis of ecdysterone was recently carried out by Japanese authors [211].

Ecdysterone is a crystalline substance, m. p. 240–242 °C. The activity of ecdysterone in the Calliphora bioassay (0.015 μg per 1 unit) is somewhat lower than that of ecdysone.

After the laborious isolation and characterisation of the two principal moulting hormones from insect and crustacean organisms, it was very surprising to isolate compounds of a similar chemical structure and of high biological activity also from some plants. The discovery of so-called phytoecdysones was anounced almost simultaneously by several groups of investigators. Nakanishi *et al.* [218] isolated the steroid ponasterone A from the plant *Podocarpus nakaii Hay*, determined its high activity in the Calliphora bioassay and elucidated its structure. In connexion with investigations on *Podocarpus elatus*, Galbraith and Horn [74] observed that this Australian tree contained a considerable quantity of a compound which was highly active in the Calliphora bioassay. On the basis of comparative chromatography, it is believed that the active substance is ecdysterone (*26*). Jizba, Herout, and Šorm [130] have shown that the earlier isolated polypodine A (from rhizomes of the Central European fern *Polypodium vulgare*) was identical with ecdysterone and that the so-called polypodine B was 5β-hydroxy-ecdysterone (with a high activity in the Calliphora bioassay). Both compounds became readily accessible, since the dried *Polypodium vulgare* rhizomes contain 1% of the bioactive material.

Takemoto *et al.* discovered ecdysterone and an additional biologically active compound (called inokosterone) from the roots of *Achyranthes fauriei* [293, 304, 305, 306, 309]. Rimpler and Schulz [248] isolated ecdysterone from *Vitex megapotamica*. The presence of ecdysone and structurally related compounds of a different biological activity has been shown in many other plants and animal organisms. These compounds will be surveyed below from a structural, rather than a chronological standpoint.

Ecdysone has been isolated from *Pteridium aquilinum*, from the fern *Blechnum minus* [33], *Polypodium vulgare* rhizomes [88], *Osmunda japonica* [298] and *Lemmaphyllum microphyllum* [292]. In addition to the plants listed, ecdysterone was isolated from *Pteridium aquilinum* [140], mulberry-tree leaves [307], *Podocarpus macrophyllus* leaves [118], *Stachyurus praecox* bark [120], *Pleopeltis thunbergiana* and *Neocheiropteris ensata* [294], *Serratula inermis* [335], *Ajuga japonica* [119], *Taxus baccata leaves* [308], *Trillium smallii* [122], *Blechnum amabile* and *B. niponicum*

[310], *Taxus cuspidata* [117] and *Vitex megapotamica* [247]. An additional compound was isolated from the latter plant, viticosterone E, and identified as the 25-acetyl derivative of ecdysterone (*40*) [247]. The biological activity of viticosterone E (m. p. 198–199 °C) in the Calliphora assay is about 14 times less than that of ecdysone.

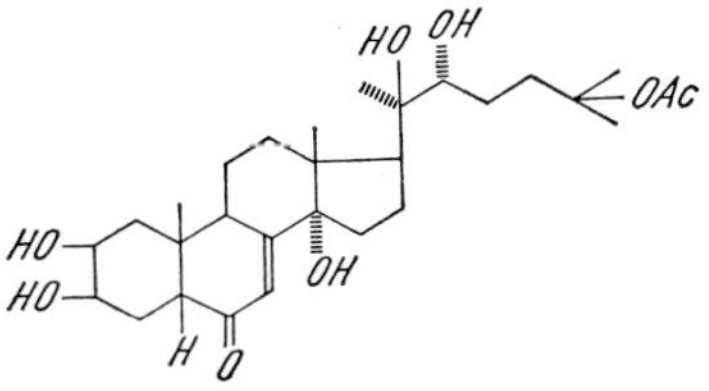

(40)

Ponasterone A is the first moulting hormone of plant origin isolated from *Podocarpus nakaii* [217, 218]. Nakanishi *et al.* isolated 2 g of ponasterone A from 4.8 kg dried leaves and established its chemical structure as 25-deoxycrustecdysone (*41*). Hüppi and Siddall [114] synthesized ponasterone A from an intermediate (*42*) of the ecdysterone synthesis by hydrogenation over prereduced platinum as catalyst (Scheme 5).

Scheme 5

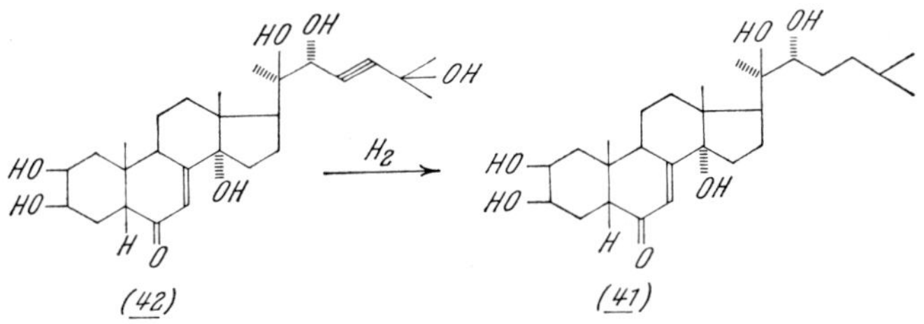

Ponasterone A was also isolated from other plants, e.g. *Podocarpus macrophyllus* [118] (0.05%), *Podocarpus chinensis* [117], *Onoclea sensibilis* [293], *Taxus cuspidata var. nana* [299], *Osmunda japonica* and *Osmunda asiatica* [298], *Blechnum amabile* and *B. niponicum* [310], and *Pteridium aquilinum* [291].

Ponasterone A is a cristalline substance, m. p. 259–260 °C, and its biological activity is about 4 times less than that of ecdysterone [215, 291].

The 3β-glucoside of ponasterone A (*43*), called ponasteroside A or warabisterone, was isolated from *Pteridium aquilinum var. latiusculum* [291] and showed a high biological activity. This glucoside is formed from ponasterone A in the fat tissue of *Calliphora erythrocephala* by the action of transglucosylase [90].

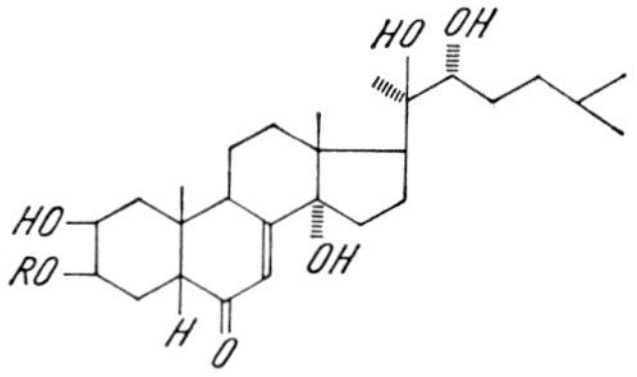

(43) R = Glucosyl

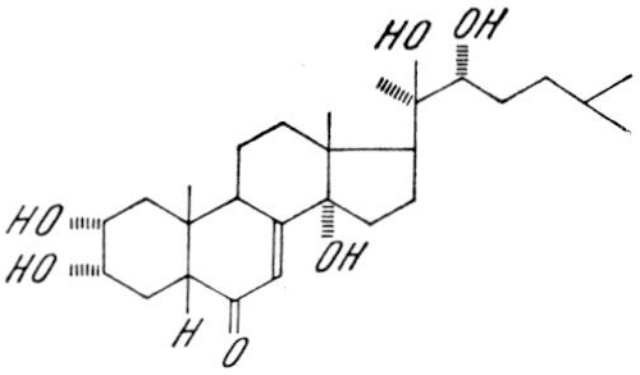

(44)

The ponasterone B (*44*), occurring in the plant Podocarpus nakaii along with ponasterone A, is remarkable in terms of α-configuration of the two hydroxylic functions in ring A in contrast to other known steroids with moulting hormone activity. The configuration of the remaining hydroxyls has not been determined [174] but is probably identical with other ecdysones.

Stachysterone C was isolated from the bark of the plant *Stachyurus praecox* and identified as the dehydro derivative (*45*) of ponasterone A [120].

Shidasterone (*46*), obtained from the plant *Blechnum niponicum* [301], is an isomer of ecdysterone. The configuration of hydroxylic functions in the side chain has not been determined. The highly active shidasterone is not identical with the synthethically obtained 22-epiecdysterone, which does not show any biological activity. The melting point of shidasterone is 257–258 °C.

Inokosterone (synonymum: Callinecdysone A) (*47*) is an isomer of ecdysterone with a shifted hydroxylic function of the side-chain. Inokosterone was obtained from the roots of *Achyranthes fauriei* [295], from the crude drug Achyranthes radix [304, 305] and from *Vitex megapotamica* [248]. Later investigations have shown the identity of inokosterone with the steroid isolated from the extract of the marine crab *Callinectes sapidus* [57]. The biological activity of inokosterone (m. p. 255 °C) is similar to that of ecdysterone.

In another ecdysterone isomer, pterosterone (*48*), the side-chain hydroxylic function is shifted from C-25 to C-24. Pterosterone has been isolated from several plants, e.g., *Matteuccia struthiopteris, Lastrea thelypteris, Onoclea sensibilis, Vitex megapotamica* [247] and *Pteridium aquilinum* [291]. Pterosterone exhibits a high biological activity. M. p. 229–230 °C.

20,26-Dihydroxyecdysterone (*49*) was isolated from 7 day old pupae of the tobacco hornworm *Manduca sexta* [315]. The configuration of the side-chain hydroxylic functions has not been determined. M. p. 149–153 °C. The biological activity of this steroid is lower than that of ecdysone.

Polypodine B (synonymum: 5β-hydroxyecdysterone, ajugasterone A

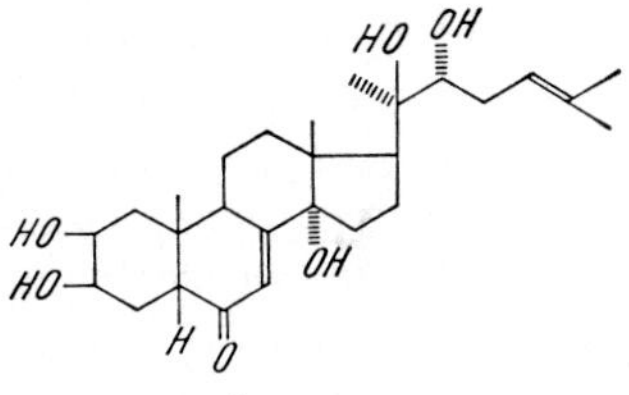

(45)

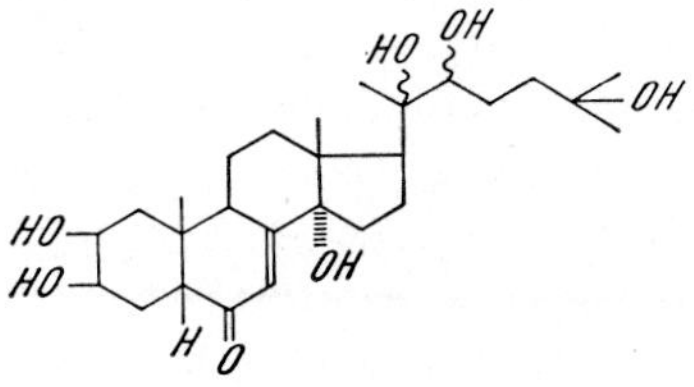

(46)

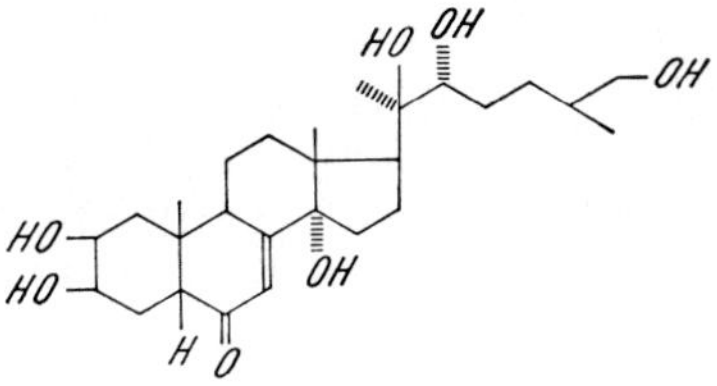

(47)

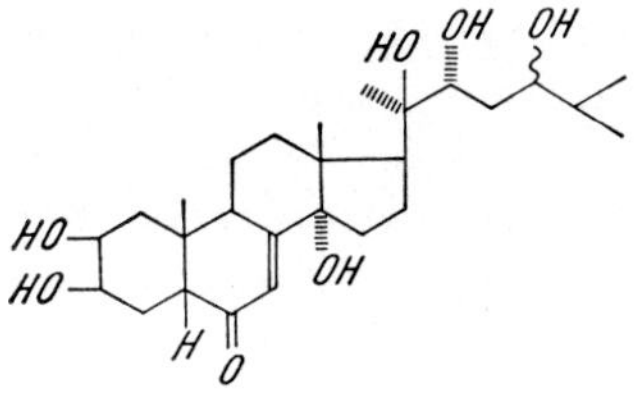

(48)

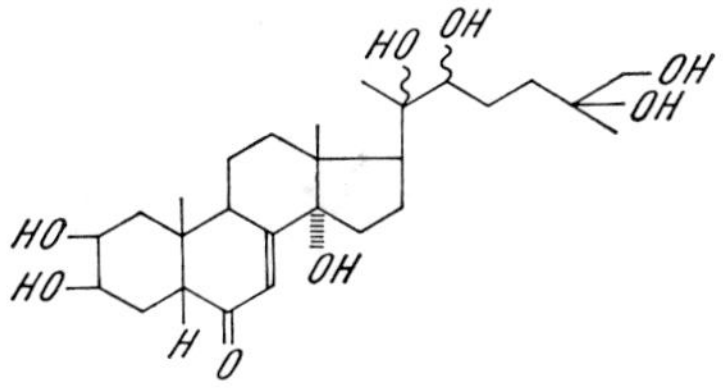

(49)

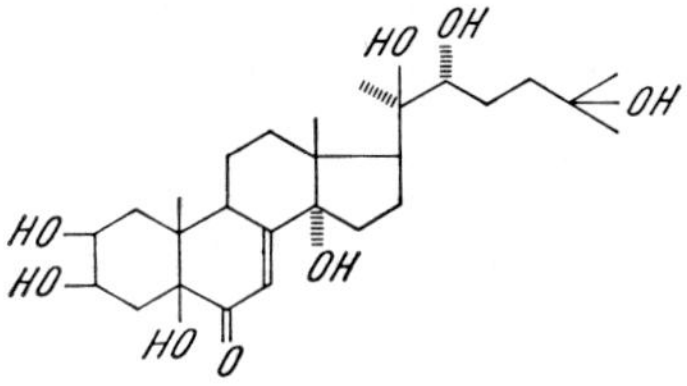

(50)

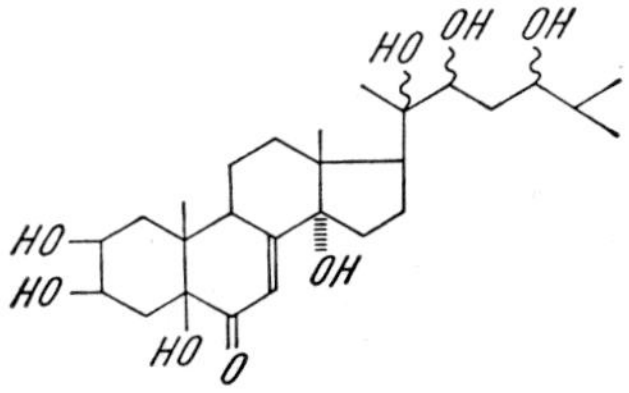

(51)

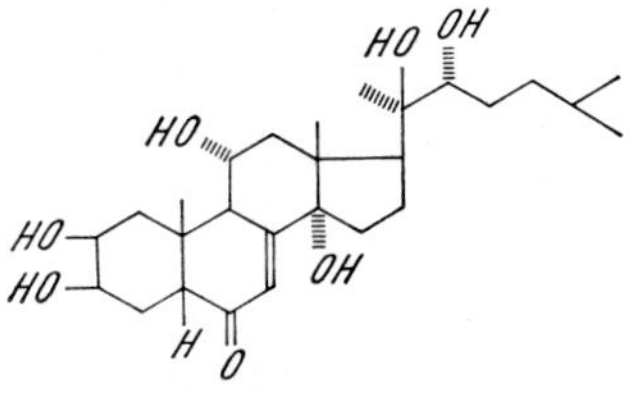

(52)

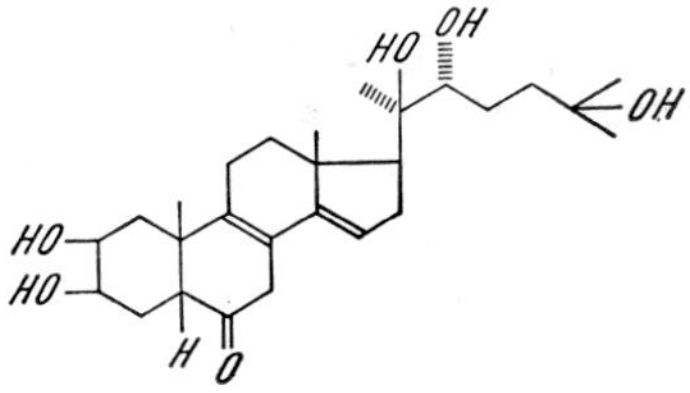

(53)

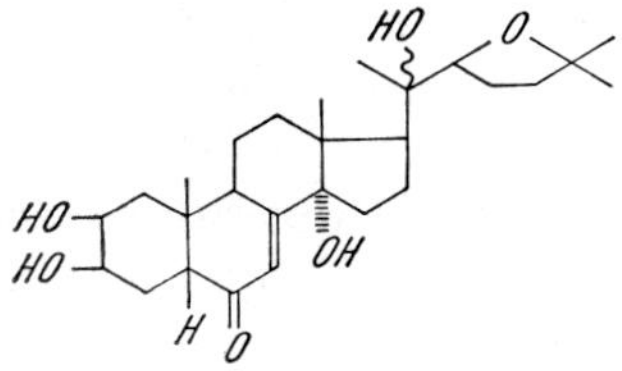

(54)

(*50*) belongs to the higher hydroxylated derivatives of ecdysterone [129]. 0.87 g of polypodine B was isolated from 2.4 kg of *Polypodium vulgare* rhizomes. Polypodine was also obtained from *Vitex megapotamica* [89, 247] and *Ajuga incisa* [119]. In the Calliphora bioassay, the biological activity of polypodine B is higher than of ecdysone. The melting point of polypodine B is 256 °C.

Ponasterone C (*51*), isolated from *Podocarpus nakaii*, is closely related structurally to polypodine B [215].

Ajugasterone C (*52*) is a cholestane derivative, the C ring of which is hydroxylated, and its biological activity is low. It was isolated from dried leaves of *Ajuga japonica* and *Ajuga decumbens* [119].

Podecdysone B (*53*) is another steroid exhibiting moulting hormone activity and possessing the characteristic arrangement of two hydroxylic functions in ring A and the configuration of the cholestane side chain. This substance was isolated from the bark of *Podocarpus elatus* [76] and was five times less active than ecdysterone in the Calliphora bioassay. M. p. of podecdysone B 125–127 °C.

Stachysterone D (*54*) is contained, along with stachysterone C, in the bark of *Stachyurus praecox* [120]. Its biological activity is negligible. M. p. 245–250 °C.

The B ring is hydrogenated in cheilanthone A (*55*) and cheilanthone B (*56*), which were isolated from the fern *Cheilanthes tenuifolia* [56]. Compounds *55* and *56* are biologically inactive. M. p. values: 235–238 °C (cheilanthone A) and 225–228 °C (cheilanthone B).

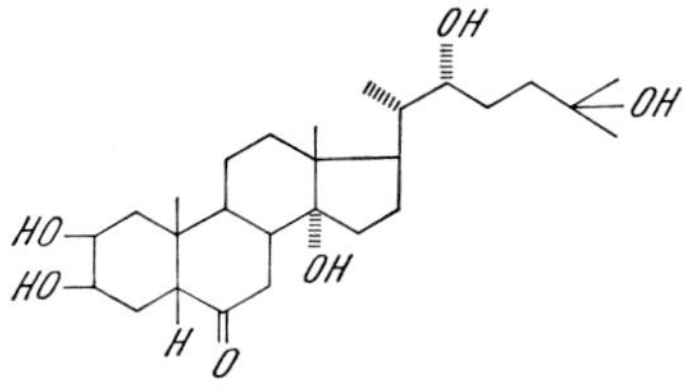

(55)

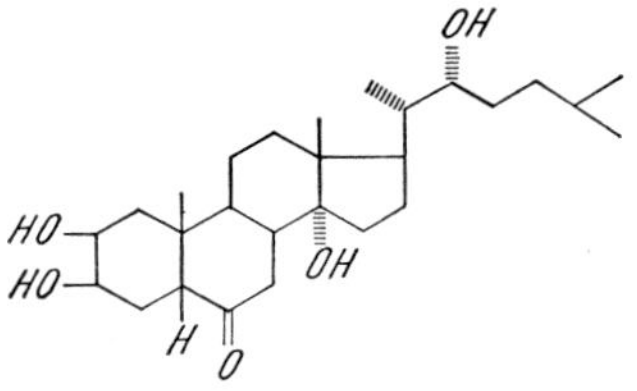

(56)

Another compound of the cholestane series, 2-deoxyecdysterone (*57*), is interesting particularly from the structural viewpoint. Compound (*57*) was obtained from the waste of the crayfish *Jasus lalandei* [75] and from the Australian fern *Blechnum minus* [33]. 2-Deoxyecdysterone appears in the Calliphora bioassay to be more active than ecdysone, despite the absence of the characteristic group at position 2 of ring A in the molecule. M. p. 250–252 °C.

2-Deoxyecdysone (*58*) is structurally related to 2-deoxyecdysterone mentioned above. It was obtained in small quantity from the Australian

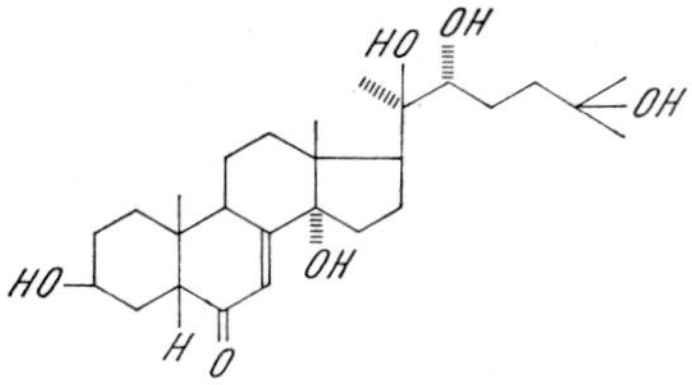

(57)

fern *Blechnum minus* [33]. The activity of 2-deoxyecdysone in the Calliphora bioassay is similar to that of ecdysone (*1*).

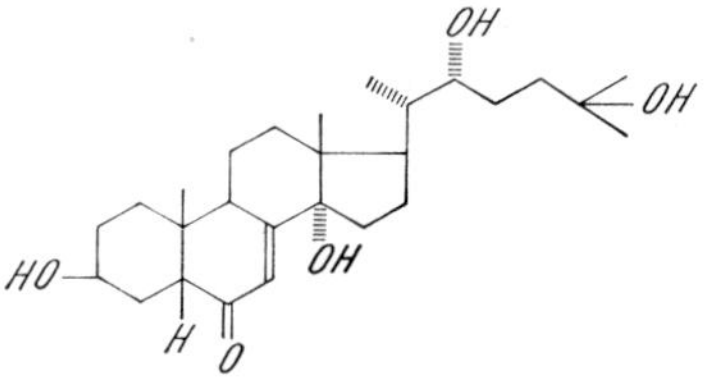

(58)

Animal and plant material also contain C_{28}-steroids which exhibit a high moulting hormone activity. Thus, makisterone A (synonymum: callinecdysone B, podecdysone D) (*59*) was isolated from the marine crab *Callinectes sapidus* [57] as well as from the leaves of the plant *Podocarpus macrophyllus* [116]. The latter plant also contains makisterone B (*60*). Compounds (*59*) and (*60*) exhibit a strong moulting hormone activity. Melting point values: 262–265 °C (makisterone A) and 172–173 °C (makisterone B).

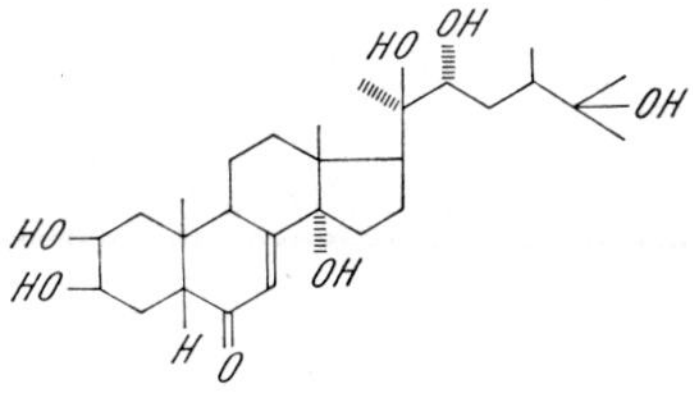

(59)

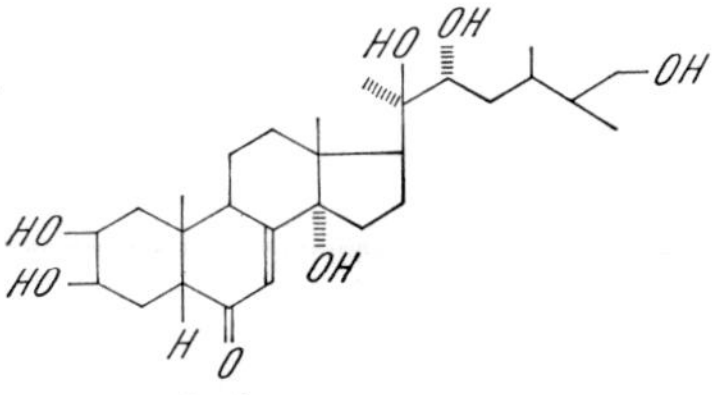

(60)

The C_{29} steroid moulting hormones are far more frequent in nature. The ecdysone ethyl derivative, lemmasterone (synonymum: podecdysone A, makisterone C) (*61*) was isolated from the plants *Lemmaphyllum microphyllum* [292], *Podocarpus elatus* [80] and *Podocarpus macrophyllus* [116], and exhibits the same biological activity as ecdysterone.

Amarasterones A (*62*) and B (*63*) were obtained along with sengosterone, capitasterone and precyasterone from the plant *Cyathula capitata* [302]. Both amarasterone A (m. p. 210–211 °C) and amarasterone B (m. p. 284–285 °C) show a high biological activity.

Makisterone D (*64*) differs from the isomeric amarasterone B by the position of the hydroxylic function on the ethyl group. Similar to other makisterones, makisterone D was isolated from the plant *Podocarpus macrophyllus* [116]. The biological activity of compound (*64*) is as high as that of other makisterones.

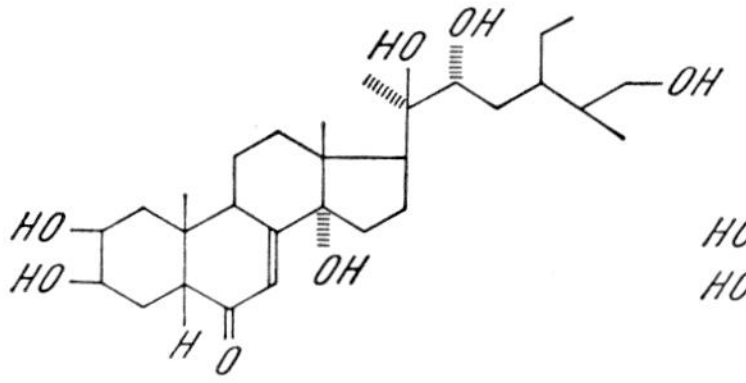

(62)

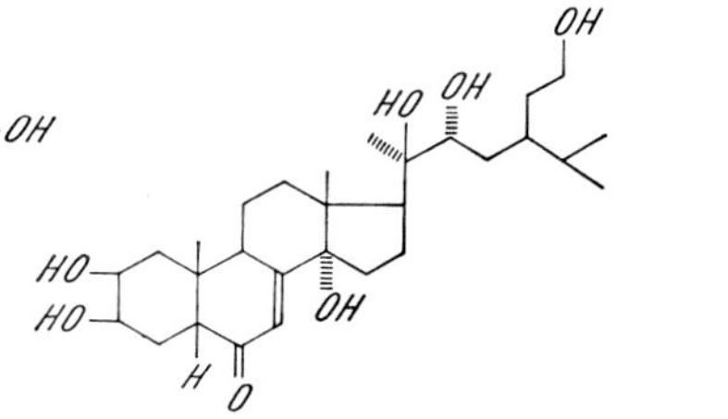

(63)

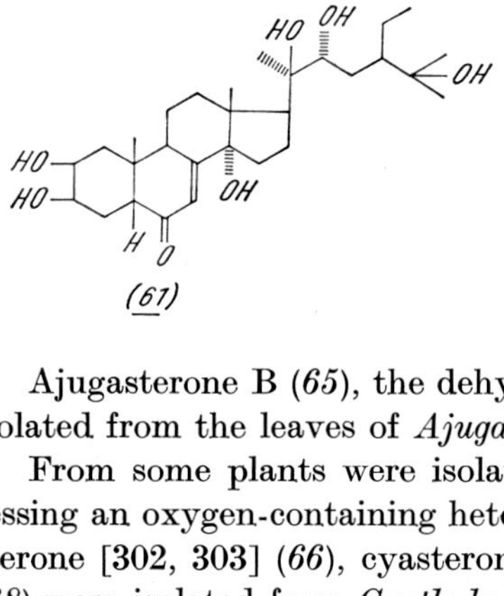

(61)

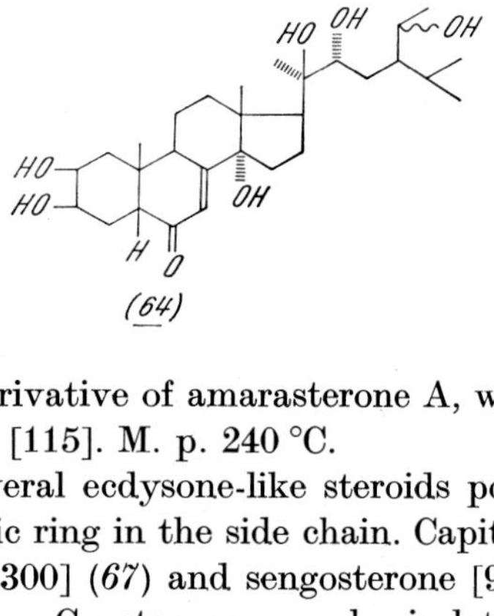

(64)

Ajugasterone B (*65*), the dehydro derivative of amarasterone A, was isolated from the leaves of *Ajuga incisa* [115]. M. p. 240 °C.

From some plants were isolated several ecdysone-like steroids possessing an oxygen-containing heterocyclic ring in the side chain. Capitasterone [302, 303] (*66*), cyasterone [91, 300] (*67*) and sengosterone [95] (*68*) were isolated from *Cyathula capitata*. Cyasterone was also isolated from *Ajuga decumbens*, *Ajuga incisa* and *Ajuga japonica* [122]. Cyasterone is highly active. According to some sources, its biological activity is higher than that of ecdysone in certain tests. The melting point values: 234–235 °C (capitasterone), 164–166 °C (cyasterone) and 159–161 °C (sengosterone).

Precyasterone (*69*) is an additional component of *Cyathula capitata* [94].

Plant material was also found to contain rubrosterone (*70*), a C_{19} steroid, the skeleton of which is arranged similarly to moulting hormones. Rubrosterone is biologically inactive and probably constitutes a degradation product of moulting hormones. The first isolation of rubrosterone

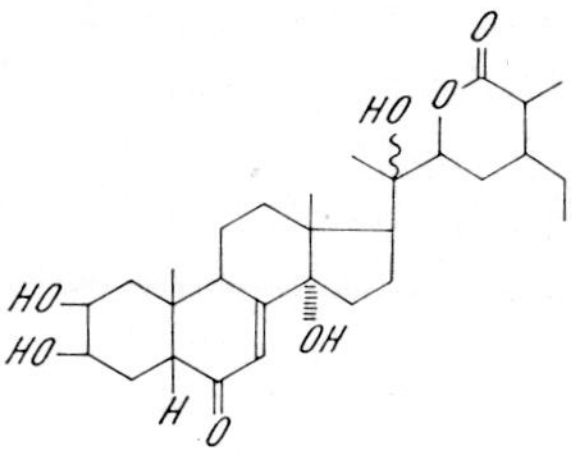

(66)

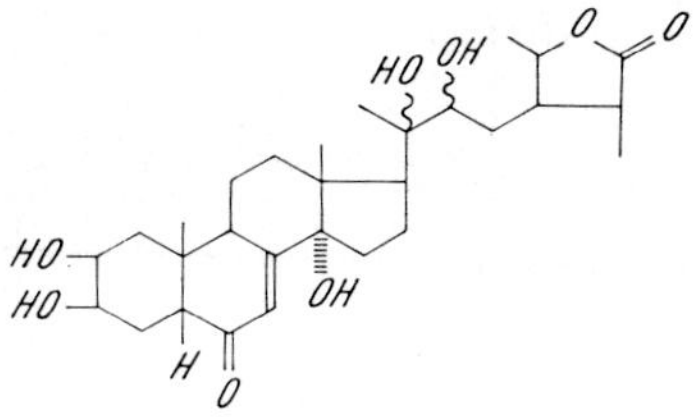

(67, 68)

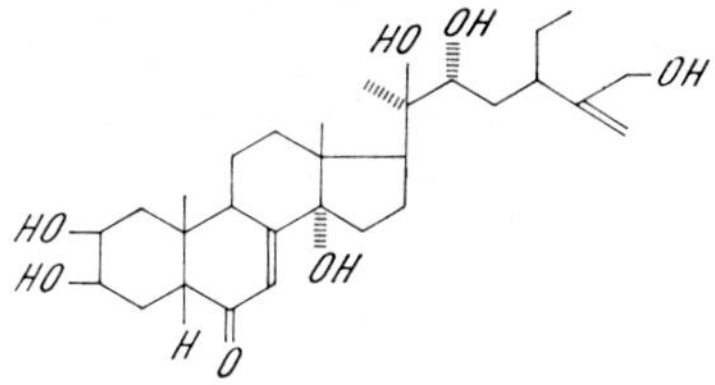

(65)

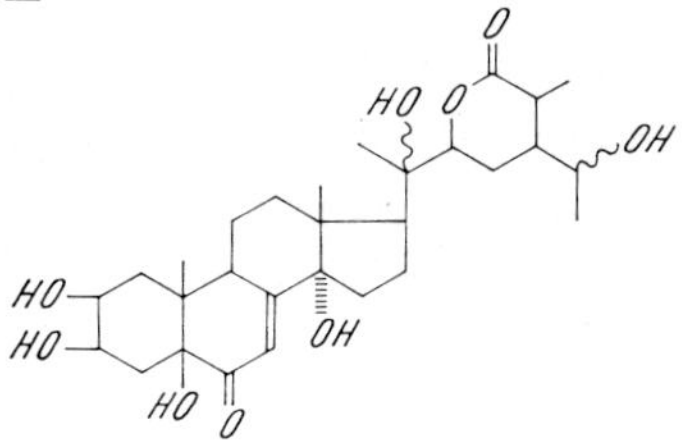

(69)

was achieved from *Achyranthes rubrofusca.* Later it was isolated from *Achyranthes fauriei* and *Achyranthes obtusifolia* [93, 296]. By chemical means, rubrosterone was prepared by degradation of the side chain of ecdysterone [92] (Scheme 6).

Scheme 6

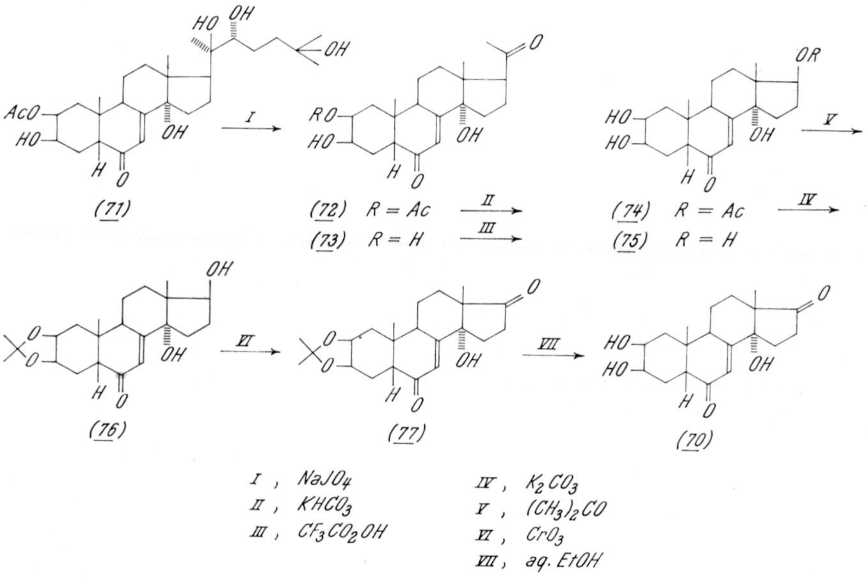

Ecdysterone 2-acetate (*71*) was oxidised with sodium periodate to afford the keto derivative (*72*) which was hydrolysed (*73*) and oxidised with perfluoroacetic acid. The acetyl group of the oxidation product (*74*) was saponified with potassium carbonate to give a tetrol (*75*), the two vicinal hydroxylic functions of which were protected by the formation of an acetonide (*76*). The acetonide was oxidised and the oxidation product (*77*) hydrolysed to rubrosterone (*70*).

A simplified procedure [93] transforms ecdysterone to the anhydro tetraacetate (*78*) on heating with acetic anhydride. Compound (*78*) was ozonolysed to afford a 55% yield of the diacetate (*79*) which was then hydrolysed to rubrosterone (*70*) (Scheme 7).

Scheme 7

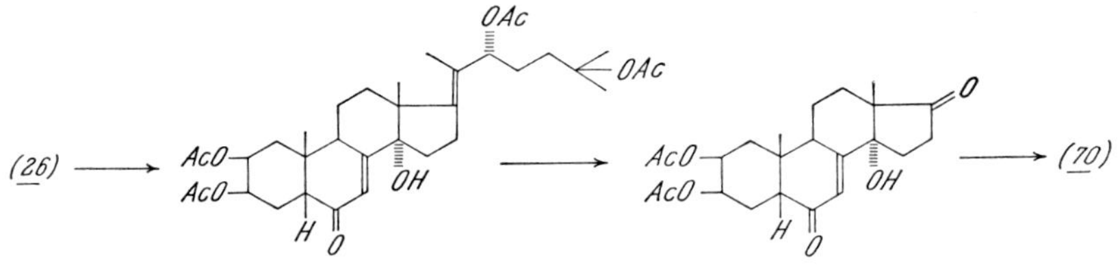

(78) (79)

Other syntheses of rubrosterone start from 17β-acetoxy-5α-androstane-3,6-dione [101] and 3β-tosyloxyandrost-5-en-17-one [19].

Poststerone [96] (*80*), a C_{21} steroid, has been recently isolated from the plant *Cyathula capitata* and shown to constitute the degradation product of ecdysterone in the insect organism.

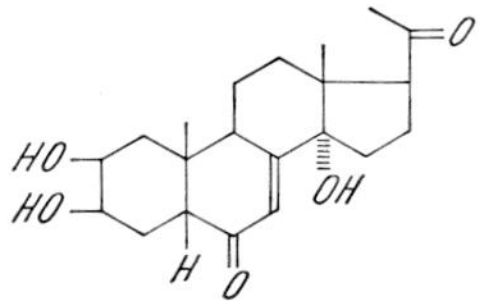

(80)

B. Relations between the Structure and Activity of Insect Moulting Hormones

In a consideration of relations between structure and activity of the insect moulting hormones it is necessary to bear in mind that different biological tests have been used in different laboratories, and that even the standard Calliphora bioassay does not appear to be strictly reproducible, as shown, e.g., by conflicting data concerning the biological activity of ecdysone (*1*) and ecdysterone (*26*). According to some authors [85, 142] ecdysone is twice as active as ecdysterone, according to others [100, 103] the activity of ecdysterone is higher than that of ecdysone. The following conclusions are therefore only qualitative in nature.

The cis fusion of rings A and B in the polycyclic skeleton appears to be of great importance, since the 5α-H compounds are biologically inactive [149]. On the other hand, replacement of the 5β-hydrogen atom by a hydroxylic function leads to an increased activity, e.g. the considerably higher activity of polypodine B (*50*) in the Calliphora test in comparison to ecdysterone. The hydroxylic function at position 3 of ring A is obviously indispensable, while the presence of the same function at position 2 does not appear necessary as indicated by the almost equal activity of 2-deoxyecdysterone and ecdysone [33]. The 3β-carbinols are far more active than the corresponding compounds of the 3α-series. Ponasterone B (*44*), the hydroxylic functions of which in ring A are α-situated, is about 30 times less active in the Calliphora bioassay than ecdysterone [215]. The presence of the keto group at position 6, and most probably, the simultaneous presence of the conjugated double bond between the positions 7 and 8, are indispensable for biological activity. Cheilanthone A (*55*) and B (*56*), which lack this double bond, are inactive [56]. On the other hand, podecdysone B (*53*), in which both the conjugation to the 6-keto group and the hydroxylic function at C-14α are absent, is only five times less active than ecdysterone [76]. It may be assumed, however, that the insect organism is capable of shifting the double bond into conjugation. In view of the high biological activity of podecdysone B, the 14α-hydroxylic function in the steroid polycyclic skeleton does not appear indispensable for biological activity. The structure of the steroid side-chain is obviously of great importance activity. Rubrosterone (*70*) which lacks the side chain, is inactive in the Calliphora test but exhibits a pronounced stimulating effect upon protein synthesis in mouse liver, similar to other moulting hormones [297]. Poststerone (*80*), one of the natural degradation products of ecdysone and ecdysterone, is also biologically inactive [278].

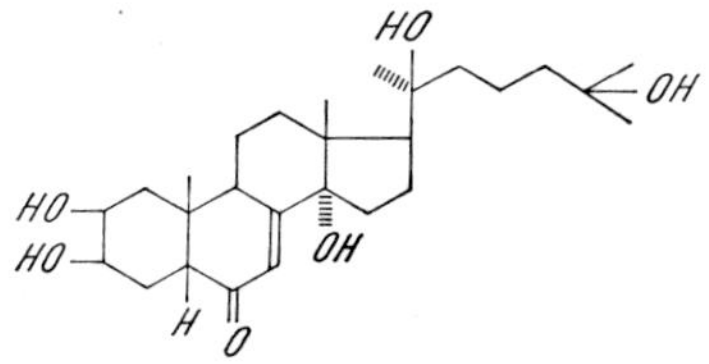

(*81*)

The hydroxylic function at C-22 is obviously indispensable for biological activity. Thus, 22-deoxyecdysterone (*81*), synthesized by GALBRAITH *et al.* [77], is about one hundred times less active than ecdysterone in Calliphora test. Biologically inactive is also the synthetically prepared 22-epiecdysterone [72]. Biological activity does not appear to depend upon the free C-22 hydroxylic function, since capitasterone (*66*) is highly active despite the esterified hydroxyl [303]. On the other hand, stachysterone D (*54*) has a low level of activity [120]. The hydroxylic function at position 20 does not seem of any importance for biological activity, as indicated by the almost equal assay results of ecdysone and ecdysterone. The biological significance of the hydroxylic function at C-25 is relatively small. Thus, ponasterone A (*41*), which is 25-deoxyecdysterone, is about four times less active than ecdysterone [215]. The biological activity of viticosterone E (*40*), which is 25-acetylecdysterone, is similar to that of ponasterone [247]. The slight significance of a hydroxylic function at position 25 is also suggested by the relatively high biological activity of stachysterone C (*45*), a dehydro derivative of ponasterone [120]. Pterosterone (*48*), the hydroxylic function of which is located at position 24 (instead of the C-25 ecdysterone position) is highly active [291]. The negligible importance of hydroxylic functions at the end of the steroid side chain is indicated by the high activity of makisterone B (*60*) and makisterone D (*64*) [116]. The properties of the latter substance also indicate that alkyl groups at C-24 (characteristic of phytosterols) play little role in biological activity.

In conclusion, the following structural features are indispensable for a high moulting hormone activity of steroid compounds:

1. cis fusion of rings A and B of the steroid skeleton;
2. β-hydroxylic function at position 3;
3. keto group at position 6 in conjugation with a Δ^7-double bond; and
4. steroid side-chain with an appropriately (R)-oriented hydroxylic function at position 22.

C. Biosynthesis and Catabolism of Ecdysoids

Despite the considerable attention paid to the biosynthesis of steroid compounds with moulting hormone activity, this problem has not been elucidated so far. Since the insect is not able to synthesize the fundamental steroid skeleton and is entirely dependent upon an exogenous supply of steroid compounds, it was necessary to determine if cholesterol, the theoretical precursor of ecdysone and ecdysterone, can be transformed to these hormones in the insect organism. This was shown for the first time by KARLSON and HOFFMEISTER [152] who administered tritiated cholesterol to the larvae of the blowfly *Calliphora erythrocephala* and isolated radioactive ecdysone. The very low conversion (0.0001%) might be explained by the circumstances that ecdysterone and not ecdysone constitutes the principal moulting hormone of this insect species. As shown by GALBRAITH *et al.* [79] on *Calliphora stygia*, the conversion of cholesterol to ecdysterone was much higher.

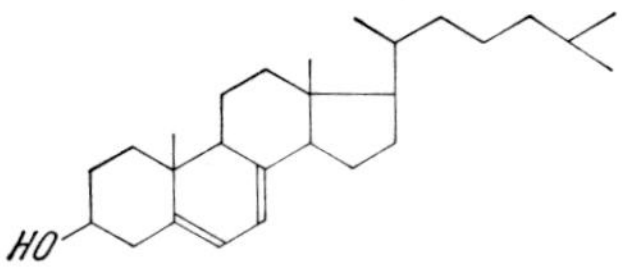

(*82*)

Transformation of cholesterol to moulting hormones appears to be initiated by dehydrogenation to cholesta-5,7-dien-3β-ol (*82*). This assumption is supported by the finding that compound [*82*] can maintain growth in some insect species [34] and that development of some insect species depends upon a supply of steroid 5,7-dienes and not on cholesterol. Thus, the larvae of the beetle *Xylobarus ferrugineus*, when reared under aseptic conditions (to exclude gut fungi as a source of steroids) can develop only with a supply of ergosterol or cholesta-5,7-dien-3β-ol [168]. The plant C_{29} steroids were also shown to be dealkylated in the insect organism to the cholesterol type [34, 289]. The possibility of cholesta-5,7-dien-3β-ol occurring as an intermediate in the synthesis of moulting hormones is also supported by the transformation of cholesterol to this dienol in some insect species under aseptic conditions, e.g. the cockroach *Blattella germanica* [251] and the housefly *Musca domestica* [209]. The formation of ecdysterone from cholesta-5,7-dien-3β-ol in the organism of *Calliphora stygia* has been recently demonstrated by GALBRAITH *et al.* [79]. According to these authors, cholesta-5,7-dien-3β-ol does not appear to serve as an intermediate in the formation of moulting hormones since it does not exert any influence on the transformation of cholesterol into ecdysterone.

On the other hand, there are several explanations for the transformation of the 5,7-diene system in ring B of the steroidal skeleton to systems containing the 6-keto group in conjugation with a double bond between positions 7 and 8. Thus, e. g. epoxidation of such a diene system is known to lead predominantly to a 5,6-monoepoxide [208] which could then be readily transformed to cholest-7-en-3β-ol-6-one. The formation of this unsaturated ketone could be accompanied by isomerisation at C-5. Such an isomerisation might be facilitated by the occurrence of an additional hydroxylic function at position 2β in view of the interaction with the C-19 methyl group of the steroid skeleton (Scheme 8). The insect organism does not seem capable of dehydrogenating ring B with a keto group at position 6 and forming a conjugated double bond. Cheilanthone A (*55*) and Cheilanthone B (*56*), which lack this double bond, are biologically inactive.

Scheme 8

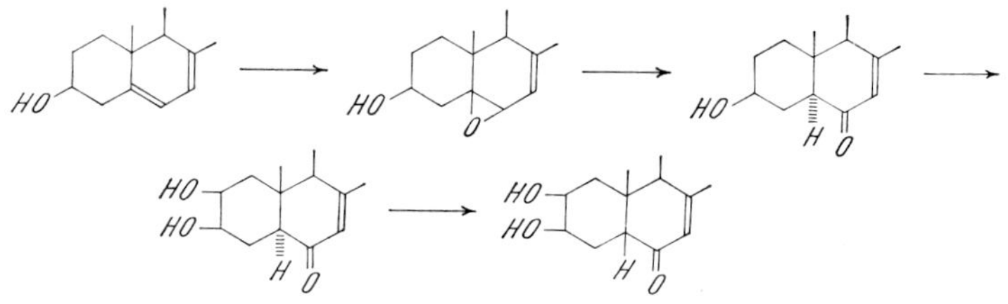

Introduction of the 14α-hydroxylic function can be readily accomplished at any stage of moulting hormone biosynthesis.

While there is little known concerning the elaboration of the polycyclic steroid skeleton, the side-chain hydroxylation is much more understood. KAPLANIS *et al.* [138] have demonstrated that [1α-^{3}H]-2β,3β,14α-trihydroxy-5β-cholest-7-en-6-one (*83*) terminated the diapause in the tobacco hornworm *Manduca sexta* and that it was converted into ecdysone, ecdysterone and (most probably) 20,26-dihydroxy ecdysone.

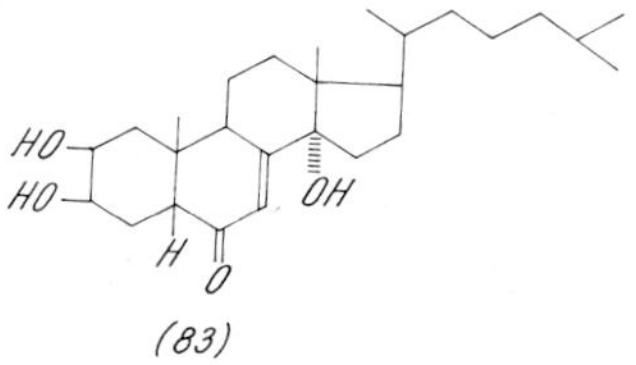

(83)

This finding suggests an enzymatic side-chain hydroxylation at all critical positions, namely, 22, 25 and 20. KING and SIDDALL have demonstrated [166] that some insect and crustacea species can effect the conversion of ecdysone to ecdysterone by hydroxylation at position 20. The most efficient conversions were observed with the moulting shrimp *Crangon nigricauda*, premoult crab *Uca pugilator* and pupating blowfly *Calliphora vicina*. Since ecdysone is not known to occur in these species, the authors suggest that it may be a short-lived intermediate in ecdysterone biosynthesis.

In another study on side-chain hydroxylation [318], 25-deoxyecdysone (*84*), labelled in the side chain with tritium, was applied to *Calli-*

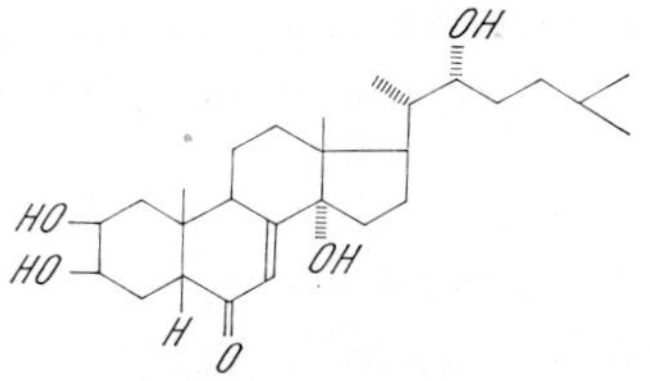

(84)

phora stygia at the time of puparium formation. The test substance was converted into radioactive ponasterone A, inokosterone and ecdysterone. The formation of ecdysone was not observed. It may be seen that position 20, 25 and 26 were hydroxylated; ecdysone alone was rapidly transformed into ecdysterone. The presence of a C-22 hydroxylic function appears critical for moulting hormones since 22-deoxyecdysterone (*81*) is almost inactive, as shown by THOMSON *et al.* [318]. It may be assumed that a hydroxylic function at C-20 strongly interferes with hydroxylation at C-22; on the other hand, the C-22 hydroxylic function does not affect hydroxylation at position 20. The hydroxylation sequence of the cholestane side chain thus appears to begin at position 22, to continue at position 20 and end at positions 25 or 26.

In contrast to insects, plants are capable of synthesizing steroids from a simple metabolite, mevalonic acid. In addition to C_{29} and C_{28} steroids, cholesterol has also been found in plants. It has been demonstrated with the use of radioactively labelled mevalonic acid that this substance can be converted into some phytoecdysones, namely, ponasterone A and ecdysterone in *Taxus baccata* seedlings [283] and ecdysterone in *Polypodium vulgare* leaves [284] (the *Polypodium vulgare* rhizomes are known as a rich source of ecdysterone). The transformation of cholesterol into ecdysterone has been demonstrated in *Podocarpus elatus* by HEFTMAN *et al.* [256]. Application of cholesterol to the leaves of *Polypodium vulgare* for periods of several weeks resulted in a rather low conversion to ecdysone, ecdysterone and 5β-hydroxyecdysterone [284]. All these experiments cannot, however, answer the question whether cholesterol is an obligatory intermediate in the biosynthesis of moulting hormone active steroids since these steroids may be formed at various stages of the biosynthesis from mevalonic acid.

Little is known of the formation of the characteristic steroid polycyclic moiety in phytoecdysones in spite of a number of hypotheses [256, 283, 284]. It may be assumed that in plants similar mechanisms operate as in the insect organism. Thus, the formation of the characteristic conjugated system in ring B from a 5,7-diene might be explained via the 5,6-epoxide. On the other hand, the 5α,6α-and 5β,6β-epoxides of cholesterol alone are not precursors of phytoecdysones [132]. Similar to insects, side-chain hydroxylation appears to proceed stepwise, e.g. the rapid hydroxylation of ecdysone to ecdysterone [233, 284].

In plant material there are a number of compounds which are structurally related to phytoecdysones but do not exhibit moulting hormone activity. The oxidative metabolic pathway of steroids appears more usual with plants than with animals. JIZBA *et al.* [127, 128] have isolated from *Polypodium vulgare* two glucosides possessing certain structural features of phytoecdysones; osladin (*85*) with the characteristic keto group at position 6 and polypodosaponin A (*86*) with the 6-keto group and the conjugated Δ^7-double bond. Viperidone (*87*), viperidinone [47] (*88*) and deoxyviperidone [167] (*89*), which occur in various cacti, are also structurally related to phytoecdysones. In all these compounds *85–89* the steroid rings A and B are trans-fused and ring A lacks the 2β-hydroxylic function which sterically facilitates isomerisation at position 5 as mentioned above (in view of the presence of a keto group at position 6, such an isomerisation does not seem improbable).

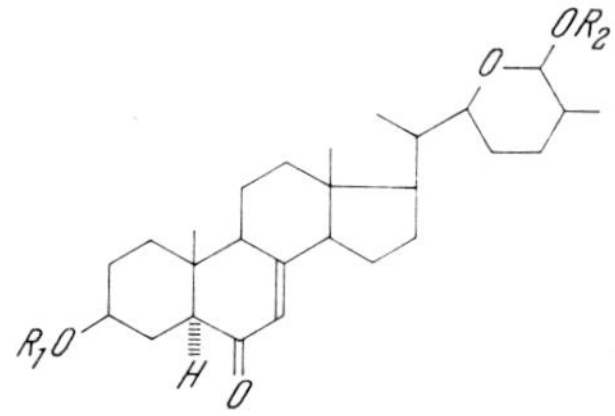

(85) R_1 = Rhamnoglucosyl
R_2 = Rhamnosyl

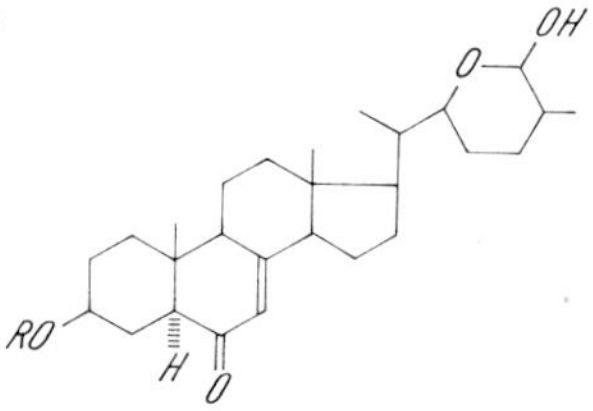

(86) R = Rhamnoglucosyl

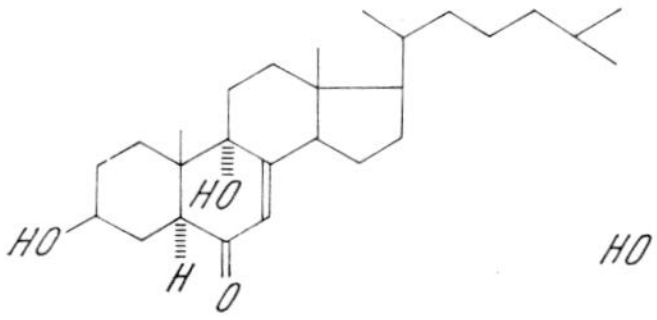

(87)

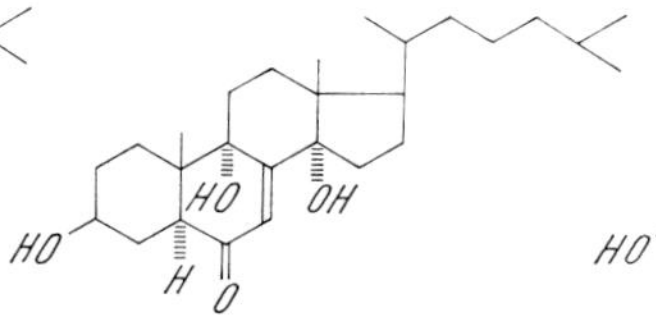

(88)

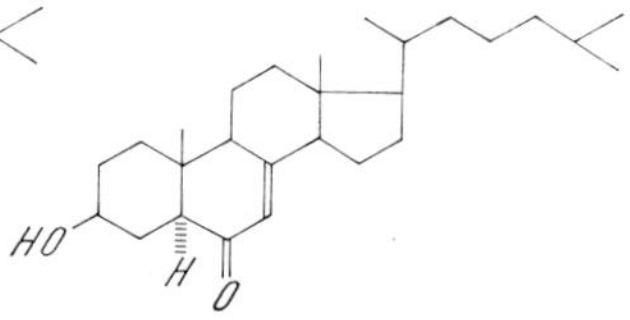

(89)

The levels of moulting hormones in the insect organism vary in relation to life cycles. These levels are obviously lowered by the action of catabolic processes which are controlled by certain enzymes. The mechanism of these breakdown processes has been partly elucidated recently. In one hypothesis, HORN *et al.* [109] suggest that ecdysterone can be catabolised by oxidation of the side-chain 20,21-diol grouping with formation of a C_{21} steroid which is known as poststerone (*80*) at present. In the case of plants, Takemoto *et al.* [93, 296] have isolated rubrosterone

(*70*) and suggest this C_{19} steroid as the product of the natural degradation of phytoecdysones in plant material and poststerone (*80*) as an intermediate of this degradation. GALBRAITH *et al.* [78] isolated a small amount of 4-hydroxy-4-methylpentanoic acid lactone (*90*) as a degradation product of ecdysterone in *Calliphora stygia* prepupae (the attempted isolation of poststerone failed). HIKINO *et al.* [96] isolated poststerone (*80*) from plant material and as the degradation product of tritium labelled ponasterone A (*41*) in the fifth instar larva of *Bombyx mori* [97]. In addition to poststerone, ecdysterone (*26*) and inokosterone (*47*) were isolated as products of primary oxidative transformation of ponasterone A. Consequently, the principal metabolic pathways in the catabolic degradation of moulting hormones appear established. Ecdysone (*1*) is obviously transformed to ecdysterone (*26*) which is then, analogously to other 20,22-diols, oxidatively degraded to poststerone (*80*); poststerone is oxidised in plants to rubrosterone (*70*) (see the following Scheme 9).

Scheme 9

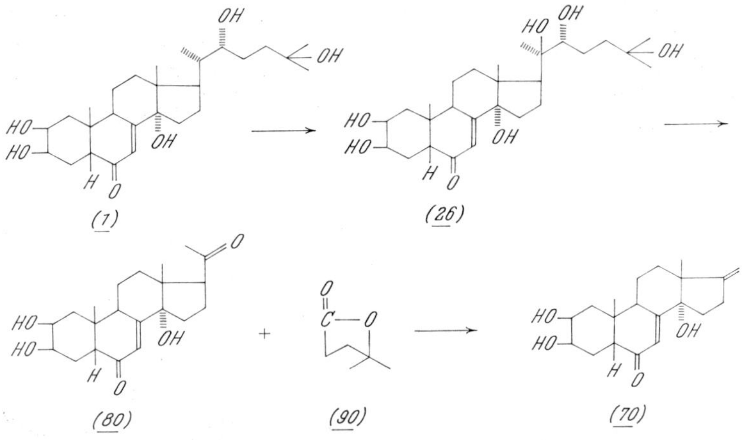

D. Steroid Compounds Interfering with Ecdysis

Some time ago it was found that some steroid compounds structurally related to moulting hormones inhibited postecdysial hardening and sclerotisation of cuticle in an Hemipteran insect, *Pyrrhocoris apterus* [107, 320]. Implantation of several microcrystals (20–50 μg) of these compounds into the body cavity of freshly moulted last instar larvae cause an atypical ecdysis, after which the insects remain soft with an unsclerotised cuticle and die within several hours. Investigation of a series of steroid compounds revealed that some parts of the steroid molecule could be substantially modified without changes in activity, whereas some other parts appeared to be quite essential [319].

All these active compounds are 6-keto steroids lacking the typical conjugated double bond between the positions 7 and 8. The presence of an hydroxylic function at position 3 (irrespective of the configuration) appears essential for the activity. When the hydroxylic function is acetylated, the corresponding compounds are inactive. The 2β,3β-dihydroxy derivatives show practically the same activity as the 3β or 3α derivatives. In contrast to moulting hormones, it is not essential whether the rings A and B are cis or trans fused, since both the 5α and 5β Δ^4-unsaturated compounds are active. The length and structure of the side chain do not exert any considerable influence on the biological activity. Derivatives of cholesterol, ergosterol, pregnane or esters of 25-homocholanic acid are highly active when other structural features of the polycyclic moiety of the steroid molecule (see above) remain intact.

The following compounds show a high antisclerotisation activity: 3β-hydroxy-5α-cholestan-6-one (*91*), 3α-hydroxy-5α-cholestan-6-one (*92*), 2β,3β-dihydroxy-5α-cholestan-6-one (*93*), 2β,3β-dihydroxy-5β-cholestan-6-one (*94*), methyl 3α-hydroxy-6-oxo-25-homo-5α-cholanoate (*95*) and 2β,3β-dihydroxy-5α-pregnane-6,20-dione (*96*). The androsterone derivatives are biologically active but are also toxic (cf. 2β,3β-dihydroxy-5α-androstane-6,17-dione (*97*).

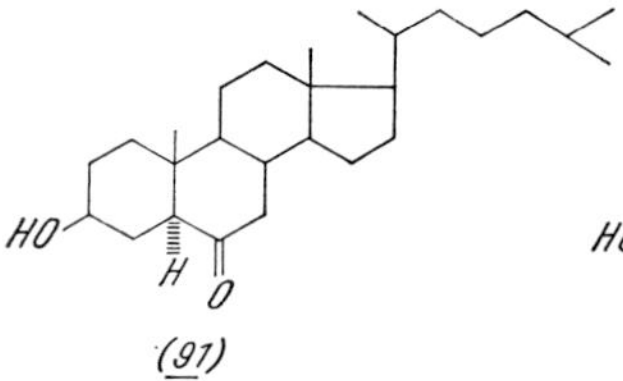

(*91*)

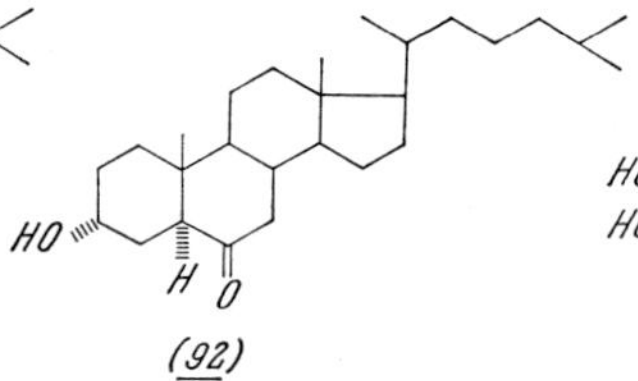

(*92*)

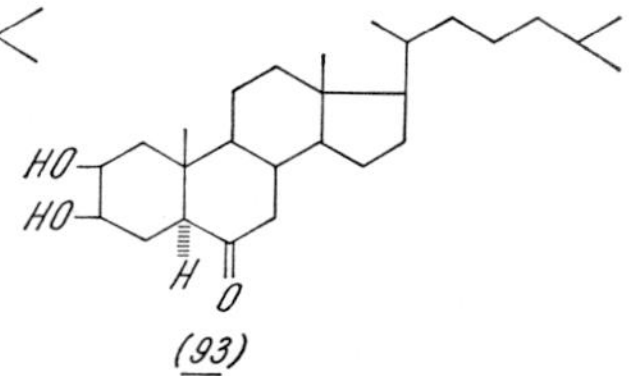

(*93*)

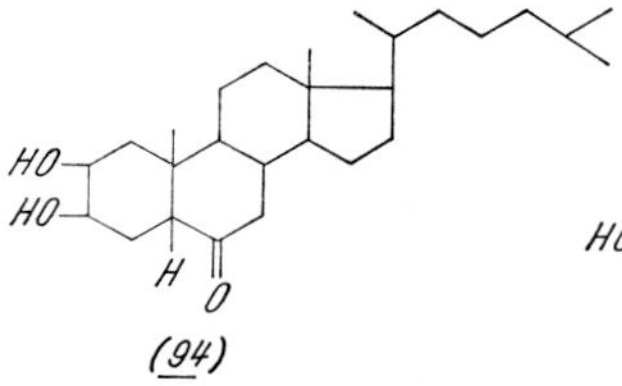

(*94*)

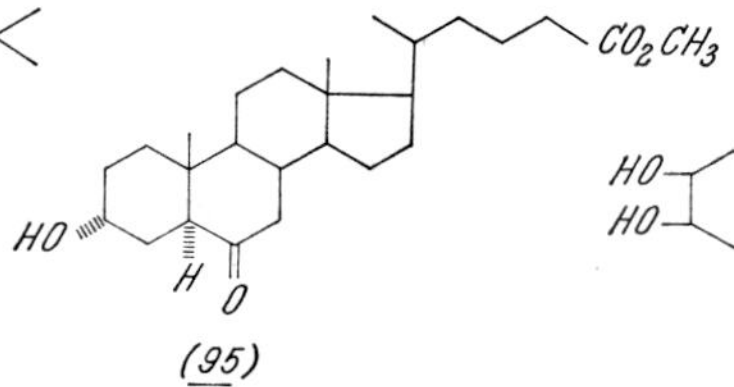

(*95*)

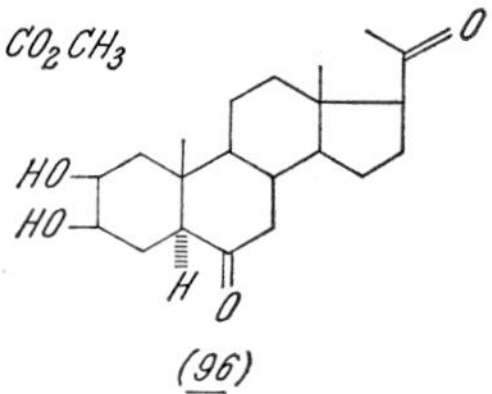

(*96*)

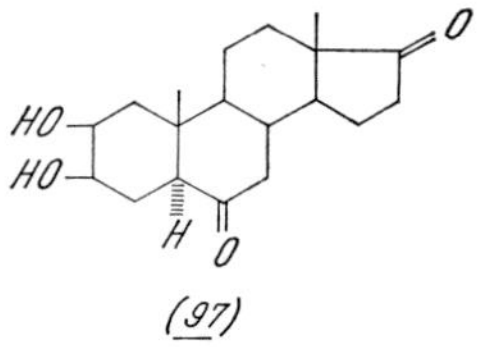

(*97*)

Some synthetically prepared steroid derivatives and their biological activities are listed in Table 16.

Table 16. *The Effect of Some Steroids on the Postecdysial Survival of Pyrrhocoris Larvae*

Compound	Activity
3β-Hydroxy-5α-cholestan-6-one	+++
3α-Hydroxy-5α-cholestan-6-one	+++
2β,3β-Dihydroxy-5α-cholestan-6-one	+++
2β,3β-Dihydroxy-5β-cholestan-6-one	+++
2β,3β-Dihydroxycholest-4-en-6-one	++
2β,3β,20α-Trihydroxy-5α-cholestan-6-one	+++
3α,25-Dihydroxy-5α-cholestan-6-one	++
3β,20α-Dihydroxy-5α-cholestan-6-one	++
3α-Hydroxy-6-oxo-25-homo-5α-cholanic acid methyl ester	+++
3α-Hydroxy-6-oxo-25-homo-5β-cholanic acid methyl ester	+++
2β,3β-Dihydroxy-5α-pregnane-6,20-dione	+++
2β,3β,20β-Trihydroxy-5α-pregnan-6-one	+++
2β,3β-Dihydroxy-17-methylene-5α-androstan-6-one	++ tox.
3β-Hydroxy-6-oxo-5α-androstan-17-carboxylic acid methyl ester	0
2β,3β-Dihydroxy-5α-androstan-6,17-dione	act. tox.
2β,3β-Dihydroxy-5β-androstan-6,17-dione	act. tox.
2β,3β-Dihydroxy-5α-androstan-6-one	act. tox.
2β,3β-Dihydroxy-5β-androstan-6-one	act. tox.
2β,3β,17β-Trihydroxy-5α-androstan-6-one	act. tox.
5α-Cholestan-6-one	0
5α-Cholest-2-en-6-one	0
5α-Cholestan-3,6-dione	0
4-Cholesten-3,6-dione	0
2β,3β-Diacetoxy-5β-cholestan-6-one	0
3β-Acetoxy-5α-cholest-7-en-6-one	0
3β-Acetoxy-5α-cholestan-6-one	0
2β,3β-Isopropylidenedioxy-5α-cholestan-6-one	0
2β,3β-Isopropylidenedioxy-5α-cholestan-5,6-diol	0

The physiological background of the effect on *Pyrrhocoris* is not clear. Some indirect evidence exists that the effect can be regarded as antiecdysone activity, e.g. the antisclerotisation activity as the first indication in this respect. The presence of the compound in the body of the last instar larva does not interfere with normal development for several days until the time of ecdysis. Then the effect is lethal. The second bit of evidence is inhibition of wound regeneration, as compared with the effect of ecdysone, the presence of which stimulates wound healing. Critical evidence would be inhibition of ecdysone action in the

Calliphora bioassay. Unfortunately, experiments so far do not permit a definite conclusion.

The 6-keto steroids also exhibit other interesting biological effects. Thus, when a series of almost fifty 6-ketones was tested by injection of a suspension into female adults of *Musca domestica*, it appeared that all compounds with a normal skeleton exhibited more or less pronounced sterilising effects, most of the ketones being strongly active [245]. Administration of these compounds disturbs the activity of the nutritive cells in egg chambers, resulting in a decreased hatching rate of laid eggs. More frequently oviposition does not occur at all owing to a more serious disturbance in vitellogenesis.

The practical use of all of these compounds is difficult since they are active only after application by injection. It would be of interest to discover compounds of the above type which would be active when applied topically.

The results obtained in Prague were later confirmed by American authors [249, 316]. It has also been shown that certain synthetic compounds containing structural features similar to the insect moulting hormones, when ingested, were effective inhibitors of larval growth and development in several species of insects, although the insect hormones ecdysone and ecdysterone were inactive or considerably less active. The most active compound was 2β,3β,14α-trihydroxy-5β-cholest-7-en-6-one (*98*) which also inhibited ovarian maturation.

In connexion with investigations on moulting hormone activity and inhibitory effects of naturally occurring, as well as synthetic, steroids, Robbins *et al.* [250] have observed that cyasterone (*67*) and ponasterone A (*41*) exhibited a pronounced inhibitory activity in the housefly ovarian test and in the confused flour beetle larvae test, while low concentrations of ecdysone and ecdysterone did not inhibit insect development.

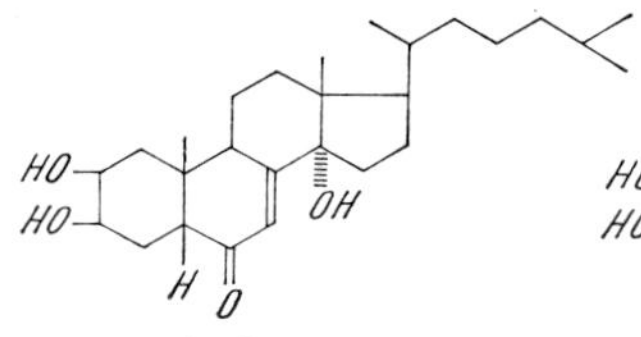

(98)

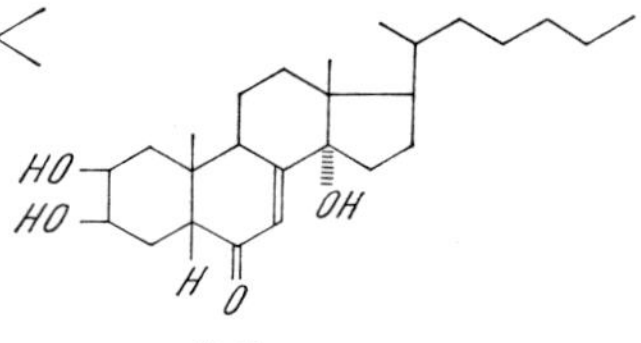

(99)

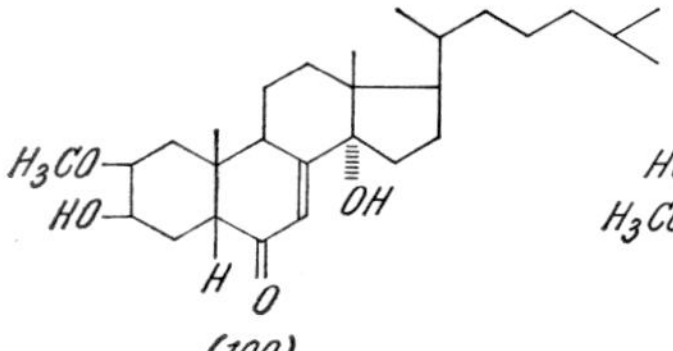

(100)

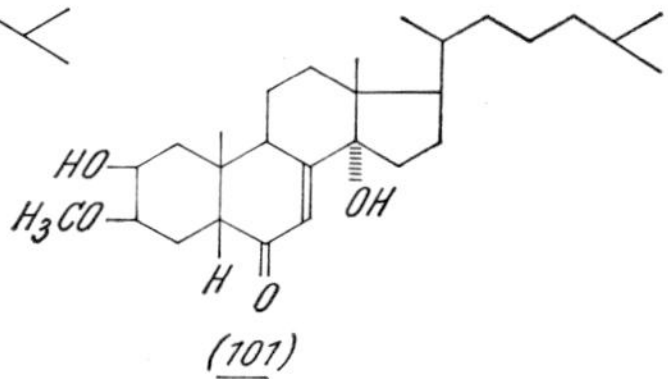

(101)

The highest activities in the tests stated were shown by some synthetic compounds, especially the above mentioned compound (*98*), its 25-demethyl derivative (*99*) as well as 2- (*100*) and 3-methyl (*101*) ethers of compounds (*98*).

NAKANISHI *et al.* [175] appear to have succeeded in isolating a virtual inhibitor of moulting hormones from the plant *Ajuga decumbens* (this plant also contains ecdysterone, ajugasterone B, ajugasterone C and cyasterone). The active substance, which was given the name ajugalactone, specifically inhibited the moulting hormone activity of ponasterone A in the so called Chilo dipping test (*Chilo suppressalis* = rice stem borer) but did not inactivate ecdysterone. The structure of ajugalactone (*102*) is closely related to C_{29} phytoecdysones. The authors claim to have isolated from the plant material an additional inhibitor which specifically interferes with the activity of ecdysterone but not that of ponasterone A.

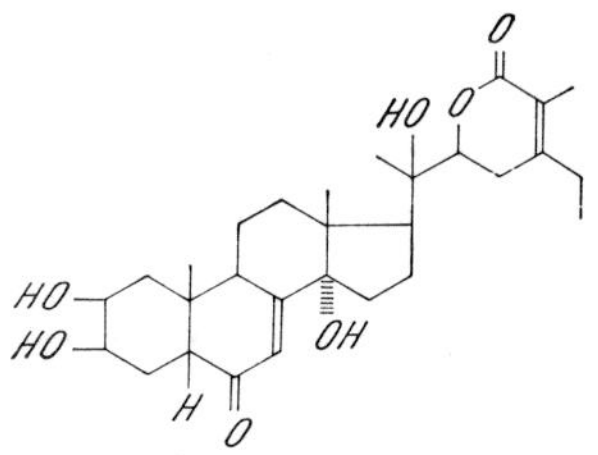

(*102*)

E. Biochemical Effects of Ecdysoids

The mechanism of action of ecdysone has been partially elucidated by KARLSON and his coworkers. The results of their investigation are reported in a number of reviews [38, 145, 146, 147, 161, 184]. The concept of the study has at first been based on the knowledge of the formation of puparia which is completed by the transformation of the original larval cuticle to the hard brown cuticle under the influence of certain quinones. These quinones react with cuticle proteins resulting in new polymeric systems with a new type of cross-linking matrix; this process is known as sclerotization. It has been established with the aid of radioactive compounds that in the larvae of *Calliphora* the process of sclerotization is closely connected with the compound derived from tyrosine, namely from its metabolite 3,4-dihydroxyphenylalanine (dopa) [143, 159]. The generally labelled tyrosine administered during the period of sclerotization is incorporated into cuticle to the same extent as the generally labelled 3,4-dihydroxyphenylalanine. In contrast, tyrosine labelled in its carboxyl moiety is not incorporated with the implication that the active sclerotizing agent is a derived amine. It has been shown further that this substance corresponds to 3,4-dihydroxyphenylethylamine (dopamine) or rather to its acetylated form which is transformed to the corresponding o-quinone reacting subsequently with cuticle proteins [155, 268]. Quinones are formed under the influence of phenoloxidase possessing a considerable affinity towards N-acetyldopamine. The formation of these compounds is shown in Scheme 10.

Scheme 10

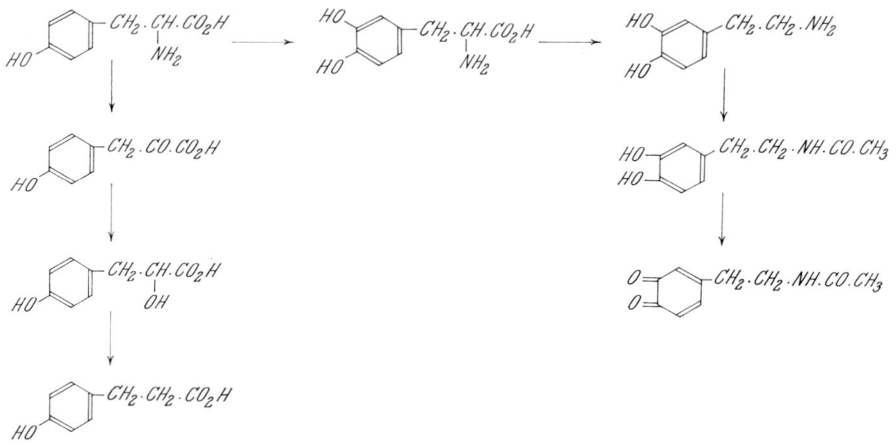

The transformation of tyrosine is characteristic for the late instar larvae [266]. At early stages of development tyrosine is prevalently transformed to p-hydroxyphenylpyruvic acid, which is further metabolised to p-hydroxyphenyllactic and p-hydroxyphenylpropionic acid. The change of tyrosine metabolism during the terminal larval stages is evidently accounted for by the synthesis of new inducible enzymes. The enzymes responsible for the transformation of tyrosine to dopamine are known and have been described by Karlson and his associates [156, 262, 271]. The key enzyme for the formation of N-acetyldopamine is dopa decarboxylase. Its level in *Calliphora erythrocephala* is considerably elevated unter the influence of ecdysone. This has been shown under physiological conditions in the intact insect when decarboxylase level rises at the same time as ecdysone is produced [275] as well as in the experimental system. Ecdysone injection into ligated abdomen of *Calliphora* causes a distinct increase of dopa decarboxylase [160]. It was also shown that this increase of dopa decarboxylase level is inhibited by inhibitors of protein biosynthesis, like actinomycin and puromycin [267].

The formation of the newly induced enzyme must be preceded by corresponding messenger RNA synthesis. This has indeed been shown in intact insects [157], and also in epidermal cells which display more than a twofold increase of P^{32} incorporation into their RNA following ecdysone [43, 263, 270]. Messenger RNA isolated from a mixture of RNA's with molecular weight corresponding to 16–18 S has a considerably lower template activity when obtained from fully grown larvae before pupation (at the time when the level of ecdysone is negligible) than its counterpart isolated from white pupae possessing high levels of ecdysone. This would suggest that ecdysone directly stimulates the synthesis of different messenger RNA species. Further work of Karlson and his associates has shown that ecdysone-stimulated RNA carries the code for dopa decarboxylase synthesis which, as mentioned before, is an active agent necessary for the sclerotization [269]. This enzyme has been obtained *in vitro* in a protein synthesizing system if messenger RNA from hormone induced larvae has been added; however, upon addition of messenger RNA from control insects practically no synthesis has occurred.

Another confirmation of this mechanism of action of ecdysone comes from the observation that the hormone causes puffing in salivary gland chromosomes of *Chironomus tentans* [42]. This effect is very specific since it takes place only at two gene loci, IR-18-C on the first chromosome and IV-2-B on the fourth chromosome. It is supposed at present in accordance with Jacob and Monod model of proteosynthesis, that ecdysone and steroids with analogous biological effects react specifically with DNA repressors (histones), and that consequently they lead to a derepression at different sites of DNA molecule resulting in the synthesis of messenger RNA at these sites [161]. The messenger is transferred to cytoplasm where the enzymes responsible for the sclerotization, especially dopa decarboxylase are formed.

This assumption indicates that the primary effect of ecdysone is bound to the cell nucleus. Using tritium labelled ecdysone it has been shown that the hormone accumulates in the nuclei of epidermis in the last instar larvae of *Calliphora* [36, 39, 162]. It has also been shown that the isolated nuclei from epidermal tissue of fully grown, not hormone induced *Calliphora* larvae are capable of performing RNA synthesis. The addition *in vitro* of ecdysone to the incubation mixture strongly enhances RNA synthesis. The newly synthesized RNA corresponds practically to messenger RNA [49, 265].

This explanation of the effect of ecdysone on insects is not the only one. The theory of Kroeger is an attempt to clarify the process of puffing and of activation of genes by the change of ion concentration in the cytoplasma and in the nuclear sap. According to this author the puffing of salivary gland chromosomes may be induced by concentrated salt solution [181, 183]. Some authors do not accept this theory [37, 123] some others support it [190, 193]. However, it seems at present that the explanation of ecdysone action as proposed by KARLSON is experimentally well documented. It is of course probable, and it has also been indicated by different observations that the direct effect of ecdysone on the process of sclerotization is only one feature of its manifestations, and that in the future the mechanisms of hormone actions during other stages of the development of insects will be discovered.

F. Bioassays for Ecdysoids

The biological assays for hormones that stimulate moult cycles in insects have an old tradition. They were initiated already at the time when WIGGLESWORTH described hormonal control of moulting in *Rhodnius*. FRAENKEL [62, 63] was the first to obtain evidence for hormonal stimulation of puparium formation in flies in 1934–1935. He ligated the larvae of *Calliphora* behind thoracic region several hours before puparium formation. The result was that only the anterior parts pupariated whereas the abdominal compartments remained larval. Pupariation of the abdominal parts was stimulated by injections of the haemolymph taken from just pupariating larvae. Later it appeared that the hormone responsible for puparium formation was secreted from the peritracheal part of the ring gland which is homologous with prothoracic glands of other insects (p. 52). The original design of FRAENKEL's assay was slightly modified by BECKER and PLAGGE [15], and by KARLSON [142] who employed this assay in his work on the isolation of ecdysone. With some additional improvements the "pupariation test for ecdysone" [66] was also utilized for discovery and further search for ecdysoids of plant origin, and it is currently used until now.

Other assays for ecdysoids were based on stimulation of metamorphosis in the last larval instars or in pupae of some Lepidoptera. These assays were originally based on WILLIAMS's studies on the prothoracic gland hormone action in Cecropia silkworms (p. 59). In connexion with this, KOBAYASHI *et al.* [171] and WILLIAMS [326] assayed ecdysone on decerebrated pupae and on isolated pupal abdomens of silkworms. Moreover, SATO *et al.* [255] and IMAI *et al.* [121] performed the tests on ligated larval abdomens of some Lepidopterans. They developed procedures for topical assays of ecdysoids.

In spite of relatively important role of these hormones in insect development, the assays for determination of their activity were conservatively restricted to the above mentioned tests in Diptera and Lepidoptera. This is an apparent contrast with a wide applicability of assays that are available for testing biological activity of the juvenoid compounds. The reasons are most probably associated with a difficulty to find suitable developmental stages for testing of the ecdysoids. For example, the prothoracic glands remain functional for considerable part of the postembryonic development. Testing of ecdysone on these stages with functional prothoracic glands is generally feasible, however, it requires to use large doses, and it needs a complicated statistical evaluation; the effects of ecdysoids on these stages are usually represented by nothing more but an acceleration of the moult cycles. Furthermore, it is well known that a removal of the prothoracic glands from the body represents one of the most difficult surgical manoeuvres in most species.

The second possibility is to assay ecdysoids on stages with inactive or degenerated prothoracic glands. Such stages are for instance diapausing larvae or pupae which were used for the assays in Lepidoptera. Other stages with inactive prothoracic glands are larvae which have been starved since the moment of ecdysis. However, such starved larvae are not the best assay animals due to their high mortality and huge doses which are required for stimulation of the moult cycles. This is associated with a lack of nutritional substrates necessary for tissue growth. Final possibility with respect to developmental stages suitable for ecdysoid assays are adult stages with degenerated prothoracic glands. But also in this case it appeared almost impossible to induce extra-adult moults, with exception of few insect species where extra-adult moults were induced by extremely large doses of ecdysoids (cf. p. 59, 341).

These data suggest that during normal development of insects it is very difficult to find a stage with a high sensitivity to ecdysoids. In fact, occurrence of susceptible stages is limited to diapausing larvae or pupae. It appeared most convenient, therefore, to perform ecdysoid assays on ligated larvae and on isolated posterior body fragments. The source of the hormone was contained in the anterior part which served eventually as a control to responses induced by the endogenous hormone. The posterior compartments may survive for a long time provided that the body has built up nutritional reserves before the moment of their isolation and, depending on time of isolation the hind compartments may become extremely sensitive to ecdysoids.

With respect to methods of ecdysoid application we can observe that, in contrast to juvenoids, there are no equivalent methods for direct topical application of ecdysoids. This is due to a limited penetration of ecdysoids through the insect integument. In certain cases, however, indirect topical application methods were successfully used, such as was the *Chilo* "dipping test" [255]. The most reliable direct application method for ecdysoid assays which is used for dose-response relationship studies, are individual injections into the haemocoele. To this end, the compounds are at first dissolved in a small amount of ethanol or acetone and then diluted 10-times with warm water or insect Ringer solution. Further dilutions are made with 10 per cent ethanolic or acetone solutions respectively. Some insects tolerate even 20 per cent solutions provided that the amount of fluid injected into the body does not exceed 5 per cent of the total body volume or weight. Dipteran insects tolerate ethanol better than acetone while in some Coleoptera and Lepidoptera we have observed reversed conditions. A limited solubility of most ecdysoids in 10 per cent ethanolic water solutions usually allows to prepare maximum concentrations of 10 mg per ml which gives a dose of 10 μg per each μl injected. When larger doses are required the compounds have to be injected in a form of suspensions or emulsions, or eventually, crystals of the pure compounds have to be implanted into the body cavity.

Evaluation of the biological activity of ecdysoids is generally based on induction of development in the epidermal cells. Developmental criteria used for evaluation of the activity are either specific, such as formation of a puparium in higher Diptera, or nonspecific, such as apolysis of the old cuticle, secretion of the new cuticle or ecdysis which are characteristic features of the moult cycles in all insects. In the following paragraphs we have described some assay procedures which were currently used or appeared convenient for determination of biological activity of ecdysoids.

a) The Pupariation Test for Ecdysoids

This assay is based on stimulation by ecdysoids of sclerotization of the old, last larval instar cuticle in higher Diptera. Formation of the puparium is brought about by phenolic tanning or calcification of the cuticle which becomes dark, solid, and contracted. The whole larval body assumes a typical barrel-shaped appearance. In addition, this process of pupariation is associated with many other physiological and biochemical changes which have been described in chapter G. It differs from the usual type of ecdysis in other insects where the old cuticle becomes digested while the new cuticle is sclerotized afterwards. The original puparitation assays [15, 63, 81, 142] were made on *Calliphora* larvae. However, some authors, e.g. Kaplanis *et al.* [139], Kobayashi *et al.* [171], and Adelung and Karlson [2] invented to use smaller larvae of *Musca* which were slightly more sensitive.

There are several important points which considerably influence economy and efficiency of these assays. Mainly, the larvae selected for the assay must be ligated at a precisely determined critical period before the expected puparium formation. Moreover, only specimens which have pupariated in the anterior part but not in the posterior abdominal part can be used for injections. Specimens which have not pupariated in either part are discarded. When ligatures were not performed at the right moment, the yield of the "sensitive" abdomens might be drastically reduced. However, this represents a considerable waist of labour involved in ligaturing because it often happens that only about 10 per cent of the "sensitive" abdomens are obtained. In order to avoid this difficulty, some authors recommended certain other species of Dipteran larvae where pupariation was inhibited for as long as the mature larvae remained in contact with moisture. The process was activated as soon as the larvae were transferred to dry conditions and this enabled to coordinate the right moment of ligaturing with the "dry treatment". This was originally found by Ohtaki *et al.* [225, 226, 227] in larvae of *Sarcophaga peregrina*, but it occurred also in some other *Sarcophaga* species [66] and in *Calliphora stygia* [313]. Further improvement which was to economize ligaturing was devised by Staal [286]. The larvae were not ligated by a thread as usual but were constricted in small apertures made in an elastic rubber strip.

Adelung and Karlson [2] have described a method in *Calliphora erythrocephala* and *Musca domestica* which has been widely used for the assays of biological activity of ecdysoids. A brief description of the method follows:

A batch of larvae among which 20 per cent (in *Calliphora*) or 10 per cent (in *Musca*) have already pupariated is selected for ligaturing of the specimens that remained larval. The ligatures are placed at one third of the body from the anterior end. After 24 hrs., all specimens which have pupariated in the anterior part but not in the posterior one are removed and incubated once more for 24 hrs. (*Calliphora*) or 12 hrs. (*Musca*) before injections. The compounds are dissolved in water or in 10 per cent ethanolic water solution and are injected into the ligated abdominal parts. The amount of solution injected per abdomen is 10 μl (*Calliphora*) or 5 μl (*Musca*) respectively. (Some authors have argued [66, 81] that this volume may be too large). Injection is performed through a loose second ligature which is applied close to the first one. After removing the injection needle the ligature is tightened to prevent bleeding.

Evaluation of the results is made after 24 (*Calliphora*) or after 20 hrs. (*Musca*) according to the degree of pupariation. Difference is made between fully pupariated abdomens (100%), strongly pupariated abdomens (75%), half pupariated (50%), or slightly pupariated ones (25%) which contain only few patches of pupariated cuticle. Toxic effects and bacterial infections which can be distinguished from brown, tanned and hard puparial abdomens by black coloration and soft cuticle, have to be eliminated from the positive counts. It seems quite satisfactory to use 10 abdomens for each dilution. The standard error is some ± 20 per cent. There exist clearcut dose-response relationships. The unit of activity, e.g. *Calliphora* unit or *Musca* unit indicates an amount of the compound which induces under the above described conditions an average of 50 per cent pupariation score. The activity units of several ecdysoid compounds in few Dipteran species range usually between 1 to 100 nanograms per abdomen.

The assay has relatively good specificity. It is reproducible, very sensitive, and gives the results in a short time. Its disadvantage is represented by a necessity of using specifically prepared, fresh larval material, and by limited yield of abdomens ready for injections. The latter can be partly avoided using *Sarcophaga* or other species where pupariation is coordinated with the dry conditions. Finally, we may advise to consult papers by OHTAKI *et al.* [227] and FRAENKEL and ZDAREK [66] who have shown that the results of these assays might be influenced by a number of factors, such as larval age at the time of ligation, waiting period between ligations and injections, dilution of the haemolymph, and other specific conditions.

b) The Chilo Dipping Test

This assay is based on indirect topical treatment of ligated abdomens of some Lepidopteran larvae. In the following lines we have described procedures for performing the assay on *Chilo*, *Ephestia*, or *Plutella*, according to SATO *et al.* [255]:

The last instar larvae are ligated behind the metathorax. The ligatures are made at certain critical moment after which the ligated abdominal parts are unable to develop spontaneously. The undeveloping ligated abdomens are dipped for 5 sec into methanolic solutions containing graded amounts of the ecdysoid compounds to be tested. After evaporation of methanol, the abdomens are incubated in the humid atmosphere at 25—28 °C. Positive responses are indicated by formation of a characteristic pupal cuticle. This is associated with changes in shape of the isolated abdomen, changes of moving responses to mechanical irritation, and rigidity of the integument due to sclerotization of the new pupal cuticle.
The effect can be easily recognized when the old cuticle has been peeled off. The range of activity is approximately between 0.01 and 0.1 per cent solutions. It was calculated that one ligated abdomen may retain about 0.01 ml of methanol during dipping. According to this, one abdomen might respond to several micrograms of the ecdysoid compound. Instead of dipping it was practically possible to evoke some positive responses by topical treatment of the ligated abdomens with 1–3 μl of methanol containing 3 to 25 μg of the ecdysoid compound.

Although the assay is less sensitive and less accurate than the pupariation test in flies, it appears quite convenient for large screenings of ecdysoid activity due to its simplicity [121].

c) The Assay on Brainless Silkworm Pupae

KOBAYASHI *et al.* [171] described this assay on *Bombyx* and WILLIAMS [326] on *Samia cynthia*. The methods were as follows:

Fresh pupae of *Bombyx* or chilled diapausing pupae of *Samia* are deprived of the brains and incubated for 57 days (*Bombyx*) or one day (*Samia*) before injections. Each pupa is injected with 10 μl of 10 per cent ethanolic water solution containing 0.1 to 10 μg of the compound (*Bombyx*), or with 50 μl of 10 per cent isopropanol solution containing approx. 0.1 to 100 μg of ecdysoids (*Samia*). The results are evaluated according to stimulation of the adult development 30 days after injection at 25 °C (*Bombyx*), or according to specific stimulation of the early development, normal development, or precocious development throughout a period of 21 days following the injection (*Samia*). In both species it appears that larger doses of ecdysoids cause abnormal precocious development associated with suppressed formation of hairs and scales; eventually there occurs retention of certain pupal patterns. In addition, unchilled diapausing *Samia* pupae which do not develop at the usual room temperatures, have also been used as assay animals as well as the isolated pupal abdomens. The range of sensitivity is usually between 1 to 10 μg per pupa of average weight 2 g.

The assay is especially suitable for comparative physiological studies of ecdysoid action. It is less convenient for routine screenings.

d) The "Anti-Juvenile" Assay on Dermestes

In the course of our investigations on ecdysoids we have developed an assay which is almost as sensitive as the Dipteran pupariation assay. Its advantage depends on a possibility to perform the tests on pre-conditioned specimens which can be stored at room temperature over a period of several weeks. The assay is made on last instar larvae of *Dermestes vulpinus* which are easily distinguished from larvae of previous instars by presence of characteristic transversal rows of tubercles on the tergites of the last 5 abdominal segments. In principle, the last instar larvae are immobilized, their metamorphosis is then inhibited by a simple topical treatment with juvenoids, and these immobile specimens are stored at room temperature until their development is again resumed by an injection of ecdysoids. It is necessary to add that the last instar larvae of this species always undergo pupation after 10 days at 27 °C, irrespective of feeding or presence of the head or thoracic endocrine glands. Description of the assay procedure according to Sláma [281] is as follows:

Freshly ecdysed last instar larvae are allowed to feed for one or two days in order to increase the amount of nutrients in the body. Then, the larvae are immobilized by a tight ligature between the head and the thorax. Metamorphosis of the immobilized larvae is inhibited by a simple topical treatment with juvenoids. Approximately 50 μg per spec. of methyl 3,7,11-trimethyl, 11-chloro,2-dodecenoate inhibits metamorphosis for about one month, however, any other juvenoid with relatively good activity on *Dermestes* (see the Appendix tables) can be used. The immobilized larvae with inhibited development are stored at room temperature until used for ecdysoid injections. When stored for a very long time, the treatment with juvenoids has to be repeated once after three weeks to prevent spontaneous recovery.

For ecdysoid injections the immobilized larvae are narcotized under water for 15 to 25 min. Injections are performed intersegmentally into the abdomen. Each specimen is injected with 1 to 5 μl of 10 to 20 per cent ethanolic water solution containing graded amounts of ecdysoid compounds.

Fig. 32. Development of pupal structures in ligated larvae of *Dermestes vulpinus* (right) after injections of small amounts (0.05 to 0.5 μg) of ecdysterone. The 1-day-old last instar larvae were ligated behind the head or thorax, metamorphosis was inhibited by topical treatment with 50 μg of 3,7,11-trimethyl-11-chloro-2-dodecen oate, and induced to metamorphose after two weeks by ecdysterone

Large doses of 1 μg or more cause precocious moult in three days. The new cuticle of the intermediate is easily visible through the old cuticle under dissecting microscope. The unit concentration for precocious moult indicates an amount of ecdysoid in μg per spec. which causes positive response in 50 per cent animals. Very low doses of 0.001 to 0.1 μg cause normal development which results in pupal moults after 6 to 7 days. The unit concentration for this effect is indicated in μg of ecdysoid per spec. which induces metamorphosis and pupal moults in 50 per cent animals. Doses which are between the unit concentrations for precocious moult and stimulation of metamorphosis cause series of intermediary forms between larva and pupa. These moult between the 3rd and 7th day at 27 °C. For each run of assays with very low doses it is necessary to set a parallell control experiment with 5 to 10 specimens in order to check for possible spontaneous initiation of metamorphosis due to consumed juvenoid supply. The immobilized larvae of *Dermestes* induced to moult by ecdysoids are shown in Fig. 32.

The larvae of *Dermestes* are very easy to breed and handle. Their ability to survive injections or even the most serious surgical treatments is enormous. The assay is suitable for routine screenings as well as for detailed analysis of juvenoid-ecdysoid interactions.

e) Some Other Assays for Ecdysoids

In connexion with the mentioned susceptibility of diapausing larvae and pupae to ecdysoids it appears possible to test these compounds in other than Lepidopteran insects. For example, effectiveness of ecdysoids has been demonstrated on larval diapause in Hymenoptera [110] or on pupal diapause in Diptera [65]. Furthermore, some specific effects of ecdysoids can serve under certain circumstances for determination of their biological activity. This may be exemplified by the effects of ecdysoids on growth of tissue cultures (p. 362) or on specific puffing patterns on the polytene chromosomes (p. 364). Some other examples have been discussed in section G.

Unfortunately, the reduced ability of ecdysoids to penetrate insect integument in most species is associated with a considerably limited use of the indirect application methods. Unlike with juvenoids, these are practically reduced to a few observations concerning peroral application of ecdysoids. Thus, Fourche [59, 60, 61] described the effects of peroral application of ecdysoids on development of *Drosophila*. Moreover, Robbins *et al.* [249, 250] described various effects of ecdysoids which were ingested with food in several species of insects.

G. Physiological Effects of Ecdysoids

Research on hormones which stimulate development and moult cycles in insects underwent tremendous progress since the time when Karlson isolated ecdysone from pupae of *Bombyx*. This was mainly associated with the availability of ecdysone analogues of plant origin and also with the availability of synthetic compounds. The basic historical facts concerning the isolation of ecdysone and the discovery of the active materials in plants have been partly mentioned in the former chapters of this section, and combined with the basic data on physiology and biochemistry of their action they were a subject of several review articles, e.g. El-Ibrashy [51], Nakanishi [216], Piechowska *et al.* [232], Rees

[242], SEHNAL [260], WILLIAMS [327], and WYATT [333]. It is clear, however, that the extent of our knowledge on physiology of action of these compounds, which we call ecdysoids, is far from being complete. Our attention has been focused to some current physiological problems and relationships between physiological functions, biochemical changes, and morphological events induced by ecdysoids in insect body. We have recorded the present state of achievements in this rapidly expanding field of science. One may notice that we have paid relatively less attention to the old problems and general informations, the full account of which is present in the above review articles.

With regard to physiological action of the prothoracic gland hormone (see part I, C) it appears that ecdysoids can duplicate almost all the physiological effects obtained with transplantation, ligaturing, or parabioses. However, there appear also certain differences in action between ecdysoids and the prothoracic gland hormone. This is most apparent when large doses of ecdysoids are used in a single injection. Most probably, the prothoracic gland never releases excessive amounts of the hormone instantaneously as the secretion of the hormone is regulated by neurohormones and by certain internal homeostatic mechanisms. Thus, we believe that many responses of insects to ecdysoids, which are seemingly patho-physiological in appearance, are merely a result of an instantaneous administration of these excessive doses.

Before going into details it may appear useful to recapitulate briefly the main developmental events which are characteristic for the moult cycles in insects. It is well known that the postembryonic development in insects proceeds in distinct cycles separated by ecdyses. The period between two ecdyses, referred to as the intermoult period, is characterized in young larval instars by periodically repeated changes in activity of the endocrine system, in feeding and digestion, cellular growth pattern, and in respiratory metabolism course, among other distinctive features. These physiological changes during the moult cycle are dependent on hormonal activity; the cycle is absent as a whole when the most important endocrine centers, e.g. the neurosecretory system which releases the adenotropic neurohormone and the prothoracic glands have been removed. Crucial targets of these hormones in the moult cycle are the epidermal cells. In response to the prothoracic gland hormone or ecdysoids, these epidermal cells undergo remarkable internal structural changes indicated earlier by WIGGLESWORTH [324] as "activation". The "activated" epidermal cells increase their nuclear and cytoplasmic volume which is associated with some structural changes in the chromatin and in the cytoplasmic organelles as well. For example, there occur changes in the mitochondrial system suggesting an increased cellular metabolism. During the stage of "activation" the epidermal cells get separated from the old cuticle. This important step in the moult cycle is called apolysis [126, 125]. After being separated from the cuticle, the epidermal cells continue to enlarge and initiate mitotic or endomitotic divisions, which gives rise to a new epithelium formed either by more numerous or larger polyploid epidermal cells. Following nuclear or cellular divisions, the cells undergo further physiological changes resulting in secretion of a new cuticle. The whole process culminates in the next ecdysis. The pharate period between apolysis and ecdysis may last only several hours in young larval instars of some species, but it may also be extended through a major part of the intermoult period. The latter is common in metamorphosis.

From a physiological point of view the prothoracic gland hormone or ecdysoids give a stimulus for initiation of the moult cycles, e.g. they cause the "activation" of the epidermal cells. We shall see, however, that there are large differences with respect to time within a moult cycle in which the presence of the hormone is essential. It seems that in certain cases the hormone serves merely as a trigger for not only one but two or more of the moult cycles. This is exemplified on metamorphosis in Hymenoptera and in many species of Coleoptera where a single administration of ecdysoids induces larval-pupal moults, which are invariably followed by pupal-adult moults. However, in other cases the hormonal stimulus is repeatedly required for further progress in a moult cycle. Such an occurrence of two distinguished sensitive periods to ecdysoids within a single moult cycle is usually associated with specific synchronization of moults with the environmental conditions. Examples to this are in special timings of puparium formation and of pupation in general. We have realized in our further discussions concerning these differences in physiological responses to ecdysoids that the developmental patterns induced by these hormones in cells of different tissues, at different ontogenetic levels, and in different species are all genetically programmed.

1. Developmental Stages Sensitive to Ecdysoid Action

In correspondence with the sensitive periods to juvenoids (p. 218), there occur certain periods in ontogenetic development of insects when a stage is increasingly sensitive to ecdysoid action. We have little information so far concerning susceptibility of the embryonic tissue to ecdysoids, except few data on *in vitro* cultivated embryonic cells. Relatively low sensitivity to ecdysoids occurs in young larval instars. This is usually explained by presence of the juvenile hormone. Also, the adult stage of insects appears to be extremely resistant to ecdysoid action. Many attempts to cause supernumerary adult moults by feeding or injections of ecdysoids were unsuccessfull [333], although this was obtained by parabiosis, transplantation of prothoracic glands, or by other endocrinological techniques with the endogenous hormone. Thus, the most sensitive developmental stages to ecdysoids are the last larval instars or the pupal instars, e.g. the metamorphosis stages.

Sensitivity of the epidermal cells to ecdysoids undergoes considerable variations throughout the course of the last larval or pupal instars. This was recognized as early as in 1957 by HANSER and KARLSON in *Ephestia* [86]. It was later confirmed by studies in other Lepidoptera, such as Saturniids [176, 210, 254, 326], or *Galleria* [200, 260, 261], and also in other insects [259, 333]. In addition, it was well known from the assays on *Calliphora* that the sensitivity to ecdysoids was drastically reduced when the compounds were administered earlier during the last larval instar [2, 63, 66, 142, 219, 220, 227, 312, 341]. These results indicate that different doses of ecdysoids may be required for the same responses in the epidermal cells when injected at various moments of the intermoult cycle.

There exist several assumptions how to explain these changes in the sensitivity of epidermal cells to ecdysoids. One of them is based on possible accumulation of some covert biochemical and physiological changes which undergo spatial and temporal summation within the target organs and finally discharge the overt developmental responses [227, 339]. Other explanations take into account possible nutritional factors which may be necessary for stimulated cellular growth [59, 60, 61]. Finally, according to some assumptions the decreased sensitivity of epidermal cells to ecdysoids may be caused by endogenous juvenile hormone which opposes metamorphosis development [86, 210, 261]. All these explanations are reasonable. However, it may depend on species or stage in question which of these alternatives will be predominating.

It was observed by POSTLETHWAIT and SCHNEIDERMAN [237, 238] that the effect of ecdysoids on metamorphosis of *Drosophila* imaginal discs cultured *in vivo* was realized as a sustained stimulus but not as a trigger to metamorphosis. Similar relationships were found in metamorphosis of *Galleria* [260]. Furthermore, SEHNAL [261] working with the ligated last instar larvae of *Galleria* induced only partial proliferation of the imaginal discs with small amounts of ecdysoids. Increased doses caused stimulation of normal larval-pupal moult cycles. The responses were dependent on time of ligaturing and on the time interval between ligatures and injections. Larvae exposed to topical treatments by ecdysoids after prolonged time since the moment of ligation were reported to bypass secretion of the pupal cuticle and seemed to develop directly from larval to adult structures.

In other instances, however, the action of ecdysoids may actually resemble a trigger to certain physiological functions and, eventually, to metamorphosis. This seems to be the case with puparium formation in flies, with stimulation of metamorphosis in *Tenebrio* and *Dermestes* [281], and with metamorphosis in certain Hymenoptera [110]. DeWilde [325] mentioned several examples from unrelated insect species where the genome was set for two successive moults in response to a single wave of prothoracic gland hormone activity. However, with respect to diversity of developmental patterns associated with metamorphosis in various insect species it may be perhaps confusing to state that ecdysoids serve as a trigger for initiation of metamorphosis. Hence, even in species where ecdysoids induce a series of two or three moult cycles, the initiation of development requires that ecdysoids are present for few hours or for as long a period as the covert biochemical changes were brought to a certain critical stage. Once this critical moment has been traversed, all further developmental steps, including a complete second moult cycle seem to proceed independently on the presence of ecdysoids. This is regulated by specific genetic instructions.

Essentially, the above relationships were derived from the effects of ecdysoids on metamorphosis. The situation with respect to the critical sensitive periods is considerably different regarding precocious induction of the highly accelerated moult cycles. These can be induced in young larval instars, as well as in metamorphosis even at a time when stimulation of normal metamorphosis is difficult to achieve.

2. Physiological Conditions for Ecdysoid Action on Growth and Development

Long before the hormones were known, entomologists observed certain deviations in insect development which were associated with precocious metamorphosis. These deviations were called prothetely in contradistinction to other developmental deviations which were conversely associated with retention of larval patterns by the metamorphosis stages and were called metathetely [cf. 221]. Whereas the main physiological effects of juvenoids have been characterized by their ability to cause artificial metathetelies, the main physiological features of ecdysoids can be briefly characterized by their ability to cause an artificial prothetely. Wyatt [333] presented a list of physiological effects of ecdysoids in various insect species. Among some rather specific effects, the most usual developmental changes caused by ecdysoids were stimulation of the puparium formation in higher Diptera, and stimulation of the moult cycles and initiation of metamorphosis development in general. Most of the investigations were performed on ligated abdomens or on decapitated larvae which was the easiest way how to prevent interventions with the endogenous prothoracic gland hormone. However, these conditions were to some extent non-physiological. They allowed to study basic developmental processes in non-feeding or immobile metamorphosis stages, but they were not suitable for studying interactions between various physiological systems. Due to this fact, our knowledge on physiological action of ecdysoids is still incomplete when compared with relatively well elaborated biochemical studies.

Stimulation of development by ecdysoids is most apparent on epidermal tissue, which is generally assumed to be a chief target tissue for these hormones. This is well documented by induction of metamorphosis by ecdysoids in diapausing larvae and pupae, further by stimulation of growth and development in regenerating imaginal discs which was extensively studied by SCHNEIDERMAN and his associates [199, 200, 237, 238], and finally, it is documented by induced growth of insect tissue cultures and explanted epidermal organs as has been described in further sections. In addition to epidermis, various internal organs are also a subject of metamorphosis rearrangements in response to ecdysoids. This proceeds hand in hand with developmental changes in the epidermal organs. Some common examples of developmental responses induced by ecdysoids in internal organs are for instance: histogenesis of the thoracic indirect flight musculature [326], development and differentiation of gonads in certain species, and structural changes in the fat body cells [333], rearrangement and metamorphosis of the intestinal epithelium [259], structural changes in the salivary glands [236], and histogenesis of the adult nerve system [233]. Thus, in a full consent with the prothoracic gland hormone, ecdysoids induce the whole series of integrated morphological, biochemical, and physiological changes which are prerequisite for insect metamorphosis. It appears that there exists hardly a biochemical or physiological process in metamorphosis which will be unaffected by ecdysoids in one or other insect species.

The best illustration of these integrated ecdysoid effects on insect metamorphosis is provided by termination of larval or pupal diapause. This effect of ecdysoids represents also an excellent confirmation of the early WILLIAMS's studies on the conditioning of pupal diapause by a lack of the prothoracic gland hormone (see p. 59). It appeared possible to interrupt the diapause developmental arrest by injections of relatively small amounts of ecdysoids at almost any moment of the diapause period. This was extensively studied in diapausing pupae of Saturniids by WILLIAMS [326], OHTAKI and WILLIAMS [228], and by JUDY and GILBERT [134]. Similar termination of pupal diapause by ecdysoids was induced by FRAENKEL and HSIAO [65] in diapausing pupae of a Dipteran insect. Furthermore, ecdysoids can induce metamorphosis and so terminate diapause even in stages where the developmental arrest has occurred at the end of larval stage. For example, BECK and SHANE [13] induced development in diapausing larvae of *Ostrinia,* although a single dose of ecdysoids was not enough to bring the adult development to its end. HSIAO and HSIAO [110] terminated larval diapause by ecdysoids in a Hymenopteran, and SCHALLER [258] succeeded in termination of larval diapause in dragonflies. Finally, we may add that WRIGHT [329] reported termination of diapause by ecdysoids even in a tick.

One of the most complex and relatively little investigated physiological problems of ecdysoid action is concerned with the effects of ecdysoids on growing and feeding larval stages. This is associated with some technical difficulties involved in extirpations of the prothoracic glands from feeding larvae. It is well known that the moult cycles are under control of the prothoracic gland hormone even in these feeding young larval stages. However, unlike with metamorphosis development, and probably due to the technical problems there was no evidence for stimulation of a physiologically perfect moult cycle by ecdysoids in a feeding larval stage. THOMSON and HORN [312] observed that injections of ecdysoids into normal last instar larvae of *Calliphora* induced accelerated puparium formation but there appeared additional physiological changes which did not occur in normal development. Moreover, STAAL [cf. 286] performed injections of ecdysoids into feeding larvae of *Locusta.* He found that this was not followed by an accelerated moult but in contrast it resulted in actual delay of ecdysis, combined with an appearance of certain physiological disturbances. Similarly ROBBINS and his associates [249, 250] reported that feeding of various insect larvae with a diet containing ecdysoids resulted in an inhibition of growth, development and reproduction (see also CARLISLE and ELLIS, 29). These few data indicate that the physiological conditions for ecdysoid action in these feeding larval instars may be considerably different from the metamorphosis instars.

At present it is almost impossible to find some reasonable interpretation for the above atypical action of ecdysoids on the feeding stages. One fact which is worth to be considered in this respect is, that almost any feeding and growing stage (except last larval instar) contains in the body effective amounts of the corpus allatum hormone. Especially in Exopterygote insects all the developmental instars represent feeding stages. This may explain partly a lack of suitable assay system for ecdysoids on these insects which are otherwise very convenient for all kinds of other hormonal studies. For many years we tried to find some physiological conditions for possible assays of ecdysoids on certain Hemipterans. We did not succeed but accidentally we found certain effects of ecdysoids which seem to be specific for feeding stages. The results have been briefly summarized below.

Most of these investigations were performed with the last instar larvae of *Pyrrhocoris*. These larvae required at least 20hrs of feeding in order to undergo metamorphosis and to moult into small adults. Larvae which were starved since the moment of ecdysis were unable to develop into adults. When injected with doses from 0.01 to 10 μg of ecdysoids the starved larvae did not develop either. But when excessive amounts of ecdysoids (approx. 100 μg) were administered into the body cavity of the starved larvae, there appeared partial moults of the epidermis mainly on the abdomen, reduced mobility and high mortality. Repeated injections of ecdysoids into feeding last instar larvae resulted in slight acceleration of ecdysis but no prothetelic forms were produced. Administration of large doses of ecdysoids to feeding larvae caused again only partial moults associated with suppressed feeding and subsequent inability of adults to shedd off the old larval cuticle. Finally, we managed to perform ectomies of the prothoracic glands in freshly ecdysed last instar larvae. These larvae were able to feed and grow normally, however, they continued to do so even at the time when all the control specimens moulted into adults. Their body became highly hypertrophic like that of the allatectomized females, they survived for long periods without any signs of the moult cycle in the epidermis. When injected with relatively small amounts of ecdysterone (0.1 μg) they underwent normal metamorphosis and moulted into adults, although these doses were completely ineffective in the starved larvae. This indicates that there exist close correlations and synchronizations between responses of epidermal cells to ecdysoids and the nutritive conditions in these feeding larval stages. The whole situation is analogous with the action of juvenoids on oocyte growth, e.g. females with completely depleted fat bodies did not develop ripe eggs in response to juvenoid stimulation.

The observed hypertrophy in larvae deprived of their prothoracic glands suggests also that feeding, digestion and food utilization, which always proceeds more or less concurrently with the moult cycles in normal development, are not under the direct control of the prothoracic gland. This was recognized earlier by WIGGLESWORTH [324] who found that development of the fat body cells in the moult cycle of *Rhodnius* was stimulated by feeding whereas stimulation of development in the epidermal cells required both nutrition and ecdysone.

In ontogeny of insects like in all animals there appear certain temporary structures which are broken down when their functions are no more required. The death of these structures is genetically programmed and the stimulus for their breakdown is provided by some internal environmental stimuli which are usually hormonal [257]. In highly specialized Endopterygote groups of insects there are many "larval" cells, tissues, or even whole organs which always disintegrate at the onset of metamorphosis. This process of histolysis is initimately bound to histogenesis of the adult organs. As a result of general inhibition of metamorphosis it can be suppressed by juvenoids. Conversely, as a result of general stimulation of metamorphosis the histolysis process is stimulated by ecdysoids. Unfortunately, most of the experimental evidence concerning the effects of ecdysoids on metamorphosis is based on histogenesis and differentiation of the adult organs. For a long time it was assumed that disintegration of the old "larval" tissues was a secondary feature conditioned by growth of the other tissues. It was rather difficult to comprehend that the same compound which stimulated intensive growth and mitotic activity in some tissues could be simultaneously used as a "death message" for other tissues.

It appeared that the breakdown of intersegmental muscles in certain adult insects was potentiated by hormonal milieu, although the main stimulus came from the central nervous system [195]. From a physiological point of view, the breakdown of the "larval" cells in insect metamorphosis appears to be an active process which brings about requirements for increased synthesis of a whole series of new hydrolytic enzymes, such as acid phosphatase, ribonucleases, esterases, cathepsins, glucuronidases, etc. The autolytic process is associated at the beginning with an increased synthesis of RNA and protein [194] before the cells start to disintegrate. Evidence that ecdysoids induce such activation of a "death message" on the genome is provided by observations of Radford and Misch [240] on intestinal tissue of *Sarcophaga*. Here injections of ecdysterone at the beginning of the last larval instar caused remarkable increase in the amount of secondary lysosomes with acid phosphatase activity. This normally occurred much later, at the beginning of metamorphosis. Further evidence for stimulation of the cell death by ecdysoids is provided by an induced breakdown of intersegmental muscles in *Calliphora* [45] and *Lucilia* [10]. These results indicate that the responses of insect tissues to ecdysoids can be as different as intensive growth and development of the cells on one hand and, disintegration and death of the cells on the other hand. It may be assumed that factors which determine these responses are related to genetical programming of morphogenesis in individual target cells.

3. The Role of Ecdysoids in Puparium Formation

Extensive studies of Karlson and his co-workers [144, 148, 161] revealed that biochemical changes caused by ecdysoids during puparium formation in *Calliphora* could be explained in the light of current theories on gene activation. Hence, ecdysone was assumed to cause a derepression of particular genes. This was followed by formation of specific mRNA required for synthesis of dopa-decarboxylase—a crucial enzyme in tyrosine metabolism which was involved in the process of tanning and sclerotization of the puparium (p. 329). In their article concerning the mode of action of ecdysone, KARLSON and SEKERIS [161] claimed that ecdysone was a hormone whose mechanism of action was best understood, and for which an influence on genetic material was best documented. However, it seems that the general mechanism of ecdysone action remains still incompletely known even with respect to puparium formation. For example, FRAENKEL and HSIAO [64] have clearly demonstrated that the above biochemical changes found in *Calliphora* cannot be regarded as a common principle for ecdysone action on puparium formation. They discovered that puparium in *Musca autumnalis* did not sclerotize on account of phenolic tanning but due to calcification of the old larval cuticle. Evidently, this process required different biochemical pathways but still it was stimulated by ecdysoids in the same way as in *Calliphora*.

Further physiological studies revealed that the process of puparium formation in flies was not as simple as it came out from the experiments with ecdysone on isolated larval abdomens. Thus, THOMSON and HORN [312] caused by large doses of ecdysoids highly accelerated pupariation without contraction to puparium. On the other hand, BERREUR and FRAENKEL [8] stimulated the contraction with small amounts of ecdysoids. Acceleration of the puparium formation process was achieved not only by ecdysoids but also as a result of electrical irritation or by scorpion venom [67]. The findings by OHTAKI *et al.* [225, 226, 227], ZDAREK and FRAENKEL [339], and THOMSON *et al.* [313] on inhibition of pupariation due to contact of larvae with moisture indicated that the release of the endogenous hormone which stimulated puparium formation was inhibited by certain internal stimuli. In analogy with regulation of the prothoracic gland activity in other species of insects it was expected that these interactions might depend on the activity of the neurosecretory system. Hence, ZDAREK and FRAENKEL [338] concluded that the neurosecretory system might potentiate effectiveness of ecdysone in puparium formation. Significance of the neurosecretory system was ascribed among other factors to possible synchronization of puparial contractions with the tanning process. However, they established later [340] that the neurosecretory system influenced puparium formation through its control over the release of the endogenous pupariating hormone from the ring gland. In more recent papers these authors have described a whole complex of physiological factors which may be involved in the process of puparium formation in flies [66, 341]. When Dipteran larvae were ligated for ecdysoid assays it sometimes happened that puriation did not occur in the anterior part as usual, but it occurred in the posterior part while the front end remained larval. This was explained by PRICE [239] as being due to some pupariation inhibitors which were present in the anterior parts of the ligated larvae.

Physiologically, the process of puparium formation represents only a terminal step of the intermoult cycle in the last larval instar of Diptera-Cyclorrhapha. It is preceded by relatively long period of feeding and development and it is most likely that the reprogramming of the genome for metamorphosis occurs in the last larval instar long before the stimulus to puparium formation has been released. It occurs that the intermoult cycle in the last instar larvae of many Endopterygote insects, including Diptera, is regulated by two successive waves of prothoracic gland hormone or ecdysoid activity. The first occurs soon after ecdysis. It is associated with initiation of feeding and initiation of the moult cycle as a whole. This period is sensitive to juvenile hormone and it is characterized by progressive accumulation of the covert biochemical and physiological changes. The second wave of the hormonal activity is then responsible for realization of pupation or puparium formation respectively. Many authors working in the competitive field of ecdysoid action on Dipteran larvae tried to induce precocious larval moults. They injected ecdysoids before the critical sensitive period for pupariation. However, the result was only precocious puparium formation or no effects whatsoever [312]. Finally, ZDAREK and SLÁMA [342] realized that due to the first wave of prothoracic gland hormone activity the developmental programming on the genome might be switched over from larval to metamorphosis programme, e.g. for puparium formation, very early in the last larval instar. They injected large amounts (10 μg per spec.) of ecdysterone into freshly ecdysed or starved last instar larvae of *Calliphora* and *Sarcophaga* and for the first time they obtained supernumerary larvae in this group of insects. They confirmed that determination of the development to puparium formation proceeds as early as 6 to 12 hrs. after ecdysis in the last instar larvae. Injections performed during this critical period caused formation of larva-puparium intermediates with mosaics of pupariated spots in the epidermis. Moreover, all injections performed later in the last instar caused only accelerated pupariation. These studies demonstrated that ecdysoids were used in last larval instars of these insects for two different things, e.g. for initiation of the moult cycle connected with reprogramming of the developmental instructions and, for stimulation of the puparium formation later in the instar.

4. Excessive versus Small Doses and the Problem of Differentiation

It is known that hormones which stimulate developmental processes are effective in certain threshold concentrations. Further increase of the hormonal concentration beyond the effective level does not cause, as a rule, any qualitative developmental changes. We have seen such an example of hormonal action among juvenoids. However, the action of ecdysoids on insect development is different. Small doses cause qualitatively different developmental responses than large ones. This results most probably from specific regulations of insect development.

Large doses of ecdysoids induce precocious moults connected with an appearance of prothetelic forms. These have been induced also by means of prothoracic gland transplantations but never in such an extent as with the large doses of ecdysoids. It appears that the moult cycle is accelerated by this treatment to such a degree that there occur physiological disturbances in secretion of the new cuticle and failures in the ability of pharate stages to digest properly the old cuticle. Furthermore, there occur some pathophysiological symptoms such as asynchronous ecdysial movements, suppression of feeding in larvae, suppression of locomotory activity and incomplete differentiation of adult organs. These effects were observed in many unrelated insects, most of them being Endopterygote. Thus, WILLIAMS [326] described the symptoms of "hyperecdysonism" in more details in pupae of *Samia*. They were commonly observed in other Lepidoptera such as *Hyalophora* [134], *Bombyx* [171, 210, 254], and *Galleria* [200, 260]. In addition, the signs of "hyperecdysonism" were found in last larval instars of Diptera [235, 312, 342], in Hymenoptera [110], and in Coleoptera [282].

The synthesis of new cuticle as well as the digestion of the old cuticle during insect moult are enzyme dependent processes which cannot be accelerated beyond certain limits dependent on reactions in intermediary metabolism. This may partly explain the above described defects in secretion and digestion of the respective cuticles during the highly accelerated moults caused by ecdysoids. One of the most characteristic feature of the "hyperecdysonism" is a formation of morphological characters which are intermediate between two developmental stages. This degree of prothetely is used as a criterion for distinction between "hyperecdysonism" and normal development stimulated by small doses of ecdysoids in insect metamorphosis. In addition, it appears that these prothetelic forms are to some extent similar with the intermediates caused by juvenoids. For example, the scaleless prothetelic adults had many morphological features in common with adultoids produced by juvenoid treatments [134, 210, 326]. Further similarities between prothetelic and metathetelic forms were observed in extra-pupae of *Tenebrio* [282], or extra-larvae of *Calliphora* [342]. However, a more detailed physiological and morphological analysis enables to make a clear distinction between both these developmental anomalies [282]. According to a general occurrence of prothetelies caused by ecdysoids in pupae we may expect that these effects were not dependent on endogenous juvenile hormone, as it was assumed by some authors [210]. More probably, as suggested by WYATT [333], in the case of "hyperecdysonism" the epidermis was forced to moult with such a rapidity that certain normal events (perhaps requiring genetic reprogramming) were by-passed.

Fig. 33. Series of intermediate forms between larva and pupa caused by injections of large amounts (10–20 μg) of ecdysterone at various time intervals of the last larval instar in *Dermestes vulpinus*. All specimens moulted 3 days after injection. The last larval cuticle was artificially removed. From upper left: freshly moulted normal last instar larva; almost perfect extra-larva, injected 0 to 20 hours after ecdysis; larval intermediate, inj. 1st day; larval intermediate retaining larval pigmentation with partly everted wing lobes, inj. 2nd day; larval-pupal intermediate with a mixture of larval and pupal patterns, inj. 3rd day; pupal intermediate with very few larval patterns and with some pupal patterns such as dorsal abdominal gintraps and fully everted wing lobes, inj. 4th day; almost perfect pupa, inj. 6th day (lower right). Usual length of the last larval instar was 10 days

In *Dermestes vulpinus* [281] the young larval instars moult in 2 to 3-day intervals. The last, 6th larval instar pupates after 10 days. Injections of large doses of ecdysterone in successive intervals of the last larval instar caused always a moult in 2 to 3 days. This was the normal duration of the previous larval-larval instars and it appeared impossible to accelerate moult beyond this limit by any amount of ecdysoids. Moreover, morphological forms of the stages produced by injections of ecdysoids at different time of the last larval instar were characterized by successive decrease of the larval epidermal patterns and by reciprocal increase of the pupal epidermal patterns in dependence on the time of injection, as shown in Fig. 33. This demonstrated that a continuous reprogramming of the developmental instructions took place throughout the first half-period of the last larval instar. Large doses of ecdysoids actually stabilized the existing situation by transforming it into the form of cuticular secretion.

One might speculate that perhaps, before the reprogramming process was initiated, an insect stage could produce several larval or pupal cuticles without morphogenetic changes when large doses of ecdysoids were applied in absence of juvenile hormone. In other words, that extreme "hyperecdysonism" might cause rapid succession of moults without differentiation even in absence of juvenile hormone. We have investigated this hypothesis on isolated abdominal fragments of freshly ecdysed last instar larvae of *Dermestes*. In contrast to most other insects, these abdomens develop spontaneously into pupal abdomens in 10 days and form also adult abdomens in another 7 days. Injections of 20 μg of ecdysterone into these abdomens induced several successive precocious moults in 2 to 3-day intervals. It appeared, however, that each of the successive cuticles contained gradually less larval and more pupal patterns. After the 3rd moult which took place approximately at the same time as the pupal moult in the controls, there was also a pupal cuticle. Thus, we have divided the larval-pupal transformation (which normally occurs in one instar) into three to four successive instars by one large dose of ecdysterone, and even under these conditions the process of metamorphosis was not completely inhibited. It remains to be noted that the abdomens which were injected with ecdysterone simultaneously with a topical treatment by juvenoids produced also several cuticles but all of them had the same morphological type.

According to WILLIAMS [326] the abnormal development associated with "hyperecdysonism" was not restricted to epidermis. It was remarkable also in the fat body, gonads, midgut, and in some other internal organs. The fact that spontaneous prothetelies are very rare in nature suggests that if any of the so far known ecdysoid were identical with the prothoracic gland hormone then its concentration in the body must have been strictly regulated to certain limits preventing spontaneous "hyperecdysonism".

When speaking about low doses of ecdysoids we have referred to certain minimum amounts of these compounds which are necessary for stimulation of metamorphosis. This treatment is usually followed by developmental features which are exactly similar to developmental changes taking place during the normal moult cycles. For example, injections of small doses of ecdysoids induced development with exactly the same timing of apolysis, initiation of mitoses, deposition of the new cuticle, pigmentation, and adult emergence, as found in normal postdiapause development or in metamorphosis [326]. Gradual increase of these low doses was followed by proportional acceleration of all these developmental events. With still increased doses there appeared increased prothetelic effects which terminated in real "hyperecdysonism". In this insect, e.g. *Samia*, the difference between large doses which caused precocious moult and the low doses was relatively small. But in *Dermestes* [281] this difference was 50 to 100-fold change in concentration depending on the ecdysoid compound used.

It is accepted by tradition that the prothoracic gland hormone or ecdysoids stimulate differentiation in absence of juvenile hormone. The process of cellular differentiation is most intensive during morphogenesis as seen in embryonic development and metamorphosis. The prothoracic gland hormone was once called a "growth and differentiation" hormone according to its effects on metamorphosis. We have seen, however, that with respect to ecdysoids the degree of differentiation is inversely proportional to the dose applied. In other words, increased doses of ecdysoids induce formation of less differentiated prothetelic forms. This suggests that under certain circumstances ecdysoids may induce accelerated moults without differentiation. Such a lack of differentiation appears also with respect to ecdysoid action on regeneration [23], and in connexion with neogenesis of certain pigments [22].

Insect development represents a special case of sequential polymorphism where the form of the body is determined mainly by the type and shape of the cuticle. In Endopterygote insects the larval-pupal ecdysis occurs at different morphogenetic stages. For example, in Lepidoptera and Diptera it occurs at the very beginning of metamorphosis, in Trichoptera and Neuroptera it is placed in about the middle, whereas in Hymenoptera it occurs at the end of metamorphosis. This shows that a moult can occur practically at any stage of differentiation, which is experimentally confirmed by results in Fig. 33. With these comparative aspects in mind we have concluded for instance, that the scaleless adults produced artificially by Williams [326] with ecdysoids in Lepidopteran pupae have been morphologically equivalent to a normal free pupa of a Hymenopteran.

During the young larval period when juvenile hormone is present, ecdysoids induce development and moult cycles but the character and amount of differentiation is completely dependent on juvenile hormone.

Furthermore it appears that at certain stage of ontogenetic development the cells may require ecdysoids for growth and differentiation, e.g. metamorphosis, whereas at some other stages such as in embryonic development ecdysoids are not required for differentiation. It thus occurs that ecdysoids have no direct control over stimulation of cellular differentiation, similarly like juvenoids have no direct inhibiting effect in this process. The hormones are used as specific messengers for selection and activation of certain alternative developmental programmes which are coded on and accomplished by the genome of each cell.

5. Effect of Ecdysoids on Reproduction

In agreement with the action of prothoracic gland hormone, edysoids stimulate metamorphosis in gonads of both sexes. In addition, there are certain insects where ecdysoids not only stimulate metamorphosis of the ovaries, but actually stimulate the whole process of ovarian growth including vitellogenesis. This occurs in certain species of Lepidoptera, Coleoptera, Diptera and Hymonoptera (see p. 37) where vitellogenesis proceeds independently from juvenile hormone. Thus, for instance Williams [326] observed stimulation of ovarian development by ecdysoids in *Samia*, Sakurai and Hasegawa [254] in *Bombyx*, Laverdure [188] in *Tenebrio*, and Spielman *et al.* [285] in *Aedes*. Ecdysoids stimulated vitellogenesis even in *in vitro* incubated ovaries [188].

Other reports suggested that ecdysoids suppressed the process of vitellogenesis in the ovaries of some insect species where this process was normally under control of juvenile hormone. However, it is difficult to interpret these findings by Robbins *et al.* [249, 250], Wright *et al.* [330, 331], and Earle *et al.* [50], on inhibition of ovarian growth by ecdysoids because of possible interferrence with suppressed feeding, side effects produced by overdosage, and the like. Interesting observations were made by Ohnishi *et al.* [224] who found the presence of ecdysoid activity in the eggs of *Bombyx*. The activity was correlated with the course of embryonic development and diapause.

6. The Content, Distribution, and Inhibition of Ecdysoids in the Body

Studies on relative content of ecdysoids in various stages and species of insects were facilitated by improved methods [158] for determination of ecdysoid content in tissue and body extracts. Relatively high ecdysone activity was found in extracts from grasshopper larvae and in their excrements by Hoffmeister *et al.* [104]. Moreover, Shaaya *et al.* [272, 273, 275] measured changes in ecdysone activity during development of *Calliphora*. They reported rather low activity in young larvae; a peak in the activity occurred several hours before puparium formation, while another peak was present in the pupal stage. Similar pattern in these ecdysone titers was later found in development of certain other Dipteran species, such as *Calliphora vicina* [20] and *Lucilia* [9].

Further it appeared that ecdysone activity underwent considerable variations in extracts from larvae and pupae of Lepidoptera and of other insects [24]. So, Shaaya and Karlson [274] found relatively low activity at the beginning of the last larval instar in *Bombyx*; there was a peak of activity in the prepupal period, a decrease of activity towards pupation, and the activity increased again in the pupal stage. Also Kaplanis *et al.* [141] observed relatively low ecdysone activity at the moment of pupation or adult ecdysis whereas there was a peak of the activity at the beginning of pupal stage in *Manduca*. Finally, in a Hemipteran *Oncopeltus*, Feir and Winkler [58] found a low ecdysone activity at the time of larval and adult ecdysis. The activity was relatively constant throughout the last larval instar and definite ecdysone activity was present also in adult stage. Most of these data on biological activity of ecdysoids in the whole body extracts indicate very low activity at the moment of ecdysis. The peaks of the activity occur either before, to stimulate pupation or puparium formation, or after ecdysis which is probably associated with stimulation of a new moult cycle. These findings provide also an indirect evidence that ecdysoids are not engaged in stimulation of the process of ecdysis, which is apparently controled by the neurosecretory system.

Endogenous ecdysoids may be rapidly inactivated in the body of certain insect stages. Shaaya [272] demonstrated inactivation of exogenous ecdysone in *Calliphora*. The biological half-life of this compound was only about 3 hrs. In addition, Ohtaki and Williams [227, 228] discovered that a velocity of such biological inactivation may be different for different ecdysoids. For example, the half-life of ecdysone was about 1 hr. whereas that of cyasterone was 3 to 4 hrs. in larvae of *Sarcophaga*. They concluded that the inactivation was a temperature dependent enzymatic process. This was confirmed by Karlson and Bode [150] who determined the activity of the ecdysone inactivating enzyme and found that its course during development of *Calliphora* was inversely proportional to the course of ecdyson content curves. This contrasted with changes in the activity of certain enzymes of tyrosine metabolism which were parallel to ecdysone titers [275]. It is still unknown how widespread this phenomenon of rapid biological inactivation of exogenous ecdysoids might be with respect to other insect species. However, evidence that it is not an universal feature is provided by already mentioned observations that ligated abdomens of *Dermestes* can undergo three successive precocious moults in response to a single injection of ecdysoids.

Initial studies on distribution of ecdysoids in the body were based on biological determination of the activity in tissue or haemolymph extracts. Thus, SHAAYA [272] observed the following relative changes in ecdysoid activity in *Calliphora*: i) haemolymph had relatively constant activity from 20 hrs. before to 5 hrs. after pupariation; ii) in the integument the activity appeared 8 to 5 hrs. before pupariation with a maximum of activity 5 hrs. thereafter; and, iii) fat body had low activity until 5 hrs. after pupariation when the activity sharply increased. This suggested that at a given moment various tissues or haemolymph may provide extracts with relatively different ecdysoid activity. In *Tenebrio*, ROMER [252] obtained extracts from oenocytes which had higher ecdysone activity than similar extracts prepared from the prothoracic glands.

Autoradiographic techniques using tritium labelled ecdysoids did not yield conclusive results on intracellular localization of these hormones, so far. WEIRICH and KARLSON [321] were unable to detect any specific chromosomal or nucleo-cytoplasmic accumulation of the label in the salivary gland cells of *Rhynchosciara* and *Chironomus*. However, slight degree of accumulation of the label at first in the nuclei (10 hr. prior to pupariation) and then in the cytoplasm (immediately before pupariation) were observed in the salivary gland cells of *Drosophila* by EMMERICH [52, 53]. Finally, THOMSON *et al.* [317] found increased binding of tritiated ecdysterone with cell membranes in the salivary glands of *Calliphora*, but did not report any localized intracellular concentration of the label. In *Antheraea*, CHERBAS and CHERBAS [31] recorded equilibrium in the distribution of ecdysone after 2 hrs. following the injection. There was no accumulation of ecdysone in any tissue nor in the excretory organs.

With regard to transport of ecdysoids, EMMERICH reported binding of labelled ecdysone with certain haemolymph proteins in *Pyrrhocoris* [55] and in *Drosophila* [54]. He assumed a presence of some temperature dependent system which might be involved in transport of ecdysone from cytoplasm to nuclei. In *Antherea*, however, the transport of ecdysone did not seem to require any rate-limiting carrier system [31]. In addition, CHINO *et al.* [32] did not find binding of labelled ecdysone with major haemolymph lipoproteins in *Philosamia*. So, it may appear that a binding of ecdysoids with protein carriers is not a necessary condition for transport of these compounds in insect body, which may be due to their considerable polarity or water solubility.

7. Developmental Relationships between Ecdysoid and Juvenoid Actions

We have already indicated (p. 260) existence of some assumptions concerning possible antagonistic action between ecdysoids and juvenoids. These views were mostly derived from investigations on some specific biochemical and biophysical effects, such as specific changes in metabolism of nucleic acids, changes in intracellular ionic balance, specific effects on puffing pattern, or antagonistic effects on growth of tissue cultures. With respect to developmental physiology, however, these hormonal agents cannot be regarded as antagonistic because each of them has different physiological functions. Ecdysoids give a general stimulus for cellular growth, multiplication and morphogenesis. Juvenoids are unable to perform these activities in immature insect stages, with one exception of prothoracotropic activity in diapausing pupae of Saturniids. A principal physiological role of juvenoids is their inhibition of the morphogenetic progress during the development induced by ecdysoids. We have seen in the initial part of the book that the same relationships apply to functions of the endogenous corpus allatum and prothoracic gland hormones. It occurs that due to some evolutionary adaptations certain Endopterygote insects reach a stage in their ontogeny where tissues and organs have no alternative developmental programme but metamorphosis into adult structures. In this case inhibition of morphogenesis by juvenoids results in a general inhibition of development stimulated by ecdysoids. And, one can easily find out that most of the assumptions on the antagonistic action between ecdysoids and juvenoids were derived from investigations on these systems.

In Exopterygote insects where metamorphosis represents a relatively simple process, juvenoids never inhibit developmental cycles stimulated by ecdysoids. In contrast, both the hormonal agents act in concert. They do not seem to compete for the same receptors. However, it appears that due to diversity of insect species and developmental habits, the genetic programming for induction of certain identical physiological function may recognize ecdysoids as the stimulus in some species while taking juvenoids as the stimulus in other species. This is exemplified by stimulation of ovarian development in certain insects by ecdysoids whereas in majority of other insect species it is tuned to juvenoids (p. 238).

With respect to these evolutionary adaptations in various species we can find numerous examples of seemingly synergistic and antagonistic actions between ecdysoids and juvenoids. This results from altered growth conditions but not from interactions between the hormones at the molecular level. For example, juvenoids cause acceleration of ecdysis in last larval instars of Exopterygotes and in the pupal instars. Superficially, this resembles a synergistic effect with ecdysoids, as may come out also from morphological similarities between prothetelic forms induced by ecdysoids and metathetelic forms induced by juvenoids. As an example of reciprocal, antagonistic action of juvenoids we may recall an inhibition of pupation which is caused by juvenoids in the last instar larvae of certain Endopterygote species. Nevertheless, these and other examples suggest that not really antagonistic, but all kinds of other, sometimes very specific developmental interactions between ecdysoids and juvenoids may exist.

8. Effects of Ecdysoids on Metabolism

Throughout the foregoing sections we have seen that ecdysoids stimulate growth and development in the epidermal cells of larval or pupal stages of insects. It is clear that the "activated" target cells or tissues require an increased amount of organic substrate for development of the internal cellular structures. Or, they need a source of energy for performing specific physiological functions to which they were subjected. This is generally associated with more or less pronounced metabolic changes. Such an availability of suitable substrates for tissue growth may be a rate-limiting factor in metabolic responses of the target cells to ecdysoids since it is known that cells or tissues cannot grow in absence of substrates for synthesis of the essential macromolecules. In general, metabolic differences induced by ecdysoids in their target cells can be regarded as differences between basal or maintenance metabolism on one hand and functional cellular metabolism on the other hand. However, development of any living cell whether stimulated by hormones, nutritional factors alone, injury, or other stimuli, is always accompanied with numerous nonspecific metabolic reactions. We have concentrated our discussions concerning metabolic consequences of ecdysoid action to changes in total body metabolism, nucleic acid metabolism, and protein synthesis.

Long before insect hormones were known it was observed that the moult cycles in insects were characterized by specific patterns in total metabolism. Later it appeared that each of the larval instars and also each of the metamorphosis instars were characterized by special course of oxygen consumption curves. When development was arrested, such as in diapause, oxygen consumption decreased to very low levels, but the typical pattern in total metabolism was again resumed with the induced development. Apparently, intensity of the total body metabolism was closely correlated with growth and developmental changes (cf. p. 256). Since ecdysoids induce development and stimulate the moult cycles in insect metamorphosis, it is highly probable that their action must be followed by changes in total body metabolism, which are always an aoutcome of certain developmental alterations. However, a direct experimental evidence to show the effects of ecdysoids on oxygen consumption in insects is still lacking, as the research on ecdysoids was mostly concerned with ligated larvae or isolated abdominal body fragments. Some general problems of ecdysoid action on insect metabolism have been discussed by WYATT [333].

More attention has been paid to ecdysoid action on RNA metabolism. Two favorite groups of insects, e.g. Diptera and Lepidoptera, have been used in these studies. In Diptera, KARLSON and PETERS [157] recorded stimulatory action of ecdysone on RNA synthesis in *Calliphora* soon after ecdysone isolation from *Bombyx* pupae. In further studies KARLSON and his associates [264, 270] demonstrated effectiveness of ecdysone on RNA synthesis in isolated nuclei of the epidermal cells. It was later confirmed that ecdysone or ecdysterone stimulated RNA synthesis in larvae of *Musca* [169], in specific tissues such as fat body of *Calliphora* [219], and also in *in vitro* incubated imaginal discs of *Drosophila* [241].

In Lepidoptera, correlations were found between ecdysone activity and some specific patterns in RNA synthesis in pupae of *Antheraea* by BARTH *et al.* [11]. Furthermore, BERRY et al. [18] extended these observations to various tissues throughout development in Saturniidae. They also observed correlations between temporal patterns in RNA synthesis and ecdysone secretion, though not in all the tissues. However, an increased RNA synthesis rate, which was similar to that induced by ecdysone, was also stimulated by a simple mechanical injury to diapausing pupae. This suggested that an increase of total RNA synthesis rate was not a specific feature which would be completely dependent on ecdysone. SAHOTA and MANSINGH [253] demonstrated that a nonspecific increase in RNA synthesis was stimulated by ecdysoids in several larval tissues in *Antheraea*, e.g. epidermis, fat body cells, silk gland cells, and in the midgut. The rise of RNA synthetic rate was temporary and occurred mainly in the nuclei. It appeared that ecdysoids caused a temporary enhancement of total RNA synthesis also in the endocrine glands, such as in corpora allata and in prothoracic glands of diapausing Saturniid pupae [83, 280]. Based on these findings assumptions were made on possible feedback effects between ecdysoids and functions of the endocrine glands.

From the above data it is clear that an increase in the total RNA synthetic rate is a general consequence of ecdysoid action on their target cells, although it may be a nonspecific feature prerequisite to any cell which was induced to develop. Several authors tried to find out whether there were some rather specific changes in RNA metabolism attributed to ecdysoid actions. We have mentioned already that KARLSON and his associates recorded induction of specific mRNA by ecdysone which was responsible for synthesis of specific enzymes of tyrosin emetabolism in the epidermal cells (p. 329, 348). However, this does not seem to be relevant to action of ecdysone on other than epidermal tissues. KARLSON *et al.* [151] working with isolated fat body nuclei of *Calliphora* concluded on basis of the nucleic acid hybridizaton that ecdysoids, as well as juvenoids, gave rise to new mRNA molecules by activating specific gene sets. Moreover, Wyatt [332] recorded stimulation of all kinds of RNA molecules early in the course of ecdysone action. Fractions extracted with hot phenol and dodecylsulphate were assayed for "template" activity in an *Escherichia coli* system. No evidence for a real template activity was obtained. Lucas and Wyatt [197] further extended these studies on hybridization capacities of RNA fractions taken from developing tissues on one hand and diapausing tissues of Cecropia pupae on the other hand. They did not obtain visible difference in hybridization abilities between both physiologically different developmental stages, indicating that a possible role of the hybridizing RNA in regulations of cell synthetic activities might be limited [333].

Pronounced correlations also exist between ecdysoids and DNA replication. We have seen throughout the text that ecdysoids stimulate histogenesis of adult tissues during insect metamorphosis, which is always associated with extensive mitotic divisions. Even in larval instars the most apparent consequence of ecdysoid action is endomitosis or mitosis, at least in the epidermal cells. It appears, therefore, that DNA replication is involved in all major developmental effects of ecdysoids. It remains to be discussed whether an increased rate of DNA synthesis is an obligatory step in ecdysoid action, and what are the chances in favour of an assumption that a mode of action of ecdysoids might depend in induced DNA synthesis.

In their extensive studies on Saturniid insects, Krishnakumaran *et al.* [176, 177] found definite temporal patterns of DNA synthesis in several tissues during the larval moult cycles and during metamorphosis. An absence of DNA synthesis, such as occurred during diapause, was correlated with absence of ecdysone and renewed DNA synthesis could be induced by ecdysone. During the larval-larval moult cycles, DNA synthesis occurred soon after secretion of ecdysone in certain tissues whereas it was not correlated with ecdysone secretion in some other tissues. In the course of these larval-larval moult cycles almost every epidermal cell synthesized DNA. However, after experiencing a moult cycle in an absence of juvenile hormone, the tissues differentiated and ordinarily did not synthesize DNA again. An important observation was that in the case of experimentally induced extra-adult moult in an adult silkmoth the epidermal cells did not synthesize DNA. The authors thus concluded that stimulation of DNA was not a primary effect of ecdysone. This conclusion was subsequently confirmed by Madhavan and Schneiderman [199] who were able to induce apolysis even in epidermal cells in which DNA synthesis had been previously blocked by mitomycine C.

The above conclusion [177] that DNA synthesis was not an obligatory step in ecdysoid action was further corroborated by studies on puparium formation in flies. For example, KOBAYASHI and AKAI [169] demonstrated that stimulation of puparium formation in *Musca* by ecdysoids was not associated with an increased incorporation of tritiated thymidine into epidermal cells or other tissues. All these data seem to indicate that the action of ecdysoids may include DNA synthesis in larval moult cycles or in metamorphosis, e.g. in systems which were genetically programmed to respond by cell or chromosomal multiplication. Conversely, the action of ecdysoids may not inlude an increased DNA synthesis in systems which were programmed for some specific physiological function, such as formation of puparium, or secretion of a new cuticle without morphogenetic changes. These differences in DNA metabolism with respect to ecdysoids were observed also on the polytenic chromosomes. We shall see later (p. 364), that ecdysoids induce specific puffs on these chromosomes which are usually the sites of increased RNA synthesis. In certain species of Diptera ecdysoids induced puffs on the chromosomes which depended on an increased DNA synthesis.

According to OBERLANDER [222] ecdysone induced DNA synthesis in the isolated wing imaginal discs of *Galleria*. The induced DNA synthesis was inhibited by ecdysterone or 22-isoecdysone, suggesting that ecdysone and ecdysterone might possibly have different functions even under physiological conditions. This interpretation is difficult to understand because in many unrelated insects both ecdysone and ecdysterone exhibit quite similar action on development and moult cycles. In addition, it was found that the pupal wing epithelium of other insects was conversely more sensitive to ecdysterone than to ecdysone when incubated *in vitro* [334].

Following KARLSON and SEKERIS [161] findings on induction of specific enzyme syntheses by ecdysone, many authors confirmed general proteosynthesis stimulating effects of ecdysoids in last instar larvae of Diptera. Working with larval fat body of *Calliphora*, NEUFELD *et al.* [219], ARKING and SHAAYA [5], and THOMSON *et al.* [314] demonstrated increased incorporation of animo acids into proteins of this tissue after addition of ecdysone. On the other hand, in *Musca* [169] there were no relationships between the rate of incorporation of tritiated leucine with respect to ecdysone presence. Also in other insects than Dipterans, such as in *Bombyx*, SHIGEMATSU and MORIYAMA [276] reported increased synthesis of proteins in the silk glands after peroral application of ecdysterone. In *Calpodes* ecdysone induced specific changes in proteosynthetic activities of the fat body cells (COLLINS, 44).

An increased protein synthesis is also rather nonspecific feature common to any cell which was induced to grow and develop. It is quite obvious that it must occur as a consequence of growth in all tissues where ecdysoids induce cell multiplication and histogenesis. However, it is also clear that ecdysoids are not essential factors for an intensive proteosynthetic activities in all the insect tissues. For example, in larvae of *Rhodnius* ecdysoids induced growth and increased proteosynthesis in epidermal cells whereas the fat body cells did not require ecdysoids for stimulation of these activites [324]. Similar conditions exist in metamorphosis of Cecropia pupae [333]. It is well known that tissues such as intestinal epithelium, fat body, and ovaries of adult females synthesize enormous emounts of proteins in a complete absence of the prothoracic gland hormone or ecdysoids. Depending on special developmental regulations ecdysoids may actually inhibit synthesis of certain proteins which also suggests their indirect effect on proteosynthesis. Thus, synthesis of muclopolysaccharide globules in the salivary glands of *Drosophila* was inhibited by application of ecdysterone to the early third instar larvae [236]. With most probability, the nature of proteins to be synthesized in the cells stimulated to grow by ecdysoids will depend on specific developmental programmes in each of the particular target tissues. Apart from a large number of nonspecific proteins, such as mitochondrial cytochrome enzymes for instance, each target cell may be programmed to synthesize specific tissue proteins, such as dopa-decarboxylase in the epidermal cells of *Calliphora* [161], sericine in the silk glands [276], lipoproteins in the fat body [44], and so on.

Ecdysoids are also reported to act on metabolism of carbohydrates in addition to other metabolic changes. Kobayashi and Kimura [170] observed conversion of blood glucose into trehalose after ecdysone injection into pupae of *Bombyx*, whereas in control specimens glucose was incorporated into glycogen. Similar effects were observed in *Musca* [169]. In agreement with some recent theories on involvement of 3,5-cyclic AMP in mediation of hormonal messages, Leenders *et al.* [189] recorded existence of competitive interactions between 3,5-AMP and ecdysterone in its effects on gene activation. Furthermore, Gilbert *et al.* [83] and Applebaum and Gilbert [4] reported stimulation of adenyl cyclase in the wing epithelium of Cecropia by ecdysterone. The effect preceded in time the changes in nucleid acid metabolism and proteosynthesis. Although this effect of ecdysterone was not universal, it represented one of the earliest metabolic changes produced by this compound. Finally, ecdysterone was also reported to enhance the activity of some enzymes of intermediary metabolism, such as epoxidase activity of microsomal fractions in the housefly [337].

9. Ecdysoids and Growth of Tissue Cultures

Before hormones were used in insect tissue cultures it was rather difficult to achieve regular growth and proliferation of the explanted tissue or organs. This was facilitated by co-incubation of the tissue cultures with prothoracic glands or, especially, by addition of ecdysoids when these became readily available. Using ecdysoids, insect tissues were not only stimulated to grow under *in vitro* conditions, but actually they were induced to undergo development which was in many respects similar to *in vivo* developmental changes. These effects of ecdysoids were reviewed by MARKS [203].

Using epidermal tissue from diapausing larvae of *Chilo*, AQUI *et al.* [3] induced a real moult under *in vitro* conditions using ecdysoids. Pronounced developmental changes were also stimulated by ecdysoids in explanted epidermal tissue of *Drosophila* by MANDARON [202], in imaginal discs of genitalia in *Tenebrio* by HUET [112], and in cockroach leg regenerates by MARKS and LEOPOLD [206]. In all these cases it was noticed that responsiveness of the explanted tissue to ecdysoids depended on developmental stage at the moment of tissue explantation. For example, epidermal tissue which was taken into culture at the end of an intermoult period developed without ecdysoids; the tissue explanted in the middle of intermoult period required relatively small amounts of ecdysoids; and, tissue which was taken at the beginning of an instar developed only in presence of relatively large amounts of ecdysoids. Moreover, JUDY and MARKS [135] demonstrated that the effectiveness of ecdysoids in tissue cultures was correlated with their relative activities in the whole body.

Using repeated additions of relatively large amounts of ecdysterone, MARKS [204] produced repeated depositions of multiple cuticular layers in *in vitro* incubated epidermal cells of a cockroach. Small doses administered at shorter time intervals resulted in a single cumulative response. Epidermal tissue explanted from the last instar larvae of *Manduca* responded to ecdysoids by specific developmental changes, e.g. by larval-pupal transformation of the cells. Furthermore, tissue explanted from diapausing pupae responded by changes which were characteristic for pupal-adult transformation [133, 203]. Apparently, under *in vitro* conditions the developmental changes induced by ecdysoids were strictly dependent on the ontogenetic stage of the explanted tissue, similarly like in the body.

Action of ecdysoids on tissue cultures was most pronounced but not restricted to the epidermal cells. So, for instance JUDY [133] observed stimulation of cell movements by ecdysterone in cultures of the hindgut cells from diapausing *Manduca*. Similar effects of ecdysoids on the cell motion were also recorded in some other tissues [244].

Extending an earlier work of WILLIAMS and his co-workers, KAMBYSELLIS and WILLIAMS [136, 137] reported that germinal cysts removed from testes of diapausing Saturniid pupae and cultured *in vitro* underwent spermatogenesis only when provided with certain macromolecular factor. It appeared that spermatogenesis within intact testes required both the macromolecular factor and ecdysoids *in vitro* as well as *in vivo*. The macromolecular factor was present in pupal haemolymph and in other sources such as in media containing bovine serum albumin. These findings were confirmed by TAKEDA [290]. Also RICHMAN and OBERLANDER [246] observed that the effects of ecdysone on morphogenesis of *Galleria* wing discs cultured *in vitro* were highly potentiated by a presence of fat body. Such dependence of growth responses to ecdysoids on macromolecular factors seems to be associated with an availability of nutritional substrates for tissue growth.

The results of these tissue culture studies indicate immediate action of ecdysoids on the epidermal tissue, provided that certain nutritional substrates essential for cellular growth were present in the medium. KAMBYSELLIS and WILLIAMS [137] further demonstrated that these effects of ecdysoids could be duplicated when an active prothoracic gland was co-incubated with the tissue in culture. This documented that the gland was a source of the hormone and that the effects were not mediated by other internal organs. In contrast, there were certain assumptions on a prothoracic gland independent synthesis of these hormones [see 82]. These assumptions were mostly derived from findings on incorporation of the labelled cholesterol into more polar fractions in insect body. However, it is known that insects, like other animals and plants may contain sterol-hydroxylating enzymes. In the light of observations of WILLIG *et al.* [328] it appears that a real identity of more or less polar sterolic compounds with ecdysoids needs to be confirmed by other than chromatographic methods.

MARKS *et al.* [205] reported that ecdysterone induced a rapid release of the neurosecretory material which was accumulated in the pericarya or in the axons of the neurosecretory cells in the brains incubated *in vitro*. Provided that this would occur also *in vivo*, it may indicate an existence of rather important feedback effects between ecdysoids and functions of the neurosecretory system.

10. Ecdysoids and Puffing Patterns on Polytene Chromosomes

Certain groups of insects, namely Dipterans, evolved specific adaptations to rapid growth and development in the larval stage. One of such adaptations is characterized by a fact that larval cells do not divide by regular mitosis but undergo an endomitotic growth. Due to this, the number of cells in the body remains more or less constant throughout the larval stage and the cells increase tremendously in size. The DNA replicates but the chromosomes do not separate. This results in an appearance of polytenic chromosomes consisting of transversal bands each of which corresponds to a normal chromatid. The cells containing polytenic chromosomes represent typically "larval" structures, similarly like the polyploid larval cells of some other Endopterygote insects. It appears that most of these cells have been programmed for a complete breakdown and decomposition during metamorphosis, or eventually, for their persistence without further morphogenetic changes.

With approaching metamorphosis the andomitotic growth is arrested in last larval instars after 8 or even more DNA replication waves. In this period there appears an increased amount of puffs representing enlargements of certain of the transversal bands or chromomeres on the polytene chromosome. CLEVER and KARLSON [42] and CLEVER [35] found that ecdysone induced some specific puffing pattern on the polytene chromosomes of the salivary glands in the last instar larvae of *Chironomus tentans*. They observed that in 15 min after ecdysone application one puff was induced, another preexisting puff was regressed and a second puff was formed later on. These findings attracted considerable attention of developmental biologists because it was assumed that the puffs represented loci of specific gene activation. Literature data pertaining to the problems are so extensive that it is impossible to give a full account for this action of ecdysoids here. We refer to numerous review articles by CLEVER [35, 39, 40, 41], KROEGER [184], KROEGER and LEZZI [185], WYATT [333], and ASHBURNER [6], among others. Our discussions to this point have been reduced to few comments.

The favorite tissue used in investigations on the effects of ecdysoids on puffing pattern have been salivary glands, Malpighian tubules, or foot pad cells of Dipteran larvae. It occurs [38] that different species of Dipteran larvae exhibit species-specific puffing pattern in response to ecdysoids. For example, BERENDES [16] found that the puffing pattern induced by ecdysone in *Drosophila hydei* was different from that observed in *C. tentans* [35, 42]. Thus, 15–20 min. following the application of ecdysone in *Drosophila*, 3 new puffs were formed, 12 puffs showed strong increase in activity, 6 puffs showed a less pronounced increase in activity, 12 puffs decreased, and 5 additional puffs appeared later on. These effects were identical *in vivo* and *in vitro* conditions.

It was generally assumed that formation of the puffs may indicate loci of activated genes on the chromosomes, in contrast to inactive normal bands. Moreover, it was expected that differences in puffing patterns on chromosomes of various tissues and developmental stages might throw some ligh into relationships between gene activation and cellular differentiation [38]. In most cases chromosomal puffing indicated the sites of intensive RNA synthesis. However, in certain species of Dipteran insects such as in *Sciara* the puffs were not associated with an increased RNA synthesis but were due to induced DNA synthesis. A whole set of these DNA-dependent puffs, which normally occurred at the end of the last larval instar, was prematurely induced by ecdysoids [46].

Ecdysoids are not the sole agents which induce formation of the puffs. We have described (p. 259) that juvenoids can also influence the induction and regression of certain puffs. In addition, juvenoids affected the Balbiani rings which represent large puffs of characteristic appearance [191, 192, 193]. Assumptions were made on possible antagonistic action between ecdysoids and juvenoids with respect to puff formation. The puffs were also induced by neurohormone extracted from the brain [26] and by many other nonhormonal factors such as high temperature, changes in oxygen supply, mechanical injury to the salivary gland cells and, especially by changes in extra or intracellular ionic concentrations [182, 183, 184, 190]. Kroeger [182, 183] discovered that certain changes in extracellular and intracellular Na^+/K^+ ratio were correlated with an induction of changes in puffing pattern including formation of puffs which were specific for ecdysone. Based on these observations and assuming that changes in intranuclear ion concentration may affect the nucleic acid—histone links associated with a derepression of certain sensitive gene loci, he put forward hypothesis [184, 185] that the action of ecdysoids on the puffing pattern was mediated by changes in electrical potentials and in the transport of ions across the membranes. Similar hypothesis on mediation of the hormonal stimuli through changes in ionic balance was later adopted for mechanism of juvenoid action by Lezzi and Frigg [191]. According to these theories ecdysoids were assumed to increase the intracellular potassium ion concentration while decreasing intracellular sodium ion concentration. Juvenoids were thought to cause reversed effects. Depending on the final ionic balance the Na^+ or K^+ sensitive gene loci were activated. A lack of these theories is that they have no reasonable physiological background; their application to hormonal control of insect development has been seriously criticized [6, 41, 151, 333].

Despite extensive cytochemical, biochemical and biophysical studies, very little is known about possible physiological significance of the hormonally induced changes in the puffing pattern. One of the interesting developmental features was that the total DNA content of the salivary gland cells decreased before pupation or puparium formation in Diptera. Conversely, there was an increased activity of DNA ase and of other hydrolytic enzymes. This suggested that some initial steps to a breakdown of these cells occurred long before puparium formation [187], e.g. at the time of intensive puffing activity. The whole gland started to disintegrate morphologically in the early pupal period [21]. It also appeared that production and transport of the secretory product in the salivary glands were not dependent on *de novo* protein synthesis, at least for several hours. Production and transport of the product was not immediately dependent on RNA synthesis as found by DOYLE and LAUFER [48] in *Chironomus*. Finally, LAUFER [186] suggested that possible correlations may exist between the puffing patterns on the salivary gland cell chromosomes and the programmed, hormonally controled self-destruction of these cells.

We have already indicated that the process of larval histolysis occurs hand in hand with the histogenesis of imaginal tissues. We have seen (p. 347) that the hormonal stimulus for breakdown of the "larval" tissues is given by ecdysoids. During accelerated metamorphosis due to ecdysoid action there occurs precocious development of some structures but also precocious disintegration of some other structures. The cells which received the final "death" message synthesize increasingly RNA and also series of new hydrolytic enzymes which are later engaged in autolysis of the individual cell particles before the cell remnants have been phagocyted completely by invading haemocytes. In some tissues, such as prothoracic glands (p. 56, 57), the initial steps of the autolytic process are visible long before the cells actually disintegrate. With respect to physiological role of the salivary glands in Diptera we can assume that changes in the puffing pattern induced by ecdysoids represent most probably the initial phases of histolysis. The most conspicuous changes in puffing pattern occur in these larval salivary gland cells long after the sensitivity to juvenoids was crossed over, e.g. after development was reprogrammed and the cells were invariably committed to metamorphosis. The general view that puffs may represent a specific gene activation was criticized by THOMSON [311]. In addition, it appears that there are some morphologically and physiologically different tissues which show the same nonspecific puffing pattern. For example, BERENDES and WILLART [17] found in the salivary gland cells identical puffing responses as in the Malpighian tubule cells, suggesting that the puffing pattern was not directly related to specific physiological functions. Moreover, it is quite impossible to correlate changes in puffing pattern on polytene chromosomes of the histolysing larval salivary gland cells with the usual situation in a normal diploid cell [311].

11. Effect of Ecdysoids in Other Animals and Plants

Selection for possible candidates of ecdysoid action in the animal kingdom was centered to Arthropods other than insects. In 1968, LOWE *et al.* [196] induced accelerated moult in a crayfish by ecdysterone. Then, KRISHNAKUMARAN and SCHNEIDERMAN [178, 179] described induction of moults by ecdysterone in other Arthropods, such as in a very primitive horseshoe crab *Limulus*, in other Crustaceans *Armadillidium* and *Procambarus*, and even in *Araneus* spiders. Ecdysterone was the most active ecdysoid on Mandibulate and Chelicerate Arthropods, whereas cyasterone had the lowest activity [180]. They concluded that these ecdysoids may be the normal moulting hormones of all Arthropods. Ecdysoid action on other Arthropods than insects was further confirmed in the fiddler crab *Uca* [288], in *Carcinus* [1], in the terrestrial or aquatic Isopods such as *Liquia* [201], *Asellus* [7] and *Idothea* [243]. Ecdysterone caused also certain biochemical changes in Crustaceans, and these changes were known to occur in insects. Thus, GORRELL and GILBERT [84] observed stimulation of protein synthesis and RNA synthesis by ecdysterone in the crayfish hepatopancreas.

It appears that the cffectiveness of ecdysoids on growth and development in animals is not limited to the Arthropod group. Possible targets are the Helminths, where ecdysoids may affect moulting in the Nematodes. In a Trematode *Schistosoma mansoni* ecdysone affected metamorphosis of the miracidia into cercaria [214]. It is possible that the plants which contain ecdysoids, especially ferns are more or less protected against continuous attack of not only insects, but mainly the soil Nematodes and perhaps also terrestrial Isopods. At present it is very difficult to estimate a real significance of ecdysoids for growth and development of plants. We have described considerable accumulation of ecdysoids in certain plants or parts of a plant system and their apparent absence in various other plants (p. 310–318). This limited occurrence of ecdysoids in certain groups or species of plants seems to indicate that these compounds represent secondary plant substances which are not used as general regulators of plant growth. This conclusion is confirmed by the results of investigations of JACOB and SUTHERS [124] who were unable to induce any qualitative or quantitative effects by ecdysone on flowering and vegetative development of the cultured shoot tips or intact plants.

Ecdysoids were reported to cause certain biochemical and pharmacological effects in mammals, including humans. BURDETTE and RICHARDS [27] observed effects of ecdysone extracts on growth of the mammalian cells *in vitro*. Moreover, BURDETTE and CODA [25] found stimulating effect of ecdysone on amino acid incorporation into hepatic proteins. This effect was later extensively studied by Japanese workers OTAKA *et al.* [229, 230, 231], who found that also other ecdysoids than ecdysone could stimulate protein and RNA synthesis rates in the mouse liver. Other reports in this respect involved a significantly increased activity of glutamic decarboxylase in the rat brain after addition of ecdysone [30]. Further it appeared that cholesterol level in tho mouse liver decreased after treatment with ecdysone [198]. Glucose incorporation into glycogen was stimulated by ecdysterone application in the rat liver [336]. This treatment in the rat was also followed by suppression of an artificially produced hyperglycaemia. Masuoka *et al.* [207] performed more detailed pharmacological analysis of the effects of ponasterone A, ecdysterone, and inokosterone in the rat. They did not observe any developmental abnormalities, anabolic androgenic, or antiandrogenic effects, nor changes in haematological and liver tests, and in growth of the organs. Thus, no special pharmacological effects of these compounds in mice were noticed. However, Hikino and Takemoto [98] stated that ecdysoids may provide dramatic terapeutic effects on patients with neuralgia. Small doses (less than 0.1 g) caused long lasting relief of pain without side effects. They mentioned that a drug called "Go-shitsu" which was prepared from plants containing high amount of ecdysoids was practically used as a remedy for 2000 years. Screenings for a possible anti-tumorigenic activity of ecdysoids did not reveal any remarcable effects on growth of the tumorous cells [99].

So far, it is not known whether ecdysoids could stimulate growth and cellular divisions in some vertebrate tissue, in analogy with their insect tissue targets. All the above effects of ecdysoids on vertebrate organs suggest rather pharmacological action than a real hormonal effect.

References

[1] Adelung, D.: Die Ausschüttung und Funktion von Häutungshormon während eines Zwischenhäutungs-Intervalls bei der Strandkrabbe *Carcinus maenas* L. Z. Naturforsch. *24*, 1447–1455 (1969).

[2] Adelung, D., and P. Karlson: Eine verbesserte, sehr empfindliche Methode zur biologischen Auswertung des Insektenhormones Ecdyson. J. Insect Physiol. *15*, 1301–1307 (1969).

[3] Agui, N., S. Yagi, and M. Fukaya: Induction of moulting of cultivated integument taken from a diapausing rice stem borer larva in the presence of ecdysterone. Appl. Entomol. Zool. *4*, 156–157 (1969).

[4] Applebaum, S. W., and L. I. Gilbert: Stimulation of adenyl cyclase in pupal wing epidermis by β-ecdysone. Develop. Biol. *27*, 165–175 (1972).

[5] Arking, R., and E. Shaaya: Effect of ecdysone on protein synthesis in the larval fat body of *Calliphora*. J. Insect Physiol. *15*, 287–296 (1966).

[6] Ashburner, M.: Function and structure of polytene chromosomes during insect development. Advan. Insect Physiol. *7*, 1–95 (1970).

[7] Balesdent, M. L.: Action d'une ecdysone de synthèse sur la mue et sur la sexualité du crustacé isopode femelle *Asellus aquaticus* L. C.R. Acad. Sci. (Paris) *273*, 1972–1974 (1971).

[8] Berreur, P., and G. Fraenkel: Puparium formation in flies: Contraction to puparium induced by ecdysone. Science *164*, 1182–1183 (1969).

[9] Barritt, L. C., and L. M. Birt: Prothoracic gland hormone in the sheep blowfly, *Lucilia cuprina*. J. Insect Physiol. *16*, 671–677 (1970).

[10] Barritt, L. C., and L. M. Birt: Development of *Lucilia cuprina*: Correlation of biochemical and morphological events. J. Insect Physiol. *17*, 1169–1183 (1971).

[11] BARTH, R. H., JR., P. P. BUNYARD, and T. H. HAMILTON: RNA metabolism in pupae of the oak silkworm, *Antheraea pernyi*: The effects of diapause, development and injury. Proc. nat. Acad. Sci. (Amer.) *52*, 1572–1580 (1964).

[12] BARTON, D. H. R., P. G. FEAKINS, J. P. POYSER, and P. G. SAMMES: A synthesis of the insect moulting hormone, ecdysone and related compounds. J. Chem. Soc. *C 1970*, 1584–1591.

[13] BECK, S. D., and J. L. SHANE: Effects of ecdysone on diapause in the European corn borer, *Ostrinia nubialis*. J. Insect Physiol. *15*, 721–730 (1969).

[14] BECKER, R.: Über Versuche zur Anreicherung und physiologische Charakterisierung des Wirkstoffes der Puparisierung. Biol. Zbl. *61*, 360 (1941).

[15] BECKER, E., and E. PLAGGE: Über das die Pupariumbildung auslösende Hormon der Fliegen. Biol. Zbl. *59*, 326–341 (1939).

[16] BERENDES, H. D.: The hormone ecdysone as effector of specific changes in the pattern of gene activities of *Drosophila hydei*. Chromosoma *22*, 274–293 (1967).

[17] BERENDES, H. D., and E. WILLART: Ecdysone-related changes at the nuclear and cytoplasmic level of Malpighian tubule cells in *Drosophila*. J. Insect Physiol. *17*, 2337–2350 (1971).

[18] BERRY, S. J., A. KRISHNAKUMARAN, H. OBERLANDER, and H. A. SCHNEIDERMAN: (see reference No. 18 on p. 282).

[19] BEVER, W. VAN, F. KOHEN, V. V. RANADE, and R. E. COUNSELL: Synthesis of the steroid rubrosterone. Chem. Commun. *1970*, 758.

[20] BÖHM, G. A.: Ecdysontiter und Blutbild bei *Calliphora vicina* Rob. Desv. unter normalen und experimentellen Bedingungen. Mitt. deutsch. Ent. Ges. *27*, 52–59 (1968).

[21] BÖHM, G. A.: Histological changes in the salivary glands of *Calliphora vicina* R. D. and their possible control by ecdyson. Endocrin. Exp. *5*, 97—100 (1971).

[22] BÜCKMANN, D.: Die Auslösung der Umfärbung durch das Häutungshormon bei *Cerura vinula* L. (Lepidoptera, Notodontidae). J. Insect Physiol. *3*, 159–189 (1959).

[23] BULLIÉRE, F., and D. BULLIÉRE: Régénération, différenciation et ecdysones chez l'embryon de *Blabera craniifer* (Insecte Dictyoptére) en culture *in vitro*. C. R. Acad. Sci. (Paris) *273*, 955–958 (1971).

[24] BURDETTE, W. J.: Changes in titer of ecdyson in *Bombyx mori* during metamorphosis. Science *135*, 432 (1962).

[25] BURDETTE, W. J., and R. L. CODA: Effect of ecdysone on incorporation of ^{14}C-leucine into hepatic protein *in vitro*. Proc. Soc. Exp. Biol. Med. *112*, 216–217 (1963).

[26] BURDETTE, W. J., and M. KOBAYASHI: Response of chromosomal puffs to crystalline hormones *in vivo*. Proc. Soc. Exp. Biol. Med. *131*, 209–213 (1969).

[27] BURDETTE, W. J., and R. C. RICHARDS: Alteration of the growth of mammalian cells *in vitro* by ecdysone extract. Nature *189*, 666–668 (1961).

[28] BUTENANDT, A., and P. KARLSON: Über die Isolierung eines Metamorphose-Hormons der Insekten in kristallisierter Form. Z. Naturforsch. *9 b*, 389–391 (1954).

[29] Carlisle, D. B., and P. E. Ellis: Bracken and locust ecdysones: Their effect on moulting in the desert locust. Science *159*, 1472–1474 (1968).

[30] Chaudhary, K. D., P. J. Lupien, and C. Hinse: Effect of ecdysone on glutamic decarboxylase in rat brain. Experientia *25*, 250–251 (1969).

[31] Cherbas, L., and P. Cherbas: Distribution and metabolism of α-ecdysone in pupae of the silkworm *Antheraea polyphemus*. Biol. Bull. Woods Hole *138*, 115–128 (1970).

[32] Chino, H., L. I. Gilbert, J. B. Siddall, and W. Hafferl: Studies on ecdysone transport in insect haemolymph. J. Insect Physiol. *16*, 2033–2040 (1970).

[33] Chong, Y. K., M. N. Galbraith, and D. H. S. Horn: Isolation of deoxycrustecdysone, deoxyecdysone, and α-ecdysone from the fern *Blechnum minus*. Chem. Commun. *1970*, 1217-1218.

[34] Clayton, R. B.: The utilization of steroids by insects. (Review). J. Lipid Res. *5*, 3–19 (1964).

[35] Clever, U.: Genaktivitäten in den Riesenchromosomen von *Chironomus tentans* und ihre Beziehungen zur Entwicklung. I. Genaktivierung durch Ecdyson. Chromosoma *12*, 607–675 (1961).

[36] Clever, U.: Von der Ecdysonkonzentration abhängige Genaktivitätsmuster in den Speicheldrüsenchromosomen von *Chironomus tentans*. Develop. Biol. *6*, 73–98 (1963).

[37] Clever, U.: Actinomycin and puromycin: Effects on sequential gene activation by ecdysone. Science *146*, 794–795 (1964).

[38] Clever, U.: The control of gene activity as factor of cell differentiation in insect development. In: Developmental and metabolic control mechanisms and Neoplasia, pp. 361–374 (1965). Published for the University of Texas by the Williams & Wilkins, Baltimore, Maryland.

[39] Clever, U.: Puffing changes in incubated and in ecdysone treated *Chironomus tentans* salivary glands. Chromosoma *17*, 309–322 (1965).

[40] Clever, U.: Chromosomal changes associated with differentiation. In: Genetic control of differentiation. Brookhaven Symp. Biol. *18*, 242–253 (1965).

[41] Clever, U.: Regulation of chromosome function. Ann. Rev. Genetics *2*, 11–30 (1968).

[42] Clever, U.: and P. Karlson: Induktion von Puff-Veränderungen in den Speicheldrüsenchromosomen von *Chironomus tentans* durch Ecdyson. Exp. Cell Res. *20*, 623–627 (1960).

[43] Congote, L. F., C. E. Sekeris, and P. Karlson: On the mechanism of hormone action. XVIII. Alteration of the nature of RNA synthesized in isolated fat body cell nuclei as a result of ecdysone and juvenile hormone action. Z. Naturforsch. *25 b*, 279–284 (1970).

[44] Collins, J. V.: The hormonal control of fat body development in *Calpodes ethlius* (Lepidoptera, Hesperidae). J. Insect Physiol. *15*, 341–352 (1969).

[45] Crossley, A. C.: The fine structure and mechanism of breakdown of larval intersegmental muscles in the blowfly *Calliphora erythrocephala*. J. Insect Physiol. *14*, 1389–1407 (1968).

[46] Crouse, H. V.: The role of ecdysone in DNA-puff formation and DNA synthesis in the polytene chromosomes of *Sciara coprophila*. Proc. nat. Acad. Sci. (Amer.) *61*, 971-978 (1968).

[47] Djerassi, C., J. C. Knight, and H. Brockmann: Neue Sterine aus dem Kaktus *Wilcoxia viperina*. Chem. Ber. *97*, 3118–3130 (1964).

[48] DOYLE, D., and H. LAUFER: Analysis of secretory processes in Dipteran salivary glands. In vitro *3*, 93–103 (1967).

[49] DUKES, P. P., C. E. SEKERIS, and W. SCHMID: On the mechanism of hormone action. II. "Increase in template activity of RNA from isolated nuclei incubated in presence of hormone". Biochem. Biophys. Acta *123*, 126–132 (1966).

[50] EARLE, N. W., I. PADOVANI, M. J. THOMSON, and W. E. ROBBINS: Inhibition of larval development and egg production in the boll weevil following ingestion of ecdysone analogues. J. Econ. Entomol. *63*, 1064–1069 (1970).

[51] EL-IBRASHY, M. T.: (see reference No. 80 on p. 285) (1972).

[52] EMMERICH, H.: Distribution of tritiated ecdysone in salivary gland cells of *Drosophila*. Nature *221*, 954–955 (1969).

[53] EMMERICH, H.: Anreicherung von tritiummarkierten Ecdyson in den Zellkernen der Speicheldrüsen von *Drosophila hydei*. Exp. Cell Res. *58*, 261–270 (1969).

[54] EMMERICH, H.: Ecdysonbindende Proteinfraktionen in den Speicheldrüsen von *Drosophila hydei*. Z. vergl. Physiol. *68*, 385–402 (1970).

[55] EMMERICH, H.: Über die Hämolymphproteine von *Pyrrhocoris apterus* und über die Bindung von Ecdyson durch Hämolymphproteine. J. Insect Physiol. *16*, 725–747 (1970).

[56] FAUX, A., M. N. GALBRAITH, D. H. S. HORN, E. J. MIDDLETON, and J. A. THOMSON: The structures of two ecdysone analogues, cheilanthones A and B from the fern *Cheilanthes tenuifolia*. Chem. Commun. *1970*, 243–244.

[57] FAUX, A., D. H. S. HORN, E. J. MIDDLETON, H. M. FALES, and M. E. LOWE: Moulting hormones of a crab during ecdysis. Chem. Commun. *1969*, 175–176.

[58] FEIR, D., and G. WINKLER: Ecdysone titres in the last larva and adult stage of the milkweed bug. J. Insect Physiol. *15*, 899–904 (1969).

[59] FOURCHE, J.: Action de l'ecdysone sur les larves de *Drosophila melanogaster* soumises au jeune. Existence d'un double conditionement pour la formation du puparium. C. R. Acad. Sci. (Paris) *264*, 2398–2400 (1967).

[60] FOURCHE, J.: Le déterminisme des mues et des métamorphoses chez *Drosophila melanogaster*: Influence du jeune et de la fourniture d'ecdysone. Arch. Anat. Micr. *56*, 141–152 (1967).

[61] FOURCHE, J.: Le déterminisme de la formation du puparium chez *Drosophila melanogaster*: Étude de la compétence a la fourniture d'ecdysone chez des larves soumises au jeune. Arch. Anat. Micr. *58*, 239–248 (1969).

[62] FRAENKEL, G.: Pupation of flies initiated by a hormone. Nature *133*, 834 (1934).

[63] FRAENKEL, G.: A hormone causing pupation in the blowfly *Calliphora erythrocephala*. Proc. Roy. Soc. (London) *B 118*, 1–12 (1935).

[64] FRAENKEL, G., and C. HSIAO: Calcification, tanning and the role of ecdysone in the formation of the puparium of the facefly, *Musca autumnalis*. J. Insect Physiol. *13*, 1387–1394 (1967).

[65] FRAENKEL, G., and C. HSIAO: Morphological and endocrinological aspects of pupal diapause in a fleshfly, *Sarcophaga argyrostoma*. J. Insect Physiol. *14*, 707–718 (1968).

[66] FRAENKEL, G., and J. ŽĎÁREK: The evaluation of the "Calliphora test" as an assay for ecdysone. Biol. Bull. Woods Hole *139*, 138–150 (1970).

[67] FRAENKEL, G., and E. ZLOTKIN: Acceleration of puparium formation in *Sarcophaga argyrostoma* by electrical stimulation or scorpion venom. J. Insect Physiol. *16*, 1549–1554 (1970).

[68] FUKUDA, S.: Induction of pupation in silkworm by transplanting the prothoracic gland. Proc. Imp. Acad. Japan *16*, 417–420 (1940).

[69] FUKUDA, S.: Induction of metamorphosis in the silkworm by transplanting the prothoracic gland. Zool. Mag., Tokyo *53*, 12 (1941).

[70] FUKUDA, S.: Role of the prothoracic gland in differentiation of the imaginal character in the silkworm pupa. Annot. Zool. Jap. *20*, 9–13 (1941).

[71] FURLENMEIER, A., A. FÜRST, A. LANGEMANN, G. WALDVOGEL, P. HOCKS, U. KERB, and R. WIECHERT: The synthesis of ecdysone. Experientia *22*, 573 (1966).

[72] FURLENMEIER, A., A. FÜRST, A. LANGEMANN, G. WALDVOGEL, P. HOCKS, U. KERB, and R. WIECHERT: Zur Synthese des Ecdysons. Helv. *50*, 2387–2396 (1967).

[73] FURLENMEIER, A., A. FÜRST, A. LANGEMANN, G. WALDVOGEL, U. KERB, P. HOCKS, and R. WIECHERT: Zur Synthese des Ecdysons. Synthesen von 2β,3β,14α-Trihydroxy-6-keto-Δ^7-A/B-cis-Steroiden. Helv. *49*, 1591–1601 (1966).

[74] GALBRAITH, M. N., and D. H. S. HORN: An insect-moulting hormone from a plant. Chem. Commun. *1966*, 905–906.

[75] GALBRAITH, M. N., D. H. S. HORN, E. J. MIDDLETON, and R. J. HACKNEY: Structure of deoxycrustecdysone, a second crustacean moulting hormone. Chem. Commun. *1968*, 83–85.

[76] GALBRAITH, M. N., D. H. S. HORN, E. J. MIDDLETON, and R. J. HACKNEY: The structure of podecdysone B, a new phytoecdysone. Chem. Commun. *1969*, 402–403.

[77] GALBRAITH, M. N., D. H. S. HORN, E. J. MIDDLETON, and R. J. HACKNEY: Moulting hormones of insects and crustaceans: The synthesis of 22-deoxycrustecdysone. Austr. J. Chem. *22*, 1517–1524 (1969).

[78] GALBRAITH, M. N., D. H. S. HORN, E. J. MIDDLETON, J. A. THOMSON, J. B. SIDDALL, and W. HAFFERL: The catabolism of crustecdysone in the blowfly *Calliphora stygia*. Chem. Commun. *1969*, 1134–1135.

[79] GALBRAITH, M. N., D. H. S. HORN, E. J. MIDDLETON, and J. A. THOMSON: The biosynthesis of crustecdysone in the blowfly *Calliphora stygia*. Chem. Commun. *1970*, 179–180.

[80] GALBRAITH, M. N., D. H. S. HORN, Q. N. PORTER, and R. J. HACKNEY: Structure of podecdysone A, a steroid with moulting hormone activity from the bark of *Podocarpus elatus*. Chem. Commun. *1968*, 971–972.

[81] GALBRAITH, M. N., D. H. S. HORN, J. A. THOMSON, G. J. NEUFELD, and R. J. HACKNEY: Insect moulting hormones: Crustecdysone in *Calliphora*. J. Insect. Physiol. *15*, 1225–1233 (1966).

[82] GERSCH, M., and J. STÜRZEBECHER: Über eine Synthese von Ecdyson-^{3}H und Ecdysteron-^{3}H aus Cholesterin-^{3}H geschnürten Abdomina von *Mamestra brassicae*-Raupen. Experientia *27*, 1475–1476 (1971).

[83] GILBERT, L. I., S. APPLEBAUM, T. A. GORELL, J. B. SIDDALL, and Y. C. SIEW: Aspects of research on insect growth hormones. Bull. World Health Org. *44*, 397–398 (1971).

[84] GORELL, T. A., and L. I. GILBERT: Stimulation of protein and RNA synthesis in the crayfish hepatopancreas by crustecdysone. Gen. Comp. Endocrinol. *13*, 308–310 (1969).

[85] Hampshire, F., and D. H. S. Horn: Structure of crustecdysone, a crustacean moulting hormone. Chem. Commun. *1966*, 37–38.

[86] Hanser, G., and P. Karlson: Über die Wirkung des Metamorphosehormons auf die Epidermis von *Ephestia*-Dauerraupen. Biol. Zbl. *76*, 129–141 (1957).

[87] Harrison, I. T., J. B. Siddall, and J. H. Fried: Synthetic studies on insect hormones. III. An alternative synthesis of ecdysone and 22-isoecdysone. Tetrahedron Letters *1966*, 3457–3460.

[88] Heinrich, G., and H. Hoffmeister: Ecdyson als Begleitsubstanz des Ecdysterons in *Polypodium vulgare* L. Experientia *23*, 995 (1967).

[89] Heinrich, G., and H. Hoffmeister: 5β-Hydroxyecdysteron, ein Pflanzensteroid mit Häutungshormonaktivität aus *Polypodium vulgare* L. Tetrahedron Letters *1968*, 6063–6064.

[90] Heinrich, G., and H. Hoffmeister: Bildung von Hormonglykosiden als Inaktivierungsmechanismus bei *Calliphora erythrocephala*. Z. Naturforsch. B *25 b*, 358–359 (1970).

[91] Hikino, H., Y. Hikino, K. Nomoto, and T. Takemoto: Cyasterone, an Insect metamorphosing substance from *Cyathula capitata*: Structure. Tetrahedron *24*, 4895–4906 (1968).

[92] Hikino, H., Y. Hikino, and T. Takemoto: Synthesis of rubrosterone, a metabolite of insect moulting substances from *Achyranthes rubrofusca*. Tetrahderon Letters *1968*, 4255–4256.

[93] Hikino, H., Y. Hikino, and T. Takemoto: Rubrosterone, a metabolite of insect metamorphosing substance from *Achyranthes rubrofusca*: Synthesis. Tetrahedron *25*, 3389–3394 (1969).

[94] Hikino, H., K. Nomoto, R. Ino, and T. Takemoto: Structure of precyasterone, a novel C_{29} insect-moulting substance from *Cyathula capitata*. Chem. Pharm. Bull. *18*, 1078–1080 (1970).

[95] Hikino, H., K. Nomoto, and T. Takemoto: Structure of sengosterone, a novel C_{29} insect-moulting substance from *Cyathula capitata*. Tetrahedron Letters *1969*, 1417–1420.

[96] Hikino, H., K. Nomoto, and T. Takemoto: Poststerone, a metabolite of insect metamorphosing substances from *Cyathula capitata*. Steroids *16*, 393–400 (1970).

[97] Hikino, H., Y. Ohizumi, and T. Takemoto: Catabolism of ponasterone A to ecdysterone, inokosterone, and poststerone in *Bombyx mori*. Chem. Commun. *1971*, 1036–1037.

[98] Hikino, H., and T. Takemoto: Arthropod moulting hormones from plants *Achyranthes* and *Cyathula*. Naturwissenschaften *59*, 91–98 (1972).

[99] Hirono, I., I. Sasaoka, and M. Schimizu: Effect of insect moulting hormones, ecdysterone and inokosterone, on tumor cells. GANN *60*, 341–342 (1969).

[100] Hocks, P., A. Jäger, U. Kerb, R. Wiechert, A. Furlenmeier, A. Fürst, and A. Langemann: Synthetische Steroide mit Häutungshormonaktivität. Angew. Chem. *78*, 680–681 (1966).

[101] Hocks, P., U. Kerb, R. Wiechert, A. Furlenmeier, and A. Fürst: Die Synthese des Rubrosterons. Tetrahedron Letters *1968*, 4281–4284.

[102] Hocks, P., and R. Wiechert: 20-Hydroxyecdyson, isoliert aus Insekten. Tetrahedron Letters *1966*, 2989–2993.

[103] Hoffmeister, H., K. Nakanishi, M. Koreeda, and H. Y. Hsu: The moulting hormone activity of ponasterones in the *Calliphora* test. J. Insect Physiol. *14*, 53–54 (1968).

[104] HOFFMEISTER, H., C. RUFER, and H. AMMON: Ausscheidung von Ecdyson bei Insekten. Z. Naturforsch. *20 b*, 130–133 (1965).

[105] HOFFMEISTER, H., C. RUFER, H. H. KELLER, H. SCHAIRER, and P. KARLSON: Zur Chemie des Ecdysons. III. Vergleichende spektrometrische Untersuchungen an α,β-ungesättigten Steroidketonen. Chem. Ber. *98*, 2361–2375 (1965).

[106] HOPPE, W., and R. HUBER: Zur Chemie des Ecdysons. II. Bestimmung des Sterin-Skeletts und seiner Orientierung mit diffuser Röntgenstreuung in Kristallen von Ecdyson. Chem. Ber. *98*, 2353–2360 (1965).

[107] HORA, J., L. LÁBLER, A. KASAL, V. ČERNÝ, F. ŠORM, and K. SLÁMA: Moulting deficiencies produced by some sterol derivatives in an insect (*Pyrrhocoris apterus* L.). Steroids *8*, 887–914 (1966).

[108] HORN, D. H. S., S. FABBRI, F. HAMPSHIRE, and M. E. LOWE: Isolation of crustecdysone (20R-hydroxyecdysone) from a crayfish (*Jasus lalandei*, H. Milne Edwards). Biochem. J. *109*, 399–406 (1968).

[109] HORN, D. H. S., E. J. MIDDLETON, J. A. WUNDERLICH, and F. HAMPSHIRE: Identity of moulting hormones of insects and crustaceans. Chem. Commun. *1966*, 339–341.

[110] HSIAO, C., and T. H. HSIAO: (see reference No. 127 on p. 288).

[111] HUBER, R., and W. HOPPE: Zur Chemie des Ecdysons. VII. Die Kristall- und Molekülstrukturanalyse des Insektenverpuppungshormons Ecdyson mit der automatisierten Faltmolekülmethode. Chem. Ber. *98*, 2403–2424 (1965).

[112] HUET, C., Différenciation en culture *in vitro* des ébauches présomptives de l'appareil génital femelle de *Tenebrio molitor* L. en présence d'ecdysone. C. R. Hebd. Séanc. Acad. Sci. (Paris) *272*, 1896–1899 (1971).

[113] HÜPPI, G., and J. B. SIDDALL: Synthetic studies on insect hormones. V. The synthesis of crustecdysone (20-hydroxyecdysone). J. Amer. chem. Soc. *89*, 6790–6792 (1967).

[114] HÜPPI, G., and J. B. SIDDALL: Synthetic studies on insect hormones. VI. The synthesis of ponasterone A and its stereochemical identity with crustecdysone. Tetrahedron Letters *1968*, 1113–1114.

[115] IMAI, S., S. FUJIOKA, E. MURATA, K. OTSUKA, and K. NAKANISHI: Structure of phytoecdysone, ajugasterone B. Chem. Commun. *1969*, 82–83.

[116] IMAI, S., S. FUJIOKA, E. MURATA, Y. SASAKAWA, and K. NAKANISHI: The structures of three additional phytoecdysones from *Podocarpus macrophyllus*, makisterone B, C, and D. Tetrahedron Letters *1968*, 3887–3890.

[117] IMAI, S., S. FUJIOKA, K. NAKANISHI, M. KOREEDA, and T. KUROKAWA: Extraction of ponasterone A and ecdysterone from *Podocarpaceae* and related plants. Stereoids *10*, 557–565 (1967).

[118] IMAI, S., M. HORI, S. FUJIOKA, E. MURATA, M. GOTO, and K. NAKANISHI: Isolation of four new phytoecdysones, makisterone A, B, C, D, and the structure of makisterone A, a C_{28} steroid. Tetrahedron Letters *1968*, 3883–3886.

[119] IMAI, S., E. MURATA, S. FUJIOKA, M. KOREEDA, and K. NAKANISHI: Structure of ajugasterone C, a phytoecdysone with an 11-hydroxy-group. Chem. Commun. *1969*, 546–547.

[120] IMAI, S., E. MURATA, S. FUJIOKA, T. MATSUOKA, M. KOREEDA, and K. NAKANISHI: Structures of stachysterones C and D. Chem. Commun. *1970,* 352–353.

[121] IMAI, S., T. TOYOSATO, M. SAKAI, Y. SATO, S. FUJIOKA, E. MURATA, and M. GOTO: Screening results of plants for phytoecdysones. Chem. Pharm. Bull. *17*, 335–339 (1969).

[122] IMAI, S., T. TOYOSATO, M. SAKAI, Y. SATO, S. FUJIOKA, E. MURATA, and M. GOTO: Isolation of cyasterone and ecdysterone from plant materials. Chem. Pharm. Bull. *17*, 340–342 (1969).

[123] ITO, S., and W. R. LOEVENSTEIN: Permeability of a nuclear membrane: Changes during normal development and changes induced by growth hormone. Science *150*, 909–910 (1965).

[124] JACOBS, W. P., and H. B. SUTHERS: The culture of apical buds of *Xanthium* and their use as a bioassay for flowering activity of ecdysterone. Amer. J. Bot. *58*, 836–843 (1971).

[125] JENKIN, P. M.: Apolysis and hormones in the moulting cycle of arthropods. Ann. Endocrin. (Paris) *27*, 331–341 (1966).

[126] JENKIN, P. M., and H. E. HINTON: Apolysis in arthropod moulting cycles. Nature *211*, 871 (1966).

[127] JIZBA, J., L. DOLEJŠ, V. HEROUT, F. ŠORM, H. W. FEHLHABER, G. SNATZKE, R. TSCHESCHE, and F. WULFF: Polypodosaponin, ein neuer Saponintyp aus *Polypodium vulgare* L. Chem. Ber. *104*, 837–846 (1971).

[128] JIZBA, J., L. DOLEJŠ, V. HEROUT, and F. ŠORM: The structure of osladin—the sweet principle of the rhizomes of *Polypodium vulgare* L. Tetrahedron Letters *1971*, 1329–1332.

[129] JIZBA, J., and V. HEROUT: Isolation of constituents of common polypody rhizomes (*Polypodium vulgare* L.). Coll. Czech. Chem. Comm. *32*, 2867–2874 (1967).

[130] JIZBA, J., V. HEROUT, and F. ŠORM: Isolation of ecdysterone (crustecdysone) from *Polypodium vulgare* rhizomes. Tetrahedron Letters *1967*, 1689–1691.

[131] JIZBA, J., V. HEROUT, and F. ŠORM: Polypodine B- a novel ecdysonlike substance from plant material. Tetrahedron Letters *1967*, 5139–5143.

[132] JOLY, R. A., C. M. SVAHN, R. D. BENNETT, and E. HEFTMANN: Investigation of intermediate steps in the biosynthesis of ecdysterone from cholesterol in *Podocarpus elata*. Phytochemistry *8*, 1917–1920 (1969).

[133] JUDY, K. J.: Cellular response to ecdysterone *in vitro*. Science *165*, 1374–1375 (1969).

[134] JUDY, K. J., and L. I. GILBERT: 1970 (see reference No. 155 on p. 289).

[135] JUDY, K. J., and E. P. MARKS: Effects of ecdysterone *in vitro* on hindgut and hemocytes of *Manduca sexta* (Lepidoptera). Gen. Comp. Endocrinol. *17*, 351–359 (1971).

[136] KAMBYSELLIS, M. P., and C. M. WILLIAMS: *In vitro* development of insect tissues. II. The role of ecdysone in the spermatogenesis of silkworms. Biol. Bull. Woods Hole *141*, 541–552 (1971).

[137] KAMBYSELLIS, M. P., and C. M. WILLIAMS: Spermatogenesis in cultured testes of the *Cynthia* silkworm: Effects of ecdysone and of prothoracic glands. Science *175*, 769–770 (1972).

[138] KAPLANIS, J. N., W. E. ROBBINS, M. J. THOMPSON, and A. H. BAUMHOVER: Ecdysone analogue: Conversion to alpha ecdysone and 20-hydroxyecdysone by an insect. Science *166*, 1540–1541 (1969).

[139] KAPLANIS, J. N., L. A. TABOR, M. I. THOMPSON, W. E. ROBBINS, and T. I. SHORTINO: Assay for ecdysone (moulting hormone) activity using the house fly, *Musca domestica* L. Steroids *8*, 625–631 (1966).

[140] KAPLANIS, J. N., M. J. THOMPSON, W. E. ROBBINS, and B. M. BRYCE: Insect hormones: Alpha ecdysone and 20-hydroxyecdysone in bracken fern. Science *157*, 1436–1438 (1967).

[141] KAPLANIS, J. N., M. J. THOMPSON, R. T. YAMAMOTO, W. E. ROBBINS, and S. J. LOULOUDES: Ecdysones from the pupa of the tobacco hornworm *Manduca sexta* (Johannson). Steroids *8*, 605–623 (1966).

[142] KARLSON, P.: Biochemical studies on insect hormones. Vitam. Horm. *14*, 227–266 (1956).

[143] KARLSON, P.: Zum Tyrosinstoffwechsel der Insekten. III. Über den Einbau von Tyrosin-Umwandlungsprodukten in das Puppentönnchen. Z. Physiol. Chem. *318*, 194–200 (1960).

[144] KARLSON, P.: Morphogenese und Metamorphose der Insekten. 14. Coll. Ges. Phys. Chem., Mosbach/Baden, pp. 101–126 (1962).

[145] KARLSON, P.: Mechanism of hormone action. A Nato Advanced Study Institute 139–148. Meersling/Bodensee, 20–26. Mai 1964. Stuttgart: G. Thieme. 1965.

[146] KARLSON, P.: Biochemistry and mode of action of ecdysone. In: Excerpta Media International Congress Series No. *83*, 516–519 (1964).

[147] KARLSON, P.: Ecdyson, das Häutungshormon der Insekten. Naturwissenschaften *53*, 445–453 (1966).

[148] KARLSON, P., and C. E. SEKERIS: Ecdysone, an insect steroid hormone, and its mode of action. Rec. Progr. Horm. Res. *22*, 473–502 (1966).

[149] KARLSON, P.: The chemistry of insect hormones and insect pheromones. Pure and Appl. Chem. *14*, 75–87 (1967).

[150] KARLSON, P., and C. BODE: Die Inaktivierung des Ecdysons bei der Schmeißfliege *Calliphora erythrocephala* Meigen. J. Insect Physiol. *15*, 111–118 (1969).

[151] KARLSON, P., L. F. CONGOTE, and C. E. SEKERIS: (see reference No. 159 on p. 289).

[152] KARLSON, P., and H. HOFFMEISTER: Zur Biogenese des Ecdysons. I. Umwandlung von Cholesterin in Ecdyson. Z. Physiol. Chem. *331*, 298–300 (1963).

[153] KARLSON, P., H. HOFFMEISTER, W. HOPPE, and R. HUBER: Zur Chemie des Ecdysons. Ann. Chem. *662*, 1–20 (1963).

[154] KARLSON, P., H. HOFFMEISTER, H. HUMMEL, P. HOCKS, and G. SPITELLER: Zur Chemie des Ecdysons. VI. Reaktionen des Ecdysonmoleküls. Chem. Ber. *98*, 2394–2402 (1965).

[155] KARLSON, P., and H. LIEBAU: Zum Tyrosinstoffwechsel der Insekten. V. Reindarstellung, Kristallisation und Substratspezifität der o-Diphenoloxidase aus *Calliphora erythrocephala*. Z. Physiol. Chem. *326*, 135–143 (1961).

[156] KARLSON, P., D. MEYENHAGEN, and C. E. SEKERIS: Zum Tyrosinstoffwechsel in Insekten. XV. Weitere Untersuchungen über das o-Diphenyloxidase-System von *Calliphora erythrocephala*. Z. Physiol. Chem. *338*, 42–50 (1964).

[157] KARLSON, P., and G. PETERS: Zum Wirkungsmechanismus der Hormone. Der Einfluß des Ecdysons auf den Nucleinsauerstoffwechsel von *Calliphora*-Larven. Gen. Comp. Endocrinol. *5*, 252–259 (1965).

[158] KARLSON, P., and E. SHAAYA: Der Ecdysontiter während der Insektenentwicklung. I. Eine Methode zur Bestimmung des Ecdysongehalts. J. Insect Physiol. *10*, 797–804 (1964).

[159] Karlson, P., and H. Schmied: Über die Tyrosinase der *Calliphora*-Larven. Z. Physiol. Chem. *300*, 35–41 (1955).

[160] Karlson, P., and C. E. Sekeris: N-Acetyl-dopamine as sclerotizing agent in the insect cuticle. Nature *195*, 183–184 (1962).

[161] Karlson, P., and C. E. Sekeris: Ecdysone, an insect steroid hormone, and its mode of action. In: Recent progress in hormone research, Vol. *22*, pp. 473–502 (1966). New York: Academic Press.

[162] Karlson, P., C. E. Sekeris, and R. Maurer: Zum Wirkungsmechanismus der Hormone. I. Verteilung von Tritium-markiertem Ecdyson in Larven von *Calliphora erythrocephala*. Z. Physiol. Chem. *336*, 100–106 (1964).

[163] Kerb, U., P. Hocks, R. Wiechert, A. Furlenmeier, A. Fürst, A. Langemann, and G. Waldvogel: Die Synthese des Ecdysons. Tetrahedron Letters *1966*, 1387–1391.

[164] Kerb, U., G. Schulz, P. Hocks, R. Wiechert, A. Furlenmeier, A. Fürst, A. Langemann, and G. Waldvogel: Zur Synthese des Ecdysons. Die Synthese des natürlichen Häutungshormons. Helv. *49*, 1601–1606 (1966).

[165] Kerb, U., R. Wiechert, A. Furlenmeier, and A. Fürst: Über eine Synthese des Crustecdysons (20-Hydroxyecdysons). Tetrahedron Letters *1968*, 4277–4280.

[166] King, D. S., and J. B. Siddall: Conversion of α-ecdysone to β-ecdysone by crustaceans and insects. Nature *221*, 955–956 (1969).

[167] Knight, J. C., and G. R. Pettit: Arizona flora: The sterols of *Peniocereus greggii*. Phytochemistry *8*, 477–482 (1969).

[168] Kok, L. T., D. M. Norris, and H. M. Chu: Sterol metabolism as a basis for a mutualistic symbiosis. Nature *225*, 661–662 (1970).

[169] Kobayashi, M., and H. Akai: Action of ecdysone on some metabolism during larval-pupal transformation of the housefly, *Musca domestica* L. (Diptera: Muscidae). Appl. Ent. Zool. *2*, 223–224 (1967).

[170] Kobayashi, M., and S. Kimura: Action of ecdysone on the conversion of ^{14}C-glucose in dauer pupa of the silkworm, *Bombyx mori*. J. Insect Physiol. *13*, 545–552 (1967).

[171] Kobayashi, M., T. Takemoto, S. Ogawa, and N. Nishimoto: The moulting hormone activity of ecdysterone and inokosterone isolated from *Achyranthis radix*. J. Insect. Physiol. *13*, 1395–1399 (1967).

[172] Kopeć, S.: Experiments on metamorphosis of insects. Bull. Int. Acad. Sci. (Cracovie) *B 1917*, 57–60.

[173] Kopeć, S.: Studies on the necessity of the brain for the inception of insect metamorphosis. Biol. Bull. Woods Hole *42*, 323–342 (1922).

[174] Koreeda, M., and K. Nakanishi: 5 β-Hydroxy-ecdysones and revision of the structure of ponasterone C. Chem. Commun. *1970*, 351–352.

[175] Koreeda, M., K. Nakanishi, and M. Goto: Ajugalactone, an insect moulting inhibitor as tested by the Chilo dipping method. J. Amer. chem. Soc. *92*, 7512–7513 (1970).

[176] Krishnakumaran, A., S. J. Berry, H. Oberlander, and H. A. Schneiderman: Nucleic acid synthesis during insect development. II. Control of DNA synthesis in the *Cecropia* silkworm and other *Saturniid moths*. J. Insect Physiol. *13*, 1–57 (1967).

[177] Krishnakumaran, A., H. Oberlander, and H. A. Schneiderman: (see reference No. 175 on p. 77) (1963).

[178] KRISHNAKUMARAN, A., and H. A. SCHNEIDERMAN: Chemical control of moulting in *Arthropods*. Nature *220*, 601–603 (1968).

[179] KRISHNAKUMARAN, A., and H. A. SCHNEIDERMAN: Induction of moulting in *Crustacea* by an insect moulting hormone. Gen. Comp. Endocrinol. *12*, 515–518 (1969).

[180] KRISHNAKUMARAN, A., and H. A. SCHNEIDERMAN: Control of moulting in mandibulate and chelicerate *Arthropods* by ecdysones. Biol. Bull. Wood's Hole *139*, 520–538 (1970).

[181] KROEGER, H.: Chemical nature of the system controlling gene activities in insect cells. Nature *200*, 1234–1235 (1963).

[182] KROEGER, H.: The mechanism of gene activation in Dipteran salivary gland chromosomes. Arch. Anat. Micr. *54*, 643–645 (1965).

[183] KROEGER, H.: Potentialdifferenz und Puffmuster. Elektrophysiologische und cytologische Untersuchungen an den Speicheldrüsen von *Chironomus tentans*. Exp. Cell Res. *41*, 64–80 (1966).

[184] KROEGER, H.: Gene activities during insect metamorphosis and their control by hormones. In: Metamorphosis: a problem in developmental biology, pp. 185–219. Published by Appleton-Century-Crofts, New York, Educational Division, Meredith Corporation. 1966.

[185] KROEGER, H., and M. LEZZI: Regulation of gene action in insect development. Ann. Rev. Entomol. *11*, 1–22 (1966).

[186] LAUFER, H.: Developmental interactions in the Dipteran salivary gland. Amer. Zoologist *8*, 257–271 (1968).

[187] LAUFER, H., B. RAO, and Y. NAKASE: Developmental studies of the Dipteran salivary gland. IV. Changes in DNA content. J. Exp. Zool. *166*, 71–76 (1967).

[188] LAVERDURE, A. M.: L'évolution de l'ovaire chez la nymphe et l'adulte de *Tenebrio molitor* (Coléoptére). La vitellogenèse. Bull. Soc. Zool. (France) *95*, 753–764 (1970).

[189] LEENDERS, H. J., G. J. WULLEMS, and H. D. BERENDES: Competitive interaction of adenosine 3',5'-monophosphate on gene activation by ecdysterone. Exp. Cell. Res. *63*, 159–164 (1970).

[190] LEZZI, M.: Induktion eines Ecdyson-aktivierbaren Puff in isolierten Zellkernen von *Chironomus* durch KCl. Exp. Cell. Res. *43*, 571–547 (1966).

[191] LEZZI, M., and M. FRIGG: (see reference No. 178 on p. 290).

[192] LEZZI, M., and L. I. GILBERT: (see reference No. 179 on p. 290).

[193] LEZZI, M., and L. I. GILBERT: Differential effects of K^+ and Na^+ on specific bands of isolated polytene chromosomes of *Chironomus tentans*. J. Cell. Sci. *6*, 615–628 (1970).

[194] LOCKSHIN, R. A.: Programmed cell death. Activation of lysis by a mechanism involving the synthesis of protein. J. Insect Physiol. *15*, 1505–1516 (1969).

[195] LOCKSHIN, R. A., and C. M. WILLIAMS: Programmed cell death. II. Endocrine potentiation of the breakdown of the intersegmental muscles of silkmoths. J. Insect Physiol. *10*, 643–649 (1964).

[196] LOWE, M. E., D. H. S. HORN, and M. N. GALBRAITH: The role of crustecdysone in the moulting crayfish. Experientia *24*, 518–519 (1968).

[197] LUCAS, K. U., and G. R. WYATT: DNA-RNA hybridization studies on ecdysone-induced wing development in the Cecropia silkmoth. J. Insect Physiol. *17*, 2301–2316 (1971).

[198] LUPIEN, P. J., C. HINSE, and K. D. CHAUDHARY: Ecdysone as a hypocholesterolemic argent. Archs. Int. Physiol. Biochim. *77*, 206–212 (1969).

[199] MADHAVAN, K., and H. A. SCHNEIDERMAN: Effects of ecdysone on epidermal cells in which DNA synthesis has been blocked. J. Insect Physiol. *14*, 777–781 (1968).

[200] MADHAVAN, K., and H. A. SCHNEIDERMAN: Hormonal control of imaginal disc regeneration in *Galleria mellonella* (Lepidoptera). Biol. Bull. Woods Hole *137*, 321–331 (1969).

[201] MAISSIAT, J.: Un ecdysis expérimentale provoqué chez l'Oniscoide *Ligia oceanica* L. et rétablissement de la mue par injection d'ecdysone ou réimplantation de glande maxillaire. C. R. Séanc. Biol. *164*, 1608–1609 (1970).

[202] MANDARON, P.: Développement *in vitro* des disques imaginaux de la *Drosophile*. Aspects morphologiques et histologiques. Develop. Biol. *22*, 298–320 (1970).

[203] MARKS, E. P.: The action of hormones in insect cell and organ cultures. Gen. Comp. Endocrinol. *15*, 289–302 (1970).

[204] MARKS, E. P.: Effects of ecdysterone on the deposition of cockroach cuticle *in vitro*. Biol. Bull. Woods Hole *142*, 293–301 (1972).

[205] MARKS, E. P., P. I. ITTYCHERIAH, and A. M. LELOUP: The effects of α-ecdysone on insect neurosecretion *in vitro*. J. Insect Physiol. *18*, 847–850 (1972).

[206] MARKS, E. P., and R. A. LEOPOLD: Deposition of cuticular substances *in vitro* by leg regenerates from the cockroach, *Leucophaea maderae* (F.). Biol. Bull. Woods Hole *140*, 73–83 (1971).

[207] MASUOKA, M., S. PRITA, A. SHINO, T. MATSUZAWA, and R. NAKAYAMA: Pharmacological studies of insect metamorphosing hormone: Ponasterone A, ecdysterone, and inokosterone in the rat. Jap. J. Pharmacol. *20*, 142–156 (1970).

[208] MAYOR, P. A., and G. D. MEAKINS: Steroids of unnatural configuration. IV. Oxidation of lumisterol and lumisterol acetate with perbenzoic acid. J. Chem. Soc. *1960*, 2792–2800.

[209] MONROE, R. E., T. L. HOPKINS, and S. A. VALDER: Metabolism and utilization of cholesterol-4-C^{14} for growth and reproduction of aseptically reared houseflies (*Musca domestica*). J. Insect Physiol. *13*, 219–233 (1967).

[210] MOROHOSHI, S., and T. IIJIMA: Induction of supernumerous ecdysis by the injection of ecdysones in *Bombyx mori*. Proc. Japan Acad. *45*, 314–318 (1969).

[211] MORI, H., and K. SHIBATA: Synthesis of ecdysterone. Chem. Pharm. Bull. *17*, 1970–1973 (1969).

[212] MORI, H., K. SHIBATA, K. TSUNEDA, and M. SAWAI: Synthesis of ecdysone. Chem. Pharm. Bull. *16*, 563–566 (1968).

[213] MORI, H., K. SHIBATA, K. TSUNEDA, and M. SAWAI: Synthesis of ecdysone. IV. A novel synthesis of the side chain structure of ecdysone. Chem. Pharm. Bull. *17*, 690–698 (1969).

[214] MUFTIG, M.: Metamorphosis of miracidia into cercaria of *Schistosoma mansoni*. Parasitology *59*, 365–371 (1969).

[215] NAKANISHI, K.: Ponasterones, compounds with moulting hormone activity. Bull. Soc. Chim. Fr. *1969*, 3475–3485.

[216] NAKANISHI, K.: The ecdysones. Pure and Appl. Chem. *25*, 167–195 (1971).

[217] NAKANISHI, K., M. KOREEDA, M. L. CHANG, H. Y. HSU: Insect hormones. V. The structures of ponasterones B and C. Tetrahedron Letters *1968*, 1105–1110.

[218] NAKANISHI, K., M. KOREEDA, S. SASAKI, M. L. CHANG, and H. Y. HSU: Insect hormones. The structure of ponasterone A, an insect-moulting hormone from the leaves of *Podocarpus nakaii* Hay. Chem. Commun. *1966*, 915–917.

[219] NEUFELD, G. J., J. A. THOMSON, and D. H. S. HORN: Short-term effects of crustecdysone (20-hydroxyecdysone) on protein and RNA synthesis in third instar larvae of *Calliphora*. J. Insect Physiol. *14*, 789–804 (1968).

[220] NOVÁK, V. J. A.: (see reference No. 219 on p. 79).

[221] NOVÁK, V. J. A.: (see reference No. 218 on p. 79).

[222] OBERLANDER, H.: Ecdysone and DNA synthesis in cultured wing discs of the wax moth, *Galleria mellonella*. J. Insect Physiol. *15*, 1803–1806 (1969).

[223] OBERLANDER, H.: α-ecdysone induced DNA synthesis in cultured wing discs of *Galleria mellonella*: Inhibition by 20-hydroxyecdysone and 22-isoecdysone. J. Insect Physiol. *18*, 223–228 (1972).

[224] OHNISHI, E., T. OHTAKI, and S. FUKUDA: Ecdysone in the eggs of *Bombyx* silkworm. Proc. Japan. Acad. *47*, 413–415 (1971).

[225] OHTAKI, T.: On the delayed putation of the fleshfly, *Sarcophaga peregrina* Robineau-Desvoidy. Jap. J. Med. Sci. Biol. *19*, 97–104 (1966).

[226] OHTAKI, T., R. D. MILKMAN, and C. M. WILLIAMS: Ecdysone and ecdysone analogues: Their assay on the fleshfly *Sarcophaga peregrina*. Proc. nat. Acad. Sci. (Amer.) *58*, 981–984 (1967).

[227] OHTAKI, T., R. D. MILKMAN, and C. M. WILLIAMS: Dynamics of ecdysone secretion and action in the fleshfly *Sarcophaga peregrina*. Biol. Bull. Woods Hole *135*, 322–334 (1968).

[228] OHTAKI, T., and C. M. WILLIAMS: Inactivation of α-ecdysone and cyasterone by larvae of the fleshfly, *Sarcophaga peregrina* and pupae of the silkworm, *Samia cynthia*. Biol. Bull. *138*, 326–333 (1970).

[229] OTAKA, T., S. OKUJI, and M. UCHIYAMA: Stimulation of protein synthesis in mouse liver by ecdysterone. Chem. Pharm. Bull. Tokyo *17*, 75–81 (1969).

[230] OTAKA, T., and M. UCHIYAMA: Stimulatory effect of ecdysterone on RNA synthesis in mouse liver. Chem. Pharm. Bull. (Tokyo) *17*, 1883–1888 (1969).

[231] OTAKA, T., M. UCHIYAMA, T. TAKEMOTO, and H. HIKINO: Stimulatory effect of insect metamorphosing steroids from ferns on protein synthesis in mouse liver. Chem. Pharm. Bull. (Tokyo) *17*, 1352–1355 (1969).

[232] PIECHOWSKA, M. J., Z. SINKIEWICZ, and M. BIELINSKA: Hormones of insect metamorphosis. Post. Biochem. *16*, 449–481 (1970).

[233] PIPA, R. L.: Insect neurometamorphosis. IV. Effects of the brain and synthetic α-ecdysone upon interganglionic connective shortening in *Galleria mellonella* L. J. Exp. Zool. *170*, 181–182 (1969).

[234] PLAGGE, E., and E. BECKER: Wirkung arteigener und artfremder Verpuppungshormone in Extrakten. Naturwissenschaften *26*, 430–431 (1938).

[235] POELS, C. L. M.: Time sequence in the expression of various developmental characters induced by ecdysterone in *Drosophila hydei*. Develop. Biol. *23*, 210–225 (1970).

[236] Poels, C. L., A. de Loof, and H. D. Berendes: Functional and structural changes in *Drosophila* salivary gland cells triggered by ecdysterone. J. Insect Physiol. *17*, 1717–1729 (1971).

[237] Postlethwait, J. H., and H. A. Schneiderman: Effects of an ecdysone on growth and cuticle formation of *Drosophila* imaginal discs cultured *in vivo*. Biol. Bull. Woods Hole *135*, 431–432 (1968).

[238] Postlethwait, J. H., and H. A. Schneiderman: Induction of metamorphosis by ecdysone analogues: *Drosophila* imaginal discs cultured *in vivo*. Biol. Bull. Woods Hole *138*, 47–55 (1970).

[239] Price, C. M.: Pupation inhibition factor in the larva of the blowfly *Calliphora erythrocephala*. Nature *228*, 876–877 (1970).

[240] Radford, S. V., and D. W. Misch: The cytological effect of ecdysterone on the midgut cells of the flesh-fly *Sarcophaga bullata*. J. Cell. Biol. *49*, 702–711 (1971).

[241] Raikow, R., and J. W. Fristrom: Effects of β-ecdysone on RNA metabolism of imaginal discs of *Drosophila melanogaster*. J. Insect Physiol. *17*, 1599–1614 (1971).

[242] Rees, H. H.: Ecdysones. In: Goodwin, T. W., ed.: Aspects of terpenoid chemistry and biochemistry, pp. 181–222. London-New York: Academic Press. 1971.

[243] Reidenbach, J. M.: Action d'une ecdysone de synthèse sur la mue et le fonctionnement ovarien chez le crustacé isopode *Idothea balthica* (Pallas). C. R. Acad. Sci. (Paris) *273*, 1614–1617 (1971).

[244] Reinecke, J. P., and J. D. Robbins: Reaction of an insect cell line to ecdysterone. Exp. Cell Res. *64*, 335–338 (1971).

[245] Řežábová, B., J. Hora, V. Landa, V. Černý, and F. Šorm: Sterilizing effect of some 6-ketosteroids on housefly (*Musca domestica* L.). Steroids *11*, 475–496 (1968).

[246] Richman, K., and H. Oberlander: Effects of fat body on α-ecdysone induced morphogenesis in cultured wing discs of the wax moth, *Galleria mellonella*. J. Insect Physiol. *17*, 269–276 (1971).

[247] Rimpler, H.: Pterosteron, Polypodin B und ein neues ecdysteronartiges Steroid (Viticosteron E) aus *Vitex megapotamica* (Verbenacae). Tetrahedron Letters *1969*, 329–333.

[248] Rimpler, H., und G. Schulz: Vorkommen von 20-Hydroxy-ecdyson in *Vitex megapotamica*. Tetrahedron Letters *1967*, 2033–2035.

[249] Robbins, W. E., J. N. Kaplanis, M. J. Thompson, T. J. Shortino, C. F. Cohen, and S. C. Joyner: Ecdysones and analogs: effects on development and reproduction of insects. Science *161*, 1158–1159 (1968).

[250] Robbins, W. E., J. N. Kaplanis, M. J. Thompson, T. J. Shortino, and S. C. Joyner: Ecdysones and synthetic analogs: Molting hormone activity and inhibitive effects on insect growth, metamorphosis and reproduction. Steroids *16*, 105–125 (1970).

[251] Robbins, W. E., M. J. Thompson, J. N. Kaplanis, and T. J. Shortino: Conversion of cholesterol to 7-dehydrocholesterol in aseptically reared German cockroaches. Steroids *4*, 635 (1964).

[252] Romer, F.: Häutungshormone in den Oenocyten des Mehlkäfers. Naturwissenschaften *58*, 324–325 (1971).

[253] Sahota, T. S., and A. Mansingh: Cellular response to ecdysone: RNA and protein synthesis in larval tissues of oak silkworm *Antheraea pernyi*. J. Insect Physiol *16*, 1649–1654 (1970).

[254] Sakurai, H., and K. Hasegawa: Response of isolated pupal abdomens of silkworm *Bombyx mori* L., to injected ponasterone A (Lepidoptera: Bombycidae). Appl. Ent. Zool. *4*, 59–65 (1969).

[255] Sato, Y., M. Sakai, S. Imai, and S. Fujioka: Ecdysone activity of plant-originated moulting hormones applied on the body surface of Lepidopterous larvae. Appl. Ent. Zool. *3*, 49–51 (1968).

[256] Sauer, H. H., R. D. Bennett, and E. Heftmann: Ecdysterone biosynthesis in *Podocarpus elata*. Phytochemistry *7*, 2027–2030 (1968).

[257] Saunders, J. W., Jr.: Death in embryonic systems. Science *154*, 604–606 (1966).

[258] Schaller, F.: Action de l'ecdysone sur des larves d'*Aeschna cyanea* Müll. (Insecte Odonate) en diapause. Mém. Soc. Sci. Nat. Math. (Cherbourg) *51*, 135–140 (1966).

[259] Schaller, F., and J. C. Andries: Role de l'ecdysone dans la multiplication des cellules de régénération de l'intestin moyenne chez la larve d'*Aeschna cyanea* Müll. (Insecte Odonate). C. R. hebd. Séanc. Acad. Sci. (Paris) *D 271*, 426–429 (1970).

[260] Sehnal, F.: Endocrines of arthropods. Chem. Zool. *6*, 307–345 (1971).

[261] Sehnal, F.: Action of ecdysone on ligated larvae of *Galleria mellonella* L. (Lepidoptera): Induction of development. Acta ent. Bohemoslov. *69*, 143–155 (1972).

[262] Sekeris, C. E.: Zum Tyrosinstoffwechsel der Insekten. XII. Reinigung, Eigenschaften und Substratspezialität der Dopa-Decarboxylase. Z. Physiol. Chem. *332*, 70–78 (1963).

[263] Sekeris, C. E.: Action of ecdysone on RNA and protein metabolism in the blowfly, *Caliphora erythrocephala*. In: Karlson, P., ed.: Mechanisms of hormone action, pp. 149–167. Stuttgart: G. Thieme. 1965.

[264] Sekeris, C. E., P. P. Dukes, and W. Schmied: The action of ecdysone on nucleic acid metabolism in insects. Biochem. J. *97*, 23–24 (1965).

[265] Sekeris, C. E., P. P. Dukes, and W. Schmied: Wirkung von Ecdyson auf Epidermiszellkerne von *Calliphora*-Larven *in vitro*. Z. Physiol. Chem. *341*, 152–154 (1965).

[266] Sekeris, C. E., and P. Karlson: Zum Tyrosinstoffwechsel der Insekten. VII. Der katabolische Abbau des Tyrosins und die Biogenese der Sklerotisierungssubstanze, N-Acetyl-Dopamin. Biochem. biophys. Acta *62*, 103–113 (1962).

[267] Sekeris, C. E., and P. Karlson: On the mechanic of hormone action. II. Ecdysone and protein synthesis. Arch. Biochem. Biophys. *105*, 483–487 (1964).

[268] Sekeris, C. E., and P. Karlson: Biosynthesis of catecholamines in insects. Pharmac. Rev. *18*, 89–94 (1966).

[269] Sekeris, C. E., and N. Lang: Induction of dopa-decarboxylase activity by insect messenger RNA in an *in vitro* amino acid incorporating system from rat liver. Life Sciences *3*, 625–632 (1964).

[270] Sekeris, C. E., N. Lang, and P. Karlson: Zum Wirkungsmechanismus der Hormone. V. Der Einfluß von Ecdyson auf den RNA-Stoffwechsel in der Epidermis der Schmeißfliege *Calliphora erythrocephala*. Z. Physiol. Chem. *341*, 36–43 (1965).

[271] Sekeris, C. E., and D. Mergenhagen: Phenoloxydase system of the blowfly *Calliphora erythrocephala*. Science *145*, 68–69 (1964).

[272] Shaaya, E.: Der Ecdysontiter während der Insektenentwicklung. VI. Untersuchungen über die Verteilung des Ecdysons in verschiedenen Geweben von *Calliphora erythrocephala* und über seine biologische Halbwertszeit. Z. Naturforsch. *24 b*, 718–721 (1969).

[273] Shaaya, E., and P. Karlson: Der Ecdysontiter während der Insektenentwicklung. II. Die Postembryonale Entwicklung der Schmeißfliege *Calliphora erythrocephala* Meig. J. Insect Physiol. *11*, 65–69 (1965).

[274] Shaaya, E., and P. Karlson: Der Ecdysontiter während der Insektenentwicklung. IV. Die Entwicklung der Lepidopteren *Bombyx mori* L. und *Cerura vinula* L. Develop. Biol. *11*, 424–432 (1965).

[275] Shaaya, E., and C. E. Sekeris: Ecdysone during insect development. III. Activities of some enzymes of tyrosine metabolism in comparison with ecdysone titer during the development of the blowfly, *Calliphora erythrocephala* Meig. Gen. Comp. Endocrinol. *5*, 35–39 (1965).

[276] Shigematsu, H., and H. Moriyama: Effect of ecdysterone on fibroin synthesis in the posterior division of the silk gland of the silkworm, *Bombyx mori*. J. Insect Physiol. *16*, 2015-2022 (1970).

[277] Siddall, J. B., A. D. Cross, and J. H. Fried: Synthetic studies on insect hormones. II. The synthesis of ecdysone. J. Amer. chem. Soc. *88*, 862–863 (1966).

[278] Siddall, J. B., D. H. S. Horn, and E. J. Middleton: Synthetic studies on insect hormones. The synthesis of a possible metabolite of crustecdysone (20-hydroxyecdysone). Chem. Commun. *1967*, 899–900.

[279] Siddall, J. B., J. P. Marshall, A. Bowers, A. D. Cross, J. A. Edwards, and J. H. Fried: Synthetic studies on insect hormones. I. Synthesis of the tetracyclic nucleus of ecdysone. J. Amer. chem. Soc. *88*, 379–380 (1966).

[280] Siew, Y. C., and L. I. Gilbert: (see reference No. 302 on p. 297).

[281] Sláma, K.: (see reference No. 311 on p. 297).

[282] Socha, R., and F. Sehnal: (see reference No. 320 on p. 298).

[283] Souza, N. J. de, E. L. Ghisalberti, H. H. Rees, and T. W. Goodwin: Studies on insect moulting hormones: biosynthesis of ponasterone A and ecdysterone from /2-^{14}C/-mevalonate in *Taxus baccata*. Biochem. J. *114*, 895–896 (1969).

[284] Souza, N. J. de, E. L. Ghisalberti, H. H. Rees, and T. W. Goodwin: Studies on insect moulting hormones: biosynthesis of ecdysone, ecdysterone and 5-β-hydroxyecdysterone in *Polypodium vulgare*. Phytochemistry *9*, 1247–1252 (1970).

[285] Spielman, A., R. W. Gwadz, and W. A. Anderson: Ecdysone-initiated ovarian development in mosquitoes. J. Insect Physiol. *17*, 1807–1814 (1971).

[286] Staal, G. B.: Plants as a source of insect hormones. Koninkl. Nederl. Akad. Wettenschap. *C 70*, 409–418 (1961).

[287] Stamm, M. D.: Estudios sobre hormones de invertebrados. II. Aistamiendo de hormonas de la metamorfosis en el *Ortóptero Dociostaurus maroccanus*. Anales real Soc. Espan. Fis. Quim. *B 55*, 171–178 (1959).

[288] Stewart, J. R., and J. P. Greeb: Ecdysone mediated events in the moulting of the fiddler crab, *Uca pugilator*. Amer. Zool. *9*, 579 (1969).

[289] Svoboda, J. A., R. F. N. Hutchins, M. J. Thompson, and W. E. Robbins: 22-trans-cholesta-5,22,24-trien-3 β-ol-an intermediate in the conversion of stigmasterol to cholesterol in the tobacco hornworm *Manduca sexta* (Johannson). Steroids *14*, 469–476 (1969).

[290] TAKEDA, N.: Effect of ecdysterone on spermatogenesis in the diapausing slug moth pharate pupa, *Monema flavescens*. J. Insect Physiol. *18*, 571–580 (1972).

[291] TAKEMOTO, T., S. ARIHARA, Y. HIKINO, and H. HIKINO: Isolation of insect moulting substances from *Pteridium aquillinum var. latiusculum*. Chem. Pharm. Bull. *16*, 762 (1968).

[292] TAKEMOTO, T., Y. HIKINO, T. ARAI, and H. HIKINO: Structure of lemmasterone, a novel C_{29} insect-moulting substance from *Lemmaphyllum microphyllum*. Tetrahedron Letters *1968*, 4061–4064.

[293] TAKEMOTO, T., Y. HIKINO, T. ARAI, M. KAWAHARA, CH. KONNO, S. ARIHARA, and H. HIKINO: Isolation of insect moulting substances from *Matteuccia struthiopteris*, *Lastrea thelypteris* and *Onoclea sensibilis*. Chem. Pharm. Bull. *15*, 1816 (1967).

[294] TAKEMOTO, T., Y. HIKINO, T. ARAI, CH. KONNO, S. NABETANI, and H. HIKINO: Isolation of insect moulting substances from *Pleopeltis thunbergiana*, *Neocheiropteris ensata*, and *Lemmaphyllum microphyllum*. Chem. Pharm. Bull. *16*, 759–760 (1968).

[295] TAKEMOTO, T., Y. HIKINO, S. ARIHARA, and H. HIKINO: Absolute configuration of inokosterone, an insect-moulting substance from *Achyranthes Fauriei*. Tetrahedron Letters *1968*, 2475–2478.

[296] TAKEMOTO, T., Y. HIKINO, H. HIKINO, S. OGAWA, and N. NISHIMOTO: Structure of Rubresterone, a novel C_{19} Metabolite of insect-moulting substances from *Achyranthes rubrofusca*. Tetrahedron Letters *1968*, 3053–3056.

[297] TAKEMOTO, T., Y. HIKINO, H. HIKINO, S. OGAWA, and N. NISHIMOTO: Rubrosterone, a metabolite of insect metamorphosing substance from *Achyranthes rubrofusca*: structure and absolute configuration. Tetrahedron *25*, 1241–1248 (1969).

[298] TAKEMOTO, T., Y. HIKINO, H. JIN, T. ARAI, and H. HIKINO: Isolation of insect moulting substances from *Osmunda japonica* and *Osmunda asiatica*. Chem. Pharm. Bull. *16*, 1636, (1968).

[299] TAKEMOTO, T., Y. HIKINO, H. JIN, and H. HIKINO: Isolation of ponasterone A from *Taxus cuspidata var. nana*. J. Pharm. Soc. Japan *88*, 359 (1968).

[300] TAKEMOTO, T., Y. HIKINO, K. NOMOTO, and H. HIKINO: Structure of cyasterone, a novel C_{29} insect-moulting substance from *Cyathula capitata*. Tetrahedron Letters *1967*, 3191–3194.

[301] TAKEMOTO, T., Y. HIKINO, T. OKUYAMA, S. ARIHARA, and H. HIKINO: Structure of shidasterone, a novel insect-moulting substance from *Blechnum niponicum*. Tetrahedron Letters *1968*, 6095–6098.

[302] TAKEMOTO, T., K. NOMOTO, and H. HIKINO: Structure of amarasterone A and B, novel C_{29} insect-moulting substances from *Cyathula capitata*. Tetrahedron Letters *1968*, 4953–4956.

[303] TAKEMOTO, T., K. NOMOTO, Y. HIKINO, and H. HIKINO: Structure of capitasterone, a novel C_{29} insect-moulting substance from *Cyathula capitata*. Tetrahedron Letters *1968*, 4929–4932.

[304] TAKEMOTO, T., S. OGAWA, and N. NISHIMOTO: Isolation of the moulting hormones of insects from *Achyranthis radix*. J. Pharm. Soc. Japan *87*, 325–327 (1967).

[305] TAKEMOTO, T., S. OGAWA, and N. NISHIMOTO: Studies on the constituents of *Achyranthis radix*. II. Isolation of the insect-moulting hormones. J. Pharm. Soc. Japan *87*, 1469–1473 (1967).

[306] TAKEMOTO, T., S. OGAWA, and N. NISHIMOTO: Studies on the constituents of *Achyranthis radix*. III. Structure of inokosterone. J. Pharm. Soc. Japan *87*, 1474–1477 (1967).

[307] TAKEMOTO, T., S. OGAWA, N. NISHIMOTO, H. HIRAYAMA, and S. TANIGUCHI: Isolation of the insect-moulting hormones from mulberry leaves. J. Pharm. Soc. Japan *87*, 748 (1967).

[308] TAKEMOTO, T., S. OGAWA, N. NISHIMOTO, and H. HOFFMEISTER: Steroide mit Häutungshormon-Aktivität aus Tieren und Pflanzen. Z. Naturforsch. *22 b*, 681–682 (1967).

[309] TAKEMOTO, T., S. OGAWA, N. NISHIMOTO, and S. TANIGUCHI: Studies on the constituents of *Achyranthis radix*. IV. Isolation of the insect-moulting hormones from *Formosan Achyranthes* spp. J. Pharm. Soc. Japan *87*, 1478–1480 (1967).

[310] TAKEMOTO, T., T. OKUYAMA, S. ARIHARA, Y. HIKINO, and H. HIKINO: Isolation of insect-moulting substances from *Blechnum amabile* and *Blechnum niponicum*. Chem. Pharm. Bull. *17*, 1973–1974 (1969).

[311] THOMSON, J. A.: The interpretation of puff patterns in polytene chromosomes. In: Currents in Modern Biology *2*, 333–338 (1969).

[312] THOMSON, J. A., and D. H. S. HORN: Effect of exogenous moulting hormones on puparium formation in *Calliphora*. Aust. J. Biol. Sci. *22*, 761–765 (1969).

[313] THOMSON, J. A., F. P. IMRAY, and D. H. S. HORN: An improved Calliphora bioassay for insect moulting hormones. Aust. J. Exp. Biol. Med. Sci. *48*, 321–328 (1970).

[314] THOMSON, J. A., J. F. KINNEAR, M. D. MARTIN, and D. H. S. HORN: Effects of crustecdysone (20-hydroxyecdysone) on synthesis, release, and uptake of proteins by the larval fat body of *Calliphora*. Life Sciences *10*, 203–211 (1971).

[315] THOMPSON, M. J., J. N. KAPLANIS, W. E. ROBBINS, and R. T. YAMAMOTO: 20,26-Dihydroxyecdysone, a new steroid with moulting hormone activity from tobacco hornworm *Manduca sexta* (Johannson). Chem. Commun. *1967*, 650–653.

[316] THOMPSON, M. J., W. E. ROBBINS, J. N. KAPLANIS, C. F. COHEN, and S. M. LANCASTER: Synthesis of analogs of α-ecdysone. A simplified synthesis of 2 β,3 β-14 α-trihydroxy-7-en-6-one-5 β-steroids. Steroids *16*, 85–104 (1970).

[317] THOMSON, J. A., D. C. ROGERS, M. M. GUNSON, and D. H. S. HORN: Developmental changes in the pattern of cellular distribution of exogenous tritium-labelled crustecdysone in larval tissues of *Calliphora*. Cytobios *6*, 79–88 (1970).

[318] THOMSON, J. A., J. B. SIDDALL, M. N. GALBRAITH, D. H. S. HORN, and E. J. MIDDLETON: The biosynthesis of ecdysones in the blowfly *Calliphora stygia*. Chem. Commun. *1969*, 669–670.

[319] VELGOVÁ, H., V. ČERNÝ, F. ŠORM, and K. SLÁMA: Further compounds with antisclerotization effect on *Pyrrhocoris apterus* L. larvae: Structure and activity correlations. Coll. Czech. Chem. Commun. *34*, 3354–3376 (1969).

[320] VELGOVÁ, H., L. LÁBLER, V. ČERNÝ, F. ŠORM, and K. SLÁMA: Some further compounds producing moulting deficiencies in an insect. Coll. Czech. Chem. Commun. *33*, 242–256 (1968).

[321] WEIRICH, G., and P. KARLSON: Distribution of tritiated ecdysone in salivary glands and other tissues of *Rhynchosciara* and *Chironomus* larvae. An autoradiographic study. Roux Arch. *164*, 170–181 (1969).

[322] Wiechert, R., U. Kerb, P. Hocks, A. Furlenmeier, A. Fürst, A. Langemann, and G. Waldvogel: Zur Synthese des Ecdysons. Synthesen von 2 β,3 β-Dihydroxy-6-keto-A/B-cis Steroiden. Helv. *49*, 1581–1591 (1966).

[323] Wigglesworth, V. B.: The physiology of ecdysis in Rhodnius. II. Factors controlling moulting and metamorphosis. Quart. J. Micr. Sci. *77*, 191–222 (1934).

[324] Wigglesworth, V. B.: The action of moulting hormone and juvenile hormone at the cellular level in *Rhodnius prolixus*. J. Exp. Biol. *40*, 231–245 (1963).

[325] Wilde, J. de: Hormones and diapause. Exc. Med. Int. Congr. Ser. *184*, 356–364 (1968).

[326] Williams, C. M.: Ecdysone and ecdysone-analogues: Their assay and action on diapausing pupae of the *Cynthia* silkworm. Biol. Bull. Woods Hole *134*, 344–355 (1968).

[327] Williams, C. M.: (reference No. 374 on p. 301) (1970).

[328] Willig, A., H. H. Rees, and T. W. Goodwin: Biosynthesis of insect moulting hormones in isolated ring glands and whole larvae of *Calliphora*. J. Insect Physiol. *17*, 2317-2326 (1971).

[329] Wright, J. E.: Hormonal termination of larval diapause in *Dermacentor albipictus*. Science *163*, 390–391 (1969).

[330] Wright, J. E., W. F. Chamberlain, and C. C. Barrett: Ovarian maturation in stable flies: inhibition by 20-hydroxyecdysone. Science *172*, 1247–1248 (1971).

[331] Wright, J. E., and J. N. Kaplanis: Ecdysone and ecdysone-analogues: effects on fecundity of the stable fly *Stomoxys calcitrans*. Ann. Ent. Soc. Amer. *63*, 622–623 (1970).

[332] Wyatt, G. R.: Macromolecular biosynthesis in insect metamorphosis. Proc. 17th Int. Congr. Biochem., Tokyo, VII-3 (1967).

[333] Wyatt, G. R.: (see reference No. 386 on p. 301) (1972).

[334] Wyatt, S. S., and G. R. Wyatt: Stimulation of RNA and protein synthesis in silkmoth pupal wing tissue by ecdysone *in vitro*. Gen. Comp. Endocrinol. *16*, 369–474 (1971).

[335] Yatschyuk, Ja. K., and G. M. Segel: On the isolation of ecdysterone. Chimia Prirodnych Soedinenij *1970*, 281.

[336] Yoshida, T., T. Otaka, M. Uchiyama, and S. Ogawa: Effect of ecdysterone on hyperglycaemia in experimental animals. Biochem. Pharmacol. *20*, 3263–3268 (1971).

[337] Yu, S. J., and L. C. Terriere: Hormonal modification of microsomal oxidase activity in the housefly. Life Sciences *10*, 1173–1185 (1971).

[338] Žďárek, J., and G. Fraenkel: Correlated effects of ecdysone and neurosecretion in puparium formation (pupariation) of flies. Proc. nat. Acad. Sci. (Amer.) *64*, 565–572 (1969).

[339] Žďárek, J., and G. Fraenkel: Overt and covert effects of endogenous and exogenous ecdysone in puparium formation of flies. Proc. nat. Acad. Sci. (Amer.) *67*, 331–337 (1970).

[340] Žďárek, J., and G. Fraenkel: Neurosecretory control of ecdysone release during puparium formation of flies. Gen. Comp. Endocrinol. *17*, 483–489 (1971).

[341] Žďárek, J., and G. Fraenkel: The mechanism of puparium formation in flies. J. Exp. Zool. *179*, 315–324 (1972).

[342] Žďárek, J., and K. Sláma: (see reference No. 391 on p. 302).

Appendix: Tables

Preface

This final part of the book is represented by tables containing the list of some selected juvenoids together with the data of their juvenile activity on certain insect species. These are data on more than 300 compounds, far less than the actual number of juvenoids known, but these compounds listed are representative of all major chemical types.

We hope that the tables will be the source of information for those interested in the activity of some juvenoids or of certain group of juvenoids, as well as for those interested in general structure-activity relations.

The compounds are arranged according to similarities of their chemical groups similar to the arrangement in section II-C. Among the acyclic juvenoids, a more or less compact group is formed by those compounds with typical isoprenoid sequences. These are represented mainly by the farnesane-type structures. The remaining part includes all acyclic juvenoids which are slightly or more different from the farnesane skeleton. This section also contains both the Cecropia compounds and their derivatives. As a main criterion of classification we have chosen the length of the basic chain and the amount of the double bonds; the acyclic oxa-analogues are divided according to the location of the ether oxygen.

Juvenoids containing an aromatic ring in the molecule are arranged according to the shorter substituent (methyl, methoxyl, chlorine, nitro group, etc.) and as a second criterion within the groups, we have used the length of the side chain and the amount of the double bonds in this chain.

In a separate group, we have placed the compounds having one or more —CONH— groupings. The majority of these compounds represent peptidic juvenoids having in addition, an aromatic ring in the molecule. They are also arranged primarily according to the smaller substituents on the benzene ring and secondly according to the character of the side chain where special attention was paid to the size of the central amino acid.

Juvenile hormone activity data are given in ID-50 Morph. in μg per specimen, as described in section II-B. A triangle indicates inactive or low activity compounds. Wherever it is present, ID-50 Morph. influence has not been obtained with up to 250 μg per spec. in the Hemipterans, 500 μg in *Tenebrio*, *Dermestes* and *Galleria*, or 1000 μg in *Gryllus* respectively. Unless otherwise stated all the applications were performed topically. In some species as in *Dermestes* and *Tenebrio*, we have given both values, topical and injection too which is indicated by (e) or (i) resp., behind the unit of activity value. In *Galleria* we have distinguished between the assays performed on larvae and on pupae by the symbol [7] or (p) respectively.

References pertaining to papers containing the data on activity of the compounds are cited at the end of part II (pp. 281–302).

Many of the values taken from the literature (bearing an approximation index) have been transformed to approximative ID-50 values from other activity data. Remaining activities represent original, partly unpublished determinations of juvenile activity by Sláma (*Pyrrhocoris*, *Dysdercus*, *Graphosoma*, *Tenebrio*, *Dermestes*, *Lygaeus*, *Gryllus*, *Triatoma*) and by Sehnal (*Galleria*).

The abbreviations in the tables are explained below. The structures of peptidic juvenoids have also been abbreviated as is common in peptide chemistry. Thus, for instance, 4-[(tert.-butoxycarbonyl-L-alanyl)amino]-1-nitrobenzene:

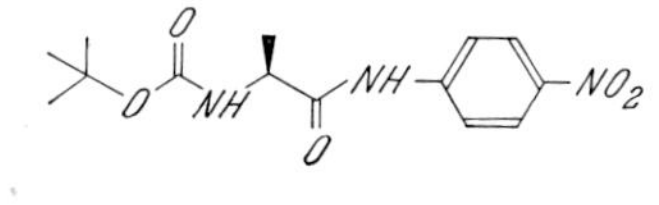

is presented in the tables as:

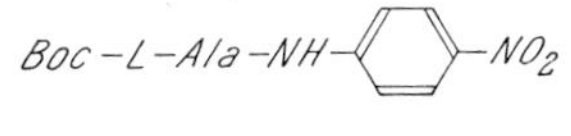

For completion we are presenting at the end of the tabular part an alphabetical list of the juvenoids in chemical nomenclature. The compounds are arranged according to the system commonly used in chemical literature. Primary aspects are the names of the main component, further there come substituents arranged also according to the alphabetical sequence. Numbers in this alphabetical list indicated by an index-T-, relate to the serial numbers of the compounds in the main tables. The same reference numbers to the particular compounds are used through the text.

We hope sincerely, that the data presented here will be useful for studies in this rapidly expanding field and will help the others to discover new more active juvenoids which could be used against insect pests.

List of Abbreviations

β-Abu	β-Aminobutyryl		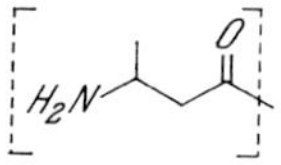
Ac	Acetyl		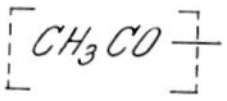
α-Aib	α-Aminoisobutyryl		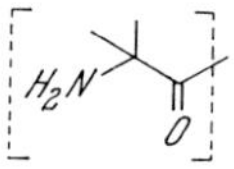
Ala	Alanyl		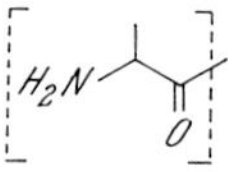
Am	Amyl	C_5H_{11}	
(t)Aoc	tert-Amyloxycarbonyl		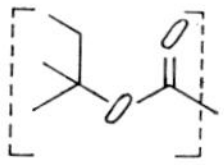
BLA	*Blatella germanica* (L.) (Blattidae)		german cockroach
(s)Boc	sec-Butoxycarbonyl		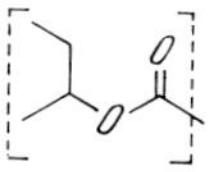
(t)Boc	tert-Butoxycarbonyl		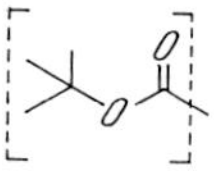
Bu	(n)Butyl	C_4H_9	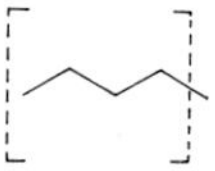
(i)Bu	Isobutyl	C_4H_9	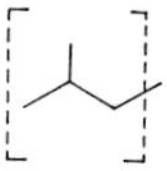
(t)Bu	tert-Butyl	C_4H_9	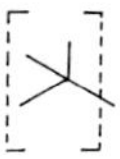
Bz	Benzyl		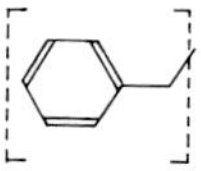
(*c*)	cis		
α-Cib	α-Chloroisobutyryl		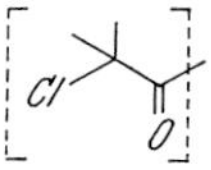

o-Cl-Ph	o-Chlorophenyl		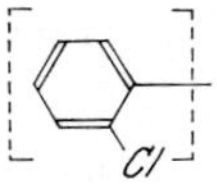
α-Cpr	α-Chloropropionyl		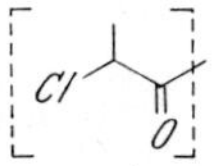
Dca	Dichloroacetyl		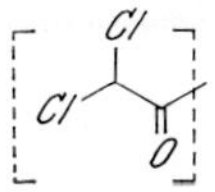
α,α-Dcp	α,α-Dichloropropionyl		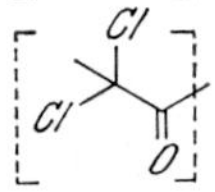
DER	*Dermestes vulpinus* (F.) (Dermestidae)		
DYS	*Dysdercus cingulatus* (F.) (Pyrrhocoridae)		Indian cotton stainer
Eoc	Ethoxycarbonyl		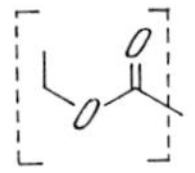
Et	Ethyl	C_2H_5	
GAL	*Galleria mellonella* (L.) (Pyralidae)		greater wax moth
Gly	Glycyl		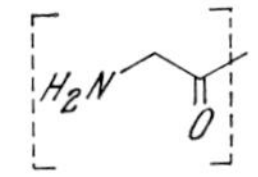
GRA	*Graphosoma italicum* (Müll.) (Pentatomidae)		
GRY	*Gryllus bimacullatus* (De G.) (Gryllidae)		
Hex	Hexyl	C_6H_{13}	
(c)Hex	Cyclohexyl		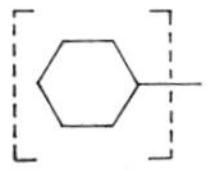
Ibu	Isobutyryl		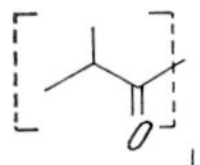
Ile	Isoleucyl		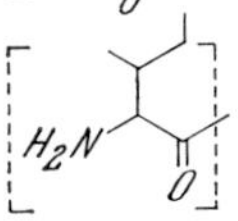

LYG	*Lygaeus equestris* (L.) (Lygaeidae)		
Me	Methyl	CH_3	
Moc	Methoxycarbonyl		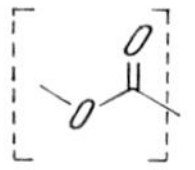
ONC	*Oncopeltus fasciatus* (Dallas) (Lygaeidae)		large milkweed bug
Ph	Phenyl		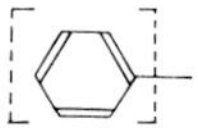
Piv	Pivaloyl		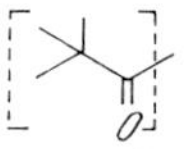
Pr	Propyl	C_3H_7	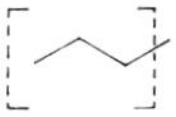
(i)Pr	Isopropyl	C_3H_7	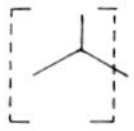
Pro	Prolyl		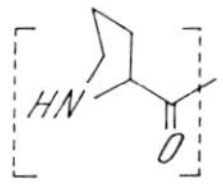
Prp	Propionyl		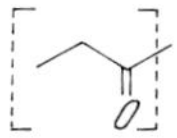
Py	Pyridine		
PYR	*Pyrrhocoris apterus* (L.) (Pyrrhocoridae)		red linden bug
RHO	*Rhodnius prolixus* (Stal) (Reduviidae)		kissing bug
STO	*Stomoxys calcitrans* (L.) (Muscidae)		stable fly
(*t*)	trans		
Tca	Trichloroacetyl		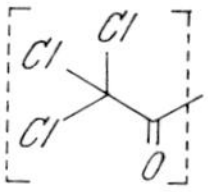
TEN	*Tenebrio mollitor* L. (Tenebrionidae)		yellow mealworm

Thf	Tetrahydrofurane	
Thp	Tetrahydropyranyl	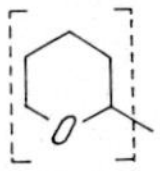
TRB	*Tribolium confusum* Jacq. du Val	confused flour beetle
Tre	Threonyl	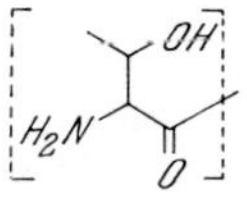
TRI	*Triatoma infestans* (L.) (Reduviidae)	southern blood sucking conenose
Ts	Tosyl	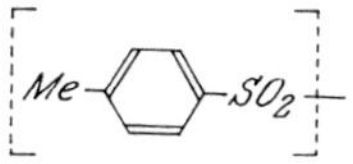
Val	Valyl	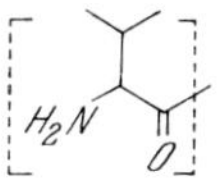

Tables

No.	Compound	PYR	DYS	GRA
T-1	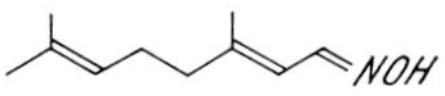	△		△
T-2	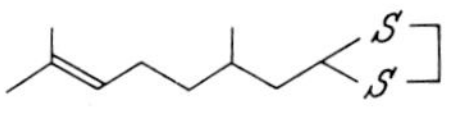		1	△
T-3	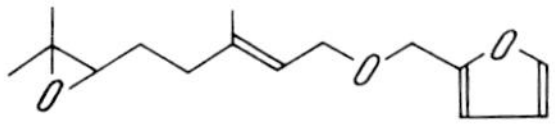			
T-4	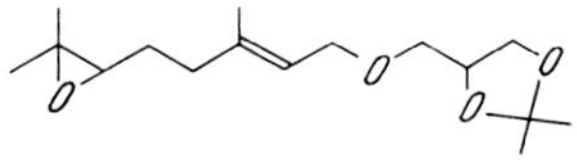			
T-5	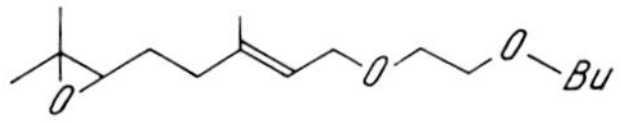			
T-6	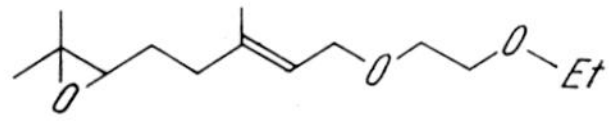			
T-7	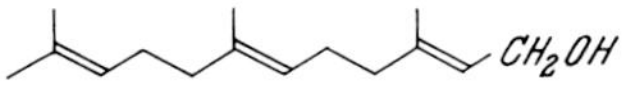	△	△	△
T-8	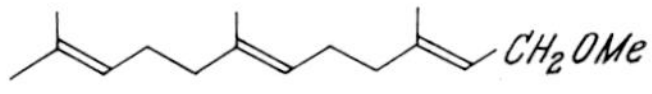	10	50	100
T-9	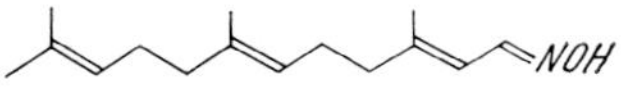	50	8	50
T-10	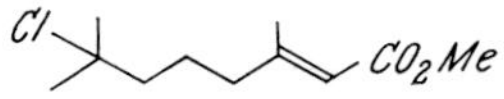	10	50	△
T-11	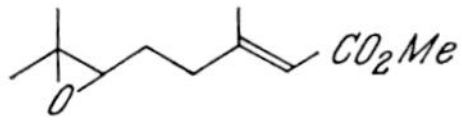	△	△	△

TRI	GRY	TEN	DER	GAL	Miscellaneous	Ref.
		△ (i)				
		100 (e)[a]				27[a]
		~ 0.1 (e)[a]				27[a]
		~ 0.1 (e)[a]				27[a]
		~ 0.1 (e)[a]				27[a]
		100 (i)		△ (7)	180 (RHO)[a]	358[a]
		10 (i)		△ (p)	~ 2 (RHO)[a]	358[a]
		60 (i)		△ (7)		
				△ (7)		

(Tables — continued)

No.	Compound	PYR	DYS	GRA
T-12	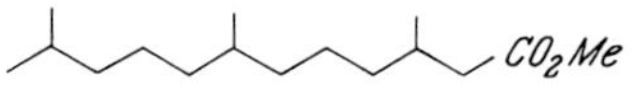	90		△
T-13	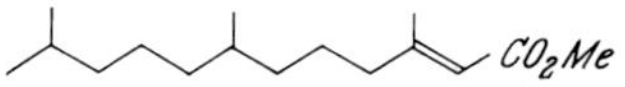	5	3	100
T-14		7	5	△
T-15	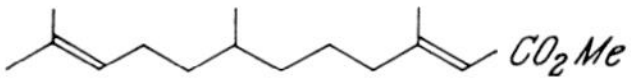	8	10	30
T-16	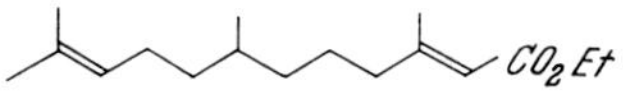	5	4	10
T-17	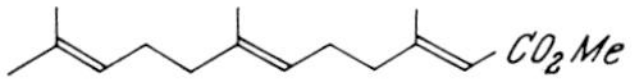	50	50	50
T-18	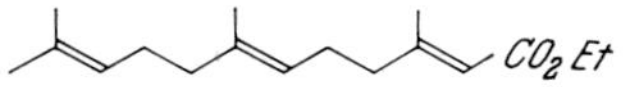	50	10	50
T-19	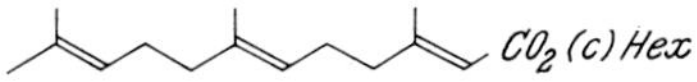	30		
T-20	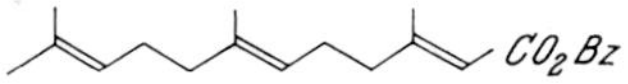	100	200	△
T-21	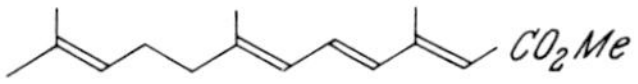	5	3	10
T-22	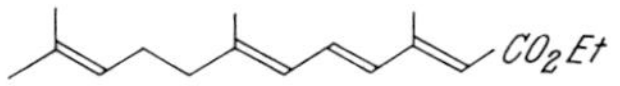	5	0.8	0.7

TRI	GRY	TEN	DER	GAL	Miscellanous	Ref.
		△ (i)		△ (p) △ (7)	△ (LYG)	
		10 (e) 1 (i)		100 (p) 20 (7)		
		△ (e) 5 (i)				
5		30 (e) 10 (i)		200 (p) 20 (7)		
		10 (i)	5 (i)	50 (p) 8 (7)	0.7 (LYG)	
30		50 (e) 5 (i)	5 (i)	△ (7)		
		5 (i)	10 (e) 2 (i)	△ (7)	40 (LYG)	
		100 (i)				
△		△ (i)		△ (p) △ (7)		
△		△ (i)		△ (p) △ (7)		
	100	△ (i)	50 (i)		5 (LYG)	

(Tables — continued)

No.	Compound	PYR	DYS	GRA
T-23	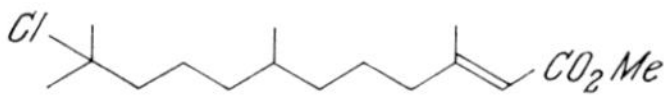	0.05		5
T-24	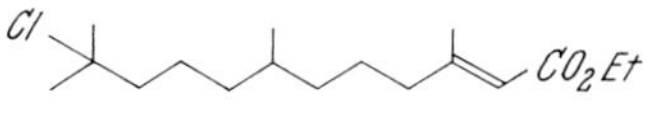	0.05	0.08	1
T-25	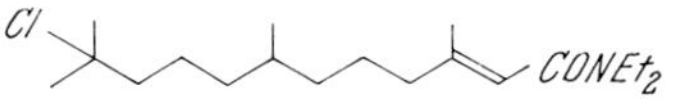	0.7	0.4	△
T-26	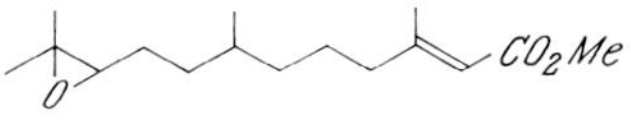	0.5	0.05	0.5
T-27	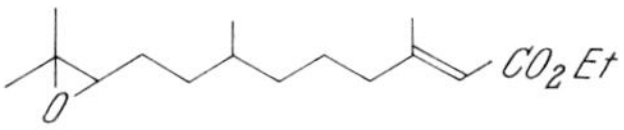	5	1	5
T-28	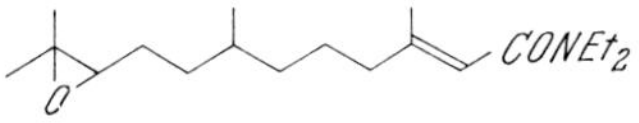	4		△
T-29	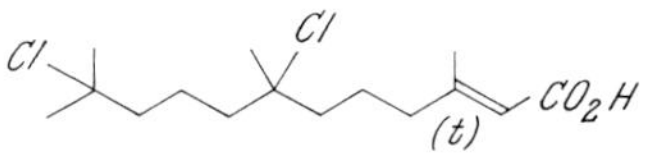	1	10	△
T-30	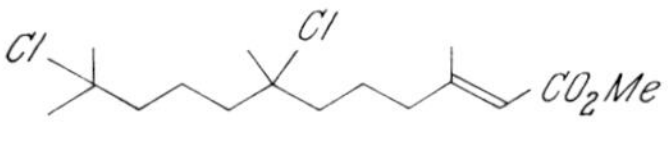	0.001		
T-31	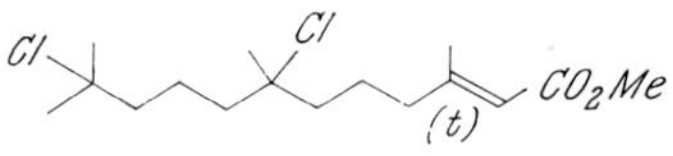	0.0008	0.008	10
T-32	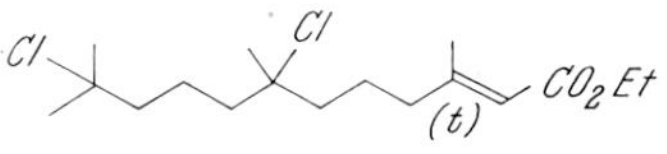	0.0005	0.001	1
T-33	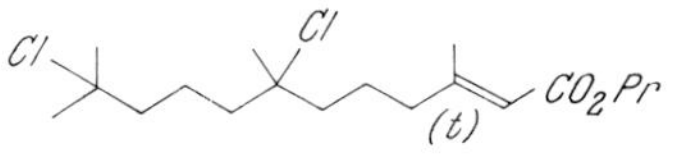	0.005	0.01	100

TRI	GRY	TEN	DER	GAL	Miscellaneous	Ref.
50		1 (e) 4 (i)	5 (e) 5 (i)	10 (7)	3 (LYG)	
	100	10 (e) 5 (i)		50 (p) 1 (7)	3 (LYG)	
	4	0.04 (e) 0.0005 (i)	0.09 (e) 0.008 (i)			
3		10 (e) 5 (i)	 50 (i)	100 (7)	0.1 (LYG)	
1		30 (i)		500 (7)	0.5 (LYG) $\sim$ 2 (RHO)[a]	358[a]
		0.05 (e) 0.001 (i)	0.5 (e) 0.04 (i)		$\sim$ 20 (RHO)[a]	358[a]
30				△ (7)	$\sim$ 9 (RHO)[a]	358[a]
30	80	100 (i)	30 (i)	△ (7)	0.01 (LYG)	
		100 (i)	10 (e) 50 (i)	△ (7)	$\sim$ 2 (RHO)[a]	358[a]
	1000	△ (i)	500 (i)		0.05 (LYG)	

(Tables — continued)

No.	Compound	PYR	DYS	GRA
T-34	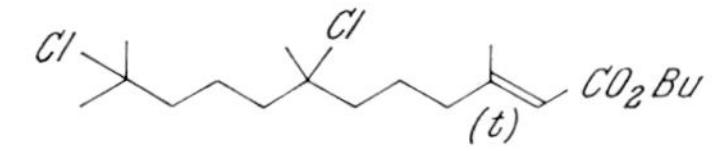	0.5	0.5	△
T-35	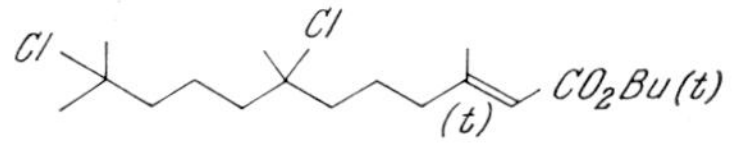	0.1	0.05	△
T-36	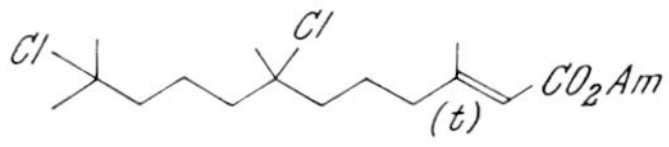	4	0.4	△
T-37	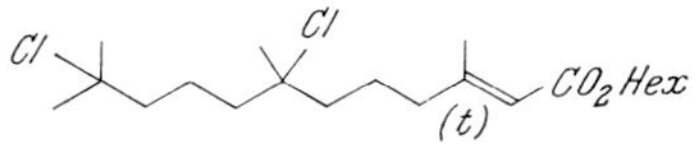	4	1	△
T-38	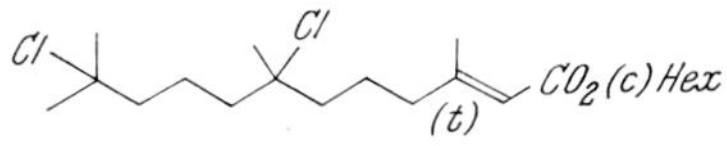	2	10	△
T-39	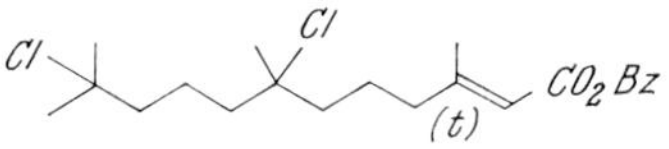	0.3	0.5	△
T-40	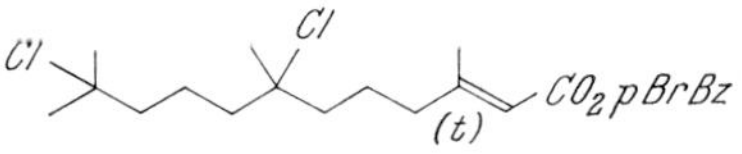	0.07	0.7	△
T-41	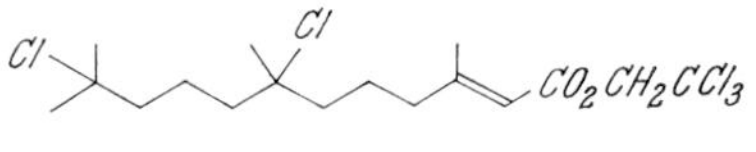	3	1	△
T-42	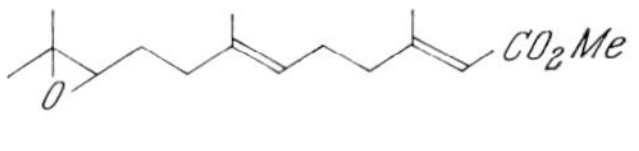	3	1	10
T-43	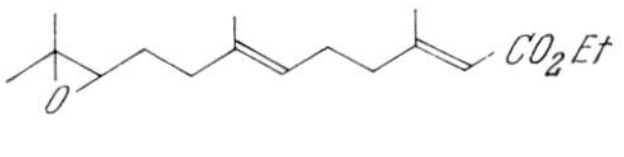			
T-44	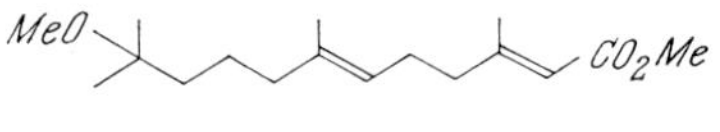	1	0.5	10

TRI	GRY	TEN	DER	GAL	Miscellaneous	Ref.
	△	△ (i)	△ (i)			
				△ (p)		
				△ (7)		
	△	△ (i)				
		△ (i)	△ (i)			
		△ (i)		△ (7)		
		△ (i)	△ (i)	△ (7)		
5		10 (i)		△ (p)	~ 1 (ONC)[a]	29[a]
				300 (7)	~ 14 (RHO)[b]	358[b]
					~ 0.05 (STO)[a]	385[a]
		100 (i)		500 (p)		
				500 (7)		

(Tables — continued)

No.	Compound	PYR	DYS	GRA
T-45	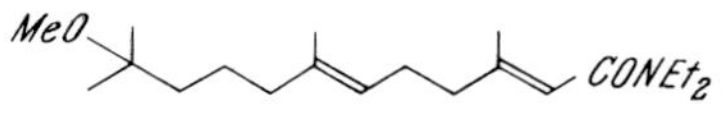	1		△
T-46	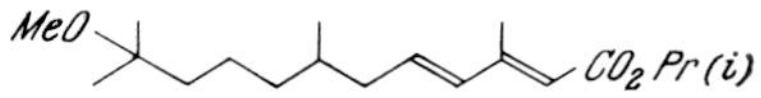		0.04	6
T-47	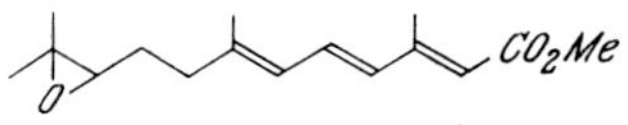	5	0.1	1
T-48	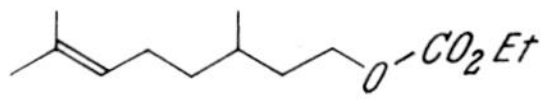			
T-49	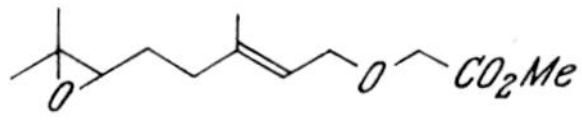			
T-50	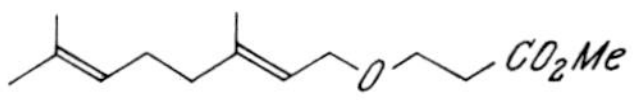	△	50	△
T-51	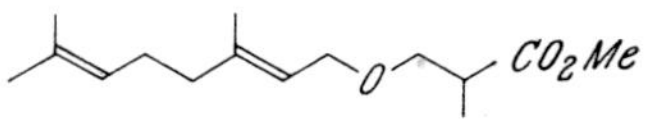	△	50	
T-52	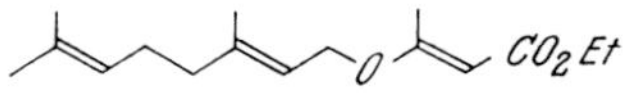			
T-53	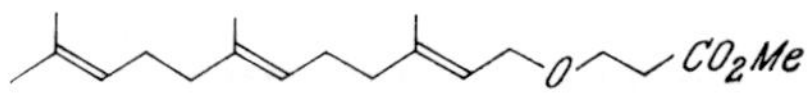	50	△	△
T-54	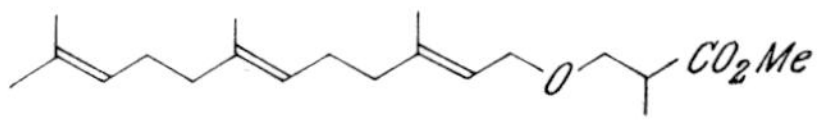	100	50	△
T-55	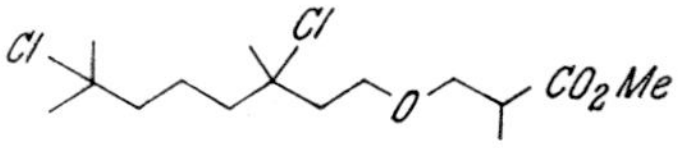	50	50	△

TRI	GRY	TEN	DER	GAL	Miscellaneous	Ref.
	250	0.07 (e) 0.05 (i)		△ (7)		
		0.01 (e)		10 (7)		
		△ (e) △ (i)		△ (p) △ (7)		
		> 300 (e)[a]				345[a]
		100 (e)[a]				27[a]
		△ (i)		△ (7)		
	△	△ (e) △ (i)		△ (7)		
		~ 25 (e)[a]				345[a]
	△	△ (i)				
		△ (i)				
	△	100 (e) △ (i)				

(Tables — continued)

No.	Compound	PYR	DYS	GRA
T-56	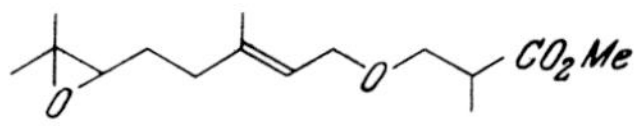	50	10	
T-57	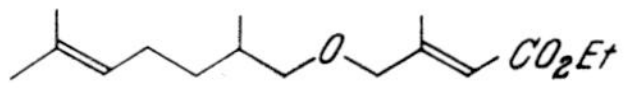	0.8	0.1	8
T-58	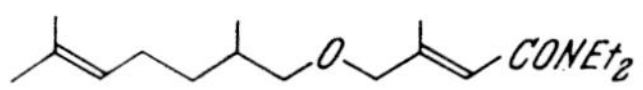		0.6	△
T-59		△	△	△
T-60	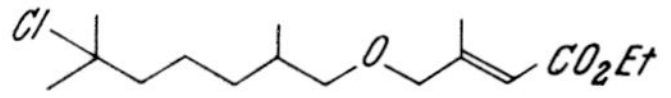	0.005	0.008	5
T-61	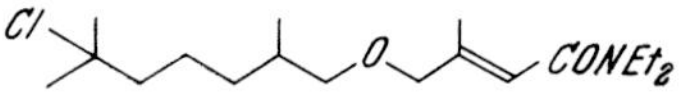	4	5	△
T-62	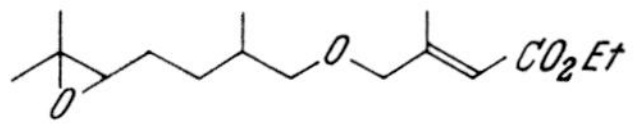	0.01	0.005	0.1
T-63	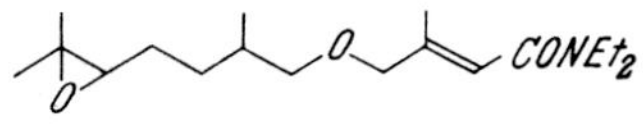	8	4	△
T-64	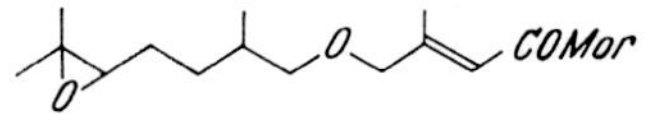	△	△	△
T-65	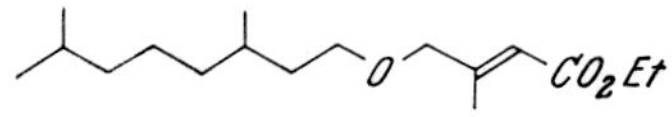	1	5	50
T-66	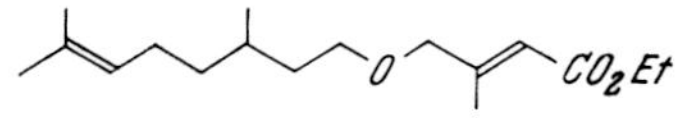	5	1	10

TRI	GRY	TEN	DER	GAL	Miscellaneous	Ref.
	△	△ (i)				
		100 (i)		△ (p) △ (7)		
		0.5 (e)				
		△ (e)				
	△	△ (e)				
		0.5 (e)				
		50 (e) △ (i)	△ (i)		0.005 (LYG)	
		0.0005 (e)				
		△ (e)				
50		△ (i)		△ (p) △ (7)		
		△ (i)		△ (p) △ (7)	5 (LYG)	

(Tables — continued)

No.	Compound	PYR	DYS	GRA
T-67	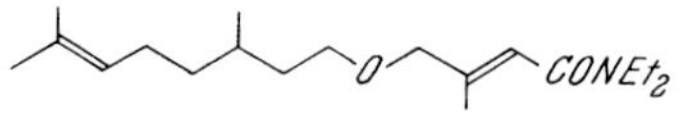		5	△
T-68	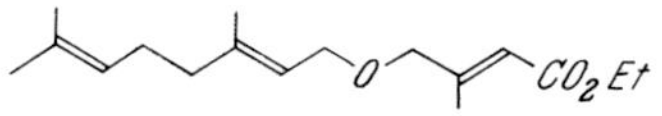	5	0.8	1
T-69	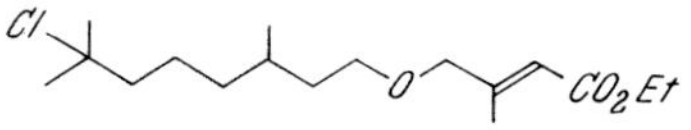	0.08	0.5	5
T-70	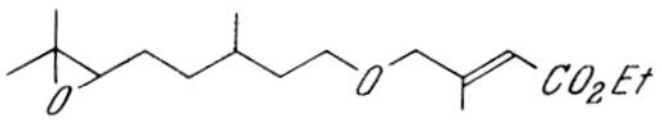	0.2	0.01	0.8
T-71	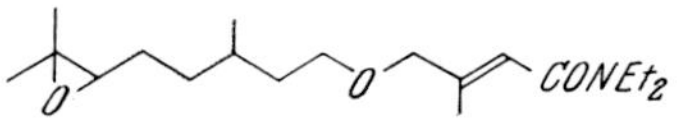		30	50
T-72	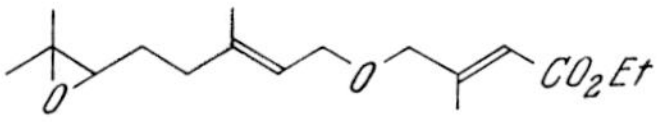	1	0.3	0.8
T-73	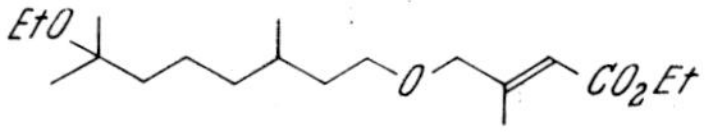	3	5	10
T-74	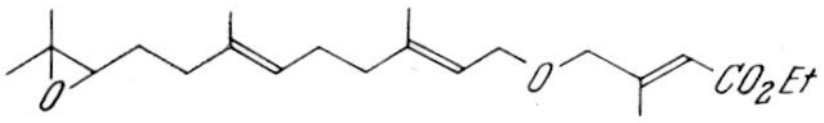	5	5	3
T-75	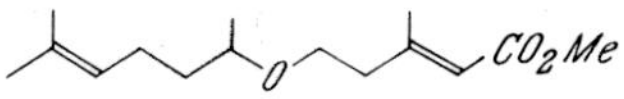		5	100
T-76	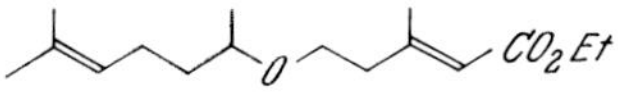	10	3	50
T-77	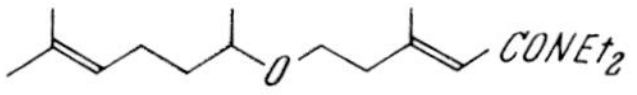	10	10	△

TRI	GRY	TEN	DER	GAL	Miscellaneous	Ref.
		△ (e) 300 (i)				
		△ (i)		△ (p) △ (7)		
	100	100 (i)				
	△	500 (e) △ (i)	△ (i)		0.1 (LYG)	
		△ (i)				
		△ (e) △ (i)	△ (i)		3 (LYG)	
	250	△ (e)				
		△ (i)				
		100 (e)		△ (7)		
	250	300 (e)		△ (7)		
		0.5 (e) 0.5 (i)				

(Tables — continued)

No.	Compound	PYR	DYS	GRA
T-78	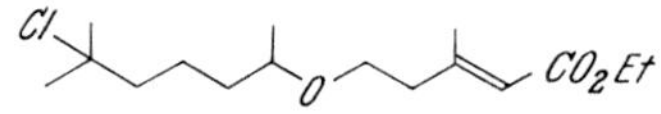		0.5	**10**
T-79	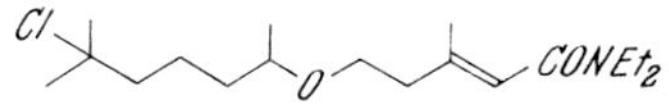		100	△
T-80	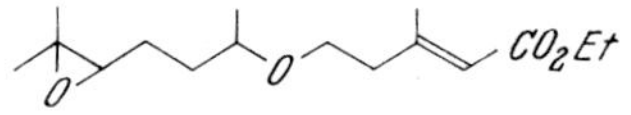	**0.08**	0.05	**4**
T-81	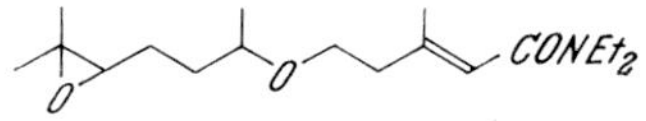	**100**	100	△
T-82	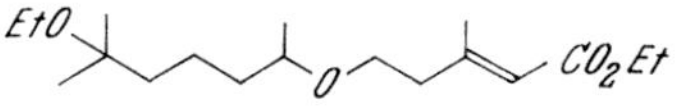		1	△
T-83	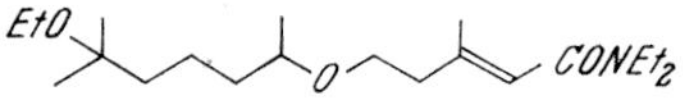		△	△
T-84	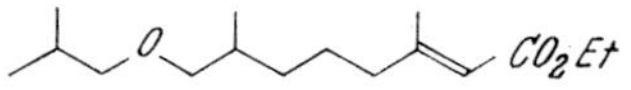		1	△
T-85	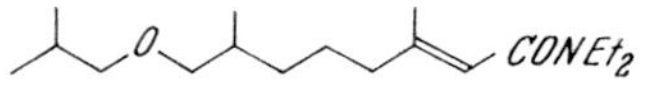		10	△
T-86	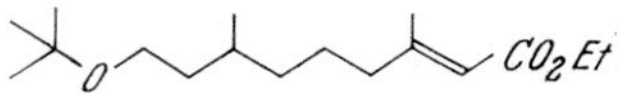		0.05	**1**
T-87	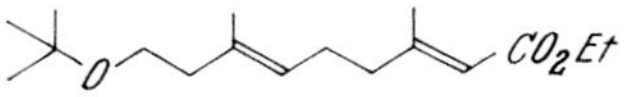		0.1	**1**
T-88	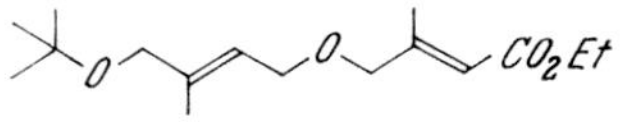		0.1	**40**

TRI	GRY	TEN	DER	GAL	Miscellaneous	Ref.
		△ (e) △ (i)		△ (7)		
	△	0.1 (e) 0.05 (i)				
		100 (e) > 100 (i)				
		0.01 (e) 0.001 (i)				
		100 (e)				
		1 (e)				
		△ (e)				
		10 (e)				
		1 (e)				
		100 (e)				
		100 (e)		△ (7)		

(Tables — continued)

No.	Compound	PYR	DYS	GRA
T-89	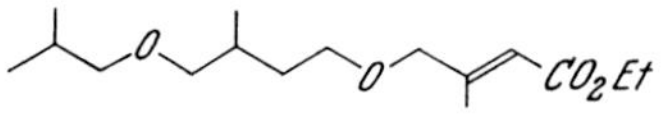	0.05	0.1	80
T-90	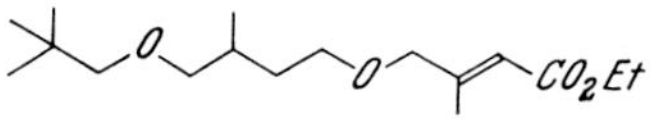		0.05	100
T-91	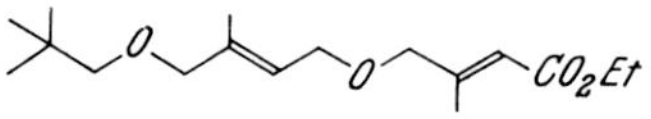		0.5	5
T-92	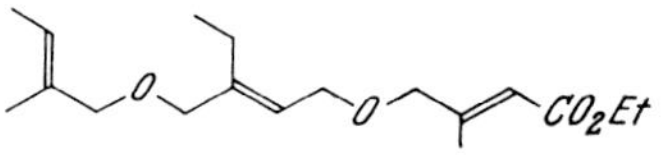	0.5	0.4	10
T-93	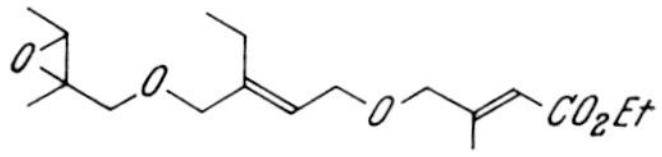		0.1	4
T-94	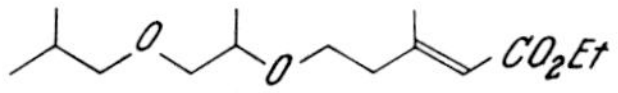	0.08	0.1	△
T-95	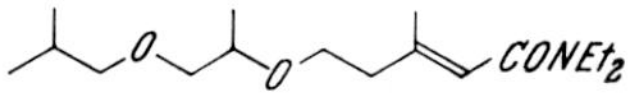		△	△
T-96	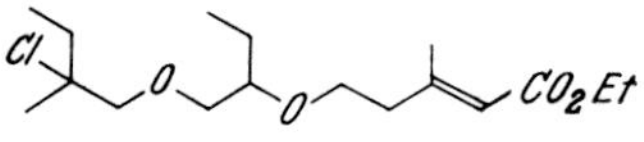	0.05	0.05	50
T-97	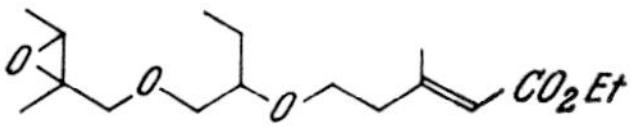	5	1	△
T-98	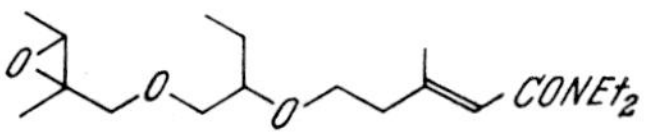		△	△
T-99	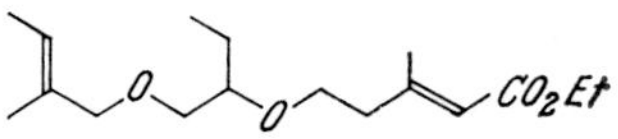	5		△

TRI	GRY	TEN	DER	GAL	Miscellaneous	Ref.
		△ (i)		△ (7)	80 (LYG)	
		△ (e)		△ (7)		
		△ (e)		△ (7)		
		△ (e)		△ (7)		
		△ (e)				
		△ (e)				
		100 (e)				
		100 (e)				
		△ (e)		△ (7)		
		△ (e)	△ (e)			
		100 (e)				

(Tables — continued)

No.	Compound	PYR	DYS	GRA
T-100	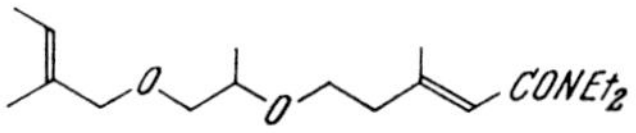		△	△
T-101	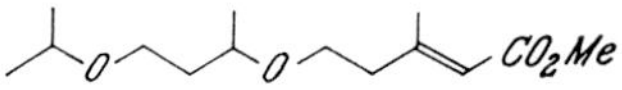		0.4	△
T-102	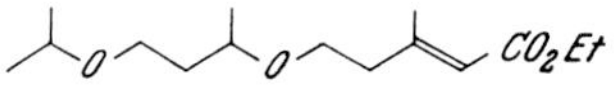		0.1	50
T-103	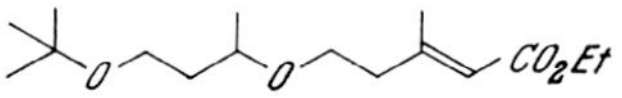	0.005		100
T-104		100	50	△
T-105	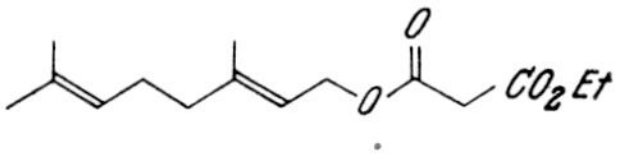	△	100	
T-106	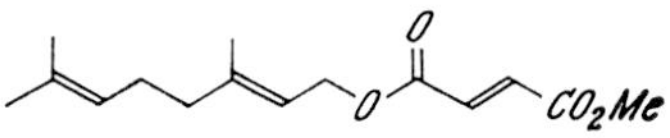	△	△	
T-107	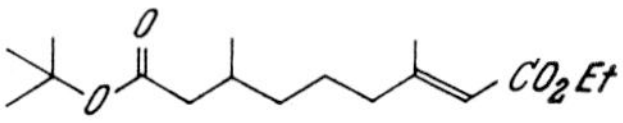	0.001	0.007	100
T-108	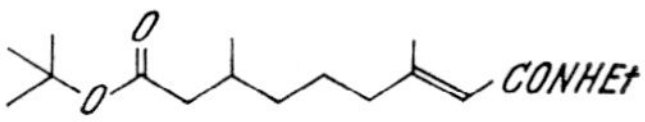		0.05	△
T-109	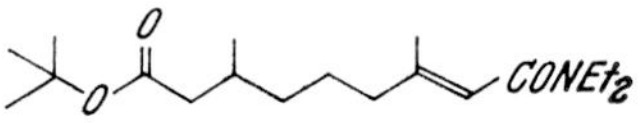	1	1	△
T-110	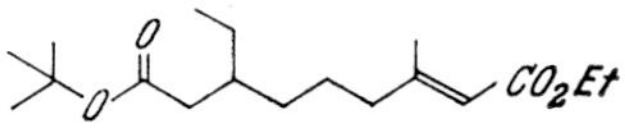		0.005	10

TRI	GRY	TEN	DER	GAL	Miscellaneous	Ref.
		500 (e)				
		500 (e)				
		△ (e)				
		△ (e)				
		1 (e)				
		△ (i)				
	500	△ (e)				
		△ (i)		△ (7)		
		10 (e)				
		10 (e)				
	250	△ (e)		△ (7)		

(Tables — continued)

No.	Compound	PYR	DYS	GRA
T-111	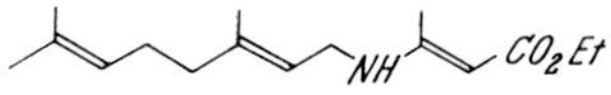			
T-112	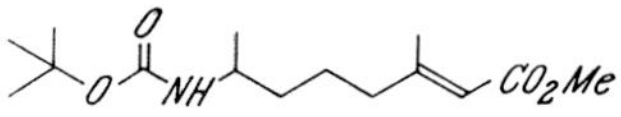		0.008	△
T-113	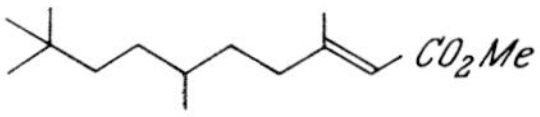	△	△	100
T-114	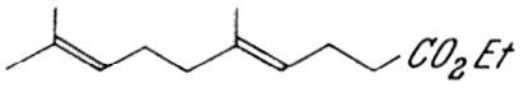			
T-115	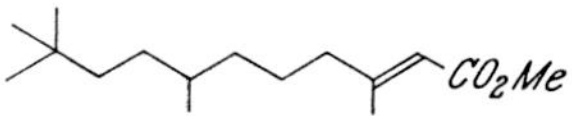	0.8	5	8
T-116	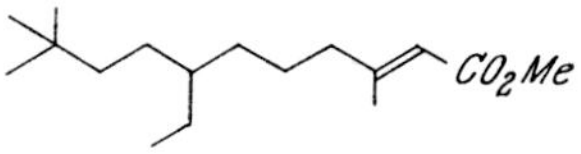	0.1	0.5	8
T-117	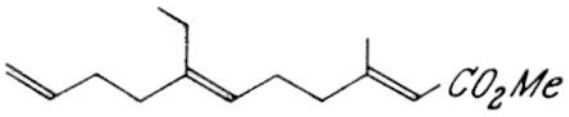			
T-118	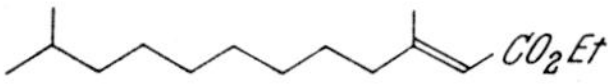		8	100
T-119	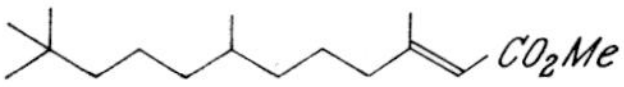	1	0.8	50
T-120	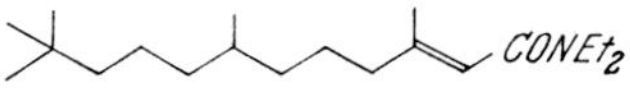		5	△
T-121	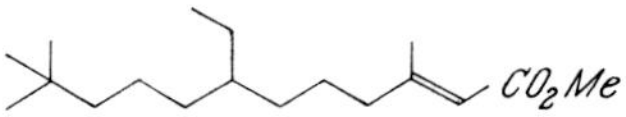	0.1	4	50

TRI	GRY	TEN	DER	GAL	Miscellaneous	Ref.
		~ 1 (e)[a]				345[a]
		△ (e)				
		△ (e) △ (i)	△ (i)	△ (7)	> 100 (LYG)	
		> 300 (e)[a]				345[a]
	5	> 100 (e) 5 (i)		80 (p) 5 (7)	> 50 (LYG)	
	5	10 (i)				
		△ (e)[a]			80 (TRB)[a]	213[a]
		80 (e)				
	5	50 (e) 1 (i)	20 (i)	20 (7)	5 (LYG)	
	100	8 (e)		△ (7)		
	0.5	10 (e) 8 (i)		10 (7)	> 50 (LYG)	

(Tables — continued)

No.	Compound	PYR	DYS	GRA
T-122	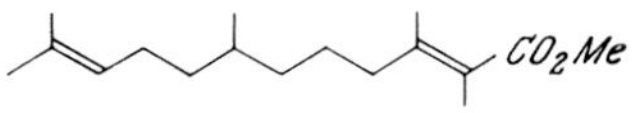	8	1	50
T-123	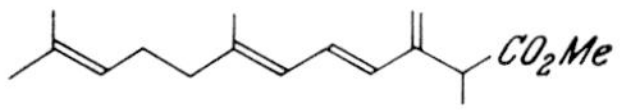	5		0.1
T-124	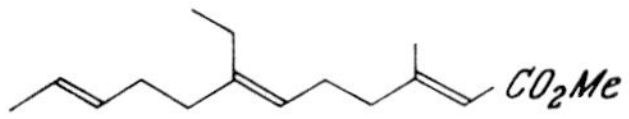			
T-125	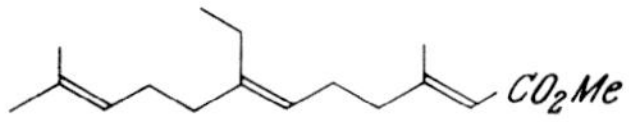			
T-126	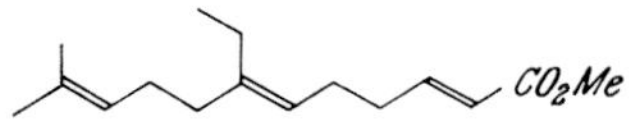			
T-127	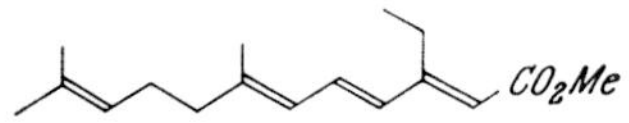	30		> 50
T-128	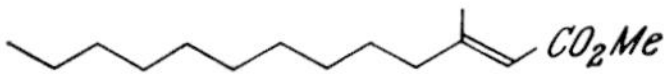	50	80	△
T-129	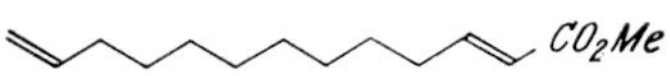	100		
T-130	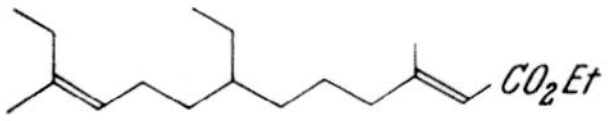	10	8	30
T-131	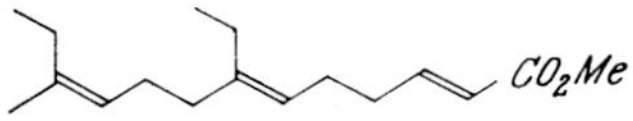			
T-132	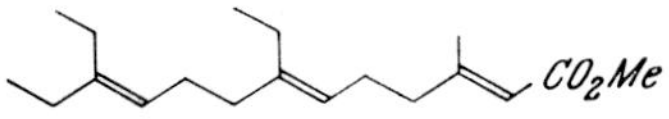			

TRI	GRY	TEN	DER	GAL	Miscellaneous	Ref.
△		△ (e) △ (i)		△ (7)		
△	△	△ (i)		△ (p) △ (7)		
		> 100 (e)[a]			~ 50 (TRB)[a]	213[a]
		~ 60 (e)[a]			~ 7 (TRB)[a]	213[a]
		~ 50 (e)[a]			~ 5 (TRB)[a]	213[a]
		△ (i)				
		△ (i)	500 (i)	△ (p) △ (7)		
	10	10 (i)	2 (i)	80 (p) 7 (7)	50 (LYG)	
		~ 6 (e)[a]			~ 3 (TRB)[a]	213[a]
		~ 8 (e)[a]			< 1 (TRB)[a]	213[a]

(Tables — continued)

No.	Compound	PYR	DYS	GRA
T-133	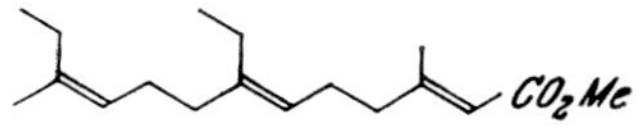			
T-134	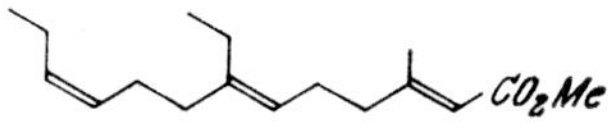			
T-135	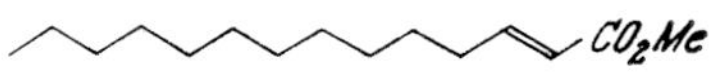	100		
T-136	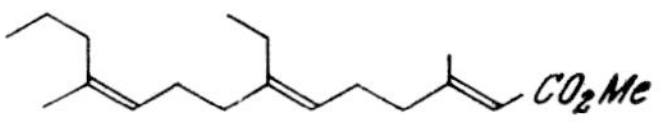			
T-137	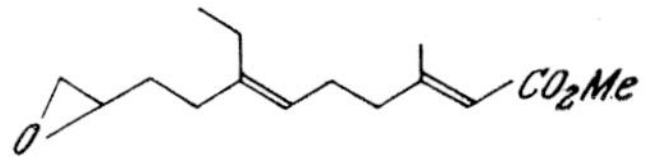			
T-138	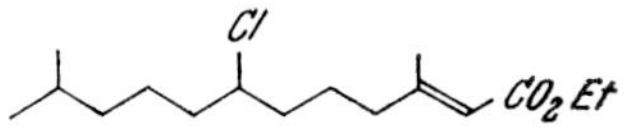		3	100
T-139	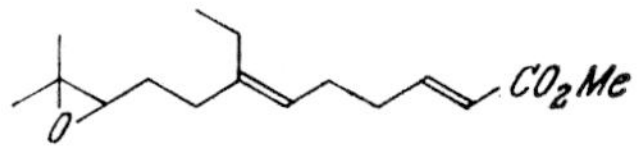			
T-140	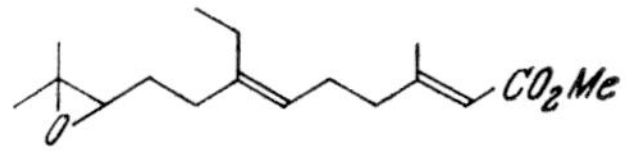			
T-141	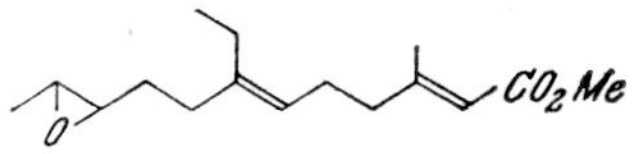			
T-142	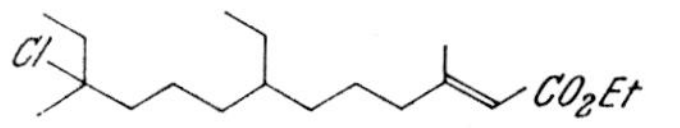	0.1	0.1	4
T-143	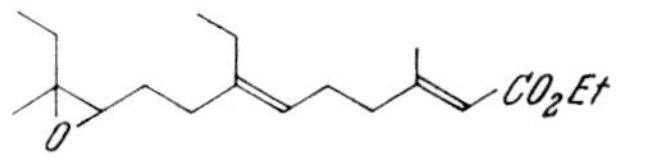			

TRI	GRY	TEN	DER	GAL	Miscellaneous	Ref.
		~ 40 (e)[a]			~ 5 (TRB)[a]	213[a]
		~ 70 (e)[a]			~ 8 (TRB)[a]	213[a]
		△ (i)		△ (p) △ (7)		
		~ 6 (e)[a]			~ 1 (TRB)[a]	213[a]
		△ (e)[a]			~ 100 (TRB)[a]	213[a]
		10 (e) 50 (i)				
		~ 6 (e)[a]			~ 1 (TRB)[a]	213[a]
		~ 6 (e)[a]			~ 1 (TRB)[a]	213[a]
		> 100 (e)[a]			~ 70 (TRB)[a]	213[a]
		0.5 (i)	0.5 (i)	10 (p) 1 (7)		
		~ 5 (e)[a]			~ 0.6 (TRB)[a] < 0.0001 (STO)[b]	213[a] 385[b]

(Tables — continued)

No.	Compound	PYR	DYS	GRA
T-144	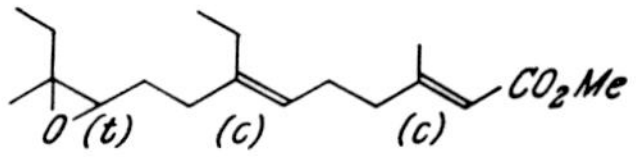		△[a]	
T-145	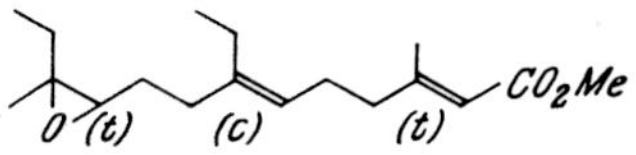		0.3[a]	
T-146	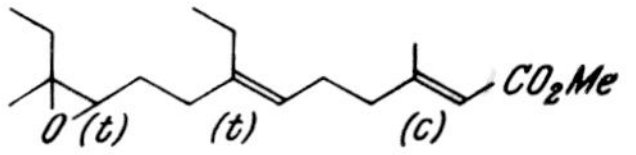		35[a]	
T-147	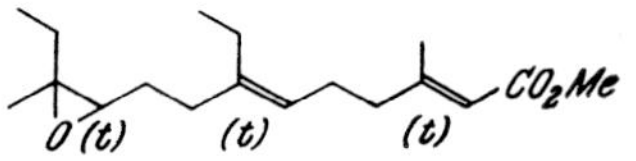		5[a]	
T-148	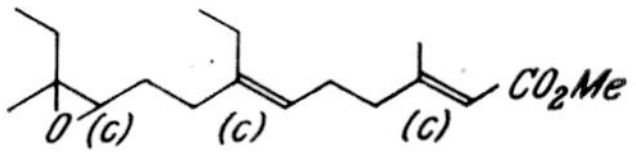		5[a]	
T-149	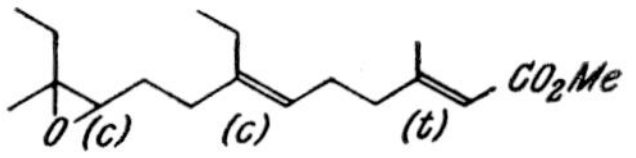		0.3[a]	
T-150	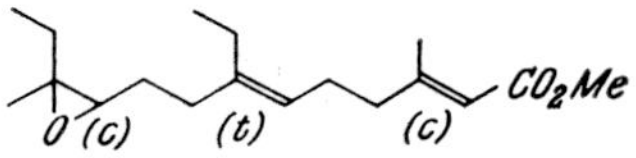		△[a]	
T-151	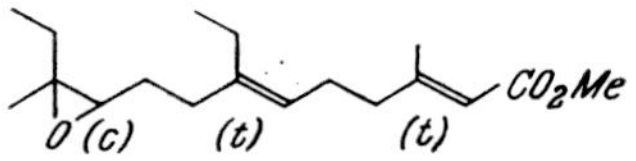	1	0.1[a] 0.5	1
T-152	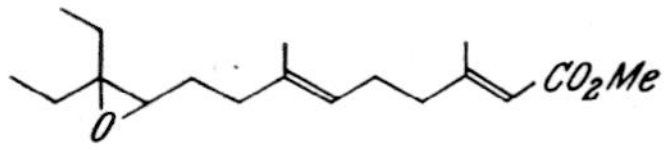			
T-153	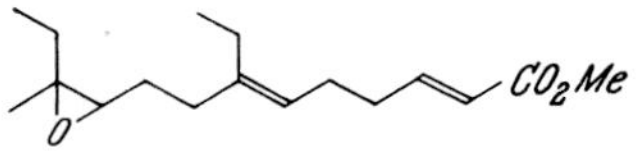			
T-154	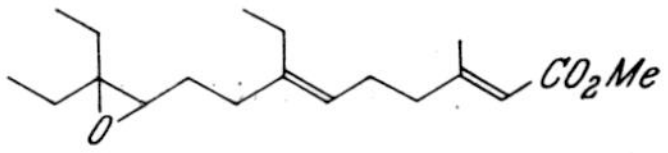			

TRI	GRY	TEN	DER	GAL	Miscellaneous	Ref.
					180 (RHO)[b]	242[a]
					> 100 (BLA)[a]	358[b]
					30 (RHO)[b]	242[a]
					50 (BLA)[a]	358[b]
					33 (RHO)[b]	242[a]
					50 (BLA)[a]	358[b]
					1.8 (RHO)[b]	242[a]
					0.5 (BLA)[a]	358[b]
					△ (RHO)[b]	242[a]
					△ (BLA)[a]	358[b]
					6.8 (RHO)[b]	342[a]
					5 (BLA)[a]	358[b]
					15 (RHO)[b]	242[a]
					10 (BLA)[a]	358[b]
		1 (e)		0.00001 (7)	0.5 (RHO)[b]	242[a]
		0.5 (i)			0.5 (BLA)[a]	358[b]
		~ 1 (e)[a]				94[a]
		~ 80 (e)[a]			~ 8 (TRB)[a]	213[a]
		~ 1 (e)[a]			< 1 (TRB)[a]	213[a]

(Tables — continued)

No.	Compound	PYR	DYS	GRA
T-155	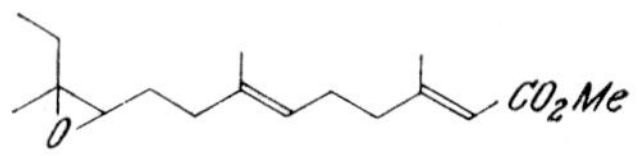	1	0.4	1
T-156	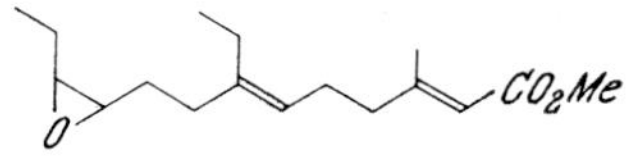			
T-157	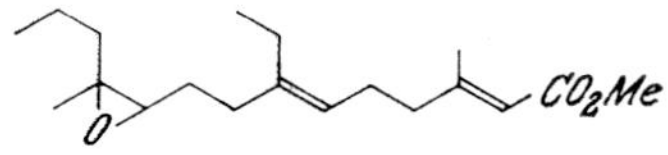			
T-158	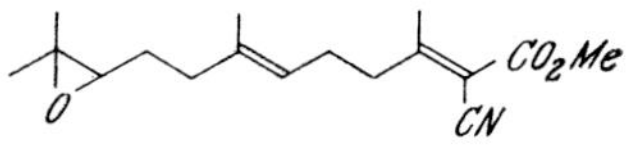			
T-159	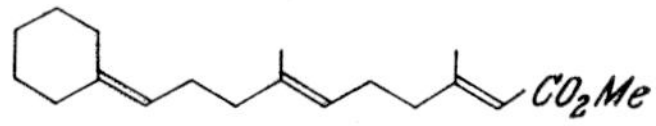			
T-160	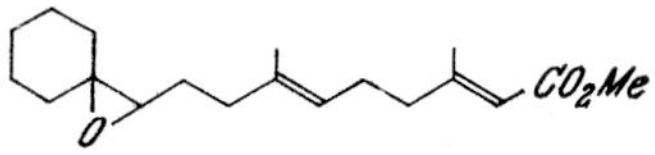			
T-161	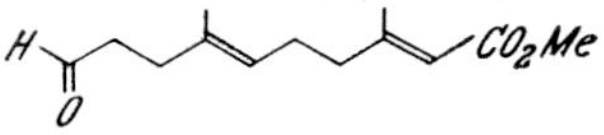			
T-162	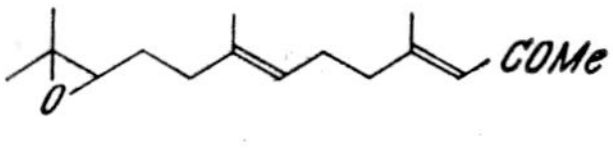			
T-163	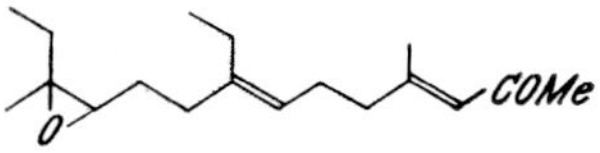			
T-164	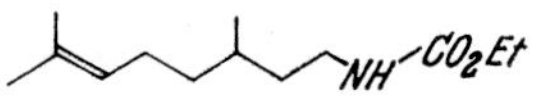			
T-165	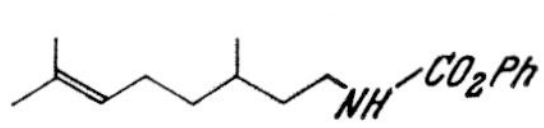			

TRI	GRY	TEN	DER	GAL	Miscellaneous	Ref.
		3 (e)		0.13 (p) 0.000001 (7)		
		~ 80 (e)[a]			~ 8 (TRB)[a]	213[a]
		~ 5 (e)[a]			< 1 (TRB)[a]	213[a]
		~ 1 (e)[a]				345[a]
		< 5 (e)[a]				94[a]
		< 5 (e)				94[a]
		~ 80 (e)[a]				345[a]
		~ 1 (e)[a]			~ 0.5 (STO)[b]	292[a] 385[b]
		~ 0.03 (e)[a]				292[a]
		~ 8 (e)[a]				345[a]
		~ 0.1 (e)[a]				345[a]

(Tables — continued)

No.	Compound	PYR	DYS	GRA
T-166	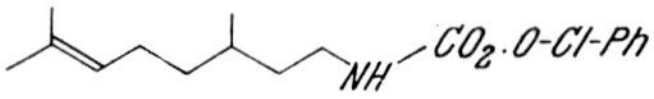			
T-167	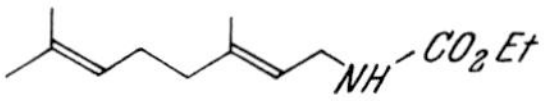			
T-168	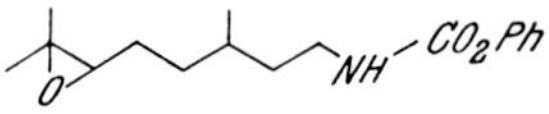			
T-169	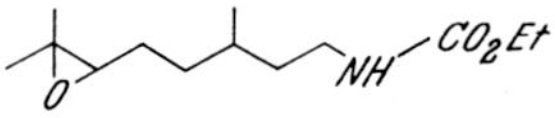			
T-170	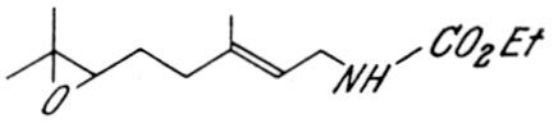			
T-171	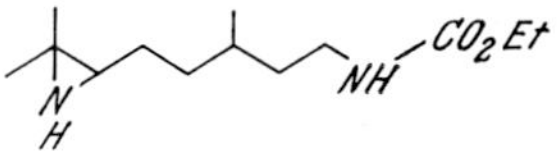			
T-172	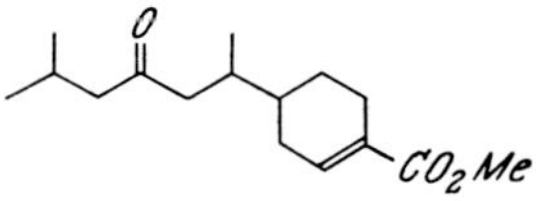	~ 1[a]	0.5	△
T-173	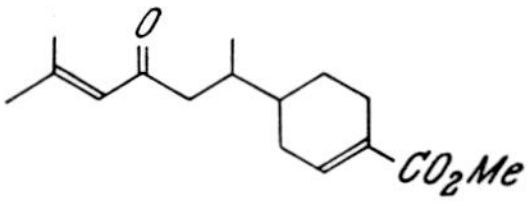	10	0.5	△
T-174	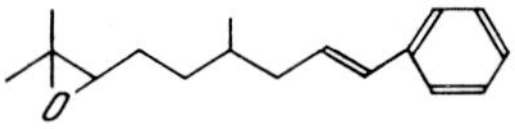			
T-175	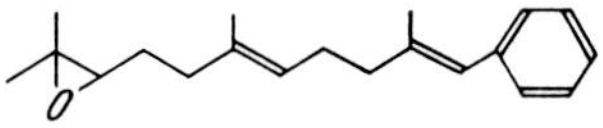			
T-176	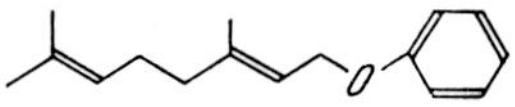	△	100	△

TRI	GRY	TEN	DER	GAL	Miscellanous	Ref.
		~ 1 (e)[a]				345[a]
		~ 7 (e)[a]				345[a]
		~ 0.3 (e)[a]				345[a]
		~ 1 (e)[a]				345[a]
		~ 1 (e)[a]				345[a]
		~ 0.8 (e)[a]				345[a]
	△	△ (e) △ (i)	△ (i)		> 50 (ONC)[a]	29[a]
	△	△ (e) △ (i)	△ (i)			
					0.003 (ONC)[a]	345[a]
		> 0.01 (e)[a]			0.03 (ONC)[a]	345[a]
		△ (e) △ (i)	△ (i)		100 (LYG)	

(Tables — continued)

No.	Compound	PYR	DYS	GRA
T-177	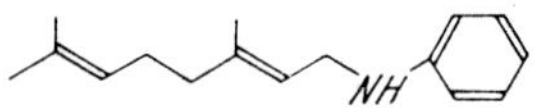	△	△	△
T-178	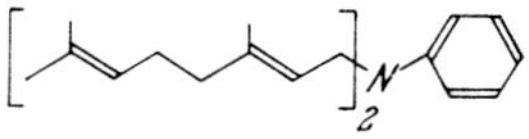	△	△	△
T-179	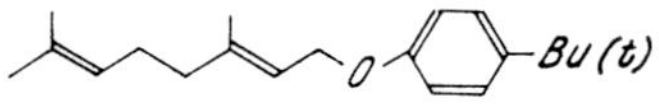	10		△
T-180	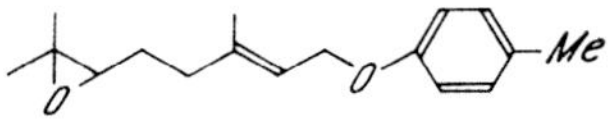			
T-181	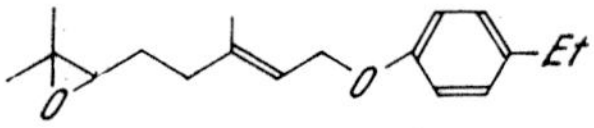		0.05	1
T-182	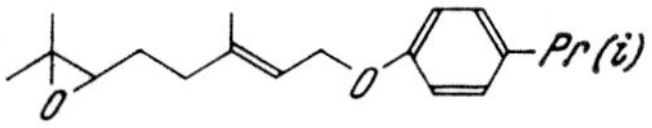			
T-183	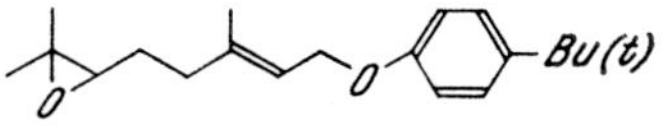			
T-184	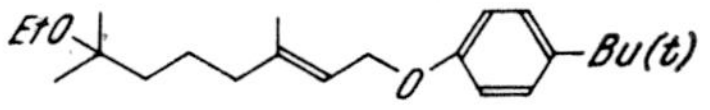	△	△	△
T-185	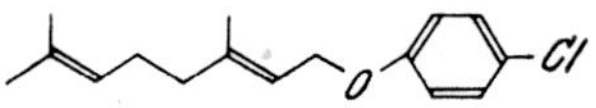	50	10	5
T-186	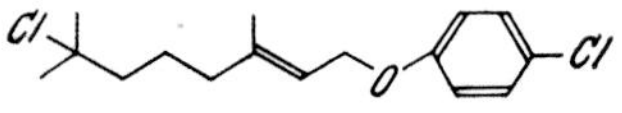		0.5	0.2
T-187	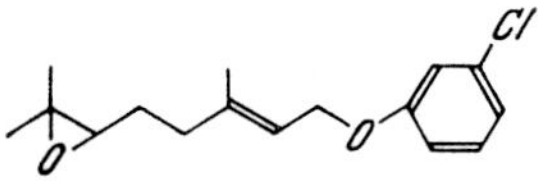			

TRI	GRY	TEN	DER	GAL	Miscellaneous	Ref.
		30 (e)				
3		△ (i)		△ (7)		
		50 (e)				
		~ 0.08 (e)[a]			~ 10 (ONC)[a]	253[a]
		0.005 (e)				
		0.04 (e)[a]			~ 10 (ONC)[a]	253[a]
		~ 0.5 (e)[a]			< 10 (ONC)[a]	253[a]
		△ (e)				
		7 (e)				
		1 (e)				
		~ 10 (e)[a]			0.01 (ONC)[a]	253[a]

(Tables — continued)

No.	Compound	PYR	DYS	GRA
T-188	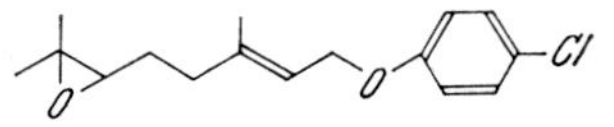	5	4	0.07
T-189	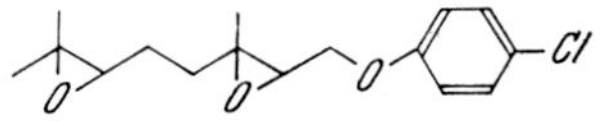	4	3	1
T-190	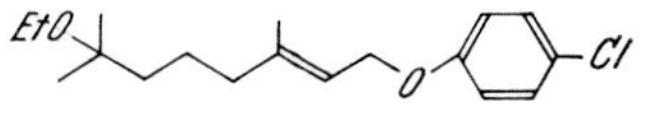	0.5	0.5	4
T-191	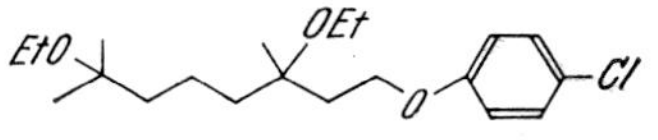	△	△	△
T-192	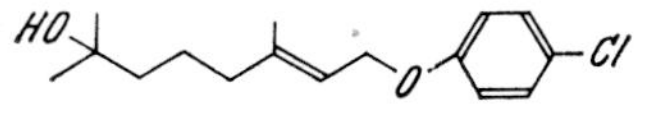		5	0.5
T-193	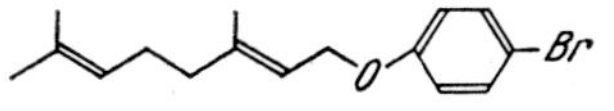	30	4	0.8
T-194	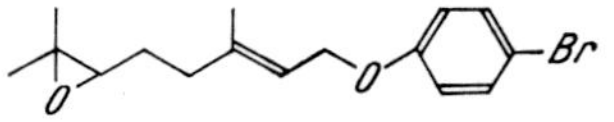	5	0.5	0.5
T-195	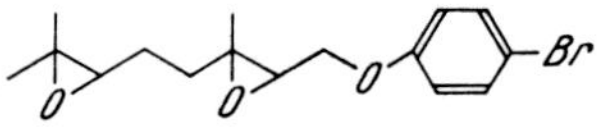	3	0.5	1
T-196	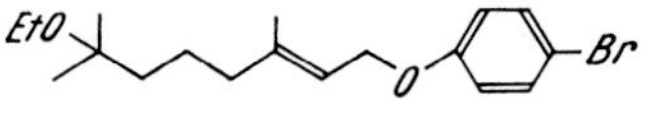	1	0.7	0.5
T-197	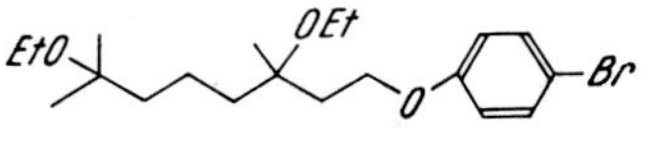	△	△	△
T-198	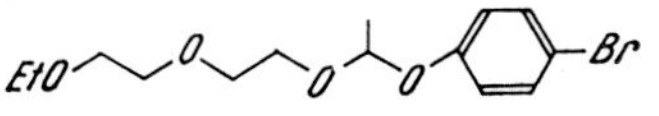			

TRI	GRY	TEN	DER	GAL	Miscellaneous	Ref.
1	100	0.5 (e)				
		5 (e)				
10		0.01 (e) 0.00005 (i)				
		5 (e)				
		0.5 (e)				
		1 (e) 0.07 (i)				
		0.1 (e)				
	~ 70	10 (e)				
8		0.005 (e) 0.00007 (i)				
		50 (e) 10 (i)				
		< 10 (e)[a]				198[a]

(Tables — continued)

No.	Compound	PYR	DYS	GRA
T-199	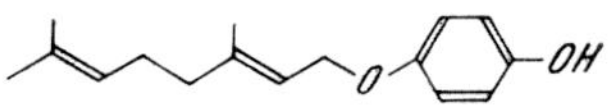	△	△	△
T-200	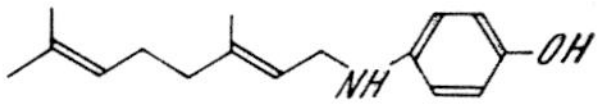		△	△
T-201	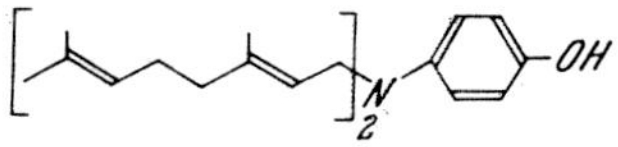		△	△
T-202	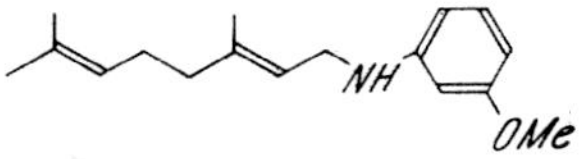		1	△
T-203	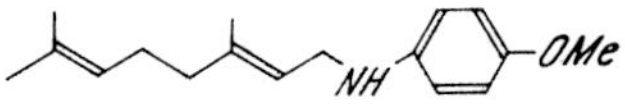		~ 100	△
T-204	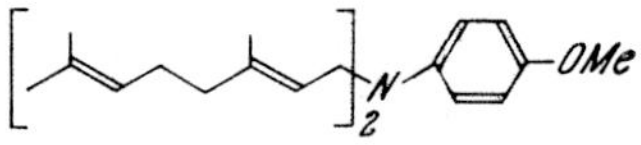		80	△
T-205	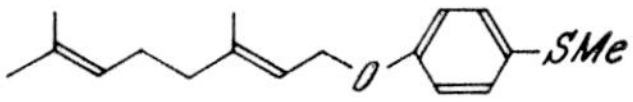	7	10	10
T-206	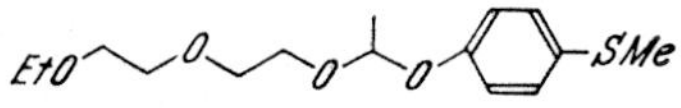			
T-207	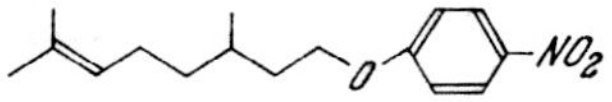	5	5	5
T-208	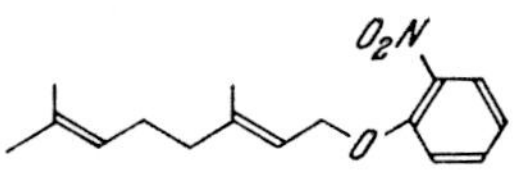	△	△	△
T-209	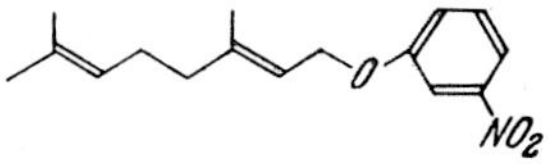	△	△	△

TRI	GRY	TEN	DER	GAL	Miscellaneous	Ref.
		△ (e)				
		△ (e)				
		△ (e)	△ (e)			
		△ (e)				
		5 (e)				
		△ (e)	△ (e)			
		3 (e)				
		10 (e)[a]				198[a]
△	△	△ (e) 100 (i)	△ (e) △ (i)			
		△ (e)				
		△ (e)				

(Tables — continued)

No.	Compound	PYR	DYS	GRA
T-210	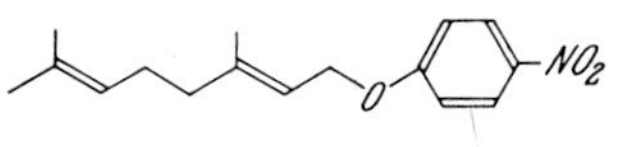	8	4	0.8
T-211	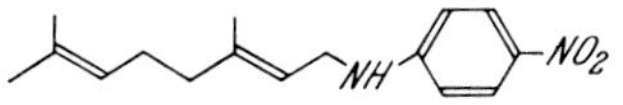	10	7	△
T-212	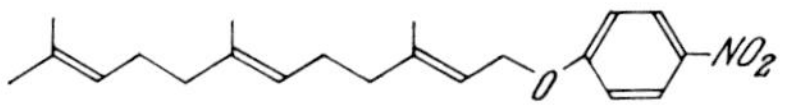	△	△	△
T-213	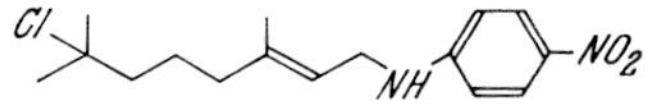	0.1	0.1	△
T-214	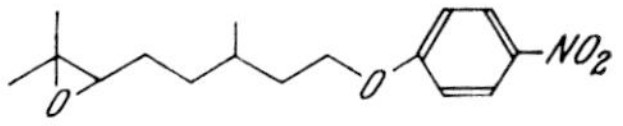	0.6	0.4	0.1
T-215	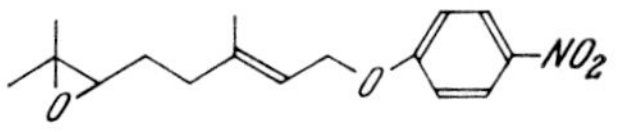	0.8	0.5	0.004
T-216	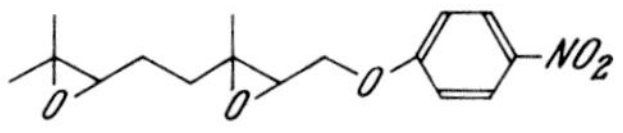	0.1	0.1	△
T-217	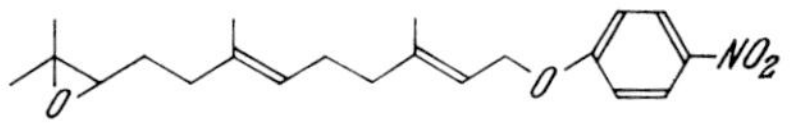		△	0.7
T-218	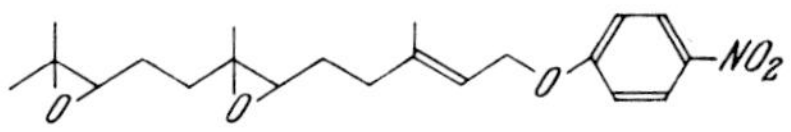		10	△
T-219	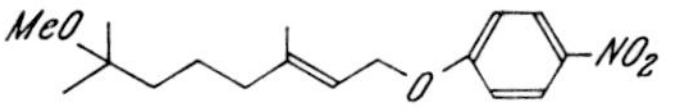	0.07	0.1	0.5
T-220	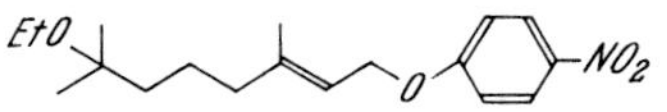	0.1	0.08	0.5

TRI	GRY	TEN	DER	GAL	Miscellaneous	Ref.
	100	30 (e) 10 (i)	100 (i)			
△	△	△ (e) △ (i)	△ (e) △ (i)			
	50	△ (i)				
	△	△ (e)				
	50	10 (e) 1 (i)	100 (e) 25 (i)	△ (7)	0.05 (LYG)	
	50	5 (e) 1 (i)	100 (e) 100 (i)		0.1 (LYG) ~ 0.04 (ONC)[a]	253[a]
	500	△ (e) △ (i)	△ (e) △ (i)			
		△ (i)		△ (7)		
		△ (i)				
		3 (e) 0.8 (i)		△ (p)		
△	50	0.005 (e) 0.003 (i)	> 10 (e) > 10 (i)			

(Tables — continued)

No.	Compound	PYR	DYS	GRA
T-221	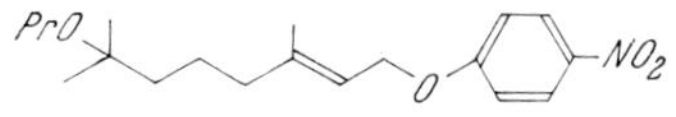	5	4	5
T-222	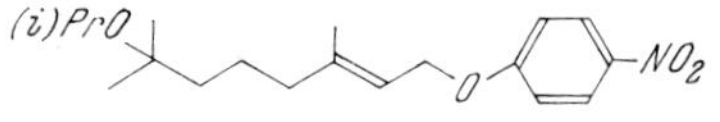	3	0.5	30
T-223	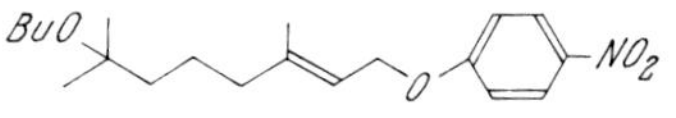	7	3	10
T-224	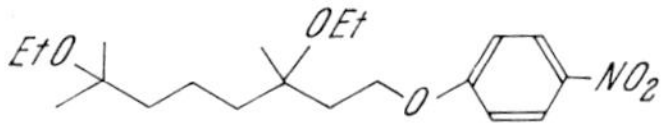	△	△	△
T-225	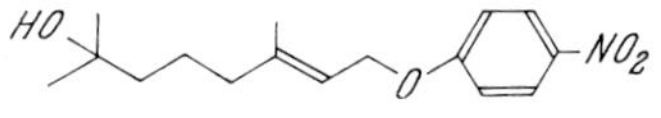	0.05	0.5	1
T-226	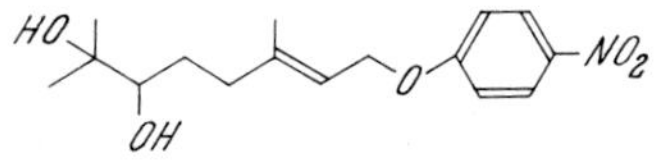	60	40	△
T-227	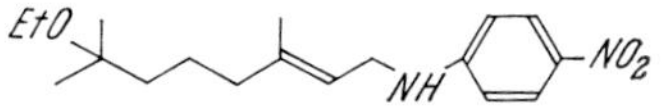	△	△	△
T-228	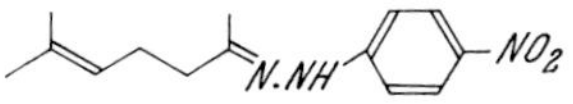	100		△
T-229	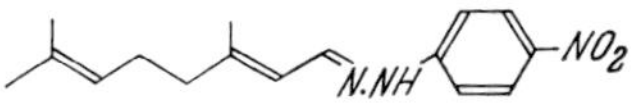	△	△	△
T-230	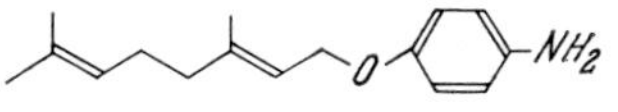	100	100	△
T-231	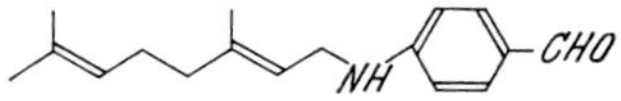	100		△

TRI	GRY	TEN	DER	GAL	Miscellaneous	Ref.
		0.005 (e) 0.003 (i)				
		100 (e) 0.1 (i)				
		500 (e)				
	△	△ (e)				
		5 (e) 5 (i)				
		△ (e) △ (i)				
		△ (i)				
		△ (e) △ (i)				
		△ (i)				
		500 (e) 10 (i)		△ (7)		
		△ (e) △ (i)				

(Tables — continued)

No.	Compound	PYR	DYS	GRA
T-232	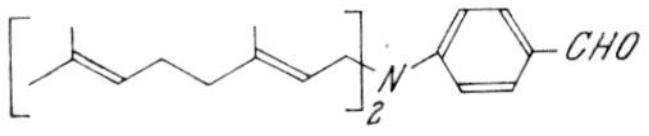		△	△
T-233	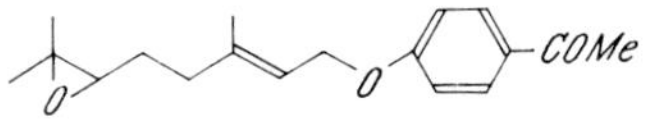			
T-234	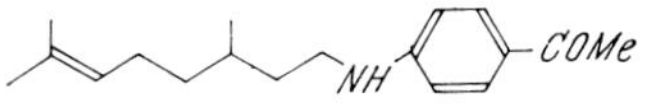	0.4		△
T-235	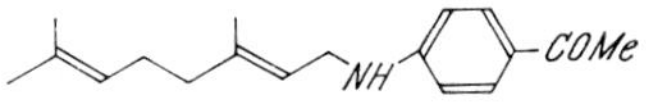	4		△
T-236	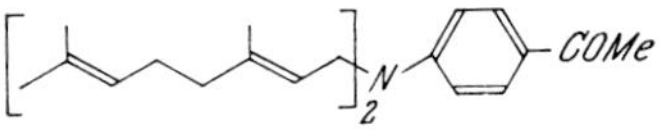	40		△
T-237	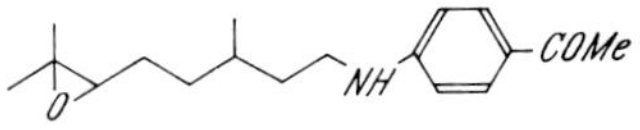	0.1	0.07	10
T-238	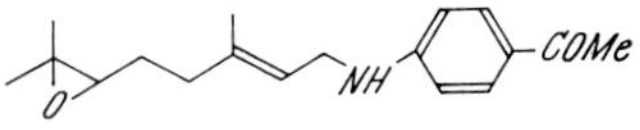	0.3	0.05	0.1
T-239	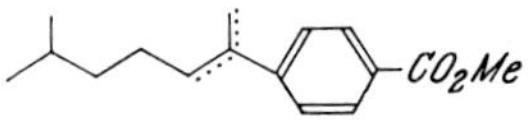	5	0.6	△
T-240	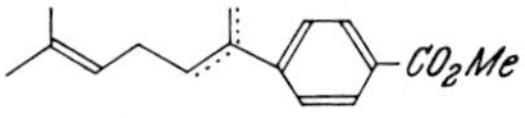	0.8	0.3	△
T-241	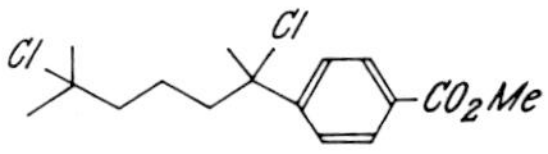	0.3	0.04	△
T-242	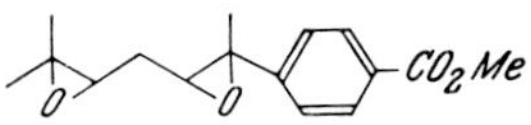	1	0.08	△

TRI	GRY	TEN	DER	GAL	Miscellaneous	Ref.
△		△ (e) △ (i)				
		~ 0.1 (e)[a]			> 0.1 (ONC)[a]	26[a]
		> 100 (e) 100 (i)	> 100 (e) > 100 (i)	△ (7)		
		△ (e) 50 (i)				
10		△ (e) △ (i)				
		500 (i)		△ (7)		
0.07		3 (e) 1 (i)				
				△ (7)		
△	△	△ (e) △ (i)		△ (7)		
	△	△ (e) △ (i)				

(Tables — continued)

No.	Compound	PYR	DYS	GRA
T-243	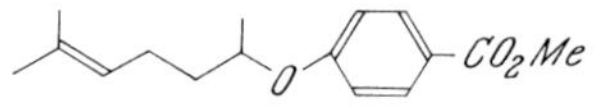	5	1	10
T-244	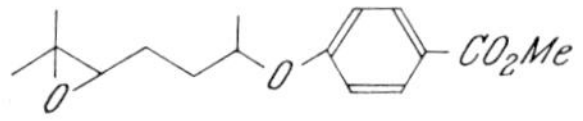	1	0.3	0.3
T-245	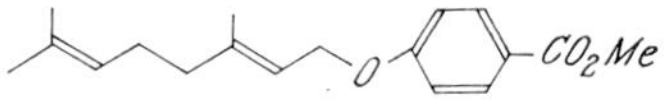	0.5	0.5	0.8
T-246	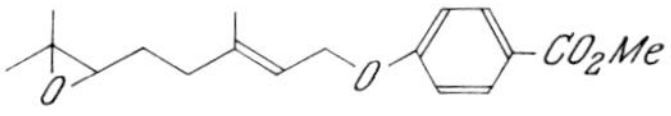	0.05	0.05	0.01
T-247	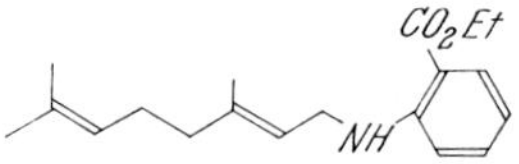		△	△
T-248	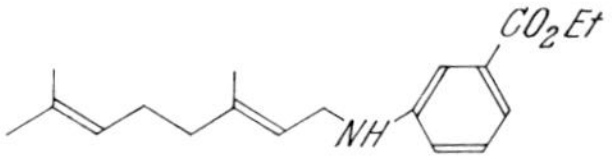		0.5	△
T-249	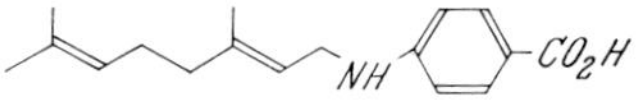		△	△
T-250	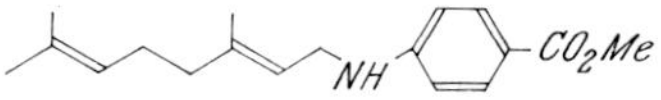	0.1	0.1	8
T-251		0.1	0.05	△
T-252	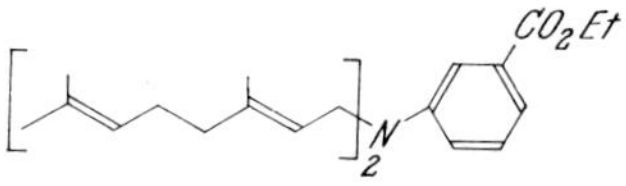		1	△
T-253	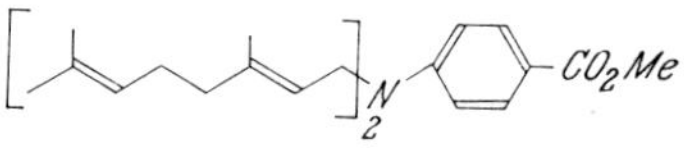	1	0.7	△

TRI	GRY	TEN	DER	GAL	Miscellaneous	Ref.
		△ (e)	△ (i)		5 (LYG)	
		△ (i)				
	△	△ (e)		△ (7)	0.01 (LYG)	
		△ (i)	△ (i)			
		100 (e)		△ (p)	1 (LYG)	
		10 (i)		△ (7)		
0.08	~ 70	100 (e)		△ (7)	0.05 (LYG)	26[a]
		100 (i)			0.05 (ONC)[a]	
		△ (e)				
		△ (e)				
△		△ (e)				
0.05	60	△ (i)		△ (7)		
10	80	△ (e)			1 (LYG)	
		△ (i)	△ (7)			
		△ (e)				
0.4	300	△ (i)				

(Tables — continued)

No.	Compound	PYR	DYS	GRA
T-254	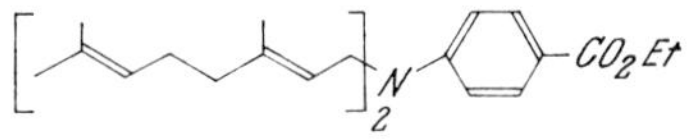		0.07	△
T-255	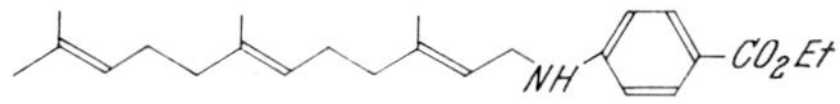		1	10
T-256	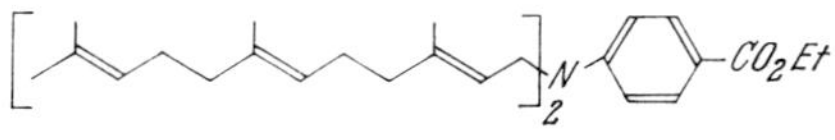		10	5
T-257		0.1	0.01	1
T-258	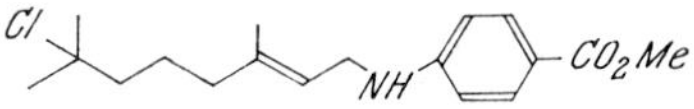	0.01	0.04	0.8
T-259	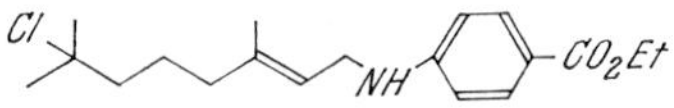	0.005	0.005	1
T-260	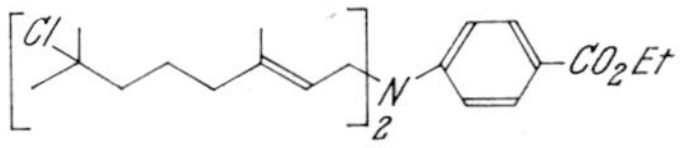	0.001	0.001	△
T-261	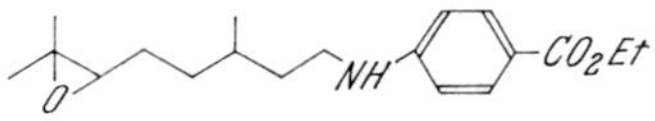			
T-262	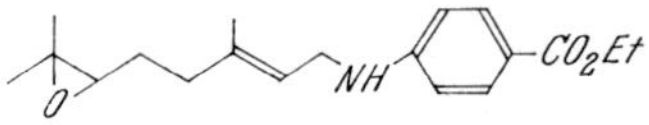		0.008	0.5
T-263	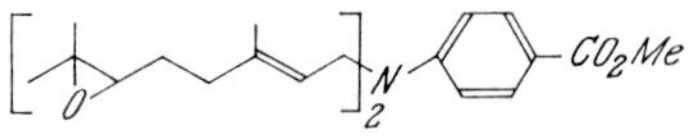	0.004	0.004	△
T-264	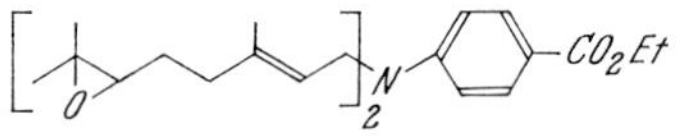	0.01	0.001	△

TRI	GRY	TEN	DER	GAL	Miscellaneous	Ref.
40	△	△ (i)				
		△ (e)				
		△ (e)				
		△ (e)				
0.03						
5	10	△ (e) △ (i)		△ (7)	< 1 (LYG)	
> 5		△ (e) △ (i)		△ (7)	5 (LYG)	
1			△ (e) △ (i)		1 (LYG)	
0.03		△ (e)				
0.2		△ (e)				
> 5		△ (e)			< 1 (LYG)	

(Tables — continued)

No.	Compound	PYR	DYS	GRA
T-265	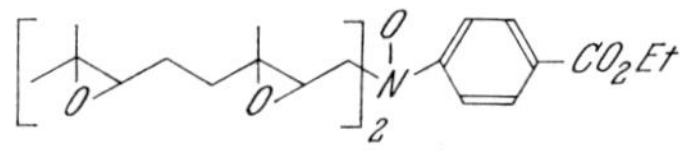	0.01	0.008	△
T-266	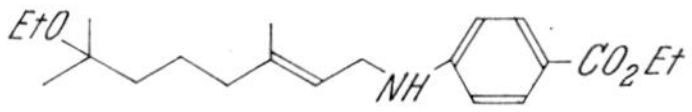	0.05	0.05	△
T-267	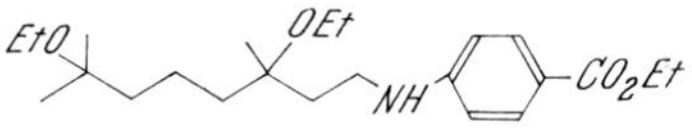	3	0.7	△
T-268	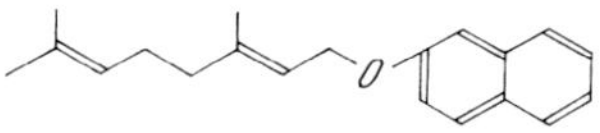		△	
T-269	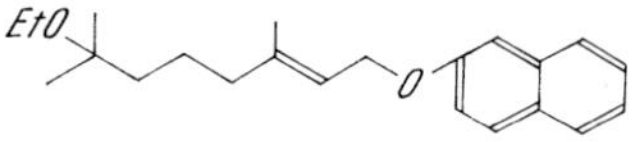	△	△	△
T-270	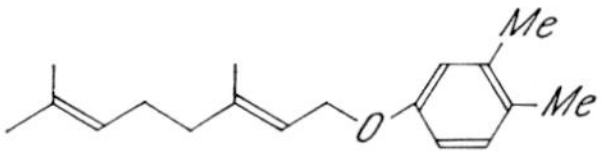		△	△
T-271	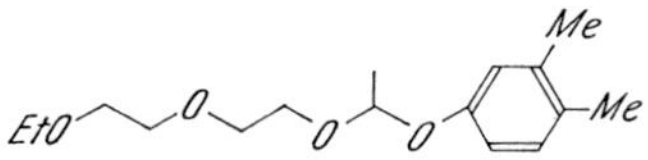			
T-272	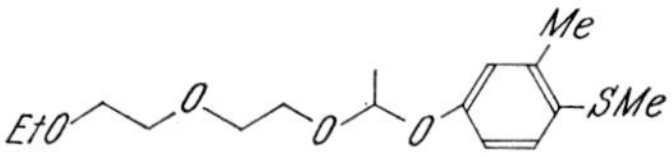			
T-273	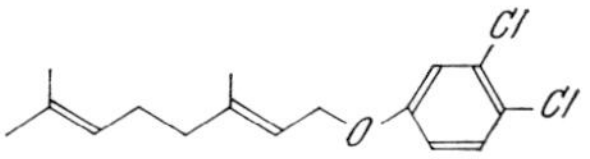		80	80
T-274	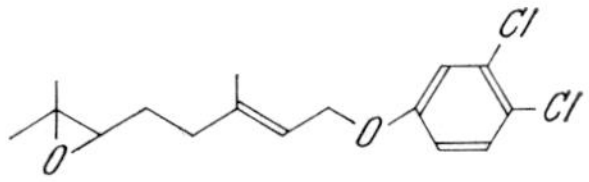			
T-275	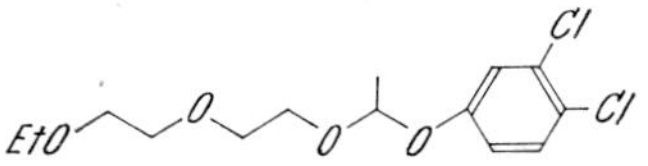			

TRI	GRY	TEN	DER	GAL	Miscellaneous	Ref.
		△ (e) △ (i)				
△	~ 70	△ (e)				
	~ 70	△ (e)				
		△ (e)				
		△ (e)				
		△ (e)				
		< 10 (e)[a]				198[a]
		< 10 (e)[a]				198[a]
		80 (e)				
		~ 5 (e)[a]			~ 0.01 (ONC)[a]	253[a]
		< 10 (e)[a]				198[a]

(Tables — continued)

No.	Compound	PYR	DYS	GRA
T-276	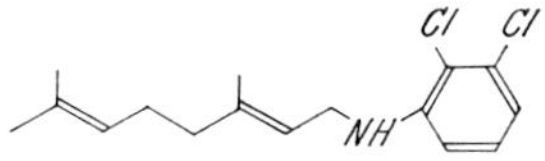			
T-277	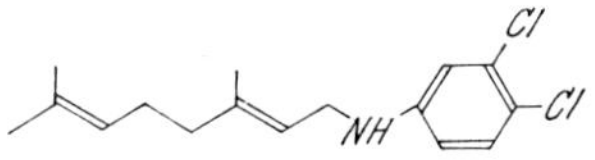		80	△
T-278	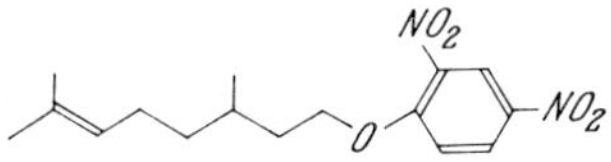		△	△
T-279	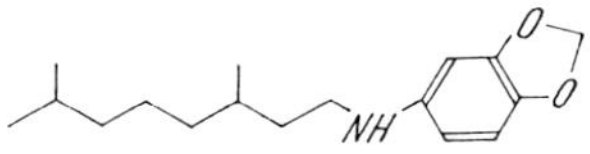	5	5	10
T-280	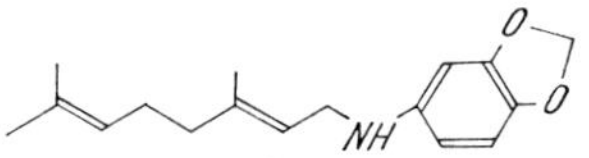	10	8	5
T-281	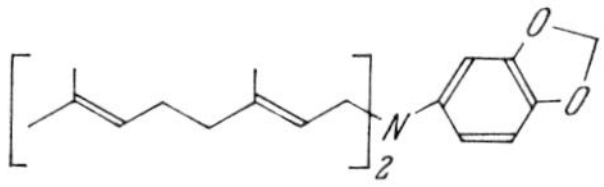	10	8	5
T-282	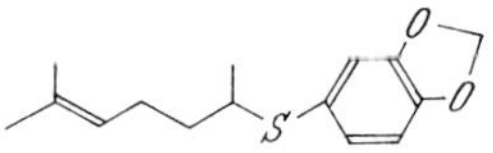	50	50	△
T-283	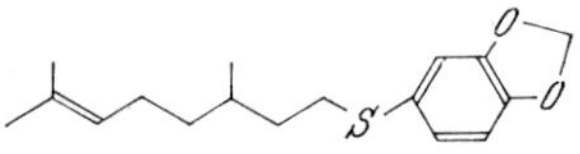	50	5	△
T-284	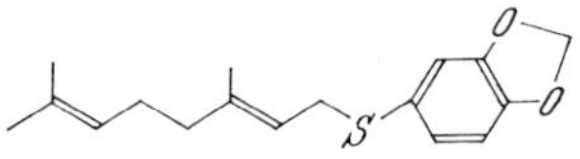	50	10	△
T-285	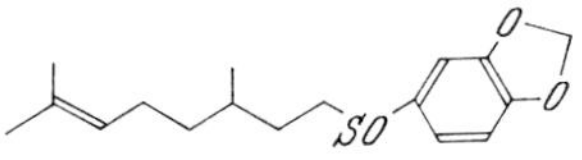		△	△
T-286	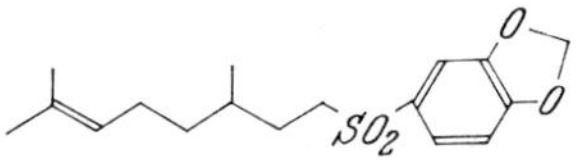		△	△

TRI	GRY	TEN	DER	GAL	Miscellaneous	Ref.
					0.003 (ONC)[a]	345[a]
		△ (e)				
		△ (e) △ (i)	△ (i)	△ (7)	250 (LYG)	
0.8		10 (e) 100 (i)	10 (e) 10 (i)	△ (7)	5 (LYG)	
0.08	> 100	10 (e) 0.8 (i)	1 (e) 0.1 (i)	△ (7)		
0.05	1000	10 (e) 5 (i)		△ (7)	3 (LYG)	
		△ (i)		△ (7)		
		△ (e) △ (i)	△ (e) △ (i)			
	△	△ (i)				
		△ (e) △ (i)				
		△ (i)				

(Tables — continued)

No.	Compound	PYR	DYS	GRA
T-287	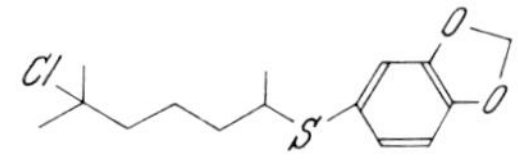	0.7	1	△
T-288	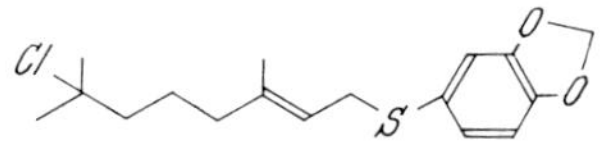	0.1	0.3	△
T-289	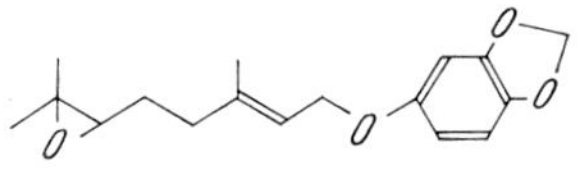	3	4	5
T-290	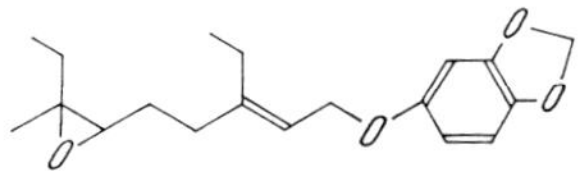		< 100[a]	
T-291	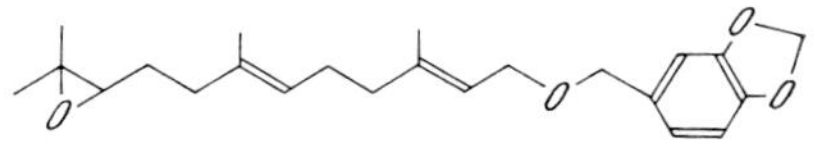			
T-292	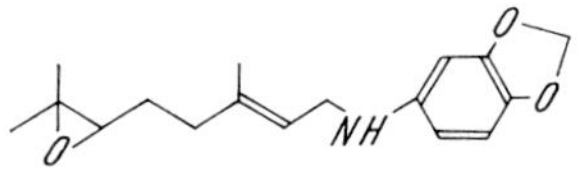	5	5	1
T-293	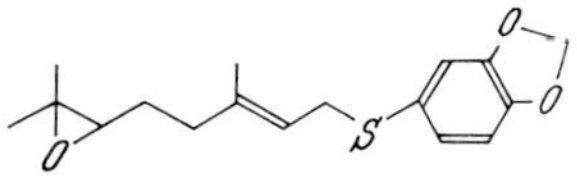	10	3	10
T-294	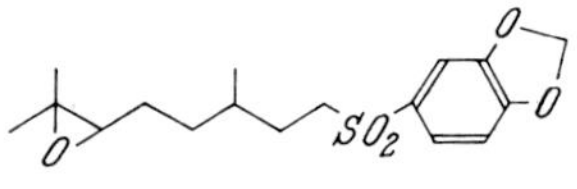		△	△
T-295	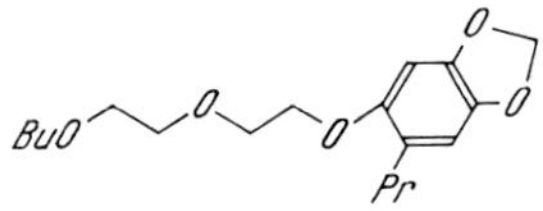			
T-296	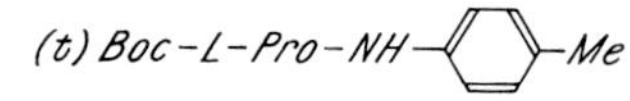	△	△	△
T-297	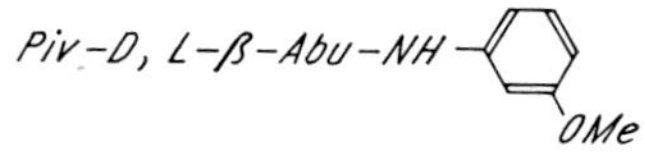		△	△

TRI	GRY	TEN	DER	GAL	Miscellaneous	Ref.
	△	△ (i)				
	△	△ (e)				
		△ (i)		△ (7)		
		0.04 (e)	0.001 (e)		5 (ONC)[a]	26[a]
0.0008		1 (i)		0.05 (7)	0.0001 (STO)[b]	385[b]
					5 (LYG)	
		~ 0.001 (e)[b]			~ 0.1 (ONC)[b]	242[a]
					10 (BLA)[a]	26[b]
		~ 50 (e)[a]			$<$ 1 (ONC)[a]	25[a]
0.008		0.08 (e)				
0.8	70	5 (i)				
		△ (i)	△ (i)			
					~ 0.001 (STO)[a]	385[a]
		△ (e)				
		△ (e)				

(Tables — continued)

No.	Compound	PYR	DYS	GRA
T-298	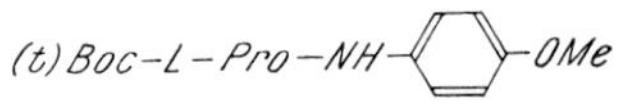	△	△	△
T-299	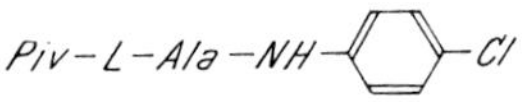	~ 100	1–100	△
T-300	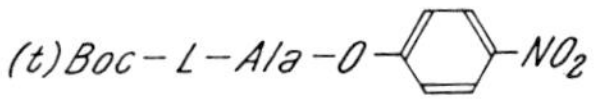	△	△	△
T-301	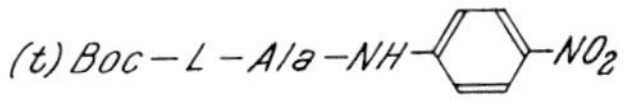	△	△	△
T-302	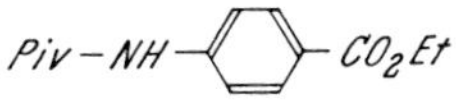	△	△	△
T-303	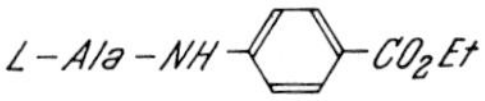	△	50	△
T-304	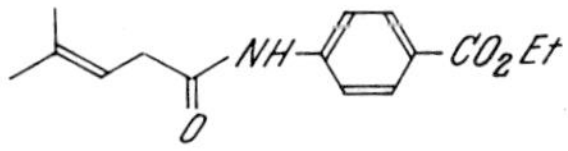		△	△
T-305	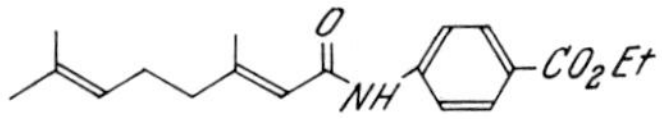		0.5	△
T-306	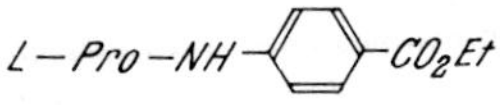	△	△	△
T-307	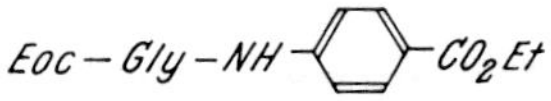		△	△
T-308	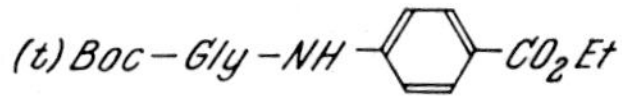	△	△	△

TRI	GRY	TEN	DER	GAL	Miscellaneous	Ref.
		△ (e)				
		△ (e)				
		△ (e) △ (i)				
		△ (e)				
		△ (e)				
		△ (e)				
		△ (e)				
		△ (e)				
		△ (e)		△ (7)		
		△ (e) △ (i)				

(Tables — continued)

No.	Compound	PYR	DYS	GRA
T-309	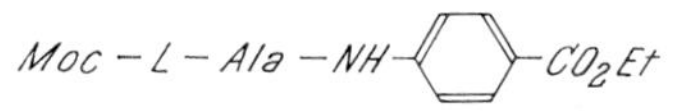		△	△
T-310	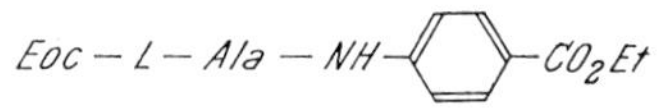		0.01	△
T-311	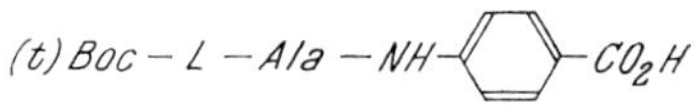	△	△	△
T-312	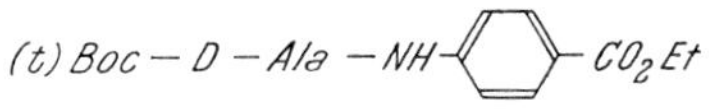	△	△	△
T-313	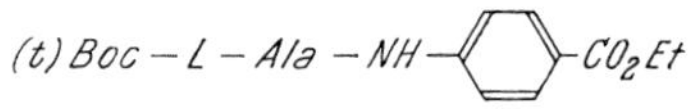	0.01	0.01	△
T-314	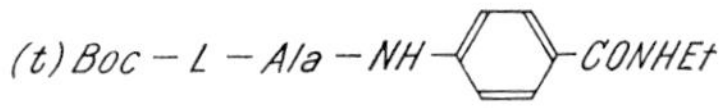	80		△
T-315	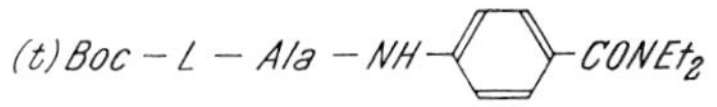		△	△
T-316	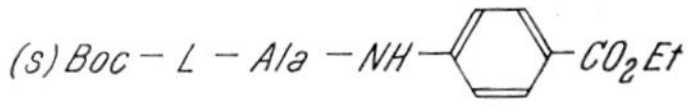	0.1	0.06	△
T-317	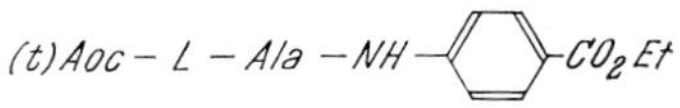		0.01	
T-318	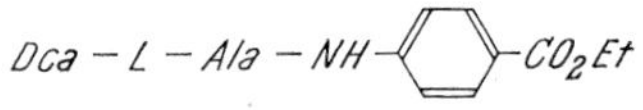		> 50	△
T-319	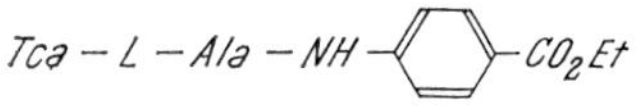	0.001	0.0008	△

TRI	GRY	TEN	DER	GAL	Miscellaneous	Ref.
		△ (e)				
		△ (e)				
		△ (e)				
		△ (e)				
		△ (e)				
		△ (e)				
		△ (e)				
		△ (e)				
		△ (e)				
		△ (e)				

(Tables — continued)

No.	Compound	PYR	DYS	GRA
T-320	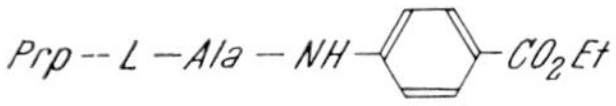		10	△
T-321	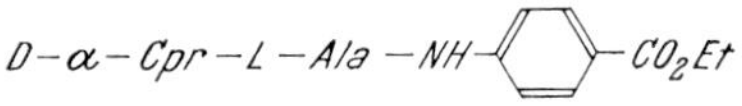	50	50	△
T-322	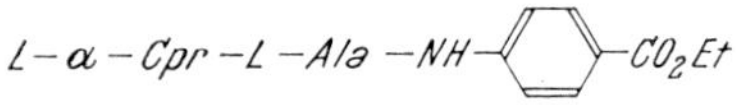	10	10	△
T-323	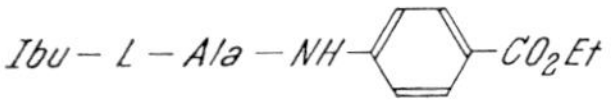		0.008	△
T-324	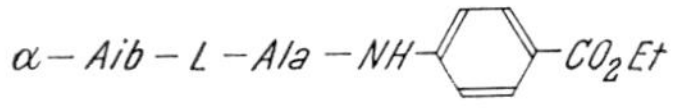		0.7	△
T-325	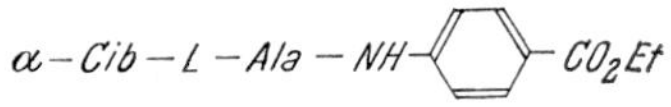	0.0001	0.00001	△
T-326	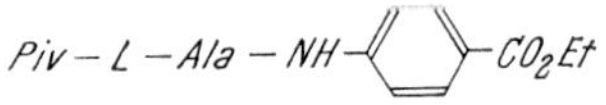	0.00004	0.00004	△
T-327	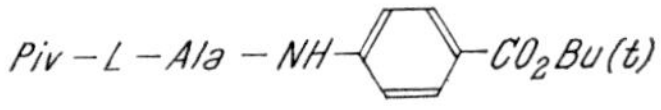		10	△
T-328			0.00008	△
T-329	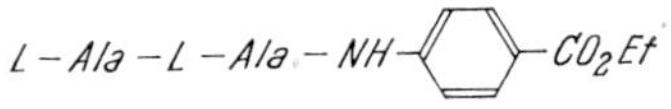	△	△	△
T-330	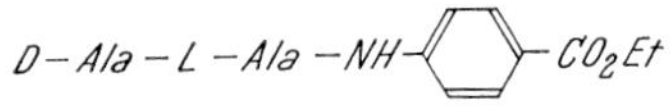	△	△	△

TRI	GRY	TEN	DER	GAL	Miscellaneous	Ref.
		△ (e)				
		△ (e)				
		△ (e)				
		△ (e)				
		△ (e)				
		△ (e)				
		△ (e)				
		△ (e)				
		△ (e)				
		△ (e)				
		△ (e)				

(Tables — continued)

No.	Compound	PYR	DYS	GRA
T-331	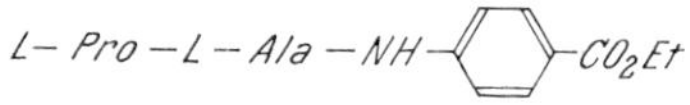		50	△
T-332	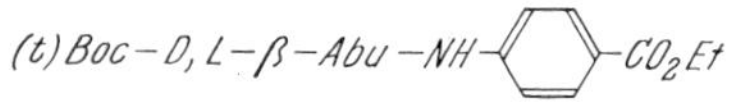	0.1	5	△
T-333	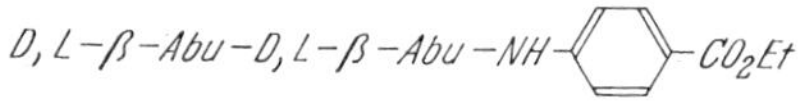	△	△	△
T-334	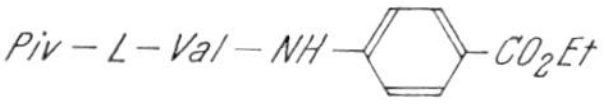		0.00004	△
T-335	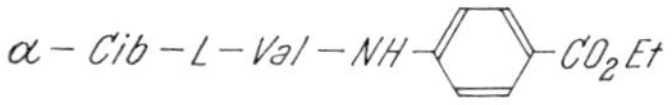		0.000002	△
T-336	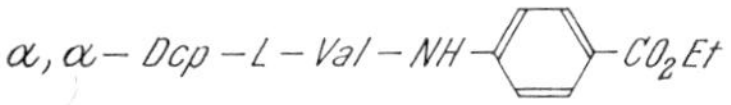		0.000005	△
T-337	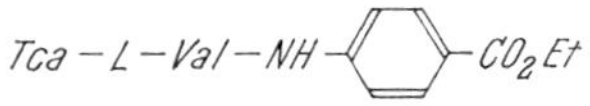		0.000009	△
T-338	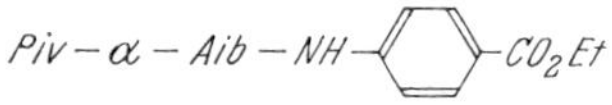	10	10	△
T-339	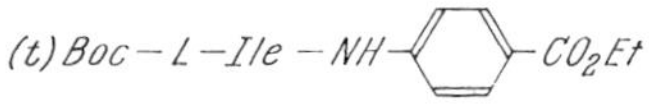	0.05	0.01	△
T-340	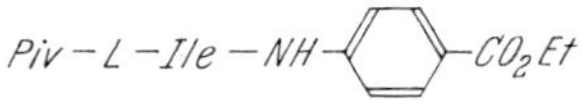	0.003	0.0001	△
T-341	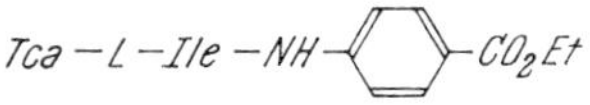		0.001	△

TRI	GRY	TEN	DER	GAL	Miscellaneous	Ref.
		△ (e)				
		△ (e)				
		△ (e) △ (i)				
		△ (e)				
		△ (e)				
		△ (e)				
		△ (e)				
		△ (e)				
		△ (e)				
		△		△ (7)		
		△ (e)				

(Tables — continued)

No.	Compound	PYR	DYS	GRA
T-342	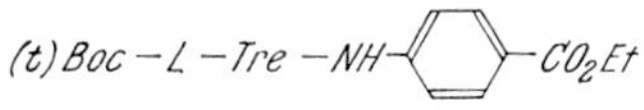		30	△
T-343	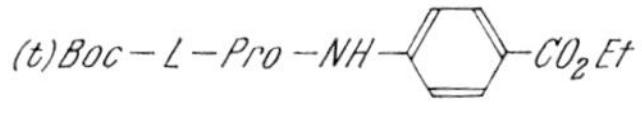	0.09	0.08	△
T-344	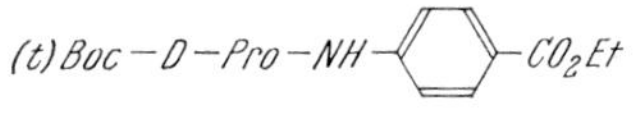	△	△	△
T-345	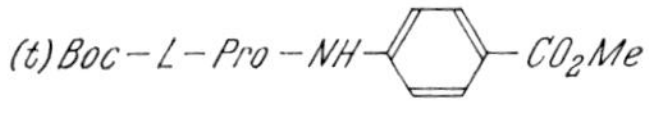	0.8	0.5	△
T-346	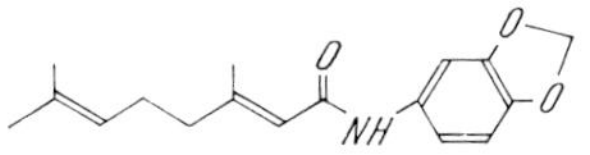		100	△
T-347	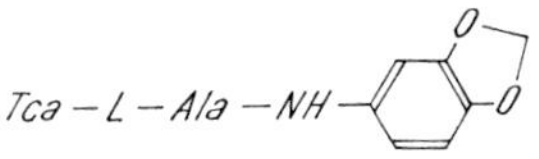		△	△
T-348	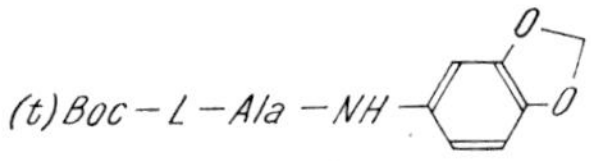	250	△	△
T-349	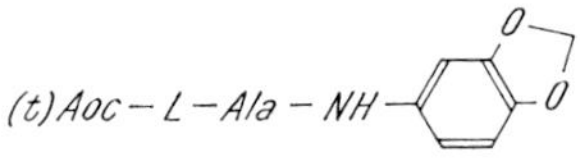	△	△	△
T-350	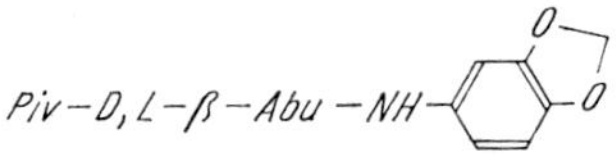		30	△
T-351	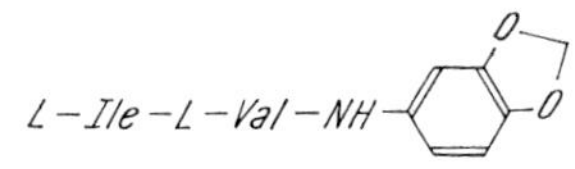		△	△
T-352	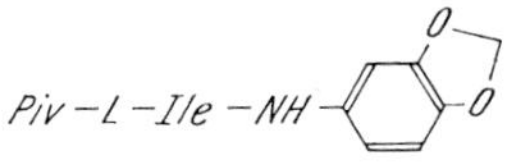		△	△
T-353	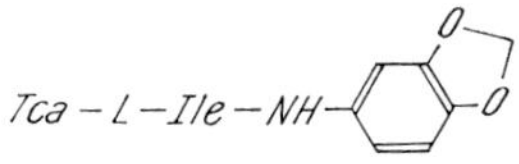		△	△

TRI	GRY	TEN	DER	GAL	Miscellaneous	Ref.
		△ (e)				
		△ (e)				
		△ (e)				
		△ (e)				
		△ (e)				
		△ (e)				
		△ (e)				
		△ (e)		△ (7)		
		△ (e)				
		△ (e)				
		△ (e)		△ (7)		
		△ (e)				

Alphabetical List of Juvenoids

Benzaldehyde,	4-[(3,7-dimethyl-2,6-octadienyl)amino]-,	T-231
	4-[bis(3,7-dimethyl-2,6-octadienyl)amino]-,	T-232
Benzene,	1-acetyl-4-[(3,7-dimethyl-6,7-epoxy-2-octenyl)amino]-,	T-238
	1-acetyl-4-[(3,7-dimethyl-6,7-epoxy-2-octenyl)oxy]-,	T-233
	1-acetyl-4-[(3,7-dimethyl-6,7-epoxyoctyl)amino]-,	T-237
	1-acetyl-4-[(3,7-dimethyl-2,6-octadienyl)amino]-,	T-235
	1-acetyl-4-[bis(3,7-dimethyl-2,6-octadienyl)amino]-,	T-236
	1-acetyl-4-[(3,7-dimethyl-6-octenyl)amino]-,	T-234
	1-amino-4-[(3,7-dimethyl-2,6-octadienyl)oxy]-,	T-230
	4-[(tert-amyloxycarbonyl-L-alanyl)amino]-1,2-(methylenedioxy)-,	T-349
	1-bromo-4-[(2,3 : 6,7-diepoxy-3,7-dimethyloctyl)oxy]-,	T-195
	1-bromo-4-[(3,7-diethoxy-3,7-dimethyloctyl)oxy]-,	T-197
	1-bromo-4-[(3,7-dimethyl-6,7-epoxy-2-octenyl)oxy]-,	T-194
	1-bromo-4-[(3,7-dimethyl-7-ethoxy-2-octenyl)oxy]-,	T-196
	1-bromo-4-[(3,7-dimethyl-2,6-octadienyl)oxy]-,	T-193
	1-bromo-4-[(1-methyl-2,5,8-trioxadecyl)oxy]-,	T-198
	4-[(tert-butoxycarbonyl-L-alanyl)amino]-1,2-(methylenedioxy)-,	T-348
	4-[(tert-butoxycarbonyl-L-alanyl)amino]-1-nitro-,	T-301
	4-[(tert-butoxycarbonyl-L-alanyl)oxy]-1-nitro-,	T-300
	4-[(tert-butoxycarbonyl-L-prolyl)amino]-1-methoxy-,	T-298
	4-[(tert-butoxycarbonyl-L-prolyl)amino]-1-methyl-,	T-296
	4-[(7-butoxy-3,7-dimethyl-2-octenyl)oxy]-1-nitro-,	T-223
	1-tert-butyl-4-[(3,7-dimethyl-6,7-epoxy-2-octenyl)oxy]-,	T-183
	1-tert-butyl-4-[(3,7-dimethyl-7-ethoxy-2-octenyl)oxy]-,	T-184
	1-tert-butyl-4-[(3,7-dimethyl-2,6-octadienyl)oxy]-,	T-179
	1-chloro-4-[(7-chloro-3,7-dimethyl-2-octenyl)oxy]-,	T-186
	1-chloro-4-[(2,3 : 6,7-diepoxy-3,7-dimethyloctyl)oxy]-,	T-189
	1-chloro-4-[(3,7-diethoxy-3,7-dimethyloctyl)oxy]-,	T-191
	1-chloro-3-[(3,7-dimethyl-6,7-epoxy-2-octenyl)oxy]-,	T-187
	1-chloro-4-[(3,7-dimethyl-6,7-epoxy-2-octenyl)oxy]-,	T-188
	1-chloro-4-[(3,7-dimethyl-7-ethoxy-2-octenyl)oxy]-,	T-190
	4-[(5-chloro-1,5-dimethylhexyl)thio]-1,2-(methylenedioxy)-,	T-287
	1-chloro-4-[(3,7-dimethyl-7-hydroxy-2-octenyl)oxy]-,	T-192
	1-chloro-4-[(3,7-dimethyl-2,6-octadienyl)oxy]-,	T-185
	4-[(7-chloro-3,7-dimethyl-2-octenyl)amino]-1-nitro-,	T-213
	4-[(7-chloro-3,7-dimethyl-2-octenyl)thio]-1,2-(methylenedioxy)-,	T-288
	1-chloro-4-[(pivaloyl-L-alanyl)amino]-,	T-299
	1,2-dichloro-4-[(3,7-dimethyl-6,7-epoxy-2-octenyl)oxy]-,	T-274
	1,2-dichloro-3-[(3,7-dimethyl-2,6-octadienyl)amino]-,	T-276
	1,2-dichloro-4-[(3,7-dimethyl-2,6-octadienyl)amino]-,	T-277
	1,2-dichloro-4-[(3,7-dimethyl-2,6-octadienyl)oxy]-,	T-273
	1,2-dichloro-4-[(1-methyl-2,5,8-trioxadecyl)oxy]-,	T-275
	4-[(2,3 : 6,7-diepoxy-3,7-dimethyloctyl)oxy]-1-nitro-,	T-216
	4-[(6,7 : 10,11-diepoxy-3,7,11-trimethyl-2-dodecenyl)oxy]-1-nitro-,	T-218

Benzene,	4-[(3,7-diethoxy-3,7-dimethyloctyl)oxy]-1-nitro-,	T-224
	4-[(6,7-dihydroxy-3,7-dimethyl-2-octenyl)oxy]-1-nitro-,	T-226
	1,2-dimethyl-4-[(3,7-dimethyl-2,6-octadienyl)oxy]-,	T-270
	(4,8-dimethyl-7,8-epoxy-1-nonenyl)-,	T-174
	4-[(3,7-dimethyl-6,7-epoxy-2-octenyl)amino]-1,2-(methylenedioxy)-,	T-292
	4-[(3,7-dimethyl-6,7-epoxy-2-octenyl)oxy]-1-ethyl-,	T-181
	4-[(3,7-dimethyl-6,7-epoxy-2-octenyl)oxy]-1-isopropyl-,	T-182
	4-[(3,7-dimethyl-6,7-epoxy-2-octenyl)oxy]-1-methyl-,	T-180
	4-[(3,7-dimethyl-6,7-epoxy-2-octenyl)oxy]-1,2-(methylenedioxy)-,	T-289
	4-[(3,7-dimethyl-6,7-epoxy-2-octenyl)oxy]-1-nitro-,	T-215
	4-[(3,7-dimethyl-6,7-epoxy-2-octenyl)thio]-1,2-(methylenedioxy)-,	T-293
	4-[(3,7-dimethyl-6,7-epoxyoctyl)oxy]-1-nitro-,	T-214
	4-[(3,7-dimethyl-6,7-epoxyoctyl)sulfonyl]-1,2-(methylenedioxy)-,	T-294
	4-[(3,7-dimethyl-7-ethoxy-2-octenyl)amino]-1-nitro-,	T-227
	4-[(3,7-dimethyl-7-ethoxy-2-octenyl)oxy]-1-nitro-,	T-220
	4-[(1,5-dimethyl-4-hexenyl)thio]-1,2-(methylenedioxy)-,	T-282
	4-[(3,7-dimethyl-7-hydroxy-2-octenyl)oxy]-1-nitro-,	T-225
	4-[(3,7-dimethyl-7-isopropoxy-2-octenyl)oxy]-1-nitro-,	T-222
	4-[(3,7-dimethyl-7-methoxy-2-octenyl)oxy]-1-nitro-,	T-219
	1,2-dimethyl-4-[(1-methyl-2,5,8-trioxadecyl)oxy]-,	T-271
	(3,7-dimethyl-2,6-octadienyl)amino-,	T-177
	bis(3,7-dimethyl-2,6-octadienyl)amino-,	T-178
	4-[(3,7-dimethyl-2,6-octadienyl)amino]-1-hydroxy-,	T-200
	4-[bis(3,7-dimethyl-2,6-octadienyl)amino]-1-hydroxy-,	T-201
	3-[(3,7-dimethyl-2,6-octadienyl)amino]-1-methoxy-,	T-202
	4-[(3,7-dimethyl-2,6-octadienyl)amino]-1-methoxy-,	T-203
	4-[bis(3,7-dimethyl-2,6-octadienyl)amino]-1-methoxy-,	T-204
	4-[(3,7-dimethyl-2,6-octadienyl)amino]-1,2-(methylenedioxy)-,	T-280
	4-[bis(3,7-dimethyl-2,6-octadienyl)amino]-1,2-(methylenedioxy)-,	T-281
	4-[(3,7-dimethyl-2,6-octadienyl)amino]-1-nitro-,	T-211
	(3,7-dimethyl-2,6-octadienyl)oxy-,	T-176
	4-[(3,7-dimethyl-2,6-octadienyl)oxy]-1-hydroxy-,	T-199
	4-[(3,7-dimethyl-2,6-octadienyl)oxy]-1-methylthio-,	T-205
	2-[(3,7-dimethyl-2,6-octadienyl)oxy]-1-nitro-,	T-208
	3-[(3,7-dimethyl-2,6-octadienyl)oxy]-1-nitro-,	T-209
	4-[(3,7-dimethyl-2,6-octadienyl)oxy]-1-nitro-,	T-210
	4-[(3,7-dimethyl-2,6-octadienyl)thio]-1,2-(methylenedioxy)-,	T-284
	4-[(3,7-dimethyl-6-octenyl)oxy]-1,3-dinitro-,	T-278
	4-[(3,7-dimethyl-6-octenyl)oxy]-1-nitro-,	T-207
	4-[(3,7-dimethyl-6-octenyl)sulfinyl]-1,2-(methylenedioxy)-,	T-285
	4-[(3,7-dimethyl-6-octenyl)sulfonyl]-1,2--(methylenedioxy)-,	T-286
	4-[(3,7-dimethyl-6-octenyl)thio]-1,2-(methylenedioxy)-,	T-283
	4-[(3,7-dimethyloctyl)amino]-1,2-(methylenedioxy)-,	T-279

Benzene,	4-[(3,7-dimethyl-1-oxo-2,6-octadienyl)amino]-1,2-(methylenedioxy)-,	T-346
	4-[(3,7-dimethyl-7-propoxy-2-octenyl)oxy]-1-nitro-,	T-221
	4-[(3,6-dioxadecyl)oxy]-1,2-(methylenedioxy)-5-propyl-,	T-295
	4-[(6,7-epoxy-3-ethyl-7-methyl-2-nonenyl)oxy]-1,2-(methylenedioxy)-,	T-290
	4-(12,13-epoxy-2-oxa-5,9,13-trimethyl-4,8-tetradecadienyl)-1,2-(methylenedioxy)-,	T-291
	4-[(10,11-epoxy-3,7,11-trimethyl-2,6-dodecadienyl)oxy]-1-nitro-,	T-217
	(9,10-epoxy-2,6,10-trimethyl-1,5-undecadienyl)-,	T-175
	4-[(L-isoleucyl-L-valyl)amino]-1,2-(methylenedioxy)-,	T-351
	1-methoxy-3-[(pivaloyl-D,L-β-aminobutyryl)amino]-,	T-297
	1,2-(methylenedioxy)-4-[(pivaloyl-D,L-β-aminobutyryl)amino]-,	T-350
	1,2-(methylenedioxy)-4-[(pivaloyl-L-isoleucyl)amino]-,	T-352
	1,2-(methylenedioxy)-4-[(trichloroacetyl-L-alanyl)amino]-,	T-347
	1,2-(methylenedioxy)-4-[(trichloroacetyl-L-isoleucyl)amino]-,	T-353
	2-methyl-1-methylthio-4-[(1-methyl-2,5,8-trioxadecyl)oxy]-,	T-272
	1-methylthio-4-[(1-methyl-2,5,8-trioxadecyl)oxy]-,	T-206
	1-nitro-4-[(3,7,11-trimethyl-2,6,10-dodecatrienyl)oxy]-,	T-212
Benzoic acid,	4-[(D-alanyl-L-alanyl)amino]-, ethyl ester	T-330
	4-[(L-alanyl-L-alanyl)amino]-, ethyl ester	T-329
	4-[(L-alanyl)amino]-, ethyl ester	T-303
	4-[(D,L-β-aminobutyryl-D,L-β-aminobutyryl)amino]-, ethyl ester	T-333
	4-[(α-aminoisobutyryl-L-alanyl)amino]-, ethyl ester	T-324
	4-[(tert-amyloxycarbonyl-L-alanyl)amino]-, ethyl ester	T-317
	4-[(tert-butoxycarbonyl-L-alanyl)amino]-,	T-311
	4-[(tert-butoxycarbonyl-L-alanyl)amino]-, N-diethyl amide	T-315
	4-[(tert-butoxycarbonyl-L-alanyl)amino]-, N-ethyl amide	T-314
	4-[(tert-butoxycarbonyl-D-alanyl)amino]-, ethyl ester	T-312
	4-[(tert-butoxycarbonyl-L-alanyl)amino]-, ethyl ester	T-313
	4-[(sec-butoxycarbonyl-L-alanyl)amino]-, ethyl ester	T-316
	4-[(tert-butoxycarbonyl-D,L-β-aminobutyryl)amino]-, ethyl ester	T-332
	4-[(tert-butoxycarbonylglycyl)amino]-, ethyl ester	T-308
	4-[(tert-butoxycarbonyl-L-isoleucyl)amino]-, ethyl ester	T-339
	4-[(tert-butoxycarbonyl-D-prolyl)amino]-, ethyl ester	T-344
	4-[(tert-butoxycarbonyl-L-prolyl)amino]-, ethyl ester	T-343
	4-[(tert-butoxycarbonyl-L-prolyl)amino]-, methyl ester	T-345
	4-[(tert-butoxycarbonyl-L-threonyl)amino]-, ethyl ester	T-342
	4-[(isobutyryl-L-alanyl)amino]-, ethyl ester	T-323

Benzoic acid,	4-[(7-chloro-3,7-dimethyl-2-octenyl)amino]-, ethyl ester	T-259
	4-[bis(7-chloro-3,7-dimethyl-2-octenyl)amino]-, ethyl ester	T-260
	4-[(7-chloro-3,7-dimethyl-2-octenyl)amino]-, methyl ester	T-258
	4-[(α-chloroisobutyryl-L-alanyl)amino]-, ethyl ester	T-325
	4-[(α-chloroisobutyryl-L-valyl)amino]-, ethyl ester	T-335
	4-[(D-α-chloropropionyl-L-alanyl)amino]-, ethyl ester	T-321
	4-[(L-α-chloropropionyl-L-alanyl)amino]-, ethyl ester	T-322
	4-[(dichloroacetyl-L-alanyl)amino]-, ethyl ester	T-318
	4-(1,5-dichloro-1,5-dimethylhexyl)-, methyl ester	T-241
	4-[(α,α-dichloropropionyl-L-valyl)amino]-, ethyl ester	T-336
	4-(1,2 : 4,5-diepoxy-1,5-dimethylhexyl)-, methyl ester	T-242
	4-[bis(2,3 : 6,7-diepoxy-3,7-dimethyloctyl)aminoxy]-, ethyl ester	T-265
	4-[(3,7-diethoxy-3,7-dimethyloctyl)amino]-, ethyl ester	T-267
	4-[(1,5-dimethyl-4,5-epoxyhexyl)oxy]-, methyl ester	T-244
	4-[(3,7-dimethyl-6,7-epoxy-2-octenyl)amino]-, ethyl ester	T-262
	4-[bis(3,7-dimethyl-6,7-epoxy-2-octenyl)amino]-, ethyl ester	T-264
	4-[bis(3,7-dimethyl-6,7-epoxy-2-octenyl)amino]-, methyl ester	T-263
	4-[(3,7-dimethyl-6,7-epoxy-2-octenyl)oxy]-, methyl ester	T-246
	4-[(3,7-dimethyl-6,7-epoxyoctyl)amino]-, ethyl ester	T-261
	4-[(3,7-dimethyl-7-ethoxy-2-octenyl)amino]-, ethyl ester	T-266
	4-(1,5-dimethyl-1,4-hexadienyl)-, methyl ester	T-240
	4-(1,5-dimethyl-1-hexenyl), methyl ester	T-239
	4-[(1,5-dimethyl-4-hexenyl)oxy]-, methyl ester	T-243
	4-[(3,7-dimethyl-2,6-octadienyl)amino]-,	T-249
	2-[(3,7-dimethyl-2,6-octadienyl)amino]-, ethyl ester	T-247
	3-[(3,7-dimethyl-2,6-octadienyl)amino]-, ethyl ester	T-248
	3-[bis(3,7-dimethyl-2,6-octadienyl)amino]-, ethyl ester	T-252
	4-[(3,7-dimethyl-2,6-octadienyl)amino]-, ethyl ester	T-251
	4-[bis(3,7-dimethyl-2,6-octadienyl)amino]-, ethyl ester	T-254
	4-[(3,7-dimethyl-2,6-octadienyl)amino]-, methyl ester	T-250
	4-[bis(3,7-dimethyl-2,6-octadienyl)amino]-, methyl ester	T-253
	4-[(3,7-dimethyl-2,6-octadienyl)N-methylamino]-, ethyl ester	T-257
	4-[(3,7-dimethyl-2,6-octadienyl)oxy]-, methyl ester	T-245
	4-[(3,7-dimethyl-1-oxo-2,6-octadienyl)amino]-, ethyl ester	T-305
	4-[(ethoxycarbonyl-L-alanyl)amino]-, ethyl ester	T-310
	4-[(ethoxycarbonylglycyl)amino]-, ethyl ester	T-307

Benzoic acid,	4-[(methoxycarbonyl-L-alanyl)amino]-, ethyl ester	T-309
	4-[(4-methyl-1-oxo-3-pentenyl)amino]-, ethyl ester	T-304
	4-[(pivaloyl-L-alanyl)amino]-, tert-butyl ester	T-327
	4-[(pivaloyl-L-alanyl)amino]-, ethyl ester	T-326
	4-[(pivaloyl-L-alanyl)N-methylamino]-, ethyl ester	T-328
	4-(pivaloylamino)-, ethyl ester	T-302
	4-[(pivaloyl-α-aminoisobutyryl)amino]-, ethyl ester	T-338
	4-[(pivaloyl-L-isoleucyl)amino]-, ethyl ester	T-340
	4-[(pivaloyl-L-valyl)amino]-, ethyl ester	T-334
	4-[(L-prolyl-L-alanyl)amino]-, ethyl ester	T-331
	4-(L-prolylamino)-, ethyl ester	T-306
	4-[(propionyl-L-alanyl)amino]-, ethyl ester	T-320
	4-[(trichloroacetyl-L-alanyl)amino]-, ethyl ester	T-319
	4-[(trichloroacetyl-L-isoleucyl)amino]-, ethyl ester	T-341
	4-[(trichloroacetyl-L-valyl)amino]-, ethyl ester	T-337
	4-[(3,7,11-trimethyl-2,6,10-dodecatrienyl)amino]-, ethyl ester	T-255
	4-[bis(3,7,11-trimethyl-2,6,10-dodecatrienyl)amino]-, ethyl ester	T-256
Carbamic acid,	N-(3,7-dimethyl-6,7-epoxy-2-octenyl)-, ethyl ester	T-170
	N-(3,7-dimethyl-6,7-epoxyoctyl)-, ethyl ester	T-169
	N-(3,7-dimethyl-6,7-epoxyoctyl)-, phenyl ester	T-168
	N-(3,7-dimethyl-6,7-imino)-, ethyl ester	T-171
	N-(3,7-dimethyl-2,6-octadienyl)-, ethyl ester	T-167
	N-(3,7-dimethyl-6-octenyl)-, ethyl ester	T-164
	N-(3,7-dimethyl-6-octenyl)-, 2-chlorophenyl ester	T-166
	N-(3,7-dimethyl-6-octenyl)-, phenyl ester	T-165
1-Cyclohexen-1-oic acid,	4-(1,5-dimethyl-3-oxo-4-hexenyl)-, methyl ester	T-173
1-Cyclohexen-1-oic acid,	4-(1,5-dimethyl-3-oxohexyl)-, methyl ester	T-172
2,6-*Decadienoic acid,*	10-cyclohexyl-3,7-dimethyl-10,1′-epoxy-, methyl ester	T-160
2,6-*Decadienoic acid,*	10-cyclohexylidene-3,7-dimethyl-, methyl ester	T-159
4,8-*Decadienoic acid,*	5,9-dimethyl, ethyl ester	T-114
2,6-*Decadienoic acid,*	3,7-dimethyl-10-oxo-, methyl ester	T-161
8-*Decenoic acid,*	5,9-dimethyl-2-oxa-, ethyl ester	T-48
2-*Decenoic acid,*	3,6,9,9-tetramethyl-, methyl ester	T-113
2,6-*Dodecadienoic acid,*	2-cyano-10,11-epoxy-3,7,11-trimethyl-, methyl ester	T-158
2,6-*Dodecadienoic acid,*	3,11-dimethyl-10,11-epoxy-7-ethyl-, methyl ester	T-140
6,10-*Dodecadienoic acid,*	7,11-dimethyl-4-oxa, methyl ester	T-50

6,10-*Dodecadienoic acid,*	7,11-dimethyl-4-oxa-3-oxo-, ethyl ester	T-105
2,7-*Dodecadienoic acid,*	5,10-dioxa-3,8,11,11-tetramethyl-, ethyl ester	T-88
2,6-*Dodecadienoic acid,*	10,11-epoxy-7-ethyl-3-methyl-, methyl ester	T-141
2,6-*Dodecadienoic acid,*	10,11-epoxy-7-ethyl-11-methyl-, methyl ester	T-139
2,6-*Dodecadienoic acid,*	10,11-epoxy-3,7,11-trimethyl-, ethyl ester	T-43
2,6-*Dodecadienoic acid,*	10,11-epoxy-3,7,11-trimethyl-, methyl ester	T-42
2,6-*Dodecadienoic acid,*	11-methoxy-3,7,11-trimethyl-, N-diethyl amide	T-45
2,6-*Dodecadienoic acid,*	11-methoxy-3,7,11-trimethyl-, methyl ester	T-44
2,4-*Dodecadienoic acid,*	11-methoxy-3,7,11-trimethyl-, isopropyl ester	T-46
2,6-*Dodecadienoic acid,*	10-oxa-3,7,11,11-tetramethyl-, ethyl ester	T-87
6,10-*Dodecadienoic acid,*	4-oxa-2,7,11-trimethyl-, methyl ester	T-51
2,10-*Dodecadienoic acid,*	5-oxa-3,7,11-trimethyl-, N-diethyl amide	T-58
2,10-*Dodecadienoic acid,*	5-oxa-3,7,11-trimethyl-, ethyl ester	T-57
2,10-*Dodecadienoic acid,*	5-oxa-3,7,11-trimethyl-, morpholide	T-59
2,10-*Dodecadienoic acid,*	6-oxa-3,7,11-trimethyl-, N-diethyl amide	T-77
2,10-*Dodecadienoic acid,*	6-oxa-3,7,11-trimethyl-, ethyl ester	T-76
2,10-*Dodecadienoic acid,*	6-oxa-3,7,11-trimethyl-, methyl ester	T-75
2,10-*Dodecadienoic acid,*	2,3,7,11-tetramethyl-, methyl ester	T-122
2,10-*Dodecadienoic acid,*	3,7,11-trimethyl-, ethyl ester	T-16
2,10-*Dodecadienoic acid,*	3,7,11-trimethyl-, methyl ester	T-15
2,10-*Dodecadienoic acid,*	3,7,11-trimethyl-, nitrile	T-14
Dodecanoic acid,	7,11-dichloro-4-oxa-2,7,11-trimethyl-, methyl ester	T-55
Dodecanoic acid,	3,7,11-trimethyl-, methyl ester	T-12
2,4,6,10-*Dodecatetraenoic acid,*	7,11-dimethyl-3-ethyl-, methyl ester	T-127
2,4,6,10-*Dodecatetraenoic acid,*	3,7,11-trimethyl-, ethyl ester	T-22
2,4,6,10-*Dodecatetraenoic acid,*	3,7,11-trimethyl-, methyl ester	T-21
2,6,10-*Dodecatriene,*	1-hydroxy-3,7,11-trimethyl-,	T-7
2,6,10-*Dodecatriene,*	1-methoxy-3,7,11-trimethyl-,	T-8
2,6,10-*Dodecatriene,*	1-oxo-3,7,11-trimethyl-, oxime	T-9
2,6,10-*Dodecatrienoic acid,*	4-aza-3,7,11-trimethyl-, ethyl ester	T-111
2,6,10-*Dodecatrienoic acid,*	3,11-dimethyl-7-ethyl-, methyl ester	T-125
2,4,6-*Dodecatrienoic acid,*	10,11-epoxy-3,7,11-trimethyl-, methyl ester	T-47
2,6,10-*Dodecatrienoic acid,*	7-ethyl-3-methyl-, methyl ester	T-124
2,6,10-*Dodecatrienoic acid,*	7-ethyl-11-methyl-, methyl ester	T-126
4,6,10-*Dodecatrienoic acid,*	3-methylene-2,7,11-trimethyl-, methyl ester	T-123
2,6,10-*Dodecatrienoic acid,*	4-oxa-3,7,11-trimethyl-, ethyl ester	T-52
2,6,10-*Dodecatrienoic acid,*	3,7,11-trimethyl-, benzyl ester	T-20
2,6,10-*Dodecatrienoic acid,*	3,7,11-trimethyl-, cyclohexyl ester	T-19
2,6,10-*Dodecatrienoic acid,*	3,7,11-trimethyl-, ethyl ester	T-18
2,6,10-*Dodecatrienoic acid,*	3,7,11-trimethyl-, methyl ester	T-17
6-*Dodecene,*	7,11-dimethyl-10,11-epoxy-1,2-isopropylidenedioxy-4-oxa-,	T-4

2-*Dodecenoic acid,*	8-aza-10-oxa-9-oxo-3,7,11,11-tetramethyl-, methyl ester	T-112
2-*Dodecenoic acid,*	7-chloro-3,11-dimethyl-, ethyl ester	T-138
2-*Dodecenoic acid,*	11-chloro-5-oxa-3,7,11-trimethyl-, N-diethyl amide	T-61
2-*Dodecenoic acid,*	11-chloro-5-oxa-3,7,11-trimethyl-, ethyl ester	T-60
2-*Dodecenoic acid,*	11-chloro-6-oxa-3,7,11-trimethyl-, N-diethyl amide	T-79
2-*Dodecenoic acid,*	11-chloro-6-oxa-3,7,11-trimethyl-, ethyl ester	T-78
2-*Dodecenoic acid,*	11-chloro-3,7,11-trimethyl-, N-diethyl amide	T-25
2-*Dodecenoic acid,*	11-chloro-3,7,11-trimethyl-, ethyl ester	T-24
2-*Dodecenoic acid,*	11-chloro-3,7,11-trimethyl-, methyl ester	T-23
trans-2-*Dodecenoic acid,*	7,11-dichloro-3,7,11-trimethyl-,	T-29
trans-2-*Dodecenoic acid,*	7,11-dichloro-3,7,11-trimethyl-, amyl ester	T-36
trans-2-*Dodecenoic acid,*	7,11-dichloro-3,7,11-trimethyl-, benzyl ester	T-39
trans-2-*Dodecenoic acid,*	7,11-dichloro-3,7,11-trimethyl-, p-bromobenzyl ester	T-40
trans-2-*Dodecenoic acid,*	7,11-dichloro-3,7,11-trimethyl-, butyl ester	T-34
trans-2-*Dodecenoic acid,*	7,11-dichloro-3,7,11-trimethyl-, tert-butyl ester	T-35
trans-2-*Dodecenoic acid,*	7,11-dichloro-3,7,11-trimethyl-, cyclohexyl ester	T-38
trans-2-*Dodecenoic acid,*	7,11-dichloro-3,7,11-trimethyl-, ethyl ester	T-32
trans-2-*Dodecenoic acid,*	7,11-dichloro-3,7,11-trimethyl-, hexyl ester	T-37
2-*Dodecenoic acid,*	7,11-dichloro-3,7,11-trimethyl-, methyl ester	T-30
trans-2-*Dodecenoic acid,*	7,11-dichloro-3,7,11-trimethyl-, methyl ester	T-31
trans-2-*Dodecenoic acid,*	7,11-dichloro-3,7,11-trimethyl-, propyl ester	T-33
2-*Dodecenoic acid,*	7,11-dichloro-3,7,11-trimethyl-, β,β,β-trichloroethyl ester	T-41
2-*Dodecenoic acid,*	3,11-dimethyl ethyl ester	T-118
2-*Dodecenoic acid,*	6,10-dioxa-3,7,11,11-tetramethyl-, N-diethyl amide	T-104
2-*Dodecenoic acid,*	6,10-dioxa-3,7,11,11-tetramethyl-, ethyl ester	T-103
2-*Dodecenoic acid,*	6,9-dioxa-3,7,11-trimethyl-, N-diethyl amide	T-95
2-*Dodecenoic acid,*	6,9-dioxa-3,7,11-trimethyl-, ethyl ester	T-94
2-*Dodecenoic acid,*	6,10-dioxa-3,7,11-trimethyl-, ethyl ester	T-102
2-*Dodecenoic acid,*	6,10-dioxa-3,7,11-trimethyl-, methyl ester	T-101
6-*Dodecenoic acid,*	10,11-epoxy-4-oxa-2,7,11-trimethyl-, methyl ester	T-56
2-*Dodecenoic acid,*	10,11-epoxy-5-oxa-3,7,11-trimethyl-, N-diethyl amide	T-63
2-*Dodecenoic acid,*	10,11-epoxy-5-oxa-3,7,11-trimethyl-, ethyl ester	T-62
2-*Dodecenoic acid,*	10,11-epoxy-5-oxa-3,7,11-trimethyl-, morpholide	T-64

2-*Dodecenoic acid*,	10,11-epoxy-6-oxa-3,7,11-trimethyl-, N-diethyl amide	T-81
2-*Dodecenoic acid*,	10,11-epoxy-6-oxa-3,7,11-trimethyl-, ethyl ester	T-80
2-*Dodecenoic acid*,	10,11-epoxy-3,7,11-trimethyl-, N-diethyl amide	T-28
2-*Dodecenoic acid*,	10,11-epoxy-3,7,11-trimethyl-, ethyl ester	T-27
2-*Dodecenoic acid*,	10,11-epoxy-3,7,11-trimethyl-, methyl ester	T-26
2-*Dodecenoic acid*,	11-ethoxy-6-oxa-3,7,11-trimethyl-, N-diethyl amide	T-83
2-*Dodecenoic acid*,	11-ethoxy-6-oxa-3,7,11-trimethyl-, ethyl ester	T-82
2-*Dodecenoic acid*,	7-ethyl-10-oxa-9-oxo-3,11,11-trimethyl-, ethyl ester	T-110
2-*Dodecenoic acid*,	7-ethyl-3,11,11-trimethyl-, methyl ester	T-121
2-*Dodecenoic acid*,	10-oxa-9-oxo-3,7,11,11-tetramethyl-, N-diethyl amide	T-109
2-*Dodecenoic acid*,	10-oxa-9-oxo-3,7,11,11-tetramethyl-, N-ethyl amide	T-108
2-*Dodecenoic acid*,	10-oxa-9-oxo-3,7,11,11-tetramethyl-, ethyl ester	T-107
2-*Dodecenoic acid*,	10-oxa-3,7,11,11-tetramethyl-, ethyl ester	T-86
2-*Dodecenoic acid*,	9-oxa-3,7,11-trimethyl-, N-diethyl amide	T-85
2-*Dodecenoic acid*,	9-oxa-3,7,11-trimethyl-, ethyl ester	T-84
2-*Dodecenoic acid*,	3,7,11,11-tetramethyl-, N-diethyl amide	T-120
2-*Dodecenoic acid*,	3,7,11,11-tetramethyl-, methyl ester	T-119
2-*Dodecenoic acid*,	3,7,11-trimethyl-, methyl ester	T-13
Furane,	2-(5,9-dimethyl-8,9-epoxy-2-oxa-4-decenyl)-,	T-3
2,7,11-*Heptadecatrienoic acid*,	15,16-epoxy-5-oxa-3,8,12,16-tetramethyl-, ethyl ester	T-74
5-*Heptene*,	6-methyl-2-oxo-, p-nitrophenyl hydrazone	T-228
6,10,14-*Hexadecatrienoic acid*,	4-oxa-2,7,11,15-tetramethyl-, methyl ester	T-54
6,10,14-*Hexadecatrienoic acid*,	4-oxa-7,11,15-trimethyl-, methyl ester	T-53
6-*Hexadecene*,	2,6-dimethyl-9,12-dioxa-2,3-epoxy-,	T-5
Naphtalene,	2-[(3,7-dimethyl-7-ethoxy-2-octenyl)oxy]-,	T-269
Naphtalene,	2-[(3,7-dimethyl-2,6-octadienyl)oxy]-,	T-268
2,6-*Octadiene*,	3,7-dimethyl-1-oxo-, p-nitrophenyl hydrazone	T-229
2,6-*Octadiene*,	3,7-dimethyl-1-oxo-, oxime	T-1
6-*Octene*,	3,7-dimethyl-1-oxo-, ethylenethioketal	T-2
2-*Octenoic acid*,	7-chloro-3,7-dimethyl-, methyl ester	T-10
2-*Octenoic acid*,	3,7-dimethyl-7,8-epoxy-, methyl ester	T-11
2,7-*Tetradecadienoic acid*,	3,12-dimethyl-5,10-dioxa-12,13-epoxy-8-ethyl-, ethyl ester	T-93
2,6-*Tetradecadienoic acid*,	3,11-dimethyl-10,11-epoxy-7-ethyl-, methyl ester	T-157

3,7-*Tetradecadien-2-one*, 4,12-dimethyl-11,12-epoxy-8-ethyl-, T-163

2,7,12-*Tetradecatrienoic acid*, 3,12-dimethyl-5,10-dioxa-8-ethyl-, ethyl ester T-92
2,6,10-*Tetradecatrienoic acid*, 3,11-dimethyl-7-ethyl-, methyl ester T-136
6-*Tetradecene*, 2,6-dimethyl-9,12-dioxa-2,3-epoxy-, T-6

2-*Tetradecenoic acid*, methyl ester T-135

2,12-*Tridecadienoic acid*, methyl ester T-129
2,6-*Tridecadienoic acid*, 7,11-diethyl-10,11-epoxy-3-methyl-, methyl ester T-154
2,11-*Tridecadienoic acid*, 3,11-dimethyl-6,9-dioxa-7-ethyl-, N-diethyl amide T-100
2,11-*Tridecadienoic acid*, 3,11-dimethyl-6,9-dioxa-7-ethyl-, ethyl ester T-99
2,6-*Tridecadienoic acid*, 3,7-dimethyl-10,11-epoxy-11-ethyl-, methyl ester T-152
2,6-*Tridecadienoic acid*, 3,7,11-trimethyl-10,11-epoxy-, methyl ester T-155
2,6-*Tridecadienoic acid*, 3,11-dimethyl-10,11-epoxy-7-ethyl-, methyl ester T-143

cis-2-cis-6-*Tridecadienoic acid*, 3,11-dimethyl-trans-10,11-epoxy-7-ethyl-, methyl ester T-144
trans-2-cis-6-*Tridecadienoic acid*, 3,11-dimethyl-trans-10,11-epoxy-7-ethyl, methyl ester T-145
cis-2-trans-6-*Tridecadienoic acid*, 3,11-dimethyl-trans-10,11-epoxy-7-ethyl-, methyl ester T-146
trans-2-trans-6-*Tridecadienoic acid*, 3,11-dimethyl-trans-10,11-epoxy-7-ethyl-, methyl ester T-147
cis-2-cis-6-*Tridecadienoic acid*, 3,11-dimethyl-cis-10,11-epoxy-7-ethyl-, methyl ester T-148
trans-2-cis-6-*Tridecadienoic acid*, 3,11-dimethyl-cis-10,11-epoxy-7-ethyl-, methyl ester T-149
cis-2-trans-6-*Tridecadienoic acid*, 3,11-dimethyl-cis-10,11-epoxy-7-ethyl, methyl ester T-150
trans-2-trans-6-*Tridecadienoic acid*, 3,11-dimethyl-cis-10,11-epoxy-7-ethyl-, methyl ester T-151

2,10-*Tridecadienoic acid*, 3,11-dimethyl-7-ethyl-, ethyl ester T-130
2,7-*Tridecadienoic acid*, 5,10-dioxa-3,8,12,12-tetramethyl-, ethyl ester T-91
2,6-*Tridecadienoic acid*, 10,11-epoxy-7-ethyl-3-methyl-, methyl ester T-156
2,6-*Tridecadienoic acid*, 10,11-epoxy-7-ethyl-11-methyl-, methyl ester T-153
2,7-*Tridecadienoic acid*, 11,12-epoxy-5-oxa-3,8,12-trimethyl-, ethyl ester T-72
2,11-*Tridecadienoic acid*, 5-oxa-3,8,12-trimethyl-, N-diethyl amide T-67
2,11-*Tridecadienoic acid*, 5-oxa-3,8,12-trimethyl-, ethyl ester T-66

3,7-*Tridecadien-2-one*, 11,12-epoxy-4,8,12-trimethyl-, T-162
2,6,10-*Tridecatrienoic acid*, 7,11-diethyl-3-methyl-, methyl ester T-132
2,6,10-*Tridecatrienoic acid*, 3,11-dimethyl-7-ethyl-, methyl ester T-133
2,7,11-*Tridecatrienoic acid*, 8,12-dimethyl-5-oxa-4-oxo-, methyl ester T-106
2,6,10-*Tridecatrienoic acid*, 7-ethyl-3-methyl-, methyl ester T-134
2,6,10-*Tridecatrienoic acid*, 7-ethyl-11-methyl-, methyl ester T-131
2,7,11-*Tridecatrienoic acid*, 5-oxa-3,8,12-trimethyl-, ethyl ester T-68

2-*Tridecenoic acid*, 11-chloro-3,11-dimethyl-6,9-dioxa-7-ethyl-, ethyl ester T-96
2-*Tridecenoic acid*, 11-chloro-3,11-dimethyl-7-ethyl-, ethyl ester T-142
2-*Tridecenoic acid*, 12-chloro-5-oxa-3,8,12-trimethyl-, ethyl ester T-69
2-*Tridecenoic acid*, 3,11-dimethyl-6,9-dioxa-11,12-epoxy-7-ethyl-, N-diethyl amide T-98
2-*Tridecenoic acid*, 3,11-dimethyl-6,9-dioxa-11,12-epoxy-7-ethyl-, ethyl ester T-97
2-*Tridecenoic acid*, 5,10-dioxa-3,8,12,12-tetramethyl-, ethyl ester T-90
2-*Tridecenoic acid*, 5,10-dioxa-3,8,11-trimethyl-, ethyl ester T-89
2-*Tridecenoic acid*, 11,12-epoxy-5-oxa-3,8,12-trimethyl-, N-diethyl amide T-71
2-*Tridecenoic acid*, 11,12-epoxy-5-oxa-3,8,12-trimethyl-, ethyl ester T-70
2-*Tridecenoic acid*, 12-ethoxy-5-oxa-3,8,12-trimethyl-, ethyl ester T-73
2-*Tridecenoic acid*, 3-methyl-, methyl ester T-128
2-*Tridecenoic acid*, 5-oxa-3,8,12-trimethyl-, ethyl ester T-65

2,6-*Undecadienoic acid*, 10,11-epoxy-7-ethyl-3-methyl-, methyl ester T-137

2,6,10-*Undecatrienoic acid*, 7-ethyl-3-methyl-, methyl ester T-117

5-*Undecenoic acid*, 6,10-dimethyl-9,10-epoxy-3-oxa-, methyl ester T-49
2-*Undecenoic acid*, 7-ethyl-3,10,10-trimethyl-, methyl ester T-116
2-*Undecenoic acid*, 3,7,10,10-tetramethyl-, methyl ester T-115

Addendum

During the last year, many papers concerning the chemistry of juvenoids have appeared which we were not able to include in full extent in our book.

We therefore consider it necessary to mention the most important ones additionally.

ANDERSON, R. J., C. A. HENRICK, and J. B. SIDDALL: Synthesis of imino derivatives of cecropia juvenile hormone. J. Org. Chem. *37*, 1266–1268 (1972).

ANDERSON, R. J., C. A. HENRICK, J. B. SIDDALL, and R. ZURFLUH: Stereoselective synthesis of the racemic C-17 juvenile hormone of cecropia. J. Amer. chem. Soc. *94*, 5379–5386 (1972).

CHANG, C. F., and S. TAMURA: Synthesis of several 3,4-methylenedioxyphenyl derivatives as inhibitors for metamorphosis of silkworm larvae. Agr. Biol. Chem. *36*, 2405–2410 (1972).

COCHRANE, J. S., and J. R. HANSON: Syntheses of the insect juvenile hormone. J. chem. Soc., Perkin Trans. *1972*, 361–366.

DRABKINA, A. A., O. V. YEFIMOVA, J. S. CIZIN, N. M. GAMPER, and E. A. PRIBANTSEVA: Sojedinenia imitiruyuscie dejstviya juvenilnovo gormona nasekomykh. Zh. Obsch. Khim. *42*, 457–459 (1972).

FAULKNER, D. J., and M. R. PETERSEN: Application of Claisen rearrangement to synthesis of trans trisubstituted olefinic bonds. Synthesis of squalene and insect juvenile hormone. J. Amer. chem. Soc. *95*, 553–563 (1973).

GRIECO, P. A.: Geraniol and nerol in organic synthesis. Useful intermediates for the synthesis of C_{17}-juvenile hormone. J. Chem. Soc. Chem. Commun. *1972*, 486–487.

HENRICK, C. A., F. SCHAUB, and J. B. SIDDALL: Stereoselective synthesis of the C-18 cecropia juvenile hormone. J. Amer. chem. Soc. *94*, 5374–5378 (1972).

KAMIMURA, H., B. D. HAMMOCK, I. YAMAMOTO, and J. E. CASIDA: A potent juvenile hormone mimic, 1-(4-ethylphenoxy)-6,7-epoxy-3,7-dimethyl-2-octene, labeled with tritium in either ethylphenyl- or geranyl-derived moiety. J. Agr. Food Chem. *20*, 439–442 (1972).

LIEDTKE, R. J., and C. DJERASSI: Mass spectrometry in structural and stereochemical problems. CCXVIII. The electron impact induced behavior of terpenoid ethers of the juvenile hormone class. J. Org. Chem. *37*, 2111–2119 (1972).

MORI, K.: Synthesis of compounds with juvenile hormone activity. Part IX. Analogues of the cecropia juvenile hormone lacking the methyl substituent at C-3. Agr. Biol. Chem. *36*, 442–445 (1972).

MORI, K.: Conversion of farnesol into methyl d,l-12-homo-10-trans-juvenate, the 10-trans-isomer of d,l-C_{17} cecropia juvenile hormone. Agr. Biol. Chem. *36*, 2563–2567 (1972).

MORI, K.: Synthesis of compounds with juvenile hormone activity — XII. A stereoselective synthesis of 6-ethyl-10-methyldodeca-5-trans,9-cis-dien-2-one, a key intermediate in the synthesis of C_{18} cecropia hormone. Tetrahedron *28*, 3747–3756 (1972).

MORI, K., M. OHKI, and M. MATSUI: A stereoselective synthesis of d,l-C_{17}-cecropia juvenile hormone. Agr. Biol. Chem. *36*, 1085–1086 (1972).

MORI, K., M. OHKI, A. SATO, and M. MATSUI: Synthesis of compounds with juvenile hormone activity — XI. New routes for the stereo-controled construction of the trisubstituted cis double bond portion of the cecropia juvenile hormones. Tetrahedron *28*, 3739–3745 (1972).

MORI, K., A. SATO, and M. MATSUI: Synthesis of compounds with juvenile-hormone activitv. 13, Synthesis of d,l-C_{17}-cecropia juvenile hormone. Agr. Biol. Chem. *36*, 1931–1936 (1972).

OHKI, M., K. MORI, M. MATSUI, and A. SAKIMAE: Synthesis of new analogues of the cecropia juvenile hormone with a cyclopentane or cyclohexane ring at the terminal position. Agr. Biol. Chem. *36*, 979–983 (1972).

ROGERS, I. H., and J. F. MANVILLE: Juvenile hormone mimics in conifers I: Isolation of (—)-cis-4-[1′(R)-5′-dimethyl-3′-oxohexyl]-cyclohexane-1-carboxylic acid from Douglas-fir wood. Canad. J. Chem. *50*, 2380–2382 (1972).

SARMIENTO, R., T. P. MCGOVERN, M. BEROZA, G. D. MILLS, JR., and R. E. REDFERN: Insect juvenile hormones: Highly potent synthetic mimics. Science *179*, 1342–1343 (1973).

Subject Index

Abies balsamea 170
abscissic acid 269
acetoacetates, in Cecropia JH synthesis 153
acetophenone derivatives 177
acetylcholine 9, 19
Achyranthes fauriei 310, 312
Achyranthes obtusifolia 317
Achyranthes rubrofusca 317
acid amides, JH activity 211
Acridiidae, JH activity 200, 216
acylmalonates, in Cecropia JH synthesis 153
Aedes, juvenoid effects on sexual instincts 271, ovicidal activity of juvenoids 271
Aeschna, hormonal effects on cell divisions 235
Ajuga decumbens 314, 316, 329
Ajuga incisa 314, 316
Ajuga japonica 310, 314, 316
ajugalactone 329
ajugasterone A 312
ajugasterone B 316
ajugasterone C 314
alar 273
alicyclic juvenoids 208
alkoxygeranyl ethers, JH activity 196
allatotropic hormone 3, 10, 13
allylic alcohols, transformation 151, 154
amarasterone A 315
amarasterone B 315
p-aminoacid derivatives 178, 192
p-aminobenzoates 215
p-aminobenzoic acid derivatives, JH activity 197
aniline derivatives 180, 206
Antheraea, distribution of labelled ecdysone 356, ecdysone in 304, ecdysone effects on RNA synthesis 359, ecdysterone in 308, juvenoid effects on esterase 258, ultrastructure of PG 53
anti-juvenile mechanisms 35, 223, 263
Aphids 8, CAH effects on polymorphism 46, ecdysial failures 231, juvenoid effects on 274, sensitive periods to juvenoids 96
apolysis 341
Apterygota, CAH effects on ovarian development 37, changes in PG 55, regulation of moults 50, 62
Araneus, moults induced by ecdysoids 368
Armadillidium, moults induced by ecdysoids 367
aromatic esters, JH activity 196
aromatic ethers, JH activity 92, 200, 241
aromatic juvabione derivatives 214
aromatic juvenoids 205
auraptene 183
azirane ring 155

behaviour, CAH effect 47
benzenesulphonic acid derivatives 188
4,5-benzo-1,3-dioxine 183
benzoic acid 177
biogenesis, of Cecropia hormones 159, of ecdysone 320, of ecdysterone 321–323, 325
Blatella, JH activity 202, 321, 390
Blatta 25
Blechnum amabile 310, 311

Blechnum minus 310, 314, 315
Blechnum nipponicum 310, 311, 312
Bombyx, assay for ecdysoids 336, 337, ecdysoid effects on carbohydrate metabolism 362, ovarian growth 354, and on protein synthesis 361, ecdysone titers in 355, ecdysterone from 308, extra-larval instars 35, farnesol effects on ovarian growth 237, growth of CA 350, hyperecdysonism 350, ultrastructure of CA 26, ultrastructure of PG 53
brain hormone 3
bursicon 22
Byrsotria 237, 271

callinecdysone A 312
callinecdysone B 315
Callinectes sapidus, ecdysterone from 308, 312, 315
Calliphora, AH effects on intestinal cells 11, biosynthesis of ecdysone 320, and ecdysterone 321–323, 325, CA of 25, CA ultrastructure 26, CAH effects on ovarian growth 36, and nucleic acids 45, distribution of labelled ecdysterone 356, ecdysoid content in the body 354 to 356, ecdysoid effects on nucleic acids 359, puparium formation 334, 335, and on protein synthesis 361, ecdysone and tissue disintegration 347, extra-larval moults 349, formation of ponasteroside A 311, inactivation of ecdysoids in the body 355, juvenoid effects on metamorphosis 272, nucleic acids 260, and ovarian growth 237, prothetelic forms 350, sensitivity to ecdysone 342, tests for ecdysoids 303, 318, 319, 334, 335
Calpodes 66, ecdysoid effects on protein synthesis 361
capitasterone 316
carbamates, JH activity 170, 201, 205, 208, 209, synthesis of 169
Carcinus, ecdysoid effects on development 368
cardioaccelerating factors 11
Carroll reaction 153
Caryedon, juvenoid effects on pupation 273
catecholamines 19
Cecropia (see *Hyalophora*), C_{18} cecropia juvenile hormone configuration 147, and *in vitro* production 159
Cecropia extracts 90, 91, 146, 148, 160, 225, effects on DNA synthesis 259, mitosis 235, ovarian growth 237, and on respiratory metabolism 255, inactivation of extracts in the body 264
Cecropia hormone, enantiomers 157, total hydrogenation 147
Cecropia hormones 140, absolute configuration 157, biogenesis 159, isolation 147, 148
Cephus, reactivation of PG 56
Cerura, juvenoid effects on behaviour 271, and pigmentation 270
Chaoborus, effect of hormones on diuresis 18, and colour change 20
Cheilanthes tenuifolia 314
cheilanthone A 314, 319, 322
cheilanthone B 314, 319, 322
Chilo, dipping test for ecdysoids 329, 336, ecdysoid effects on tissue cultures 363
Chironomus, distribution of labelled ecdysoids 356, ecdysoid effect on puffing 365, juvenoid effects on puffing 260
cholesterol, in ecdysoid biosynthesis 320, 321
cholesteryl farnesyl ether, JH activity 197
Chrysocornis, CA cytology 27
Cimex, farnesol effects on ovarian growth 237
citronellol, derivatives 142
citronellylacetone 143
Claisen rearrangement 151, 152, 158
Cockroaches, changes in PG 55, juvenoid effects on ovarian growth 92, 237, and on vitellogenic proteins 258
Coleoptera, adult diapause 39, CAH effects on respiration 41, hyperecdysonism 350, JH activity 200, 216, juvenoid effects on the delay of pupation 273, on development

273, and respiration 240, hormonal control of ovarian growth 37, PG and their degeneration 51, 57
colour change, CAH effects 46
coumarone derivatives 182, 183
Crangon nigricauda 322
crustecdysone 307
Culex, CA of 25
cyasterone 316
Cyathula capitata 316, 318

Danaus, juvenoid effects on ovarian growth 238, 272, and on vitellogenic proteins 258
dauerlarvae 58
dehydrojuvabione 92, 170
demethyljuvabione 175
2-deoxyecdysone 314
2-deoxyecdysterone 314, 319
22-deoxyecdysterone 320
deoxyviperidone 324
Dermaptera, CAH effect on ovarian development 37
Dermestes, anti-juvenile mechanisms 223, 263, assay for ecdysoids 337, CAH effect on pupation 35, ecdysoid effects on development 351 to 353, extra-pupal instars 222, 231, JH activity on 169, 389, 391, juvenoid applications 100, juvenoid effects on delay of pupation 273, and on last larval instars 223, larval intermoult periods 226, "pupal" differentiation 232–235
Dermestidae, JH activity 200, 216
diapause, effect of CAH 36, ecdysoids 339, 344, 345, juvenoids 243, neurohormones 14, and PGH 59
diapause hormone 22
7,11-dichlorofarnesoate, absence of mutagenicity 272, effect on body composition 258, ecdysial failures 229, embryonic development 248, metamorphosis 131, 133, and on ovarian growth 241
7,11-dichlorofarnesoyl chloride 144
Dictyoptera, CAH effects on ovarian growth 37, and respiration 41, prothoracic glands 51
dienoates, JH activity 201
diesters, synthesis 161, 162
N,N-digeranyl derivatives 207
dihomojuvenates (see Cecropia hormones), JH activity 203
20,26-dihydroxyecdysterone 312
Diptera, adult diapause 39, CAH effects on ovarian growth 37, and on respiration 41, ecdysoid effects on nucleic acids 361, and on puparium formation 348, 349, hyperecdysonism 350, juvenoid effects on metamorphosis 271, 272, reproduction 240, and on sexual behaviour 271, larval intermoult periods 226, sensitive periods to juvenoids 96, 97
Dixippus, CAH effect on pigmentation 47, growth of allatectomized larvae 40, neurohormones and colour change 20
Dociostaurus, ecdysone in 303, ecdysterone in 308
1-dodecanol, JH activity 198
n-dodecyl alcohol derivatives 210
dodecyl methyl ether, JH activity 196, 198
DOPA, bursicone effects on metabolism 23, ecdysone effects 330
Doralis, farnesol effect on wing development 274
Drosophila, CA ultrastructure 26, developmental changes in PG 55, distribution of ecdysoids 356, ecdysoid binding with proteins 356, ecdysoid effects on puffing 365, growth of imaginal discs 342, metamorphosis 339, nucleic acids 359, protein synthesis 362, and on tissue cultures 363, juvenoid effects on metamorphosis 272, PG ultrastructure 53
Dysdercus, CAH effects on musculature 42, dose-response relationships with juvenoids 108, JH activity 173, 188, 189, 191, 193, 195, 202, 205, 211, juvenoid effects on body composition 258
Dytiscus, gonadotropic effects of CAH 36

ecdysial failures 229
ecdysial glands 50
ecdysone 303, 319

ecdysterone 307, 319
effective doses of juvenoids 104
Ephemera 25
Ephestia, CA size 29, ecdysoid assays 336, sensitivity to ecdysone 342
22-epiecdysterone 320
epijuvabione 173, 213
Epilachna 182
epoxidation 145
10,11-epoxyfarnesoate 91, effect on adult diapause 243, hydrolysis of juvenoid esters 197
ethyl farnesoate diacetate 155
Eurygaster, JH activity 177, juvenoid effects on diapause 243, ultrastructure of CA 26, 27
Euthystira, farnesol effects on ovarian growth 237, 271

farnesal 140
farnesoic acid 142, derivatives 142, 143, 199, JH activity 142
farnesoic acid amides 146
farnesol 138, 140, 195, 197, biological degradation 265, and oxidation 265, content in *Samia* 159, conversion of mevalonate in 265, derivatives of 138, 141, effects on development in Lepidoptera 272, growth in plants 274, mortality 273, ovarian growth 237, 241, and on wing development 274, epoxidation of 141, 145, isomers 138, JH activity 210, JH activity of isomers 138, 201, synthesis 138, 140
farnesyl acetate 154, diepoxide 154
farnesyl alkyl ethers 91, 141, 210, 241, effect on Lepidoptera 272, on respiration 255
farnesylamines 141
farnesyl esters 141, 143
farnesyl methyl ether, epoxidation 141
farnesyl pyrophosphate 138
fatty acids, JH activity 197, 253, 254
feed-back mechanisms 38, 240, 265

Galleria, CAH effects on behaviour 47, larval growth 40, metamorphosis 36, respiration 41, cecropia extract effects on larvae 225, ecdysone effects on DNA synt. 361, extra-larval moults 35, hyperecdysonism 350, JH activity 195, 198, 216, 389, 391, JH activity of ethers 196, 198, 200, juvenoid assay 94, 124, juvenoid effects on membrane permeability 263, and metamorphosis 272, larval intermoult periods 226, juvenoid versus ecdysoid effects 218, sensitivity to ecdysone 342
geometrical isomers, of Cecropia JH 154
geraniol 142, 197, derivatives 142
geranyl-p-aminobenzoate 207
Gomphocerus, CAH effect on sexual instincts 48
Graphosoma, JH activity 177, 179, 185, 186, 187, 195, 198, 211, 391, juvenoid effects on diapause 243, and pigmentation 270
Gryllus, CAH effects on musculature 42, and pigmentation 47, dose-responses with juvenoids 108, JH activity 161, 198, 389, 391

halophenols, derivatives 185
Heliothis, inhibition of ovarian growth by juvenoids 272
Hemiptera, adult diapause 39, CAH effects on alary polymorphism 46, ovarian growth 37, and on respiration 41, dose-responses with juvenoids 108, JH activity 200, 209, 214, 216, 389, PG 51, scoring systems for juvenoids 105
hexahydropseudoionone 143
homeostatic mechanisms 48, 64, 340
homojuvenates (see Cecropia hormones), JH activity 198, 201, 216
homologization, of farnesane skeleton 154, 155
Hyalophora, CA ultrastructure 26, ecdysoid effects on adenyl cyclase 362, and fat body cells 362, hyperecdysonism 350, juvenoid effects on embryogenesis 246, 249, PGH effects on metamorphosis 56, 59

hydrogen chloride, addition to unsaturated compounds 141, 144
Hydrous 25
hydroxy acids, derivatives 141
hydroxy nitriles, derivatives 141
p-hydroxyacetophenone derivatives 209
Hymenoptera, hyperecdysonism 350, juvenoid effects 273, 274, ovarian growth 37, 238, PG 51, sensitive periods to juvenoids 96
hyperecdysonism 350, 351

Idothea 368
imino analogues, of Cecropia hormones 155, 156
Imperatoria ostruthium 183
inhibition doses, of juvenoids 105, 109
inokosterone 310, 312
insecticide synergists 92, 207, 252
iodination 138
Ips, juvenoid effects on pheromone production 271
isoinokosterone 307

Japyx 25
Jasus lalandei, ecdysterone from 308, 314
juvabione, chemistry 170–173, effects on ecdysis 230, 269, 277, and on embryogenesis 244, JH activity 196, 197, 199, 208, 213

Kolbe anodic synthesis 160

larvalization 236
Lasioderma 182
Lastrea thelypteris 312
Law-Williams mixture 91, 144
Lemmaphyllum microphyllum 310, 315
lemmasterone 315
Lepidoptera, assay for ecdysoids 332, ecdysoid effects on nucleic acids 359, ecdysone titers 355, hormones and ovarian growth 37, 38, JH activity 198, 200, 216, 272, juvenoid effects on ecdysis 244, prothoracic glands 50, sensitivity to ecdysone 342
Leptinotarsa, adultoids 226, CAH effects on behaviour 47, blood proteins 43, and on musculature 42, hormonal control of diapause 29, juvenoid effects on homogenates 255, pupation 227, 273, and on vitellogenic proteins 258
Leucophaea, CA activation 28, gonadotropic effects of CAH 36, juvenoid effects on ovarian growth 237, and on vitellogenic proteins 258, ultrastructure of CA 26, 27
Limulus 367
Liquia 368
lithium azide 156
Locusta, CAH effects on blood proteins 43, fat body 41, mitosis 34, nucleic acids 45, pigmentation 47, polymorphism 46, regeneration 47, and on sexual instincts 48, ecdysoid effects on larvae 345, JH activity 200, 216, juvenoid effects on dose-responses 108, reproduction 237, and on respiration 255, PGH effects on pigmentation 65, PG survival 57, developmental changes 55, and ultrastructure 53, ultrastructure of CA 26
Lucilia, ecdysone titers 355, ecdysterone effect on disintegration of muscles 347
Lygaeidae, JH activity 215
Lygaeus, JH activity 177, 391, juvenoid effects on diapause 243

macromolecular factor 364
makisterone A 315
makisterone B 315, 320
makisterone C 315
makisterone D 316, 320
Malacosoma, effects of farnesol 272, 273
Manduca, ecdysoid effects on tissue cultures 363, ecdysoids in 304, 308, 312, 322, ecdysone titers 355, juvenoid effects on ovarian growth 238, and on vitellogenic proteins 258
manganese dioxide 151, 154
Matteucia struthiopteris 312
Melanoplus 36
Melolontha 47

metathetely 33, 223, 267, 343, 350
methionine, in Cecropia JH biosynthesis 159, 160
methyl acrylate 141
methyl 12,14-dihomojuvenate (see Cecropia hormones) 139, 148
methyl 9,10-epoxyhexadecanoate 146
methyl farnesoate, hydrogenation of 143
methyl geranylate 98
methyl 12-homojuvenate (see Cecropia hormones) 139, 148
methyl juvenate 146, 154, total hydrogenation 147
methyl methacrylate 142
methyl senecionate, photooxidation 151
methylenedioxyphenyl ethers 196, 209
methylenetriphenyl phosphorane 151
mevalonate 159
Mosquitoes, AH effect on glycogen 17, Cecropia extract effect on ovarian growth 237, juvenoid effects on development 271, 272
Musca, biosynthesis of ecdysoids 321, ecdysoid effects on carbohydrate metabolism 362, nucleic acids and protein synthesis 359, 361, and on puparium formation 334, 335, 348, effect of 6-ketosteroids 325, farnesol effects on ovarian growth 237, juvenoid effects on sexual instincts 271

narcosis 98, 99
Nauphoeta, CAH effects on blood proteins 43
Neocheiropteris ensata 310
nerol 141
neurohumours 9
neurosecretory axons 5, 8
neurotransmitters 9
p-nitroaniline derivatives 205, 209
nitrophenol derivatives 185
Nomadacris, CAH effects on sexual instincts 48
Nomia, inhibition of metamorphosis by juvenoids 274
noradrenaline 8, 9
Nosema, JH activity from 91

Oediboda, CAH and sexual instincts 48
olefinic ketal Claisen reaction 151, 153
Oncopeltus, ecdysone titers 355, gonadotropic effects of CAH 36, intermediate cell types 221, JH activity 177, 180, 181, 188, 195, 208, 392, neurosecretory material release 8
Onoclea sensibilis 311, 312
optical rotation, in Cecropia hormones 157
Orthoptera, CAH effects on ovarian development 37, and respiration 41, JH activity 200, juvenoid effects on embryogenesis 244, and pigmentation 270
osladin 324
Osmunda asiatica 311
Osmunda japonica 310, 311
Ostrinia, ecdysoid effect on diapause 345
ostruthin 183
Oulema, juvenoid effect on diapause 243
oxa-esters, JH activity 142, 164, 204, 205, synthesis 142, 162, 164
oxirane ring 145

"paper factor" 92, 170, 269, 275
Pentatomidae, JH activity 200
peptidic juvenoids 92, 190, 199, 209
pericardial glands 50, 52
Periplaneta, bursicon in 22, CAH effects on blood proteins 43, juvenoid effects on accessory sexual glands 242, and ovarian growth 237, rhythms in locomotory activity 20
perisympathetic neurohaemal organs 7, 18, 19
peritracheal glands 50, 52
permanent larvae 58
pharate periods 341
pheromones, CAH effects 47, and juvenoid effects on production of 271
photooxidation, of ethyl farnesoate diacetate 155, and methyl senecionate 151
phytoecdysones 310, 323

Pieris, CA and PG 55, farnesol effect on pigmentation 270, juvenoid effect on ovarian growth 238
piperonylbutoxide 180, 252
Pleopeltis thunbergiana 310
Plutella, ecdysoid assay 336
podecdysone A 315
podecdysone B 314, 319
podecdysone D 315
Podocarpus chinensis 311
Podocarpus elatus 310, 314, 315, 323
Podocarpus macrophyllus 310, 311, 315, 316
Podocarpus nakaii 310, 311, 312, 314
Polyphemus, CAH effects on blood proteins 43
polypodine A 307, 310
polypodine B 310, 312, 319
Polypodium vulgare 310, 314, 323, 324
polypodosaponin A 324
ponasterone A 310, 311, 320
ponasterone B 312, 319
ponasterone C 314
ponasteroside A 311
poststerone 318, 319
potentiation of JH activity 156, 251, 252
precyasterone 316
Procambarus, induction of moults by ecdysoids 368
propargylic alcohols 138, 150, 151
prothetely 33, 253, 343, 350
prothoracotropic hormone 3, 10, 131
pseudoionone 143
pseudo-juvenilizing effects 253
Pteridium aquilinum 310, 311, 312
pterosterone 312, 320
Pterostichus, juvenoid effect on ovarian growth 237
Pyralidae, JH activity 200
Pyrrhocoridae, JH activity 198, 199, 200, 204, 207, 213, 214
Pyrrhocoris, activation of CA 12, CA 25, CAH effects on blood proteins 43, mitosis 34, respiration 42, and on sexual instincts 48, dose-responses with juvenoids 105, 108, ecdysoid binding with proteins 356, JH activity 170, 176, 177, 188, 191, 193, 195, 204, 209, 211, 214, 392, juvenoid applications 99, 100, juvenoid effects on allatectomized larvae 225, ecdysial failures 229, 230, embryogenesis 244, 245, 246, 249, reproduction 116, 237, 239–241, respiration 255, 256, and on transaminase 258, PG ectomy 60, 63, and survival 57

quaternary alkyls, in juvenoids 160

regeneration 47
reversal of metamorphosis 236
Rhodnius, assay of juvenoids 94, CAH effects on blood proteins 43, mitoses 34, nucleic acids 45, and on ovarian development 36, effect of Cecropia extracts 90, effects of injury 47, growth of PG 55, hormonal effects on adult moults 59, diuresis 18, and regulation of moult cycles 50, 61, JH activity 141, 195, 198, 200, 202, 203, 214, 392, juvenoid applications 100, juvenoid effects on metamorphosis and reproduction 92, 117, 237, 249, larval differentiation 232, parabiosis and transplantations 31, 33, PGH effects on epidermal cells 64, and fat body 63, sensitive periods to juvenoids 96, 97
Rhynchosciara, distribution of ecdysoids 356
rubrosterone 316, 317, 319, 324

Samia, assay for ecdysoids 336, 337, ecdysoid effects on nucleic acids 44, and ovarian growth 354, hyperecdysonism 350
Sarcophaga, ecdysoid effects on disintegration of tissue 347, and puparium formation 335, 336, extra-larval moults 349, inactivation of ecdysoids 355, juvenoid effects on metamorphosis 272
Schistocerca, CAH effects on blood proteins 43, and sexual instincts 48, ecdysoid effect on development 368, juvenoid effects on embryogenesis 245, and ovarian growth 237, PGH effect on pigmentation 65, sensitive periods to

juvenoids 136, ultrastructure of CA 26
Sciara, ecdysoid effect on puffing 365
sengosterone 315, 316
serotonine 8, 19
Serratula inermis 310
sesamex 208
sesamoline 180
sesoxane 180
shidasterone 312
Sialis 25
sirenin 177
solubility, of juvenoids 195
solvomercuration 145
Sphinx 25
squalene 98
stachysterone C 312, 320
stachysterone D 314, 320
Stachyurus praecox 310, 312, 314
stereospecific reactions 138
stereospecific reduction 140
Stomoxys 392, juvenoid effect on metamorphosis 272
systemic effects of juvenoids 93

Taxus baccata 310, 323
Taxus cuspidata 311
Tenebrio, ecdysoid effects on ovarian growth 354, and tissue cultures 363, ecdysoid activity in oenocytes 356, effect of antibiotics 253, growth of PG 55, JH activity 140, 141, 158, 162, 169, 170, 177, 180, 182, 185, 186, 188, 195, 198–202, 204–206, 208–215, 389, 392, JH activity in excrements 91, juvenoid assays 94, 124, application 99, 100, concentration-activity curves 200, dose-responses 107, scoring systems 106, juvenoid effects on cuticle formation 236, enzymatic activity 257, nucleic acids 259, ovarian growth 237, pherome production 271, and on respiration 255, larval-pupal intermediates 35, prothetelic pupae 350
Tenebrionidae, JH activity 200
Termites, juvenoid effects on caste differentiation 274
terpenoid aniline derivatives 92
1-tetradecanol, JH activity 197
tetrahydrogeraniol derivatives 142
thiophenol derivatives 184, 206
Tortricids, ovicidal effects by juvenoids 272
transesterification 141
Triatoma, JH activity 179, 182, 214, 215, 392
Tribolium, JH activity 162, 216, 392
Trillium snialii 310
tritium chloride 144
Trogoderma 273

Uca, ecdysoid effect on development 322, 368
unsaturated compounds, reaction with hydrogen chloride 141, 144
unsaturated systems, effect on JH activity 143, 144, 201

ventral glands 50, 51
viperidinone 324
viperidone 324
Vitex megapotamica 310–314
viticosterone E 311, 320

Wadsworth-Emmons reagent 143
warabisterone 311
wax wound test for juvenoids 94
Wittig reaction 143

Xylobarus ferrugineus 321